Environmental Earth Sciences

Series Editor:
James W. LaMoreaux

For further volumes:
http://www.springer.com/series/8394

Nicolaos Lambrakis
George Stournaras
Konstantina Katsanou
Editors

Advances in the Research of Aquatic Environment

Volume 1

Springer

Editors
Prof. Dr. Nicolaos Lambrakis
University of Patras
Department of Geology
Laboratory of Hydrogeology
Patras
Greece
nlambrakis@upatras.gr

Prof. Dr. George Stournaras
University of Athens
Department of Geology and
Geoenvironment
Athens
Greece
stournaras@geol.uoa.gr

Konstantina Katsanou
University of Patras
Department of Geology
Laboratory of Hydrogeology
Patras
Greece
katsanou@upatras.gr

ISBN 978-3-642-19901-1 e-ISBN 978-3-642-19902-8
DOI 10.1007/978-3-642-19902-8
Springer Heidelberg Dordrecht London New York

Library of Congress Control Number: 2011936434

Cover design: deblik, Berlin

Printed on acid-free paper

Springer is part of Springer Science+Business Media (www.springer.com)

Preface

These two volumes contain the proceedings of the 9th International Congress of Hydrogeology and the 4th MEM Workshop on Fissured Rocks Hydrology, organized by the Hellenic Committee of Hydrogeology in collaboration with the Cyprus Association of Geologists and Mining Engineers.

The number of the manuscripts submitted to the Organizing Committee throughout 15 countries all over the world reflects the rapidly increasing interest that Hydrology gains nowadays worldwide. The papers cover more or less all fields, such as mathematical modeling, statistical, hydro-chemical methods, etc., focusing on the environmental aspect.

Aquatic environment, the main topic of the Congress, as it is shown by the title of the Proceedings "Advances in the research of aquatic environment" is covered by articles mostly dealing with ecological impacts *versus* water requirements, climate change implications on groundwater, anthropogenic impacts on the groundwater quality, groundwater vulnerability, and more.

Both volumes follow the general structure of the Congress topics. Moreover the keynote lectures are also included.

On behalf of the International Scientific Committee I would like to take this opportunity to thank all the authors for their contributions, as well as all participants for their cooperation, which made this Congress possible. Additionally, I would like to express my gratitude to the staff of Springer and especially Christian Witschel and Agata Oelschlaeger for their hard work, patience and support.

Last but not least, I would like to thank my wife Aggeliki and my children Athina and Ioannis for their patience and love.

Prof. Nicolaos Lambrakis
President of the Organizing Committee
University of Patras
Laboratory of Hydrogeology
Rio – Patras, Greece

Acknowledgements

The 9th International Hydrogeological Congress of Greece would not have been possible to be carried out without the active engagement of many persons and the financial support of Institutions and Organizations. I would like to express my gratitude to all of them. Moreover, I would like to thank in particular Mr Dimitris Dalianis for his perfect work in the construction and maintenance of the Congress website. I would also like to acknowledge Mrs Konstantina Katsanou, Panagoula Kriempardi and Katerina Karli, for their valuable contribution during the organization.

Sponsors

BANK OF GREECE

UNIVERSITY OF PATRAS

MARIOLOPOULOS KANAGINIS FOUNDATION

AGRICULTURAL UNIVERSITY OF ATHENS

DR C.J.VAMVACAS LTD HIGH-TECH PRODUCTS-CONSULTANTS

GEOTECHNICAL CHAMBER OF GREECE

OLYMPIC BREWERY S.A.

FIX HELLAS

TERNA ENERGY SA

ΔΕΥΑΠ M.E.W.S OF PATRAS

SCIENTACT S.A.

RIGAS LABS S.A.
when details lead to excellence

METRICA
When it's a matter of trust

Marathon Data Systems
e-mail: marathon@otenet.gr
www.marathondata.gr

LOUX
Natural Greek Pleasure

ZAGORI®
natural mineral water

ΤΕΤΡΑΜΥΘΟΣ WINERY
TETRAMYTHOS

KLEOPATRA GOUVA – CHEMICALS PARTNERSHIP LAZARIDIS

Organizing Committee

The 9th International Hydrogeological Congress and the 4th MEM Workshop of Fissured Rocks Hydrology of the Hellenic Committee of Hydrogeology in collaboration with the Geological Society of Greece and the Cyprus Association of Geologists and Mining Engineers, was organized by the Laboratory of Hydrogeology, Department of Geology, University of Patras, with the cooperation of colleagues from several universities and authorities. The Organizing Committee consists by the following members:

President:

Nikolaos Lambrakis, University of Patras, Laboratory of Hydrogeology

Vice President:

Evangelos Nikolaou, IGME Greece

General Secretary:

Anastasia Pyrgaki, Region of Western Greece

Executive Secretary:

Konstantina Katsanou, University of Patras, Laboratory of Hydrogeology

Koumoutsou Eleni Chelmos Vouraikos Geopark

Treasurer:

Eleni Zagana, University of Patras, Laboratory of Hydrogeology

Members:

Christos Petalas, Department of Environmental Engineering, Democritus University of Thrace

Constantinos Constantinou, Geological Survey of Cyprus

Grigorios Krestenitis, Region of Western Greece

Markos Sklivaniotis, M.E.W.S.P of Patras

Georgios Soulios, Aristotle University of Thessaloniki, Department of Geology

Georgios Stamatis, Agricultural University of Athens, Laboratory of Mineralogy-Geology

Georgios Stournaras, Faculty of Geology and Geoenvironment, University of Athens

Leonardos Tiniakos Region of Western Greece

Scientific Committee

Mohamed Aboufirass (N. Africa)
Ian Acworth (Australia)
Apostolos Alexopoulos (Greece)
Bartolome Andreo (Spain)
Athanasios Argyriou (Greece)
Alicia Aureli (Italy)
Giovanni Barrocu (Italy)
Konstantinos Chalikakis (France)
Antonio Chambel (Portugal)
Massimo Civita (Italy)
John Diamantis (Greece)
Alexandros Dimitrakopoulos (Greece)
George Dimopoulos (Greece)
Romeo Eftimi (Albania)
Christophe Emblanch (France)
Dolores Maria Fidelibus (Italy)
Björn Frengstad (Norway)
Michael Fytikas (Greece)
Michael Galabov (Bulgary)
Jacques Ganoulis (Greece)
Panagiotis Giannopoulos (Greece)
Vasileios Kaleris (Greece)
George Kallergis (Greece)
George Koukis (Greece)
John Koumantakis (Greece)
Andre Kranjc (Slovenia)
Jiri Krasny (Czech Republic)
Ioannis Kyrousis (Greece)
Patrik Lachassagne (France)
Nikolaos Lambrakis (Greece)
John Leontiadis (Greece)
Michael Leotsinidis (Greece)
Argyro Livaniou (Greece)
Manuel José Margues (Portugal)
Paulos Marinos (Greece)
Henrik Marszalek (Poland)
Boris Mijatovic (Serbia)
Jacque Mudry (France)
Konstantinos Nikolakopoulos (Greece)
Euagelos Nikolaou (Greece)
Andreas Panagopoulos (Greece)
George Panagopoulos (Greece)
George Papatheodorou (Greece)
Didier Pennequin (France)
Christos Petalas (Greece)
Fotios Pliakas (Greece)
Maurizio Polemio (Italy)
Antonio Pulido Bosch (Spain)
Dimitris Rozos (Greece)
Kim Rudolph-Lund (Norway)
Nikolaos Sambatakakis (Greece)
Allen Shapiro (USA)
George Soulios (Greece)
George Stamatis (Greece)
George Stournaras (Greece)
Luigi Tulipano (Italy)
Peter Udluft (Germany)
Sotirios Varnavas (Greece)
Konstantinos Voudouris (Greece)
Qin Xiaoqun (China)
Eleni Zagana (Greece)
Hans Zojer (Austria)
Nikolaos Zouridakis (Greece)

Water bodies and ecosystems

Groundwater in integrated environmental consideration

G. Stournaras

Professor of Hydrogeology and Engineering Geology, University of Athens

Introduction

The co-existence and thus the association and mutual interaction between the geological world (abiotic phase) and the biological world (biotic phase) and the role of the water, especially of the subsurface water, is a very old scientific field. This subject the last decades became particularly interest in the frame of the ecosystem's study. The previous predominant consideration declared that the geological phase is the material "on which" or "in which" the living world survives and evolutes. To-day, our knowledge goes beyond this simplified consideration. The discussed co-existence seems to be more complicated, the detected interactions seem to be more crucial and intensives and it is about a dynamic symbiosis of the two worlds, where one main bridge is the water, mainly the subsurface water. Initially the science of ecology, the concepts of ecosystems and the environmental maintenance and improvement, formed a scientific field of Biology. Although biologists did not forget the role of the geological environment, their main target became the living beings. The last decades, Geology assumed to reappoint its targets, oriented to the Environmental Geology, which includes the study of the ecosystems from the geological point of view. At the same time, ecosystems reminded themselves that they dispose a geological side equally important as the biological side*. The abiotic and the biotic world (cosmos) although are separated by inviolate borders, they present an active and mutual interaction in the frame of their symbiosis and association. From one part, the geological environment, the lithosphere, the hydrosphere and the atmosphere and, from the other part, the living beings and in the middle, the world of flora which accomplished to be placed among the living beings according the definition of life or at least according the common characteristics of life in animals and plants. This accession is due to a rather expansive, although correct, reading of the definition of life (Gohau 1987, Pavlides 2007). Moreover, in the geological world, its characteristics are comparable to the characteristics of life. The structural geological, hydrographical, gases and fluid components are coming into being by preexisted material, are developed, combined, feed each other and, finally are transformed. Hence, according a more expansive reading of life the accession of geological components in living beings could be presented. The classical biological saying "*the ontogenesis is a brief repetition of* *phylogenesis (*the controversial recapitulation theory "*ontogeny recapitulates phylogeny*") is something that can be cited for the mineral, the rock, the mountain, the pore water, the stream, the aquifer, the lake or the sea.

The use of the dialectic for the nature world offers the element of the analogy (analogon) to both, abiotic and biotic world and consequently to the study of the evolutionary processes within the nature according Engels (Mazis 2010). The addition of an atom in an organic substance forms another substance, entirely different from the initial one. In the same way, the addition of an element to a mineral creates another mineral. The more essential "prove" that the earth is an active (living) planet comes from the procedures in the abiotic environment (movement of the tectonic plates, earthquakes, volcanoes, external factors such as erosion, degradation etc), tides, wave action, winds, orogenesis, isostasy, eustatic movements and many other procedures in man and geological time. In the light of an environmental consideration, the "living" planet becomes a living planet and the living beings could include the geological environment. If biology, compared with geology, has been placed at the epicenter of human interest, connected with the philosophy, theology, and man is why dealing directly with these concepts is and not indirectly as is geology does. Finally, the concept of the ecosystem is based both on geological (abiotic) and biological (biotic) environment and any research, study, analysis, management, protection or any other consideration should take into account both concepts (Stournaras 2010).

Brief historical approach (Stournaras 2007)

Thales, naming the water as *origin of everything* sets him free from the mythological thinking, which was the heritage of Homer and the Theogony by Hesiodus. The experience of the water in many forms becomes the basis for the Thales philosophic approach concerning the unity and entity of the natural world. Thales with Anaximander and Anaximenes founded the subsequent philosophy related to a simple and unified natural world. Heraclitus preceded one more step, in the given subject, among the pre-Socratic philosophers who described a frame very closed to a symbiotic planet. He underlined that *one is everything*, which means that the beings are in a continuous argument, expressing their uniform antithetical tendency. Analyzing these terms with the present philosophical and scientific consideration some concepts could vaguely detected, such as coexistence, symbiosis, living space, food chain, parasitism, life cycle, photosynthesis, mutual action and reaction between living beings and abiotic and biotic environment. Zeno says something similar…*not any part of a being will be last and without a relation with the other...* Anaxagoras formulates one thought, which open the gate to the present scientific thoughts saying… *Nothing could be done from not being and the being cannot be ruined to a not being... We get bread, water, and many other foods and all these are transformed to arteries, hairs, flesh, bonds etc. Since all these happen, we must accept that all beings exist in the offered food and because of these beings everything is increased...* It is obvious that the presocrates philosophers, disposing a complete consideration of the natural world and of the philosophy what this world creates or is governed by, approached the concept of the coexistence and they almost described it without giving a special name.

In the middle age period, the ancient doctrines of science dominated in case they were agreed with the religion dogmas. Among the ancient philosophers, Aristotle has been used since his sayings could easily adapt to the church. It is about a period of guided systematization and criticism. The taxis of the creation elements could not be called in question. Not even in the most complete and popular books, as they had been the *Natural History* by Plinius and the *Questiones* (*Book IV of Questions on the eight books by Aristotle*) one can find reports or mentions related to the discussed subject. During the last decades of the 16th century and the initial decades of the 17th century, a peculiar cosmic theory dominates, saying that...*the nature is absolutely unalterable and invariable....* This doctrine concerned the geologic and biologic parts of the nature and their overlapping the climate. Only some scientists resisted as it was Linné, Kant, Laplace, and Herschel and, finally Lyell, based on the discoveries of Geology and Paleontology. At the same time, in the field of Biology, where the scientific expeditions, the paleontology, the anatomy, the comparative anatomy, the physiology, the discovery and use of the microscope, the discovery of the cell, confirmed the evolution, irrespective of the nature of its mechanism. In the frame of the comparative physical geography, the life conditions of different beings and their relation with the abiotic environment become clear, the appearance of skilled organs is explained and a new science is founded, which later will get the name Ecology, with its basic laws, as it is the law of Liebeg (law of minimum) and the law of Shelford (law of tolerance). Finally, the discovery that the protoplasm and the cell, structural units of all beings, exist also as self-existent lower forms of life put an end to the controversies concerning the origin of beings. At the same time, the chasm between the abiotic and the organic material is waning and *the nature is moving in an eternal and circular course (Engels).*

The histories of the two worlds are parallel, where the biologic world appears when the geologic world gives the necessary conditions for such an appearance. The formation of the crust, the water accumulations on the earth's surface, the continuous isolation from the high temperature of the core resulting from the new formed layers, and the reduction of the ultraviolet radiation transform the atmosphere to an almost similar to the present. The photosynthesis is the specific bridge, after the initial and primordial bridge of the water, between the two worlds, the exit of the life from the sea and the production of amino acids from electrical discharges are different views of a permanent and mutual co-existence and association.

New considerations

The notion of Ecology has been established in 1886, by Ernst Haeckel, one of the scientists who carried on the Work of Darwin. Haeckel gave three definitions of Ecology (Stournaras 2007) in his three volumes of «Generelle der Organismen» *...the science of economy, of living way, of vital external relations of organisms......the totality of science of relations between the organisms with the environment, including all the conditions of life...ecology or the geographical distribution of organisms (...) the science of the totality of relations between the organisms with the surrounding external cosmos, with the organic and inorganic conditions of life. What we have called economy of the nature, the mutual relations of all the living beings in a unique space, their adaptation in the environmental space, their transformation due to their fight to survive, the phenomena of parasitism etc....*In the above definitions, it is obvious the separation of the inorganic and organic world within the notion, the concept of Ecology.

During the last decades, all the sciences have been resumed and adapted in order to include ecology within their concept. Classic sciences presented a new face to accomplish the nature description and analysis starting from hydrology, hydrogeology, hydrometeorology, and generally, all the water sciences, since the water is fundamental element of the natural environment, consequently fundamental element of ecology. The radical theory of Mandelbrot* proposed a geometric version of the natural structures. The theory of chaos, natural continuity of the meteorological physics, has been applied in the ecologic studies with an orientation to the dynamics. The theory of biodiversity is deeply impregnated by the principles of ecology. New fields have been appeared, such as economic ecology, urban ecology, ecological engineering, ecology of the systems, all considering the two worlds as a unity, therefore as an undivided environment.

Looking at the main principles of ecology, two of them are revealed about the mutual active symbiosis of the two worlds. The *law of Liebeg* (law of minimum) declares that for each living being there is a minimum limit of necessary conditions of substratum, mainly of the water since they are beings living without air but not without water (i.e. the water in a desert, the oxygen for the water of a lake etc.). *The law of Shelford* (law of tolerance) declares that the tolerance of an organism, related to the environmental parameters modification, is not unlimited. This is the reason of the organisms' classification as narrow ecological range and wide ecological range species.

The role of the water is similar as previously. The history determines ecological conditions, which dispose as a fundamental element the relation of living beings systems with the water as substratum and mean of surviving and evolution or as supply element. This is why the biotic societies cannot be reformed *de novo,* the evolution is not inverted and the ecosystems cannot be divided to segments.

Related to the above considerations, the first conclusion is that the natural systems in both, abiotic and biotic environment, developed in a not inverted direction.

Seasonal episodes, the days-nights and summer-winder alternations, the climatic changes, the geologic procedures insinuate in the development of living beings, modifying the natural environment, including the aquatic environment, creating new conditions. The second conclusion is that development is connected to the form of the hierarchic structures. For example, a lake with fishes contains much more information related to a lake with only bacteria. In case that this is true, another conclusion – surprise is that the elements of an ecosystem demand the information existence in the form of a classified hierarchy. On the contrary, the biotic elements do not demand the information hierarchy; even they are adapted to a more complicate system. The main abiotic factors of the aquatic or terrestrial ecosystems, having a direct or indirect relation with the water are:

- The water in a liquid or aerial form. *The water itself.*
- The total of dissolved salts (salinity, ionic strength). *The role of the water in solubility and configuration of the salts concentrations.*
- The sunlight. *The role of the sunlight is connected with the depth in seas and lakes.*
- The temperature. *It is about the water temperature and the environmental temperature affected by the water temperature.*
- The ground or seas bottom – riverbeds sediments nature. *The role of the water becomes obvious.*
- The oxygen and other gases (i.e. atmospheric air). *The role of the water becomes obvious, both as containing and contained.*
- The pH. *It is connected with the water presence.*
- The nutritious elements. *The same as their concentration.*
- The trace elements. *The same as their concentration*

The substratum nature and the suspended load of an aqueous receiver have a very important meaning in the frame of ecological and physicochemical procedures. The terrestrial ecosystems cannot develop when the soil is not fertile or it is instable. Therefore, the erosion, both as natural and anthropogenic action denudate the land from the soil layer, even thin, prevents the nourishing materials accumulation and the flora development, just as it happens in the case of the sand's aeolic mobilization (dunes).

The temporary consideration about the mutual coexistence of the two worlds starts from the Russian geoscientist Vladimir Vernadsky (Vernadsky, 1926 republished 1997). It is true that Vernadsky was preceded of his era and scientists were not ready to understand his view*. Collecting two points of his works, related to our target, we should note his consideration of the *life as a main geological factor* on the earth. Furthermore, he called the living matter as *vitalized water*, which is a perfect definition of the life. It is, also, correct that the participation of the water in the appearance, evolution, and maintenance of the life is not limited to a passive role, as substratum or nourishing or conservative element.

---

* B. Mandelbrot "The fractal geometry of nature", W.H. Freeman, 1982

This is obvious after the role of the water in *symbiosis*, which is the bridge between the two worlds, the inorganic and organic matter, and the abiotic and living world. Within this frame, life in our planet is characterized by the direct or indirect, visible or presumed, permanent or temporary symbiosis, which permitted to the biologist Lynn Margulis to call our planet, a *symbiotic planet* (Margulis 1999). A common denominator in all forms of symbiosis, consequently in all forms of life, as source or environment of life, is the water. The concept of Gaia, name of mother earth in ancient Hellas (Greece) contains a postulate. *The earth is alive.* This is well known among the geologists procedures mentioned above. In the same frame, the water is the main component of the theory of Gaia, proposed by J.E. Lovelock (Lovelock J. (1988) *Translated to Greek 2007,* 2006), which underlines, among other, that *...the characteristics of the gases of the atmosphere, of the rocks and of the waters on the surface and in the interior of the earth, are fixed by means of the evolution, the death, the metabolism and the other activities of living beings...* The action of the environment, common action of the two worlds, resulted in the partial exit of the life from the sea and the conquest of the initially hostile land. The ultraviolet solar radiation, the dearth of nourishing mater, and the devastating dryness 500 millions years ago, have been surmounted by the active co-existence with its unique action as it is the *photosynthesis, the biological degradation* etc. described in the Theory of Hypersea by Marc and Dianna Mac Menamin (Mc Menamin and Mc Menamin, 1994).

The self organized critical character is a new consideration of the nature (Bak, 1999). According the fundamental principle of the above consideration, the nature is always in a non balance state, but organized in a meta-instable state, the crucial state, where everything can happen in the frame of well defined statistical laws. The science of the self organized critical character is oriented to the examination of the question: Why nature is not limited simply to express the physics laws, but it appears much more complicate. The obvious next question is whether this complication is exclusively limited to the living world, where is evident, or is extended to the abiotic geological world. An initial and eventually final answer is that this complication is not extended to geology, because of the procedures of evolution, development, genesis, and consummation in the living world are unlimited, while, in the geology, the modifications are limited in a typical frame of chemical reactions and physical laws. The passage from the simple laws of physics to the complexity remains a severe question considering its profundity, range and the induced mechanisms. Geology, Biology, Atmosphere Science, Climatology and other sciences gave already partial answers, but a complete answer still is demanded.

"...When Lynn Margulis and I introduced the Gaia Hypothesis in 1972 neither of us was aware of Vernadsky's work and none of our much learned colleagues drew our attention to the lapse. We retraced his steps and it was not until the 1980s that we discovered him to be our most illustrious predecessor... (declaration by J. Lovelock)". "The Biosphere" by V. Vernadsky (1926) republished in 1977, revised and annotated by M. Mc Menamin

Bak gives some examples such as geophysists speak about plate tectonics without any mention to astrophysics. Biologists describe the characters and the evolution of species without any reference to geophysics. Economists analyze the monetary exchanges without any connection with biology etc. Later, is quite possible that these references will be considered as necessary, but the symbiosis of the two worlds is to be described and analyzed in their real dimensions now.

It is cited that, in many sciences, the use of a narrative account is not enough, since the outcome of actions and conditions can be a result of different experiments. All these are useless in evolution (biology) or in paleontology, where the actions and conditions cannot be exactly reproduced. Another consideration, referred t the nature, is the concept of the *fractal structures*. These structures dispose geometrical forms, with characteristics of all the scales, which constitute the nature, according Mandelbrot (Bak). The research on the geometrical characters of the fractal structures has already proceeded, but the question of their origin still remains unanswered. The accession of the nature in the frame of the fractals has been attempted by Bak in a piecemeal way, especially regarding certain geoforms. Their full accession is an open matter, while there is no mention about Geology or Biology.

About the organization on the earth surface

Looking at the beginning of the Geology, which means the initial stage of the crust formation, Gautier (Gautier 1853), cited that, according the law of the rotation, the freezing started from the poles, where should be the origin of the oceans. At the poles, also, the first surface elevations existed and the formation of the initial relief where the first organization of flora or fauna appeared. Step to step the freezing mobilized to the equator. If this is true, it is evident that the triggering point for the appearance of life was the diversification of the geological and climatic conditions (Lyell 1830-33, republished 1997). Within the above evolution, the different zones acquired different climate, mainly in the base of the temperature. In this case, the earlier mentioned laws of Liebeg and Shelford start arrange the survival, development and evolution of living beings. The role of the sunlight becomes crucial. Being a part of the abiotic environment, the solar energy created the second bridge between the two worlds, after water, the photosynthesis procedure, the transformation of inorganic matter to organic. The structure of an inorganic molecule is solid and it is maintained by a cohesion forces in the surface. The structure of an organic molecule is cellular, it is consisted by real molecules but the disconnection is easier, due to the most mobile nature of the given elements. Although the discussed differentiations exist, the molecular chemical combinations seem the same in both cases. The inorganic and organic different forms are visible only after successive modifications, where the combinations of the organic molecules are more complex (Rossi et al. (Eds) 1999, Schoch et al. 1999).

Hydraulic connections

Since the water is a fundamental environmental element and, at the same time, the first bridge between the two worlds, the internal water bodies (surface and ground flow) and the hydraulic relations among them obtain a special interest. The water is the essential element in the frame of the ecosystems sustained and development, in association with the activity of the geological environment. The last ecological conquest, offered by the geological side, is the legislatively adapted *ecologic flow*, which means that the totality of the water running to the sea or to a lake is not lost water, since it ensures the maintenance of the ecosystems. From technical point of view, the ecologic flow means a certain minimum river flow and a minimum lake level that must be maintained.

The initial shape concerning the water distribution on and I the earth is the hydrologic water circle P (Precipitation) = R (Runoff) + E (Evapotranspiration) + I (Infiltration) ± RV (Reserves Variations). The hydrologic water circle has a meaning only considered in certain period and in certain area. The above general distribution is modified because of secondary, tertiary etc. phenomena as they are the secondary runoff of the springs water the secondary infiltration during the runoff and so on.

The climatic change in the frame of the water economics define another very important field to be analyzed and been used in the environmental maintenance and improving, as a new factor which, directly or indirectly, intrudes itself in the relevant considerations (Alexandersson 1976; Caldecott 2007, Holst-Warhaft et al. (Eds) 2010; Kolbert 2006; Lane 2009; Orsenna 2008; Raymond (Ed) 1998; Stern 2007; Weizsäcker et al. 1997).

Beyond and additionally to the above water balance, there are many special types of hydraulic connection between surface and ground water bodies, in different geoforms (Stournaras 2007).

Mountainous areas The main characteristics here are the high level of precipitation, the ice formation and the long distance run in a big hypsometric difference. The high slope helped by the coarse bed materials favor the secondary infiltration. The complexity of the aquatic exchanges in the mountainous areas, the micro- and megaclimat, the flora and fauna and the evolution of the geoforms, acquire remarkable intensity and visible in the man scale life.

Riverine zones The exchanges between surface and groundwater depend on the presence of local and regional aquatic systems, the permanent or seasonal floods and the evapotranspiration index. Generally, the aquifers are in reduced depth, related to the river level.

Coastal zones The discussed exchanges in these zones are affected by the discharge of the surface runoff and of the groundwater flow to the sea, which is the base level for both. These actions are accomplished by the evapotranspiration and the action of the sea mechanisms, as they are the sea currents, the tide action and the wave action (Stournaras 1999).

Glacier and sandy areas Because of the alternation of hills and ground depressions the drainage of these areas is very often incomplete, since a great part of the runoff is engaged in lakes and wetlands.

Karstic areas They represent the most complicate aquatic system, while the runoff can be limited or even negligible (dry valleys etc.). The hydraulic exchanges between the surface and ground waters include a considerable number of physical, chemical, geological, and biological procedures actualized in an extended range of physiographic and climatic conditions. Some of them are products of the natural geological and biological evolution, some other are products of the man made pollution and contamination. Such procedures are the eutrophication, the attack of the water bodies by the atmosphere, the desertification, the wetlands destruction etc. The same consideration adapted to necessary particularities, can be cited for fissured rocks (Stournaras 2008).

Regarding the consideration of the aquatic exchanges between the different water bodies, an extended range of exchanging procedures can be recognized.

Exchanges between surface runoff and groundwater aquifers The main procedures in this subject are the development of the hydrologic circle by secondary, tertiary etc. phenomena as they are the secondary runoff by the groundwater coming out from a spring, the secondary infiltration during the runoff, etc.

Exchanges between lakes and groundwater aquifers The regime of these exchanges is settled by the hydraulic charges of the lake and the surrounded aquifer in the different points of the lake's perimeter.

Exchanges between lakes and groundwater wetlands The procedures of the exchanges depend on the wetland type and the groundwater hydraulic characteristics. A particular case is the total or partial discharge of the groundwater aquifer in the wetland, before its arrival at the sea (Stournaras et al 2007).

The role of water and groundwater in the environment and the ecosystem

Considering the water as an environmental element, its fundamental role becomes obvious since

- There are organisms living without air, but not without water.
- The water is a bridge between the abiotic (geological) world and the biotic (biological world).
- The water maintains another bridge between the two worlds, the photosynthesis, which means transformation of inorganic (abiotic) groundwater ecology matter to organic (biotic) one.
- The water keeps the essential role within the relation between the life and the environment: land uses (the water is definitive for each land use), natural resources (the water is itself an essential resource, and a determinate factor for

the genesis, evolution and use of all natural resources), sub products disposal (the water is the final receiver of all waste sub products of the natural and human activities).

Considering the groundwater origin, flow, storage, and its hydraulic connections with surface water bodies, including the sea water, there is a multiple role in the above described frame.

- The groundwater is one of the constituents of the hydrologic circle
- The groundwater takes place at the secondary phenomena of the hydrologic circle development (exit from springs and added to the runoff, receiving secondary infiltration from runoff etc.)
- The groundwater maintains the top soil and sub soil moisture, indirectly, giving a priority to the soil moisture, and directly, by the capillary water rising
- The groundwater forms itself a field of ecological conditions (groundwater ecology) where its chemical properties contribute to the general environmental formation. Furthermore, the groundwater ecology takes place at the food chain which is another important environmental contribution.

Groundwater ecology

Groundwater ecology is the study of the interactions between groundwater organisms and their external environment, be it the immediate aquifer or connected terrestrial system (Danielopol et al. 1994). Groundwater animals are collectively known as *stygofauna* after the Styx spring connecting the existing world and the underworld, where Thetis put Achilles to make him permanently uninjured.

The decade of 1990s was a significant one in the field of groundwater ecology, commencing with the publication on the surface water/groundwater ecotone concept. An *ecotone* is a transition zone between two ecosystems that displays characteristics of both. Gibert et al. (Gibert et al. 1990) argued that the hydrological interaction between surface waters and groundwaters mediated an ecotone that consisted of high intensity, oxygen saturation, and amplitudes in variables such as water temperature, along with biogeochemical gradients in micronutrients such as nitrogen. The acetones is especially evident in the hyporheic zone (see below), which can act as a physical, chemical, and biological filter. At the same time it is able to immobilize or transform nutrients or pollutants, preventing or reducing their transfer between groundwater and the surface. Another significant advance in the 1990s was the synthesis of detailed reach-scale studies concerning the central role of hydrological exchange between groundwater and river water in governing the chemistry and fauna of the hyporheic zone. The third major development was the incorporation of specific elements of spatial scale into understanding groundwater ecosystems, especially hyporheic zones. In brief, this explored the different trends that emerge when groundwater-surface water interactions are assessed at a sediment scale. As the hydrology of all surface water bodies relates to their groundwater flow systems, geological characteristics of their beds, and their climatic settings, these interactions must be understood and are likely to influence groundwater organisms at a variety of scales (Baxter et al. 2000; Boulton et al. 2002).

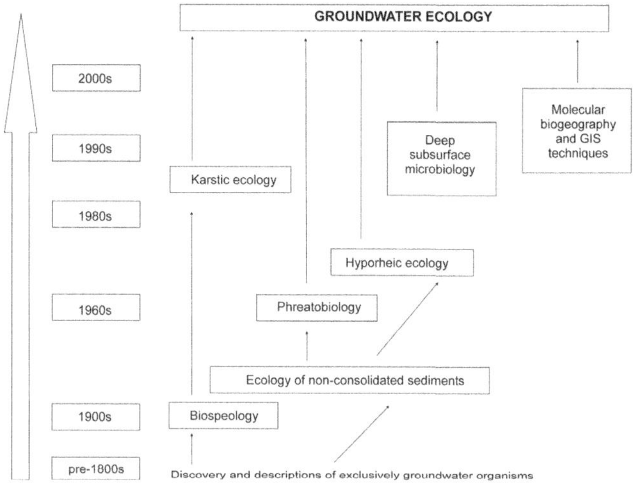

Fig. 1. The evolution of researches perspectives in the development of groundwater ecology (Gibert 1994, in Hancock and Boulton 2005).

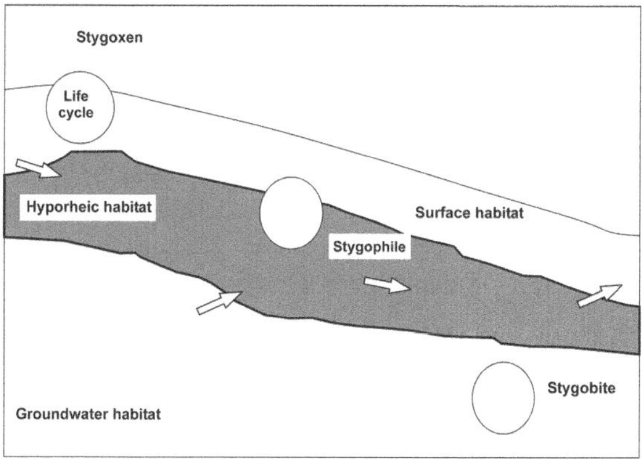

Fig. 2. A functional classification of hyporheic fauna based on their habitat affinity for groundwater (Marmonier et al. 1993, modified by Hancock and Boulton 2005).

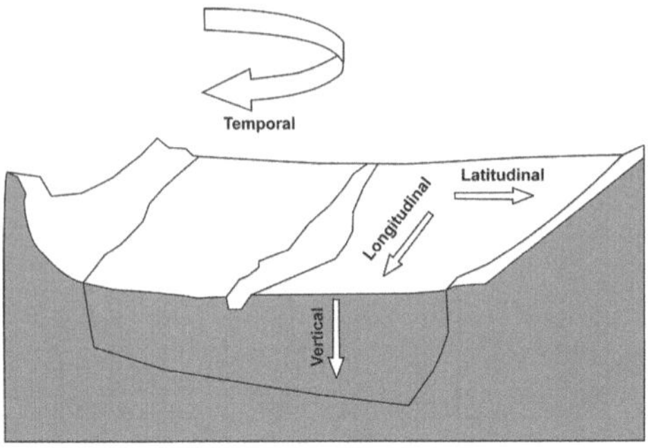

Fig. 3. The four-dimensional nature of alluvial groundwater ecosystem (Dole-Olivier et al. 1994, modified by Hancock and Boulton 2005).

Groundwater-surface water interactions

The biogeochemical processes within the upper few centimeters of sediments beneath nearly all surface water bodies (*hyporheic zone*) (Hancock et al. 2005) have a profound effect on the chemistry of the water interchange, and here is where most of the recent research has been focusing is. Groundwater and surface water are not isolated components of the hydrologic system, but instead interact in a variety of physiographic and climatic landscapes. Besides that the origin of the groundwater is the surface water, development of contamination of one commonly affects the other. Therefore, an understanding of the basic principles of interactions between groundwater and surface water is needed for effective management of water resources. To understand these interactions it is necessary to understand the effects of what Tóth (Tóth 1970) called *hydrogeological environment* on groundwater flow systems – that is the effects of topography, geology, and climate. Focusing at the aquatic relations, the interactions of streams, lakes, and wetlands with groundwater are governed by the positions of the water bodies with respect to groundwater flow systems, geologic characteristics of their beds and their climatic settings.

Considering these interactions in a larger scale, a geomorphologic perspective is also helpful. For example, Larkin and Sharp (Larkin et al, 1992 classify stream-aquifer systems (based on the predominant regional groundwater flow component) as (1) *underflow-component dominated* (the groundwater flux moves parallel to the river and in the same direction as the streamflow; (2) *baseflow-component dominated* (the groundwater flux moves perpendicular to or from the river depending on whether the river is effluent or influent, respectively, or (3) mixed, according Sophocleous (Sophocleous 2002).

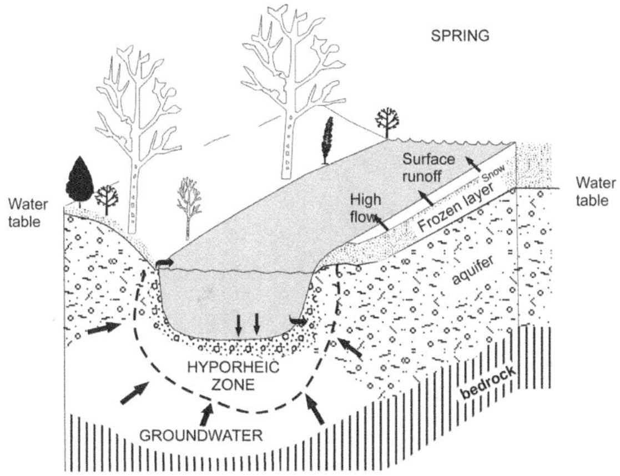

Fig. 4. Descriptive model of the dynamics of the hyporheic zone and the surrounding water systems. The arrows size indicates the relative magnitude of flow (Williams 1993; in Sophocleous 2002).

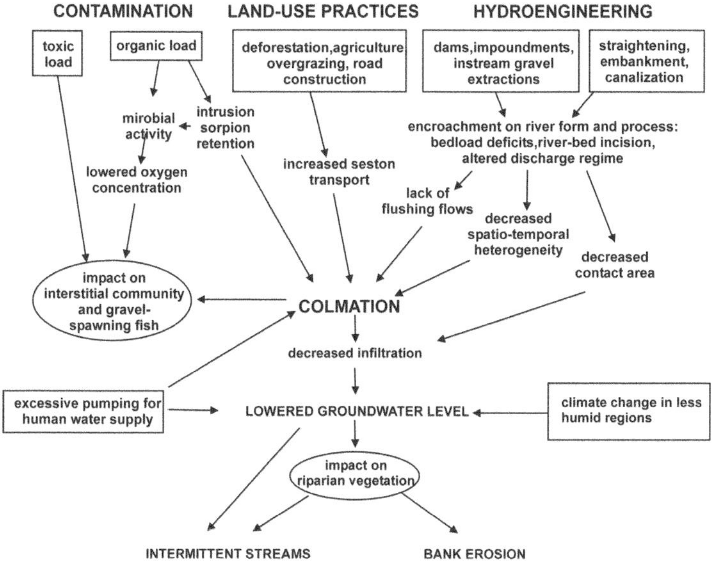

Fig. 5. Human-induced impacts promoting clogging of stream-bed sediments and their ecological consequences (Brunke and Gonser 1997; in Sophocleous 2002).

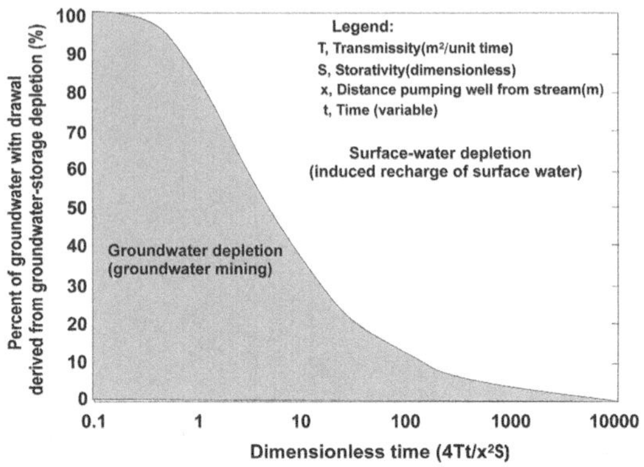

Fig. 6. Transition from reliance upon groundwater storage to induced recharge of surface water (Balleau 1988; in Sophocleous 2002).

Hyporheic zone

Water flows not only in the open streams, but also through the interstices of stream channel and bank sediments, creating a mixing zone with subsurface water. The zone of the discussed mixing is the *hyporheic zone.* This zone is obviously a zone of intensifies biogeochemical activity. The time travel estimated for the water in a stream channel might be too short to permit significant mineralization of organic nutrients. However, if hyporheic exchange is an important process, residence time within a reach and contact with subsurface sediments may result in dramatic alterations in material transported from the catchment to the receiving water body. According Maddock et al. (Maddock et al, 1995) the flow paths within the river bed are a function of the geomorphology of the bed.

A homogeneous and isotropic streambed is assumed to exist between the stream and the underlying hyporheic zone (Fox et al, 2002). It is also assumed that the streambed has a lower hydraulic conductivity than the underlying soil and that leakage from the stream to the aquifer represents a steady, vertical, one-dimensional flow. Consequently, three regimes of stream/aquifer hydrologic states exist.

- Regime A – *Saturated flow.* Stream leakage rate depends on the location of the water table.
- Regime B – *Transition zone.* Stream leakage rate governed by unsaturated flow hydraulic conditions.
- Regime C – *Unsaturated gravity-driven flow.* Stream leakage rate is not a function of the location of the water table.

Mechanisms of interactions

Hydrologic interactions between surface and subsurface waters occur by subsurface lateral flow through the unsaturated soil and by infiltration into or exfiltration from the saturated zones. In the case of karst or fractured terrain, interactions occur through flow in fracture/solution channels. Water that enters a surface-water body promptly, in response to such individual water input events as rain or snowmelt, is known as *event flow, direct flow, storm flow, or quick flow.* This water is distinguished from *base flow*, or water that enters a stream from persistent, slowly varying sources and maintains stream flow between water-input events. Although some base flow is derived from drainage of lakes or wetlands, or even from the slow drainage of relatively thin soils on upland hill slopes, most base flow is supplied from groundwater flow. Subsurface flow can also enter streams quickly enough to contribute to the event response. Such flow is called *subsurface storm flow* or *interflow.* Interflow involves both unsaturated and saturated flows. If interflow encounters a seepage face, the interflow process may grade into *return flow* by which subsurface water can contribute to *overland flow.* The mechanisms by which subsurface flow enters streams quickly enough to contribute to stream flow responses to individual rainstorm and snowmelt seem to be the following: *translatory flow, macropore flow, groundwater ridging and return flows.*

Ecological significance of interactions

The time travel estimated for steam water might be too short for significant mineralization of organic nutrients. I the case of the hyporheic zone, exchange is an important process. An important aspect of groundwater-surface water interchange is that surface water in streams, lakes, and wetlands repeatedly interchanges with nearby groundwater. Thus, the length of time water is I contact with mineral surfaces in its drainage basin is extended after the water first enters a stream, lake, or wetland. An important consequence of these continued interchanges between surface water and groundwater is their potential to further increase the contact time between water and chemically reactive geologic materials.

References

Alexandersson O ((1976), "Lebendes Wasser", (Translated to Greek, 2008), ΕΙΣΕΛΙΞΗ Eds
Bak P. (1999) How Nature Works – The Science of the Self-organized Criticality", (Translated to Greek), Katoptro Eds, Athens
Baxter CV, Hauer FR (2000) Geomorphology, hyporheic exchange, and selection of spawning habitat by bull trout (Salvelinus confluentus), Can. J Fish Aquat. Sci. 57:1470-1481

Boulton AJ, Hakenkamp C, Paslmer M, Strayer D (2002) Fresh-water meiofauna and surface water-sediment linkages: A conceptual framework for cross-system comparisons, In Rundle S.D., Robertson A.L., Schmid-Araya J.M. (Eds) Fresh-water meiofauna biology and ecology, Backhuys, Leiden, The Netherlands, p 241-259
Caldecott J (2007), Water. The Causes, Costs and Future of a Global Crisis", Virgin Books Eds
Danielopol DL, Creuzé des Châtelliers, Moeblacher F, Pospisil P, Popa R (1994) Adaptations of Crustacea to interstitial habitats : a practical agenda for ecological studies, Groundwater Ecology (Eds J. Gibert, D.L. Danielopol, and J.A. Stanford), pp 218-243, Academic Press, San Diego
Fox GA, Du Chateau P, Durnford DS (2002) Analytical model for aquifer response incorporating distributed stream leakage, Ground Water 40:378-384
Gautier A (1853), "Introduction Philosophique à l' étude de la Géologie", Masson Eds
Gibert J, Dole-Olivier MJ, Marmonier P, Vervier P (1990) Surface water – groundwater ecotones, In Naiman R.J., Décamps H. (Eds) The ecology and management of aquatic – terrestrial ecotones, UNESCO / Parthenon Publishing, London, pp 199-225
Gohau G. (1987) "Histoire de la Géologie, La Découverte Eds
Hancock PJ, Boulton AJ (2005) Aquifers and hyporheic zones: Towards an ecological understanding of groundwater, Hydrogeological Journal, 13:98-111
Holst-Warhaft G, Steenhuis T (Eds) (2010) Loosing Paradise. The Water Crisis in Mediterranean", Ashgate Eds
Kolbert E (2006) "Field Notes from a Catastrophe", (Translated to Greek, 2007), AVGO Eds
Lane N (2009) Life Ascending, Profile Books Eds
Larkin RG, Sharp JM Jr (1992) On the relationship between river-basin geomorphology, aquifer hydraulics, and groundwater flow direction I alluvial aquifers, Geol. Soc. Am. Bull. 104:1608-1620
Lovelock J (1988) The Ages of GAIA, (Translated to Greek 2007), Octavision Eds, Athens
Lovelock J (2006) The revenge of GAIA, (Translated to Greek, 2007), A.A. Livanis Eds, Athens
Lyell Ch (1830-33, republished 1997) Principles of Geology, Penguin Books Eds
McMenamin M, McMenamin D (1994) "Hypersea – Life on Land", Columbia University Press, New York
Maddock IP, Petts GE, Evans EC, Greenwood MT (1995) Assessing river-aquifer interactions within the hyporheic zone, In Brown A.G. (ed) Groundwater and Geomorphology, Wiley, Chichester, pp53-74
Margulis L (1999) Symbiotic Planet, (Translated to Greek, 2001), Katoptro Eds, Athens
Mazis ITh (20010 Water Geopolitics in Middle East", Papazisis Eds (in Greek)
Orsenna E (2008) L' Avenir de l' Eau, Fayard Eds
Pavlides S (2007) Pangaea. Leader Books, Athens (in Greek)
Raymond S (Ed) (1998) Science, Technology and the Economic Future", New York Academy of Sciences Eds
Rossi C, Bastianoni S, Donati A, Marchettini N (Eds) (1999) Tempos in Science and Nature. Structures, Relations, and Complexity, New York Academy of Sciences, Vol. 879
Schoch RM, McNally RA. (1999) Voices of the Rocks, Thorsons Eds
Sophocleous M. (2002) Interaction between groundwater and surface water: the state of the science, Hydrogeological Journal, 10: 52-67
Stern N (2007) The Economics of Climatic Change", Cambridge Eds
Stournaras G (1999) Correlating morphometric parameters of Greek Rhone-type deltas. Hydrogeologic and environmental aspects. Environmental Geology, V 38/1, No 1
Stournaras G (2007) Water. Environmental Aspect and Route, Tziolas Eds, Thessaloniki (In Greek)
Stournaras G, Leonidopoulou D, Yoxas G. (2007) Geoenvironmental approach of Tinos Wetlands (Aegean Sea, Hellas), 35th International Congress of IAH, Groundwater and Ecosystems, PORTUGAL, Lisbon

Stournaras G. (2008) Hydrogeology and vulnerability of limited extension fissured rocks islands, Ecohydrology and Hydrobiology, Vol. 8, N0 2-4, pp. 391-399
Stournaras G. (2010) The groundwater bodies in ecology and ecosystems Conference of Interacademy Panel Towards Engineering Harmony between Water, Ecosystem and Society. Strengthening the Collaboration between European Academies of Science in the IAP Water Programme, Zakopane, Poland
Tóth J (1970) A conceptual model of the groundwater regime and the hydrogeologic environment, J Hydrol 10: 164-176
Vernadsky V I (1926 republished 1997) The Biosphere, Copernicus, Springer Verlag Eds
Weizsäcker Von E., Lovins ZA.B., Lovins I.H. (1997) Factor Four - Doubling Wealth, Halving Resources Use, Earthscan Eds, London

Ecological requirements (Habitats Directive) versus water requirements (Water Framework Directive) in wetland ecosystems in Spain

A. de la Hera[1], J.M. Fornés[1], M. Bernués[2], J.J. Durán[1]

[1] Geological Survey of Spain. Ríos Rosas 23. 28003 Madrid. Spain. a.delahera@igme.es; jm.fornes@igme.es; jj.duran@igme.es;

[2] Ministry of Environment. Ríos Rosas 24. 28003 Madrid. Spain. mbernues@mma.es

Abstract The Water Framework Directive (WFD) and the Habitats Directive (HD) have a number of similarities in relation to the conservation of wetland ecosystems. The WFD aims to maintain the ecological health of aquatic ecosystems. The concept of "good status" by this directive is set to be equivalent to the concept of "good conservation status" of habitats of community interest (HCI) established by the HD. Both concepts are based on maintaining the structure and functions of ecosystems in the long term. However, the transposition of the WFD in Spain has introduced a new concept "water requirements" of the surface water body type lake, to the concept of "environmental requirements" of the habitats of community interest raised by the HD. In Spain, many of Sites of Community Important (SCI) as part of the Natura 2000 network are surface water bodies such as lakes or Protected Areas as results of the implementation of the WFD, identified and subject to protection as stated in the Basin Water Plans. Therefore, protection elements are common in both directives. The purpose of this paper is directed to conduct a conceptual analysis to define the similarities and differences, and determine the scope of the meaning of both concepts. A superficial analysis might point that the two concepts are equivalent; however, a more detailed analysis reveals significant differences that must be taken into account to avoid misunderstandings from scientists and technicians.

1 Introduction

The Habitats Directive (HD), along with the Birds Directive (BD) is the backbone of biodiversity policy in the EU. Protected areas under these Directives constitute the so-called Natura 2000 network, which covers about 20% of the EU (Carcavilla et al. 2008). Moreover, the Water Framework Directive (WFD) aims to establish a framework for the protection of all surface water and groundwater, by achieving good ecological status by 2015. The above mentioned three directives, aim to ensuring the health of aquatic ecosystems while, at the same time, ensure a balance between water/nature and sustainable use of natural resources. Among these guidelines there are a number of synergies, such as the implementation of measures under the WFD to benefit, in general, the HD, BD and WFD objectives. However, the implementation of these policies in practice has opened a number of issues, and although there has been a large number of guidance documents to assist and coordinate this practical implementation, a number of them remain on the table pending further solve.

This document is addressed to "aquatic ecosystems and, with respect to their water needs, terrestrial ecosystems and wetlands directly depending on aquatic ecosystems" (article 1 (a) WFD).

2 Similarities and differences between the Water Framework Directive and the Habitats Directive

The work done for the implementation of the WFD, has been carried out by the River Basin Authorities and concluded with the identification of surface water bodies type lake, according to the criteria defined in the so-called Instructions for Water Planning of Spain. Generally, they are recognized by the WFD as "lake" (art. 2, (5)) are those inland water bodies, which are shallow and quiet, but also some "artificial water bodies" and some "heavily modified water bodies".

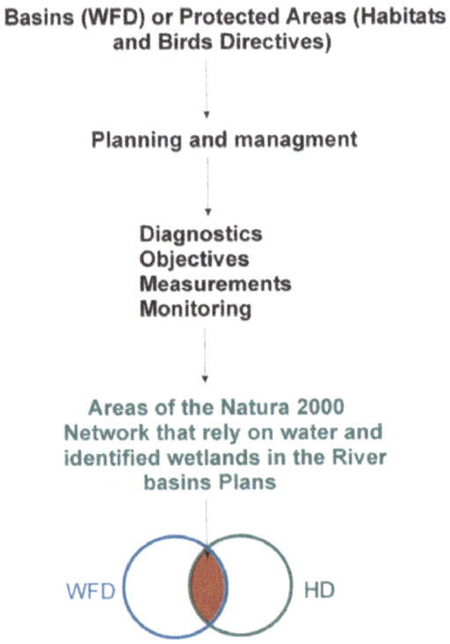

Fig. 1. Overlapping responsibilities between planning and watershed management (WFD) and protected areas (as the Birds and Habitats Directives) (SEO / BirdLife 2010).

On the other hand, the article 6 (1) of the WFD stipulates the establishment of a register of protected areas "that have been designated as requiring special protection under specific Community legislation for the protection of surface and groundwater, and conservation of habitats and species directly dependent on water". These areas are identified as "Natura 2000 sites dependent on water" (Fig. 1). For these Natura 2000 sites, the targets of HD, BD and WFD directives must be applied. Many of them overlap with sites identified as Sites of Community Importance (SCI) or Special Areas of Conservation (SAC) of the Natura 2000 network. This network, in turn, depending on the habitat types of Community interest defined in Annexes I and II of the HD, initially identify, in a first step, the SCI, and in a second step, the SAC. In Spain, are the Autonomous Communities who define these spaces, known as SCI according to the criteria in Annex III of the HD.

Within the requirements of the HD, each Member State must perform an upgrade and evolution of the conservation status of habitats every 6 years (article 17.1, Directive 92/43/EC), while according to the WFD, River Basin Management Plans should be updated every 6 years (article 13.7, Directive 2000/60/EC). A comparison between some elements of both directives allows identifying some other synergies of interest (Fig. 2). However, one of the main questions to be addressed is: what is the relationship between good ecological status (WFD) and a favorable conservation status (HD)?

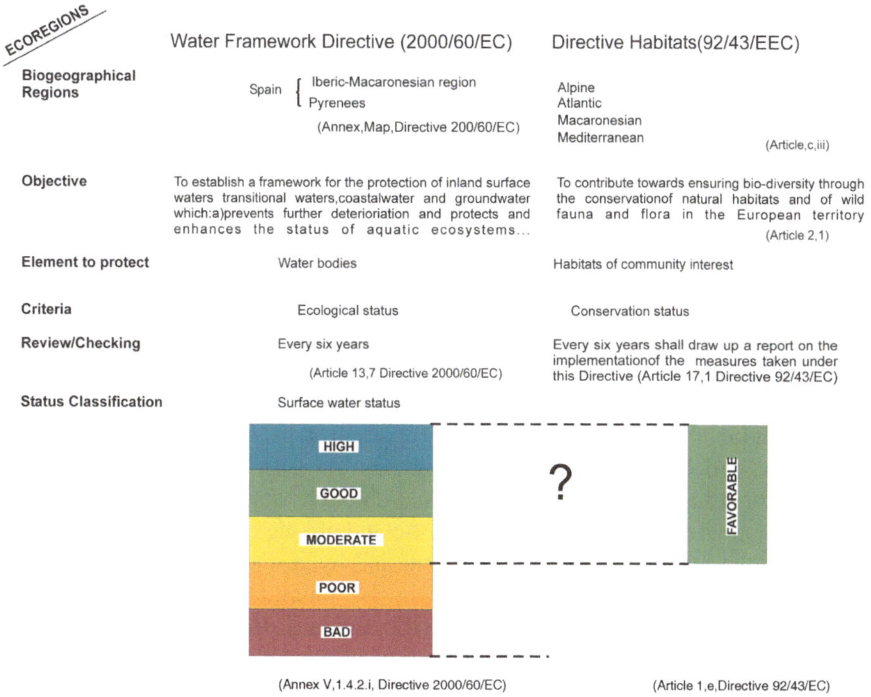

Fig. 2. Comparative analysis of some issues of Water Framework Directive (WFD) and Habitat Directive (HD).

The concept of "good ecological status" is defined as the status of a body of surface water, which is classified as such under the criteria of Annex V of the

WFD (art. 2, 22). This concept is built on two additional concepts: good "qualitative" and "quantitative" state. These are set based on comparisons with undisturbed reference conditions set for the type of water body. The assessment of ecological status is based, ultimately, on the deviation of state of its specific type of reference conditions and relies on physical-chemical and hydromorphological elements selected (Fig. 3). The ecological status is classified according to the normative definitions of high, good and acceptable state for each of the quality hydromorphological, biological and physical-chemical quality elements (Fig. 2). In this sense, it is noteworthy that the WFD does not look to the presence or absence of certain species (as the Habitats Directive does), unless their presence or absence is essential to determine the ecological status of such water body.

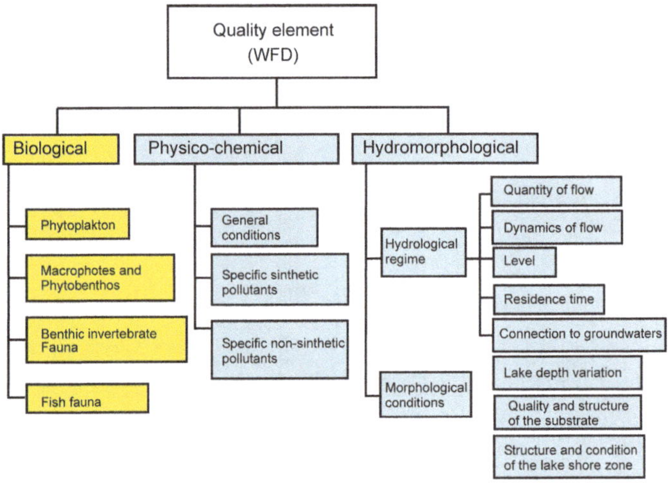

Fig. 3. Definitional elements of the state of the water body type "lake" established by the Directive 2000/60/EC.

In addition, the HD stated that the conservation status of habitat refers to keeping quality reflected in its structure and function (Fig. 4). This structure and functioning are described in terms of biotic and abiotic characteristics, against which the question arises again: could these biotic and abiotic characteristics be identified with the biological, physical-chemical and hydro-morphological assessment of the status of water body of the WFD (Fig. 3)?. Interestingly, the WFD provides indicators of the water body, but not the HD. This is partly logical, given that water is an element found only in certain types of habitats, not in all, and because the conservation status of a habitat, is also related to its relationship with the environment (air, water, soil, vegetation, etc.) (EC 2000).

The surface water body type "lake" (WFD) was established based on a set of descriptors defined in Annex II, 1.2.2 WFD, while the identification of SCI and SAC is based on the presence of types of habitats and species identified in Annexes I and II of the HD. Therefore, the identification criteria are different.

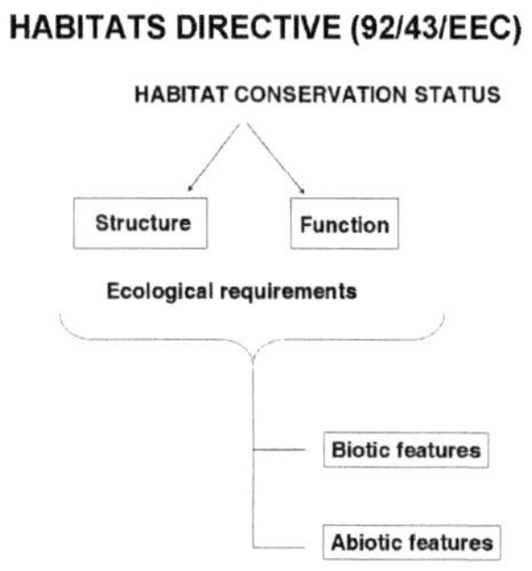

Fig. 4. Definitional elements of the state of conservation established by the Directive 92/43/EC.

There are other aspects to be noted between the two directives, such as differences in terminology (Table1). The WFD uses a language eminently limnological, while HD does it with a dominantly botanical language in order to define the same types of water bodies (SEO / BirdLife, 2010).

Table 1. Terminology differences between the WFD and the HD for the types of lakes (SEO / BirdLife 2010).

Types of lakes WFD	Types of lakes HD
• Marsh types Littoral	• Oligotrophic waters containing very few minerals (Littorelletalia uniflorae)
• Interior sedimentation basin, karst, mixed contribution, small	• Oligo-mesotrophic calcareous waters with vegetation of Chara spp.
• Interior sedimentation basin, karst, hypogene, small	• Natural eutrophic lakes and ponds with vegetation Magnopotamion o Hydrocharition
• Interior sedimentation basin, karst not permanent, shallow, non-saline	• Natural dystrophic lakes and ponds
• Interior basin sedimentation, floodplain type, low to medium mineralization	• Mediterranean temporary ponds and lakes
• Littoral dune complex, ongoing	• Lakes in gypsum karst lakes

3 IGME work for the Habitats Directive and the Water Framework Directive

The work recently published by the Ministry of Environment of Spain (MARM) (WAA 2009), has several thousands of pages devoted to the ecological characterization of habitats of Annex I of the HD, related to water in Spain. This eminently scientific-technical work describes what is the favorable conservation status for each type of habitat in Annex I of the HD in Spain. In this study, the MARM contacted the main companies and scientific institutions in Spain related to environmental conservation, among which the Geological Survey of Spain (IGME). Reports produced are focused on describing each of the habitats, placing emphasis on six areas: (1) definition of habitats, name and geographical distribution; (2) ecological characterization of habitats, control biophysical, ecological requirements and subtypes; (3) parameters and procedures for assessment and monitoring of the condition and future prospects; (4) recommendations for conservation; (5) goods and services offered by the habitat; (6) literature and photographic documentation. It is intended that the document not only describes the habitat, but also provides specific guidelines to assess their conservation status and evolution over time (Carcavilla et al. 2008).

The HD also includes geology, geomorphology, among other reasons for conservation, because some of the habitats are of geological and geomorphological importance (such as natural cavities), while in others (such as wetlands) the geological conditions play an important role (Carcavilla et al. 2008). In particular, the HD defines 9 categories of habitats. In turn these are subdivided into 21 others, from which the 117 types of habitats of Community interest emerge (Table 2). Wetlands are distributed among categories 1, 3 and 7 of that table, although there are many other habitats present in them.

Table 2. Habitat classification (Habitats Directive 92/43/EEC).

Code and name of the major categories of habitat
Coastal and halophytic vegetation
Coastal and inland dunes
Habitats of freshwater
Health and scrub warm of template zone
Sclerophyllous shrublands
Semi-natural grassland formations raised bogs and lower marsh areas
Rocky habitants and Caves
Forests

More recently, the MARM contacted IGME to carry out a series of scientific and technical work to support the sustainability and protection of groundwater, namely the identification and characterization of the interaction that occurs between groundwater and wetlands, and other natural ecosystems of special hydrological interest (MARM 2011). Within this work, it has been made a conceptual model of wetland-aquifer relationship that places special emphasis on the following points: (1) Identification and conceptual model for the existing wetlands in each intercommunitary river basin, i.e., one that includes several regions; (2) the relations between the wetland and the groundwater bodies. It is intended that the document is addressed to current and future protection of aquatic ecosystems dependent on groundwater. Within this work, we have analyzed, in those cases where the available data have allowed it, the quantification of wetland-aquifer relations. These findings exist for specific cases of total wetlands studied. But what relationship does this quantification with the water requirements of wetlands? These estimates are looking for a quantification of the transfer rate between the wetland and the aquifer to which it relates. In no case was an assessment of the ecological needs of the wetland.

4 Conclusions

Although the WFD and the HD have common elements, a detailed analysis reveals striking differences in their objectives, scope and focus. The work done so far for the implementation of both policies lead us to conclude that the water requirements (understood as quality and quantity of water for wetlands) although necessary, seem to be not sufficient to ensure the ecological requirements of the habitats. Water is a fundamental factor of wetland ecosystems, but not the only one. The quantifications made in the work of the WFD in Spain can be a useful basis for further research aimed at what minimum water volumes can guarantee the good state of habitats preservation.

References

Carcavilla L, De la Hera A, Durán Valsero JJ, Gracia FJ and Pérez Alberti A (2008) Geomorfología y geomorfología en la Directiva Hábitats de la Unión Europea [Geology and Geomorphology in the European Habitats Directive]. Trabajos de Geomorfología en España, 2006-2008. X Reunión Nacional de Geomorfología. Cádiz, Spain

EC (2000) Gestión de Espacios Natura 2000 [Management of Nature 2000 Sites]. Apartado 1 del Artículo 6. Disposiciones del artículo de la Directiva 92/43/CEE sobre hábitats. 73 pp. http://ec.europa.eu/environment/nature/natura2000/management/docs/art6/provision_of_art6_es.pdf

Marm (2011) Actividad 4 de la Encomienda de Gestión IGME-DGA. Identificación y caracterización de la interrelación que se presenta entre aguas subterráneas, cursos fluviales, descargas por manantiales, zonas húmedas y otros ecosistemas naturales de especial interés hídrico [Identification and characterization of the interaction that occurs between groundwater, streams, downloads and streams, wetlands and other natural ecosystems of special water]. DVD. Ministry of Environment and Ministry of Science and Innovation

SEO/BirdLife (2010) La Directiva Marco del Agua y la conservación de los humedales y los espacios de la Red Natura 2000 que dependen del agua [The Water Framework Directive and the conservation of wetlands and areas of the Natura 2000 network that rely on water]. 60 pp. with annexes

WAA (2009) Bases ecológicas preliminares para la conservación de los tipos de hábitat de interés comunitario en España [Preliminary ecological basis for conservation of habitat types of Community interest in Spain]. Dirección General de Medio Natural y Política Forestal, Ministerio de Medio Ambiente, y Medio Rural y Medio Marino. Madrid, Spain

Ecological impacts due to hydraulic technical projects to ecosystems near Natura 2000 network

Th.M. Koutsos[1,2], **G.C. Dimopoulos**[1], **A.P. Mamolos**[2]

[1] Laboratory of Engineering Geology and Hydrogeology, Department of Geology, School of Geology, Faculty of Sciences, Aristotle University of Thessaloniki, 54124, Greece
[2] Laboratory of Ecology and Environmental Protection, Faculty of Agriculture, Aristotle University of Thessaloniki, 54124, Greece

Abstract The construction of hydraulic technical projects plays an important role in preserving water resources for the continuously increase of water needs especially due to the last climate changes. Though water reservoirs contribution in local economic prosperity is substantial, questions rise about the possible impacts on protected areas near the construction site. This study established an environmental information system in combination with an eco-environmental vulnerability assessment method using Analytical Hierarchy Process and GIS tools in order to reveal areas of possible threats for protected species of Natura 2000 sites due to a hydraulic technical project construction. Using the proposed methodology it became possible to assess and map the expected ecological impacts on endangered species of a Natura 2000 site, due to the technical project activities at the construction and operation phase of Akropotamos dam (Kavala, Greece).

Introduction

Last decade climate changes along with increased human activities constitute a growing menace for protected species and areas especially for those susceptible to changes. Water is the determinative factor for the survival of every ecosystem particularly in regions with relatively abundant rainfall where evapotranspiration is expected to be very high (Isendahl and Schmidt 2006). To cope with the growing water needs the construction of dams and other hydraulic technical projects are the answer to water scarcity and to the increased frequency of drought events (White 2005). The construction of hydraulic structures may alter stream downflows or upflows (Hu et al. 2008), affecting the nearest ecosystems (Santos et al. 2008; Koutsos et al. 2010).

The lack of mitigation measures along with the increased human activities (due to the growth of settlements) may be crucial for the future of environmentally 'fragile' regions such as the protected areas of Natura 2000 network. In order to promote the maintenance or restoration of biodiversity the European Union issued a conservation directive (2009/95/EC) taking into account the economical, social, cultural and regional requirements.

Geographical Information Systems have much to contribute in environmental assessment applications (Segurado and Araϊjo 2004), especially in the case of rare or endangered species and their habitats, such as Natura 2000 Network. Recent studies (Koutsos et al. 2010) have shown that combining simple procedures for impact assessment and mapping we could form an idea about the geographic location where negative impacts will take place and which species or habitats will be affected and to what extent due to a technical project construction. Impacts on protected or endangered species can be mapped in a GIS with the assumption that their habitats can be linked to specific types of vegetation that represents their living environment (DeMaynadier and Hunter 1995).

The implementation of the proposed method leads to a thematic map presenting the risk of influence and the areas where additional mitigation measures need to be taken in order to efficiently minimize all possible threats. The evaluation scheme includes four steps: (a) determination of the areas at which the impact evaluation scores will be assigned (grouped habitat regions), (b) evaluation procedure using environmental criteria and use of an equation for calculating totals, (c) registration of evaluation scores to polygon areas (habitats) in the GIS system and (d) creation of thematic maps to show the range of plotted values as regions with different degree of impact on species near Natura 2000 network sites.

Methods

Proposed methodology

(a) Determination of areas where the impact evaluation scores will be assigned

The IUCN Red List of Threatened Species along with bibliographical data offers comprehensive information about the status of the most threatened, vulnerable and endangered species and particularly about the geographic range of their population and the variety of their habitats (grasslands, woods, forests, rivers etc.). Using the embedded SQL support in the GIS system we can select and merge (combine) the different types of vegetation (as polygon areas) and save them as one separate layer (habitat) for each of the species. This layer represents the geographic limits where each of the species live, move and reproduce. As a result layers for different species may overlap each other as expected because species share many times the same type of habitat. In order to assign impact evaluation scores, the different local habitats types included in the vegetation map, must be organized into groups of habitats (zones) subjected to the same degree of impacts due to the technical project activities. Intersection among these layers can reveal the degree of influence due to human activities or the risk for each of the species.

(b) Evaluation procedure

The main steps for the evaluation process are: (1) use of an evaluation criteria pack, (2) calculation of totals via impact equation, (3) prioritizing impact values. The pair-wise comparisons are supposed to be carried out taking advantage of the expert's knowledge and experience.

Evaluation criteria pack

For each habitat region a set of 10 criteria (IAIA 2002; Koutsos et al. 2008) is proposed to be used to evaluate the possible impacts to one or more species of the nearest Natura 2000 sites. As it is common in many studies these criteria have simple integer values (Koutsos et al. 2008) in order to facilitate the impact assessment and to avoid possible bias errors. The number of values for each criterion differs according to its nature.

Calculation of totals using an impact equation

To calculate a total value representing the final impact score for each zone and for all species separately, the following equation has been proposed to merge all individual criteria evaluations (Koutsos et al. 2008):

$$E=5 \times \text{type} \times \text{risk} \times (\text{extent} + \text{duration} + \text{reversibility} + \text{significance})$$

where number 5 is used for data normalization (scale -100 to 100).

Prioritizing impact values

For achieving a more objective impact ranking the methodology proposes the use of the Analytical Hierarchy Process (Saaty 1980) to make the pair-wise comparisons on the basis of the question "which of the two compared species will be more affected due to human activities?" As usually expected, judgment errors are likely in pair-wise comparisons and therefore we had to check the result consistency of the comparison matrix by calculating the consistency index (Cl) and consistency ratio (CR).

(c) Registration of evaluation scores to areas in a GIS system

The calculated impact total values were assigned to the corresponding zones (grouped habitat regions) in the GIS system (Gillenwater et al. 2006). The GIS program proposed for the data manipulation and mapping is MapInfo professional.

(d) Creation of thematic maps

As a result, instant thematic maps can be derived using the thematic map capability of the GIS system according to custom needs. This final step gives the visual image of the expected impacts on different species due to the construction of a specific technical project. GIS combined with the assessment procedure can now present the degree of impact on different species according to their habitats taking into account the characteristics of their living environment, the distance from the construction site, the degree of their susceptibility to changes etc.

Data set

In order to test the effectiveness of the method, we used a study area of a dam which is currently under construction (Akropotamos Dam, River Marmaras, Pieria basin, Kavala, Greece) having at the north of the dam basin, a Natura 2000 site (GR1150005). According to a previous study (Koutsos et al. 2010), it became possible to assess and plot impacts on species due to the dam construction using a project activity zone classification. Each zone (consisting of different types of vegetation) was supposed that it would be subjected to the same degree of impact due to the technical project activities. For that purpose the different local vegetation types, included in the vegetation map, were organized into the following five zones: (i) zone A (1), areas near the river with riparian vegetation and cultivating fields, (ii) zone B (2), coppices and abandoned fields, (iii) zone C (3), pastures and village areas, (iv) zone D (4), oak forests and beeches, (v) zone E (5), the part of Natura 2000 site, which intersects with the dam basin (Koutsos et al. 2010).

This methodology attempts to go further one step by combining the habitats for each of the species, assessing the expected impacts and mapping them for each one separately calculating the area and the percentage of the impact while giving also a qualitative value in order to make it easier for the decision makers to evaluate the results.

Results

Using the proposed methodology for assessing the impacts for each of the protected species a table with the evaluation scores has been constructed (Table 1, 2).

Table 1. Evaluation scored for each zone and all species at construction phase.

Species	Zones				
	A	B	C	D	Mean
Lutra lutra	-20.00	-22.50	-11.25	-3.75	-14.38
Triturus cristatus	0.00	0.00	0.00	-3.75	-0.94
Bombina variegata	-27.50	-15.00	0.00	-10.00	-13.13
Testudo	-20.00	-6.25	-22.50	-15.00	-15.94
Mean	**-16.88**	**-10.94**	**-8.44**	**-8.13**	**-44.38**
Weighted	**-21.48**	**-17.28**	**-8.10**	**-6.94**	**-13.45**

Table2. Evaluation scored for each zone and all species at operation phase.

Species	Zones				
	A	B	C	D	Mean
Lutra lutra	-30.00	-22.50	-27.50	-3.75	-20.94
Triturus cristatus	0.00	0.00	0.00	-2.50	-0.63
Bombina variegata	20.00	-15.00	0.00	-3.75	0.31
Testudo	6.25	-8.75	-12.50	-6.25	-5.31
Mean	**-0.94**	**-11.56**	**-10.00**	**-4.06**	**-26.56**
Weighted	**-8.38**	**-17.53**	**-15.55**	**-3.94**	**-11.35**

In the GIS system, topographical, hydrological and vegetation data were used as derived from digitized maps, presented at the dam study (Dimopoulos et al. 2001), in order to be the graphical background where impacts will be presented. All assessment data were assigned to the polygon areas representing the habitats of the species and they were also grouped according the technical project activities (zones A-D). For each of the species the area of their habitats has been calculated in the GIS system (Table 3). To make the comparison among species possible, the percentage of the area for each zone was also calculated (Table 4, 5).

Table 3. Area of habitats (sq km).

Species	Zones				
	A	B	C	D	Totals
Lutra lutra	38.699	28.65	6.1442	0	73.4925
Triturus cristatus	0.4716	0	0	12.72	13.1916
Bombina variegata	38.699	29.141	9.5385	0	77.3784
Testudo	37.678	29.141	9.54	3.6141	79.9731
Totals	**115.55**	**86.932**	**25.223**	**16.334**	**244.036**

Table 4. Percentage of impact area at construction (%).

Species	Zones			
	A	B	C	D
Lutra lutra	-34.78	-39.13	-19.57	-6.52
Triturus cristatus	0	0	0	-100
Bombina variegata	-52.38	-28.57	-19.05	0
Testudo	-31.37	-9.8	-35.29	-23.53

Area of the expected impacts for each zone has also been calculated and then grouped using a qualitative classification (Koutsos et al. 2010) in order to emphasize the regions and the degree where each of the species is more vulnerable (Table 6, 7). Finally eight thematic maps where constructed for each or the species at construction and operation phase of the technical project (Fig. 1, 2).

Table 5. Percentage of impact area at operation (%).

Species	Zones			
	A	B	C	D
Lutra lutra	-35.82	-26.87	-32.84	-4.48
Triturus cristatus	0	0	0	-100
Bombina variegata	51.61	-38.71	0	-9.68
Testudo	0	-31.82	-45.45	-22.73

Table 6. Qualitative assessment of impacts at construction.

Species	Zones			
	A	B	C	D
Lutra lutra	Low (-)	Low (-)	Neutral	Neutral
Triturus cristatus	Neutral	Neutral	Neutral	Neutral
Bombina variegata	High (-)	Medium (-)	Neutral	Neutral
Testudo	Low (-)	Neutral	Low (-)	Low (-)

Table 7. Qualitative assessment of impacts at operation.

Species	Zones			
	A	B	C	D
Lutra lutra	Medium (-)	Low (-)	Medium (-)	Neutral
Triturus cristatus	Neutral	Neutral	Neutral	Neutral
Bombina variegata	Low (+)	Low (-)	Neutral	Neutral
Testudo	Neutral	Neutral	Neutral	Neutral

As expected *Lutra lutra* is susceptible to water quality and quantity changes and possible loss of the riparian vegetation and will be affected more during the construction phase where many areas of its habitats will be in danger. *Triturus cristatus*, as a typical forest amphibian, lives in the north part of the study area in the forests away from any human activity and will almost not be influenced by the dam construction. *Bombina variegata* lives in a wide range of environment types and will be influenced negatively during the dam construction phase but it will take advantage of water abundance at the operation phase and receive a positive bonus.

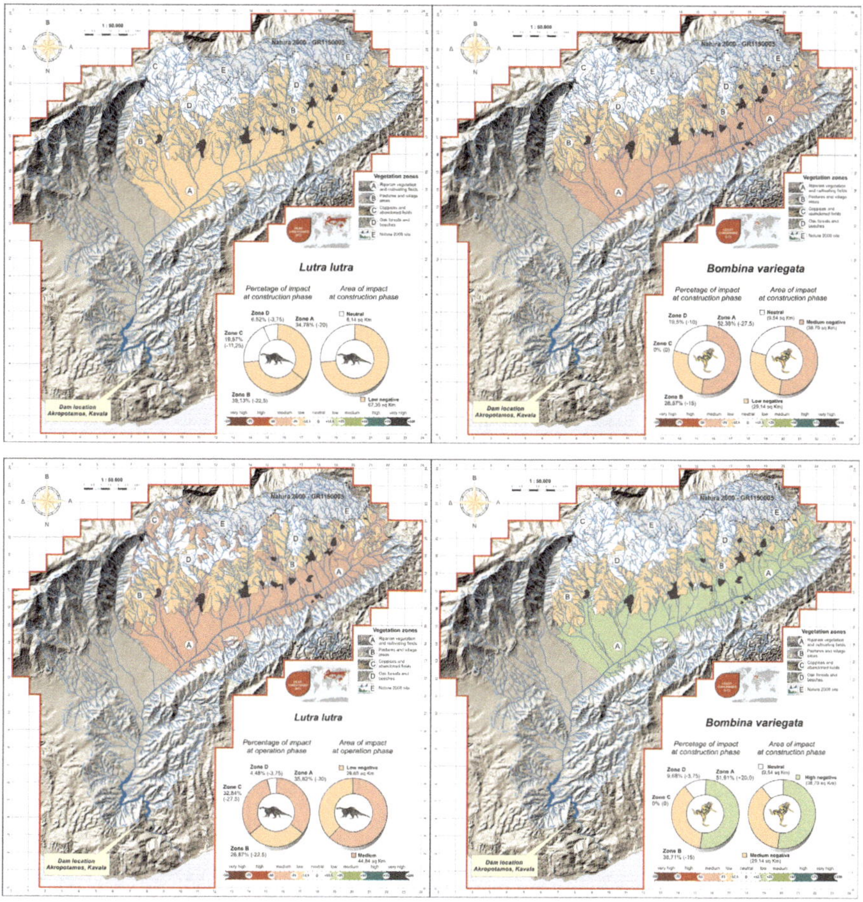

Fig. 1. Thematic maps of evaluation scores at both construction and operation phase of the dam for *Lutra lutra* and *Bombina variegata*.

Contrary to *Lutra lutra* this specie is more adaptive to environment changes caused by human activities and it is expected to gain profit of the dam construction after its completion. *Testudo* species are also adaptive to changes in their environment but they have to deal with different types of threats basically concerning the possible increase in human population or traffic, as they are usually being killed by humans and vehicles. It is expected to be influenced more during the construction phase as shown in the thematic maps, where machinery and vehicle circulation expected to be increased.

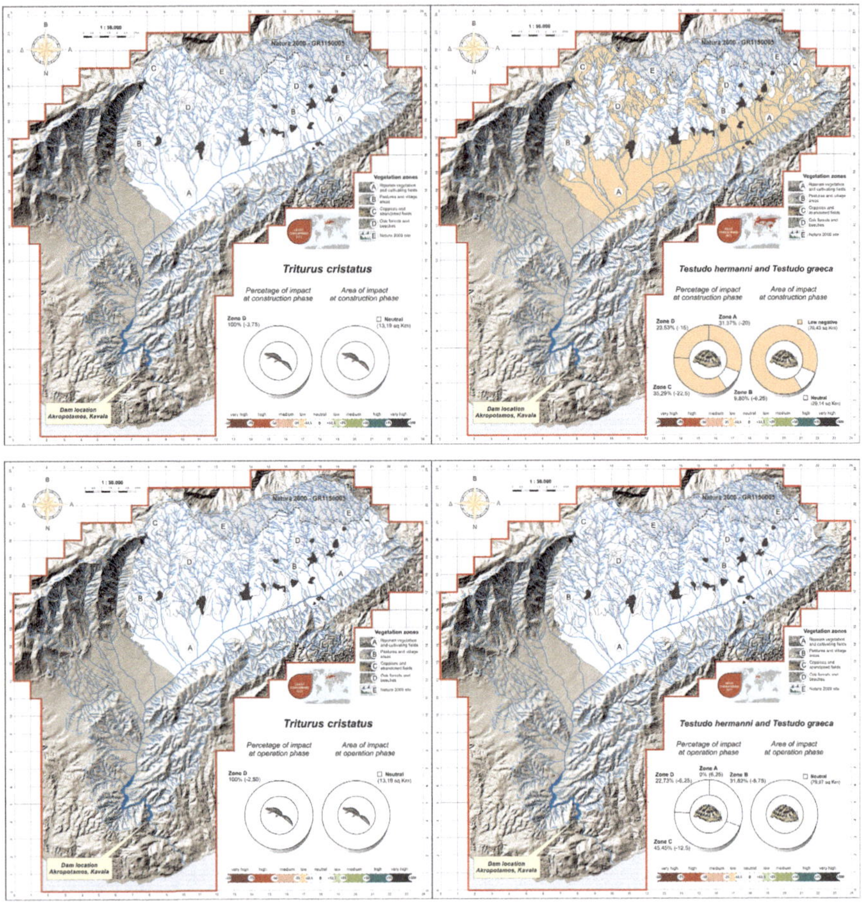

Fig. 2. Thematic maps of evaluation scores at both construction and operation phase of the dam for *Triturus cristatus*, *Testudo hermanni* and *Testudo graeca*.

With the proposed methodology it was finally possible to assess and map ecological impacts on endangered species due to human activities, showing the areas where species are more vulnerable and to what extent. It is also feasible to make comparisons among species on which will be more or less influenced and what degree. Refining the assessment procedure using a group of experts for improving evaluation scores and plotting the results separately for each of the species in thematic maps decision makers could now redesign mitigation measures and practices in order to protect endangered species from possible threats at specific areas of interest.

References

DeMaynadier PG, Hunter ML (1995) The relationship between forestmanagement and amphibian ecology: a review of the North American literature. Environ. Rev. 3, 230-261

Dimopoulos G, Papacharisis N, Papacharalampous C (2001) Study of Akropotamos dam construction in the region of village Akropotamos, Marmaras River. Research Report, Research Project Auth., Kavala, Greece (in Greek)

Gillenwater D, Granata T, Zika U, (2006) GIS based modeling of Spawning Habitat Suitability for Walleye (Sander vitreus) in the Sandusky River, Ohio, and implications for dam removal and river restoration. Ecol. Eng. 28 (3), 311-323

Hu WW, Wang GX, Deng W, Li S (2008) The influence of dams on ecological conditions in the Huaihe River basin, China. Ecol. Eng. 33, 233-241

International Association for Impact Assessment (IAIA) (2002) Topic 6-Impact analysis. EIA Training Resources Manual, 2nd edition, pp 251-302. Available at: http://www.iaia.org/publicdocuments/EIA/ManualContents/Sec E Topic 6.PDF

Isendahl N, Schmidt Q (2006) Drought in the Mediterranean: WWF policy proposals - A WWF report. WWF-World Wide Fund for Nature, Germany, 45

Koutsos T (2008) Contribution in Assessment and Evaluation of Environmental Impacts at Technical Projects with the use of Geographical Information Systems (GIS). PhD Thesis, School of Geology, Aristotle University of Thessaloniki, Greece (in Greek, with English abstract)

Koutsos TM, Dimopoulos GC, Mamolos AP (2010) Spatial evaluation model for assessing and mapping impacts on threatened species in regions adjacent to Natura 2000 sites due to dam construction, Ecol. Eng. 36, 1017-1027

Saaty TL (1980) Analytic Hierarchy Process, Planning, Priority Setting, Resource Allocation. McGraw-Hill, New York

Santos MJ, Pedroso NM, Ferreira JP, Matos HM, Sales-Luvs T, Pereira N, Baltazar C, Grilo C, Candido AT, Sousa I, Santos-Reis M (2008) Assessing dam implementation impact on threatened carnivores: the case of Alqueva in SE Portugal. Environ. Monit. Assess. 142, 47-64

Segurado P, Araïjo MB (2004) An evaluation of methods for modeling species distributions. J. Biogeogr. 31, 1555-1568

White WR (2005) World Water Storage in Man-made Reservoirs. Foundation for Water Research, Marlow, UK

Climate change

Approaches for increasing and protecting fresh water resources in light of climate change

G.A. Kallergis

University of Patra, Rio, 261 10 Patra, Greece

geokal@hol.gr

Abstract Available fresh water resources are shrinking and periods of droughts and water shortage are becoming more frequent. Such water Shortage is mainly man-induced and therefore should be treated as such. The impacts of Water shortage are ecological disruptions, general decrease of water levels in lakes, ponds, streams, reservoirs and aquifers, loss of topsoil and increased airborne dust, crop shortages or losses, increases in insect activities and predators, multiplying of weeds, replacement of useful plant species by less useful species, increased fire hazard, transformation of land uses, adverse effects to livestock such as reduction of herd size, increase of diceases and problems of vectror control, reduction or loss of wildlife resources.

Measures for mitigating water shortage, such as, *inter alia*, supply oriented, impact minimization, interbasin and within-basin water transfer and exchange, drought forecasts and early warning, surface and groundwater storage, transfer of surplus water in time must be reliably designed and implemented.

1 Aridity, Drought, Desertification and Water shortage

The primary determinant of *Aridity* is the nature-induced low average annual Precipitation, leading to a permanent imbalance of water availability. In arid areas the rainfall on a given parcel of land is not adequate for regular crop production and in semi-arid areas the rainfall is sufficient for short-season crops and the grass is an important element of the natural vegetation (Correia, 1999). Aridity is a function of the relationship P/E_p between the amount of *rainfall* (P) and the potential *evapotranspiration* (E_p). It is exacerbated when rainfall decreases and evapotranspiration increases. The latter is a function of the air temperature, among other factors, which is again a climatic feature. Aridity, which is a quasi-permanent climatic feature leads to nature-induced quasi-temporal *Drought* (Fig. 1). The latter is such a lack of rainfall as to adversly affect the plant and animal life of a region and to deplete water supplies, both for domestic purposes and for the operation of irrigation schemes. Lack of rainfall in certain areas does not necessarily indicate a drought, if the streamflows and ground-waters in the area are derived from rainfall in distant places (Baloutsos et al. 1999).

Droughts may be local, confined to a single river basin or area, or they may be widespread, extending over many states and parts of the world. When droughts are widespread, however, they are likely to be very severe in only a limited part of the total area affected.

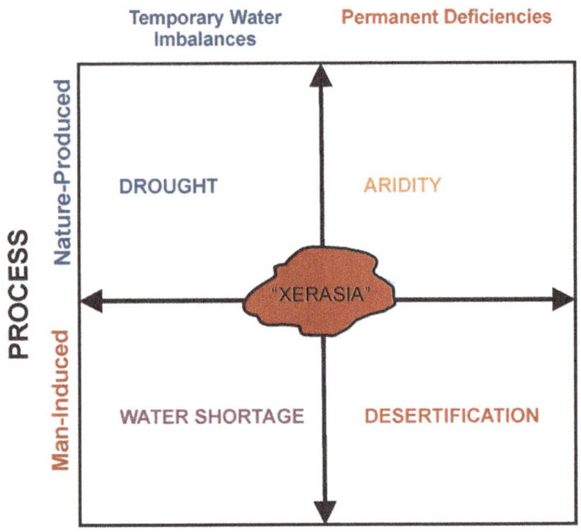

Fig. 1. Sketch definition of Aridity, Drougt, Water Shortage and Desertification (Adapted from Vlachos-Douglas 1983).

The environmental impacts of drought are, among others, (Vlachos and James 1983), ecological disruptions, general decrease of water levels in lakes, ponds, streams reservoirs and aquifers, loss of topsoil and increased airborne dust, crop shortages or losses, increases in insect activities and predators, multiplying of weeds, replacement of useful plant species by less useful species, increased fire hazard, transformation of land uses, adverse effects to livestock such as reduction of herd size, increase of diceases and vector control problems, reduction or loss of wildlife resources.

Factors such as topography, geology, and land grading influence the severity of droughts (Yassoglou 1999). With like amounts of rain, a steep area or an area having soil with low permeability and little underground storage may fail to supply vegetative needs and may yield zero streamflow at critical periods. On the other hand, a highly permeable area with ground storage may support sizable streamflows throughout dry areas, except in the more elevated portions of the area. Man-made changes such as grading, farming, drainage works, and residential improvements, aggravate drought conditions by speeding up water runoff and by reducing surface pondage, infiltration, and underground storage.

Measures for coping with droughts are *inter alia* (Yevjenich-Vlachos 1983):

- Supply-oriented measures that are intended to augment supply during droughts, i.e. surface and groundwater storage, interbasin and within-basin water transfer and exchange.

- Demand-oriented measures that are intended to decrease demand during droughts, such as active measures discouraging reduction, i.e. legal restrictions and penalties, economic incentives, recycling etc.
- Impact-minimization measures that are basically related to water users, such as drought forecast and warning.

Aridity does not necesserily lead to *Desertification* (Thornes 1999,Van der Leeeuv 1999) a situation which is an alteration of the ecological regime often associated with Aridity and/or Drought but always caused by human activities. Properly speaking, Desertification is the product of nature induced aridity and man- induced imbalance of water availability leading to *Water Shortage* (Fig.1).

2 Climate change and Water Resources

Climate is the dominant factor leading to aridity and drought.. On the other hand land becomes naturally desertified when the ratio: P/E_p (annual rainfall P to potential evapotranspi-ration E_p) acquires values below a certain threshold, regardless of the state of the other components. In contrast, when the ratio exceeds an upper threshold, desertification does not advance. But does indeed climate change? There are indications but not clear evidence that climate becomes warmer, but no statistically significant indications that it becomes drier.

Recent years have been among the warmest since 1860 - the period for which instrumental records are available. Warming is evident in both sea-surface and land-based surface, air temperatures. Urbanization in general and desertification could have contributed only a small fraction of the overall global warming, although urbanization may have been an important influence in some regions. Indirect indicators such as borehole temperatures and glacier shrinkage provide independent support for the observed warming. It should also be noted that the warming has not been globally uniform. The recent warming has been greatest between 40°N and 70°N latitude, though some areas such as the North Atlantic Ocean have cooled in the recent decades (WMO-UNEP 2008).

From the dawn of the Earth's history, over 4.6 billion years, the climate has changed constantly from warm to very cold and vice-versa (Fig. 2).

Over the last 400,000 years the Earth's climate has been unstable, with very significant temperature changes, going from a warm climate to an ice age in only a few decades (WMO-UNEP 2008). These rapid changes suggest that climate may be quite sensitive to internal or external climate forcings and feedbacks. Temperatures have been less variable during the last 10,000 years. Based on the incomplete evidence available, it is unlikely that global mean temperatures have varied by more than 1°C in a century during this period. Figure 3 shows temporal changes of precipitation and temperature in the Athens region (data from the meteorological station of Athens of National Observatory of Athens) during the last century

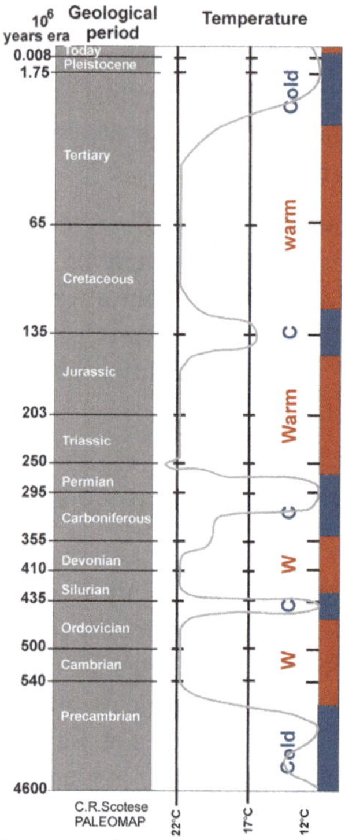

Fig. 2. Climate instability from the dawn of the Earth's history to the present.

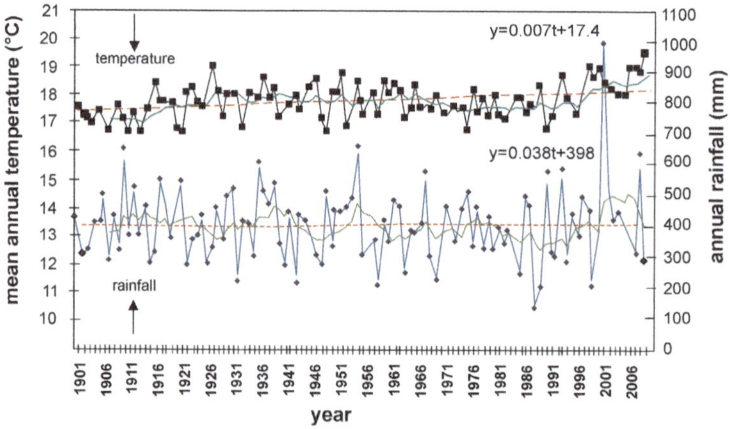

Fig. 3. Mean annual temperature (°C) of the period 1901-2010, trend line (dashed red lines) and 7-years moving average (solid green lines) from the meteorological station of National Observatory of Athens.

After Figure 3, there is clear evidence that air-temperature rises but statistically no significant indications that rainfall decreases. Founda, Giannakopoulos (2009) pose the rhetorical question of whether the exceptionally hot summer of 2007 (44,8 °C on 26 June) in Athens Greece is "a typical summer in the future climate", concluding that "the abnormally hot summer of 2007 is perhaps not the proof but a strong indicator of what eastern Mediterranean summers could resemble in future". Founda (2011), in a review paper on the evolution of the air-temperature in Athens, concludes that the long term linear trend of the entire time series (1890-2009) is positive and statistically significant.

3 Drought and water resources

The first and most powerful effect of drought is to reduce agricultural production over a wide area; other adverse economic impacts follow from this loss in production. During droughts the quantity of moisture drawn from storage by transpiration increases, resulting in the exhaustion of soil moisture early in the growing season. This is reflected in lower water levels in shallow wells and in deep wells subject to recharge in the drought area. High temperatures aggravate the situation by increasing the transpiration and evaporation requirements. Because of the size of underground storage, the depletion of groundwater reservoirs during a single drought is not common. However, depletion may occur when draft rates greatly exceed recharge capabilities or where the groundwater reservoir is relatively shallow. Because droughts are the result of a cumulative deficiency, *records for individual days, months, and in some cases even years are not significant.*

It is usually assumed that the severety of a drought is independent of the size of the area involved. This amounts to assuming that the low-flow unit discharge (discharge per unit drainage area) will not be affected by changes in area, an assumption that is known to be incorrect for flood flows. While mean flows are assumed to be proportional to drainage areas, drought-frequency studies indicate that unit discharges during drought periods are smaller from small areas when compared with like unit discharges from larger areas. Low-flow studies indicate that the discharge during a 7-day drought was proportional to the drainage area raised to the 1.25 power. Expressed in another way, the unit discharge varies as the 0.25 power of the drainage area.

The severity of droughts may be measured by various parameters such as deficiencies in rainfall and runoff, decline in soil moisture, reduction in groundwater levels, and the storage required to meet prescribed drafts or demands.

4 Man made causes of Watershortage

Water shortage is primarily human induced and less dependent on climatic changes.

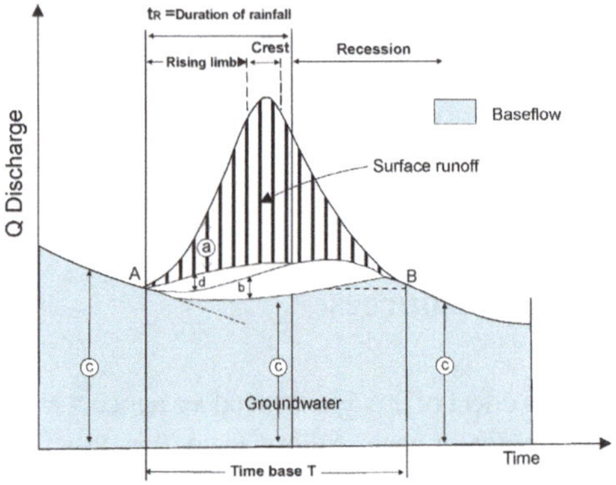

Fig. 4. The base flow in the Hydrograph (Adapted from De Wiest 1965).

An indicative but not restrictive list of the main man-made causes of water shortage are: failure to establish appropriate "capacity building", the lack or inefficient implementation of water policy and of management of water resources on a national and awatershed scale, the lack of conjuctive development and use of both surface and ground water, the over-exploitation of aquifers resulting either in depletion or in sea water contamination, the "over-use" or "unwise" use of water resources in order to "produce more agricultural goods" or to irrigate home grass-plots, especially in arid or semiarid regions, the minimization, if not elimination, of the "base flow"(Fig. 4) - due to important decrease of piezometry in aquifers - contributing to runoff, the intentional or unintentional pollution of surface or groundwater bodies (misuse or excessive use of agrochemicals, uncontrolled discharges of solid and liquid wastes, accidental pollution of water bodies, polution fom irrigation returns etc.), the lack of effectively implemented antipollution preventive measures and of decontamination policies in order to renovate the polluted water bodies, the lack of incentives or mechanisms to recycle water or to use water of lower quality where feasible, the inefficient adoption, or failure to adopt, "polluter pays" and "recovery of costs for water services, environmental and resource costs included" policies, the important water losses due to leakages from the distribution systems as well as due to adoption of inappropriate irrigation methods, especially in windy and temperate regions, the water losses due to lack of impoundment or of regulation of runoff especially in the upstream part of mountainous creeks or ephemeral torrents resulting in a small time lag until the epheneral runoff enters the sea, the lack of early warning systems on water resources depletion or degradation due to lack of effectively designed, implemented and functioning monitoring system on the quantity and quality changes of water resources.

5 Drought Impacts to the Baseflow

The cost of developing ground water is lower in most areas than the cost of developing surface water. Ground water may be developed on a "pay-as-you-go" or "pay-as-you-grow" basis. Surface water development generally involves a large initial capital investment.The above doctrines have led to overdevelopment of groundwater in relation to surface water.

Overpumping of aquifers, especially heavy in periods of drought, causes heavy drawdowns of watertables in the aquifers. This leads to minimization and even to depletion of the baseflow (Fig. 4), expressed as depletion of spring discharges and decreases the runoff over the whole watershed. Many springs do not recover even in wet periods and the water resources deficit could be perennial.

6 Effects of sea-level rise on coastal water resources

The water resources of small islands and low-lying coastal areas are very susceptible to sea-level rise. Figure 5 illustrates the direct impacts on the water resources sector, as well as the plethora of higher-order impacts which affect not only that sector but most, if not all, other sectors including health, transport and agriculture.

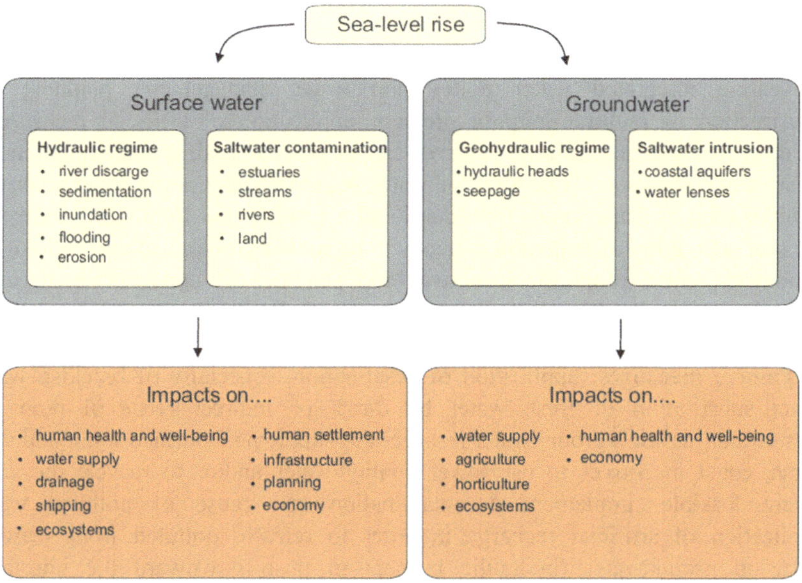

Fig. 5. Effects of sea-level rise on water resources (adapted from Hay. and Mimura 2006).

7 Impacts of Drought on Water Supply and Use

Vlachos and James, (in V. Yevjevich, L. Da Cunha, E.Vlachos ed. 1983) classified the Drought - Related Impacts as:

- *Economic,* namely, economic loss from drought-impacted-cattle breeding, loss from drought-impacted crop production, loss from drought-impacted timber production, loss from drought-impacted fishery production, loss to industries directly dependent on agricultural production, unemployment from drought-related production declines, strain on financial institutions, revenue losses to state and local governments, revenues to water supply firms etc.
- *Environmental,* namely, damage to animal species,damage to fish species, damage to plant species, water quality effects, air quality effects, visual and landscape quality.
- *Social,* namely, public safety from forest and range fires, health-related, low flow problems, inequity in the distribution of drought impacts/relief.

8 Towards Sustainable Water Resources Management

In order to match water availability and water needs, in quantity and quality, in space and time, at a reasonable cost and with the minimum acceptable environmental impacts, there is an urgent need to apply *active* (preventive) and *passive* (curative) measures such as:

Active measures: establishing institutional infrastructure, application of preventive integrated water policy and water management, planning and construction of multiple-purpose projects, application of artificial recharge in order to increase groundwater reserves and to prevent aquifer contamination due to seawater intrusion,"Wise use" of water resources integrating strict priorities, minimization of water losses from distribution systems and from water courses to the sea, adaptation of irrigation methods to regional hydrological and pedological conditions, enact economic incentives and penalties so conservation becomes attractive to consumers, information of farmers on the need to adopt the "Codes of good agricultural practices," and water resources monitoring.

Passive measures: application of desalination, especially of brackish water, enact substitution of fresh water by direct or indirect reuse of renovated wastewater, develop groundwater resources to supplement surface water and vice-versa, enact incentives to use lower quality water and/or to recycle the water where feasible, implement decontamination and reuse of polluted water, application of artificial recharge in order to refresh, polluted from seawater intrusion groundwater (hydraulic barriers to push downward the interface), minimization or prohibition of the use of agrochemicals (pesticides, fertilizers), in sensitive areas, by adopting appropriate incentives and disincentives and even penalties, revision of the traditional categories of "potential" water users and even of existing land uses.

Conclusions

In order to mitigate the shortage of water resources, water resources policy, management and practices should assure that water is used wisely and in conformity with appropriate planning of land use, is conserved, that water quality is protected, and that new technologies are adopted to augment available water quantities and enhance water supplies. Changing the way water is used is a difficult matter politically, legally, and economically. But as supplies of water shrink or become contaminated in many regions, the appropriate laws must be reliably implemented and when nessecary appropriately adopted to cope with regional problems so that water can be used most effectively. Specific actions should assure ,“ inter alia”, that:

- Water used for irrigation is used wisely and efficiently.
- Urban water-supply systems are well maintained, water loss is reduced, and all users have individual meters.
- Prices for water are consistent with the cost of supplying the water.
- Administrative or legislative methods are developed that permit higher beneficial uses of water as economic or social conditions change.
- Augmentation of available water resources by desalination of brackish water
- Conservation of water resources
- Reuse and recycling of municipal and industrial wastewaters where feasible
- Land-use priorities are established so that water quality is protected.

References

Baloutsos G et al. (1999). Characteristics of Meteorological Droughts of the last 132 years in the plain of Attica, Greece. In Greek with Abstract in English . Geot. Epist. Themata, 4:2, 5-11

Bolle J H (1999). Impact of climate variability on desertification process. In Balabanis P., D. Peter, A. Ghazi and M.Tsogas (ed.) Mediterranean Desertification, Research results and policy implications, Proc. Int.Conf. pp. 41-86, Dir. Gen. Res. EUR

Correia N (1999). Water Resources under the threat of Desertification. In Balabanis P., D. Peter, A. Ghazi and M.Tsogas (ed.) Mediterranean Desertification, Research results and policy implications, Proc. Int. Conf. pp. 215-242, Dir. Gen. Res. EUR

Founda D. (2011) Evoloution of the air temperature in Athens and evidence of climatic change: A review Advances in building energy research, 5, 7-41, Earthscan

Founda D, Giannakopoulos C (2009) The exceptionally hot summer of 2007 in Athens, Greece-A typical summer in the future climate? Global and Planetary Change, 67, 227-236, Elsevier

Hay, JE and Mimura, N (2006). Sea-level rise: Implications for water resources management. Mitigation and Adaptation Strategies for Global Change, 10, 717-737

Thornes, J (1999). Mediterranean desertification: the issues. In Balabanis P., D. Peter, A. Ghazi and M.Tsogas (ed.) Mediterranean Desertification, Research results and policy implications, Proc. Int. Conf. pp. 9-16, Dir. Gen. Res. EUR

Van der Leeuv S (1999). Degradation and Desertification:some lessons from the long-term perspective. In Balabanis P, Peter D, Ghazi A and Tsogas M (ed.) Mediterranean Desertification, Research results and policy implications, Proc. Int. Conf. 17-32, Dir. Gen. Res. EUR

Vlachos E, James L, Douglas (1983). Drought Impacts. In V. Yevjevich, L. Da Cunha, E.Vlachos (ed.) Coping with Droughts, pp. 44-76. Water Resources Publ

Yassoglou N (1999). Land, Desertification vulnerability and management in Mediterranean landscape. In Balabanis P, Peter D, A, Ghazi and Tsogas M (ed.) Mediterranean Desertification, Research results and policy implications, Proc. Int. Conf. pp. 87-114, Dir. Gen. Res. EUR

Yevjevich V, Vlachos E (1983). Strategies and measures for coping with Droughts, In V. Yevjevich, L. Da Cunha, E.Vlachos (ed.) Coping with Droughts, pp.77-101. Water Resources Publ

WMO-UNEP (2008) Climate Change and Water, IPCC Technical Paper VI, 1-200 Geneva

Estimation of hourly groundwater evapotranspiration using diurnal water table fluctuations

L.H. Yin[1], G.C. Hou[1], D.G. Wen[2], H.B. Li[3], J.T. Huang[1], J.Q. Dong[1], E.Y. Zhang[2], Y. Li[1]

[1] Xi'an Center of Geological Survey, No. 438, Youyidong Road, Xi'an, 710054, China, ylihe@cgs.cn

[2] China Geological Survey, Beijing 100035

[3] China University of Geosciences (Beijing), Beijing 100083

Abstract Groundwater is a valuable resource in the arid environments. Evapotranspiration (ET_G) is the major component of the groundwater balance in these regions. In this study, the Loheide-method was used to estimate hourly ET_G using the diurnal water table fluctuations and the variation ET_G of throughout the day was analyzed. The methods mentioned above were applied to a pre-existing data set that was collected in a well (SB1) that is screened across the water table in the Subei Lake watershed in NW China. It is obvious that the pattern of estimated ET_G and air temperature and solar radiation are very similar. The distinct feature of estimated ET_G is the diurnal fluctuations. The results present a wide range of diurnal ET values ranging from 10 to 22 mm/d.

1 Introduction

Groundwater is a valuable resource in the arid environments. Evapotranspiration (ET_G) is the major component of the groundwater balance in these regions. ET_G is notoriously difficult to measure directly due to its inherent complexities and small scale spatial and temporal variations (Mould et al. 2010). ET_G is commonly quantified by first quantifying near-surface ET; then ET_G is estimated using a simple linear decrease of ET as a function of the depth of the water table from the land surface (Banta 2000). ET is driven by incoming solar radiation (Oke 1978); therefore it has a strong diurnal variation caused by evaporation or transpiration of phreatophytic vegetation or both. In response to water loss, the water table fluctuates diurnally. There are several practical advantages to using diurnal water-table fluctuations. They are a direct measurement and accurate over daily time steps and relatively simple to calculate and monitoring diurnal water-table fluctuations requires less intensive fieldwork than monitoring evaporation pans or lysimeters

(Lautz 2008). White (1932) firstly proposed a method of estimating ET_G from groundwater hydrographs. Recently, Loheide (2008) formalized a general methodology for estimating subdaily ET_G. In this study, the Loheide-method was used to estimate hourly ET_G and the variation ET_G of throughout the day was analyzed.

2 Description of the study area

The methods mentioned above were applied to a pre-existing data set that was collected in a well (SB1) that is screened across the water table in the Subei Lake watershed in NW China. This site (39°20′N, 109°00′E) was established to investigate plant water use in an ecohydrologic setting common to the Ordos Plateau region of NW China. The climate is characterized by semi-arid, receiving about 350 mm of precipitation annually, with potential evapotranspiration rates up to 2500 mm. Rainfall occurs from March to October, but approximately 60 -80 % of the rainfall occurs from July to September.

The mean monthly temperature ranges from -4.3°C in January to 19.6°C in July, the mean annual temperature being 6.8°C. The study site is located in the discharge area of the Subei watershed, where the soil is mainly eolian sand at the top and fluvial sand underlying with unknown thickness. The baserock is the Cretaceous sandstone that is the most productive aquifer in the study site. The vegetation is dominated by Willow (Salixmatsudana), bushes (Salix psammophila) and grasses, a mix found in many watersheds in the Ordos Plateau. Meteorological data, including air temperature, precipitation, relative humidity, global irradiance, pressure and with speed and direction, are collected every 1 hour at a weather station (Hobo Weather Station, Onset Computing Corp.) about 1 km east to the study site. In the study site, water tables were measured using pressure transducers and date logger unit submerged in the water column (Minidivers, Van Essen Instruments, Delft, The Netherlands). Pressure readings are taken at 1-hour intervals and checked periodically with manual measurements.

3 Methods

Loheide (2008) modified the original White-method and estimated subdaily evapotranspiration values of riparian zones. In his analysis, the rate of change of head at the recovery source is assumed to be equal to the overall rate of water table change at the observation location. Firstly, he removed the trend (assuming it linear with a slope m_T and intercept b_T from the groundwater-level time series, WT(t)). The detrended water table depth (WT_{DT}) is obtained as follows:

$$WT_{DT(t)} = WT_{(t)} - m_T \times t - b_T \tag{1}$$

Application of this detrending procedure yields a relationship to predict d WT_{DT} /dt as a function of the detrended water table depth, Γ (WT_{DT}), using periods of zero ET_G. Then the net inflow rat can be estimated for the period of analysis using:

$$r(t) = S_y \lfloor \Gamma(WT_{DT(t)} + m_T \rfloor \qquad (2)$$

Once r(t) is estimated in this fashion, ET_G is calculated directly through Equation (3):

$$ET_{G(t)} = r(t) - S_y \frac{dWT}{dt} \qquad (3)$$

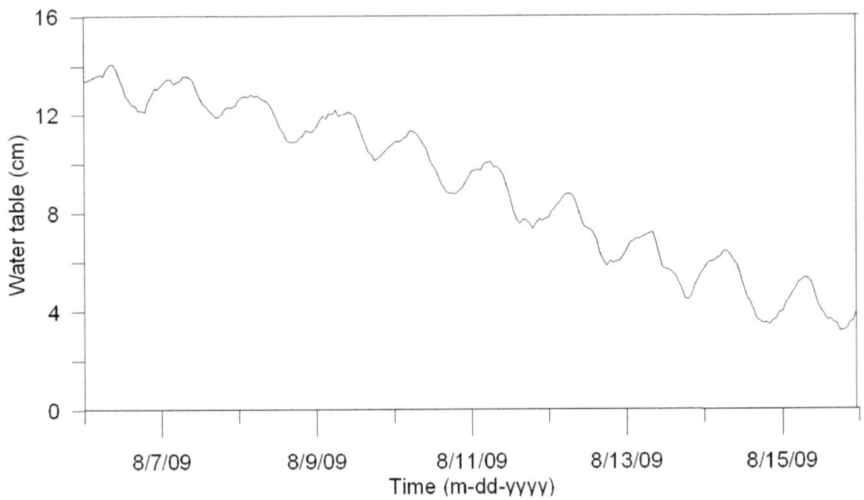

Fig. 1. Water table fluctuations in a well.

4 Results and discussion

The water table fluctuations collected at this well are shown in Figure 1 for a 9-day period at the beginning of August. Many factors control the phreatophyte-induced fluctuations in the water table and the major ones are cyclical pumping of nearby wells, changes in stream stage or barometric pressure (Butler et al. 2007). Around the study site, there are no pumping wells and rivers. The effect of barometric pressure on the water table was removed using the pressure data collected in the meteorological station. Water table elevations are generally at a minimum around 18:00 or 19:00 each day as shown in Figure 1, due to consumptive water use by vegetation. Water table elevations reach the peak between 07:00 and 09:00, following a period of recharge by a net influx of groundwater during night, when transpiration is minimal. These diel fluctuations typically have a magnitude of ~2 cm and are very regular on all days.

ET_G for this record was calculated using the Loheide-method. A readily available specific yield, 0.25, was assumed, resulting the range of estimates plotted in Figure 2 for comparison, air temperature and solar radiation from the weather station were also plotted. It is obvious that the pattern of estimated ET_G and air temperature and solar radiation are very similar. The distinct feature of estimated ET_G is the diurnal fluctuations. The results present a wide range of diurnal ET values ranging from 10 to 22 mm/d. During a single day's fluctuation the following sequence is observed: immediately after sunrise around 05:00, air temperature started to rise, follow by an increase of ET from vegetation. As a result, groundwater level reaches a maximum with equilibrium of ET and net groundwater inflow. During the day, solar radiation reaches a maximum around 13:00, whilst the groundwater level is still falling. During night time ET is zero and so groundwater levels increase again to the next daily maximum. However, this is a lower maximum than that of the previous day, as the daily total ET was higher than the total daily groundwater recharge.

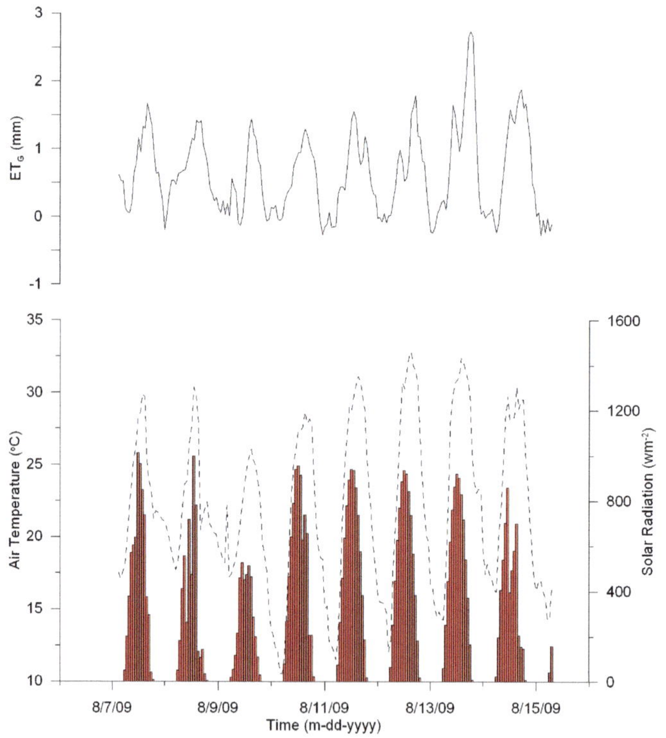

Fig. 2. Estimated ET_G, air temperature and solar radiation.

5 Conclusions

Groundwater is a valuable resource in the arid environments. Evapotranspiration (ET_G) is the major component of the groundwater balance in these regions. In this study, the Loheide-method was used to estimate hourly ET_G using the diurnal water table fluctuations and the variation ET_G of throughout the day was analyzed. The methods mentioned above were applied to a pre-existing data set that was collected in a well (SB1) that is screened across the water table in the Subei Lake watershed in NW China. It is obvious that the pattern of estimated ET_G and air tempera-ture and solar radiation are very similar. The distinct feature of estimated ET_G is the diurnal fluctuations. The results present a wide range of diurnal ET values ranging from 10 to 22 mm/d.

Acknowledgments This research was funded by the Groundwater Circulation and Rational Development in the Ordos Plateau project (Project Code: 1212010634204) and the National Natural Science Foundation of China (Grant No. 41002084)

References

White WN (1932) A method of estimating ground-water supplies based on discharge by plants and evaporation from soil: results of investigations in Escalante Valley, Utah. US Geol Surv WaterSuppl Pap 659-A

Mould DJ, Frahm E, Salzmann Th, Miegel K, Acreman MC (2010) Evaluating the use of diurnal groundwater fluctuations for estimating evapotranspiration in wetland environments: case studies in southeast England and northeast Germany. Ecohydrology, 3: 294-305

Banta ER (2000) MODFLOW 2000, the US Geological Survey modular ground-water model: documentation of packages for simulating evapotranspiration with a segmented function (ETS1) and drains with return flow (DRT1), US Geol Surv Open-File Rep 00-466, 127 pp

Oke TR (1978) Boundary Layer Climates. London

Lautz LK (2008) Estimating groundwater evapotranspiration rates using diurnal water-table fluctuations in a semi-arid riparian zone. Hydrogeol J 16: 483-497

Loheide SP. (2008) A method for estimating subdaily evapotranspiration of shallow groundwater using diurnal water table fluctuations. Ecohydrology 1: 59-66

Estimation of precipitation change over Greece during the 21st century, using RCM simulations

J. Kapsomenakis[1], P.T. Nastos[2], C. Douvis[1], K. Eleftheratos[2], C.C. Zerefos[1,2]

[1]Research Centre of Climatology and Atmospheric Environment, Academy of Athens, Greece

[2]Laboratory of Climatology and Atmospheric Environment, Faculty of Geology and Geoenvironment, University of Athens, Athens, GR 15784, Greece. nastos@geol.uoa.gr

Abstract In the present study, an estimation of the future change in precipitation within the 21st century was carried out for the 12 climatic zones of Greece. The estimation is based on the results of 12 Regional Climate Models (RCM) simulations of the ENSEMBLES Program. The precipitation is projected to decrease mildly in the near future (2021-2050) and moderately in the far future (2071-2100) by 5% and 20% respectively for the area of Greece. The decrease seems to be most prominent during spring and summer, less prominent during winter. The autumn precipitation is projected to increase mildly in the near future and to decrease mildly in the far future. The projected precipitation change in Greece is expected to result in drastic changes on the ground water balance and on several sectors of human activity and environment parameters, as well.

1 Introduction

In recent decades, the Earth has experienced abrupt climate changes, including decrease of precipitation within the Mediterranean (Amanatidis et al. 1993, Piervitali et al. 1998, Türkes et al. 1998, González-Rouco et al. 2001, Xoplaki et al. 2002, Dünkeloh and Jacobeit 2003). Scientific research has revealed similar conclusions while focusing on the area of Greece. Kandylis et al. (1989) and Hatzioannou et al. (1998) have detected statistically significant decrease of precipitation over Greece, especially over the southern parts for the second half of the 20th century. Feidas et al. (2007) show that the decrease of the annual precipitation over Greece appears mainly after the year 1984. Besides, Nastos and Zerefos (2009) concluded that the temporal variability of consecutive wet days shows statistically significant (confidence level of 95%) negative trends, mainly in the western region of Greece, characterized by large orographic precipitation amounts (Metaxas et al. 1999) against insignificant positive trends of consecutive dry days appear almost all over the country with emphasis in the southeastern region.

It is very likely that the abrupt climate changes result from the increase of the greenhouse gases (GHG) concentration (IPCC 2007). Given that the concentration of the GHGs is raising, it is plausible that more abrupt climate changes may occur in the future. The modern tool used to project the future climate change is General Circulation Models (GCMs). Due to computational resources limitations, the horizontal resolution of present day GCMs is quite low, usually in the order of hundreds of kilometres. In such a crude resolution many local aspects of the climate are unable to be represented. In addition, the topographical input is equally crude, thus excluding important local features of the topographic forcing. For these reasons downscaling methods have been developed, which input the GCM results producing high resolution localized climate information. Dynamical downscaling is achieved using Regional Climate Models (RCMs) that increase the resolution of the GCMs to even less than 10 km.

According to most GCM-RCM results, the Mediterranean is one of the regions where the impacts of human-induced climate change are estimated to be high (Hulme et al. 1999, Giorgi 2006, IPCC 2007). All climate simulations in the literature project a rise in temperature throughout the planet but less agreement between projections is found for precipitation. However, it is considered that in the area of Mediterranean strong precipitation decreases for both winter and summer will occur in the 21st century (Pal et al. 2004, Tselioudis et al. 2006, Zanis et al. 2009). Giorgi (2006) considers the area on of the two primary climate change hot-spots of the planet while IPCC (2007) reports that the decrease is considered especially robust and very likely to occur, in annual precipitation.

In the present study, the results of simulations (25 km horizontal resolution) made by the European Project ENSEMBLES are used in order to produce state of the art projections of the future change of mean precipitation over the 12 climatic zones of Greece.

2 Data and Analysis

The precipitation datasets of a total of 12 simulations from Regional Climate Models (RCM) carried out by the European program ENSEMBLES (http://ensemblesrt3.dmi.dk/) are analyzed in this study. More specifically, we work on the results of the following simulations: RCA3 (HadCM3Q), RM5.1 (ARPEGE), HIRHAM5 (ARPEGE), HIRHAM5 (ECHAM5), HIRHAM5 (BCM), CLM (Hadcm3Q0), RegCM3 (ECHAM5), RACMO2 (ECHAM5), REMO (ECHAM5), RCA (BCM), RCA (ECHAM5), RCA (HadCM3Q3); the name of the General Circulation Model (GCM), from which data were used as input in the simulation process for each RCM, appear in brackets, while the name of RCM used for the mentioned simulations appear in front of the bracket. The spatial resolution of the 12 RCMs is 0.22° longitude x 0.22° latitude. The assessment of the climatic conditions within the 21st century was based on SRES A1B (Nakicenovic et al. 2000), while the performed analysis used the average of 12 simulations (ensemble mean). Initially, the precipitation derived from the ensemble mean of the 12 RCMs was compared to the gridded precipitation from stations for the reference period (1961-1990).

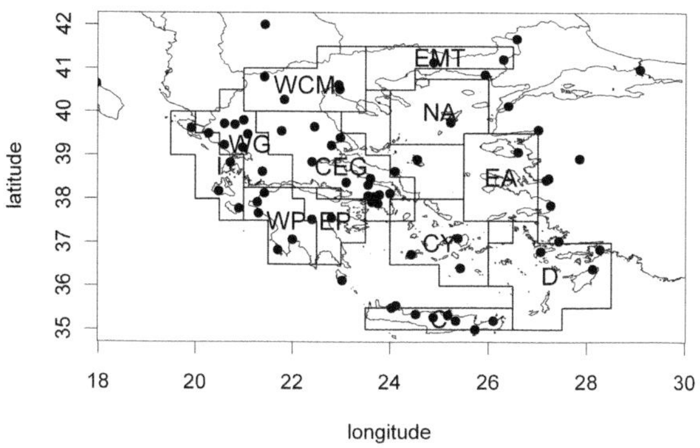

Fig. 1. The network of meteorological stations, along with the 12 climatic zones of Greece, used in the analysis.

The gridded station precipitation data were derived from the data processing of a network of 57 available meteorological stations from the Hellenic National Meteorological Service (Fig. 1), using the 2-D Kriging methodology, coupled with a routine which calculates the elevation component. The interpolation was done in the CRU like grid (spatial resolution 0.25° longitude × 0.25° latitude). The same grid concerns the 12 RCMs simulations. Then, the ensemble mean precipitation simulations were calculated for two future periods, one concerning the near future (2021-2050) and the second the far future (2071-2100). In the process, the precipitation changes (%) were estimated with respect to the reference period 1961 - 1990. The analysis was performed for each one of the 12 climatic regions of Greece (Fig. 1), extracted by using climatic and geographic criteria, as well as for the wide area of Greece, in seasonal and annual basis. The 12 climatic regions concern: Western Greece (WG), Central and Eastern Greece (CEG), West and Central Macedonia (WCM), Eastern Macedonia-Thrace (EMT), West Peloponnese (WP), Eastern Peloponnese (EP), Crete (C), Dodecanese (D), Cyclades (CY), Eastern Aegean (EA), North Aegean (NA) and Ionian (I) (Zanis et al. 2009).

3 Results and Discussion

The precipitation derived from the ensemble mean simulations of the 12 RCMs was compared with the gridded station observations for the reference period (1961-1990). In general, the RCMS reproduce satisfactorily the spatial distribution of mean annual and seasonal precipitation in the region of Greece (not shown).

Additionally, the RCMs reproduce satisfactorily the mean annual and seasonal precipitation totals fallen in the different climatic zones of Greece with the exception of Central and Eastern Greece (CEG) and the Eastern Peloponnese (EP), where a significantly underestimation of the precipitation appears (not shown).

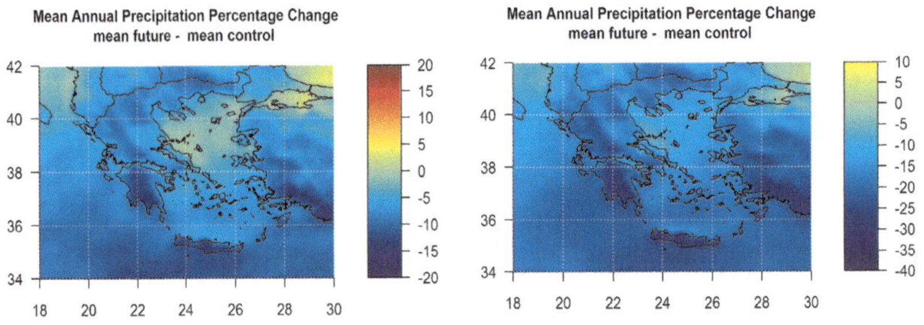

Fig. 2. Changes (%) in mean annual precipitation totals between future periods 2021–2050 (left), 2071–2100 (right) and the control period 1961–1990.

Taking into account the mentioned findings, we conclude that, despite any errors of the RCMs from the ENSEMBLES project, in general they reproduce satisfactorily the mean annual and seasonal precipitation totals for the various Greek regions. Therefore, future simulations of these RCMs are a useful tool for assessing possible climatic changes within the different climatic zones of Greece.

In the process, the estimated changes (%) in mean annual and seasonal precipitation totals, based on the ensemble mean of the 12 simulations, between the future periods 2021-2050 (Fig. 2, 3), 2071-2100 (Fig. 2, 4) and the control period (1961-1990) were estimated.

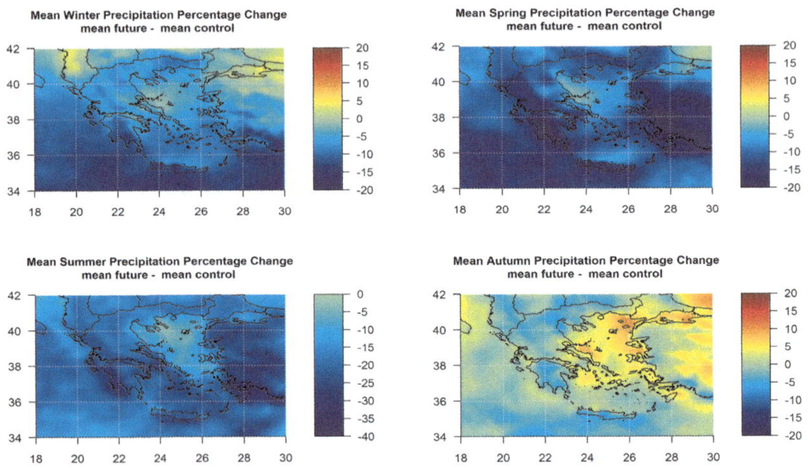

Fig. 3. Changes (%) in seasonal precipitation totals between the future period 2021-2050 (SRES A1B) and the control period 1961–1990.

Table 1. Mean seasonal and annual precipitation totals (mm) for the thirty year periods 1961-1990, 2021-2050, 2071-2100 and their changes (%) between the periods 2071-2100 and 1961-1990 as well as between the periods 2021-2050 and 1961-1990 (mean and standard deviation of the studied 12 RCM simulations, SRES A1B).

Region	Time Period	Winter		Spring		Summer		Autumn		Annual	
		PRE	ΔPRE%	PRE	ΔPRE%	PRE	ΔPRE%	PRE	ΔPRE%	PRE	ΔPRE%
	1961-1990	209.9±52.7		188.4±40.3		91.3±73.8		168.3±37.2		658.9±143.7	
WCM	2021-2050	195.3±51.6	-7.1±3.4	167.5±34.1	-10.6±8.7	76.4±62.9	-18.6±12.6	165.5±31.9	-1.0±6.7	605.8±126.3	-7.8±4.1
	2071-2100	187.0±42.3	-10.3±7.6	139.8±34.5	-25.7±10.0	60.4±51.6	-37.5±16.7	150.5±38.2	-10.8±11.7	539.0±114.5	-18.0±4.9
	1961-1990	266.2±89.5		186.5±44.5		78.9±63.9		176.3±71.2		709.8±184.7	
EMT	2021-2050	243.4±82.7	-8.1±7.8	162.9±40.1	-12.4±9.0	65.5±57.8	-20.3±12.8	177.5±67.2	2.0±9.8	651.2±169.4	-8.2±2.9
	2071-2100	233.3±73.2	-11.6±9.2	142.8±41.0	-23.8±9.8	50.9±44.2	-39.8±15.2	151.4±48.4	-12.6±10.0	580.4±155.6	-18.3±4.7
NA	1961-1990	217.1±83.2		102.9±32.8		31.5±33.1		155.8±88.4		509.7±205.6	
	2021-2050	207.7±76.0	-3.3±9.4	94.9±36.6	-8.9±9.3	29.5±30	-10.0±25.5	167.0±88.4	9.2±12.2	501.4±198.8	-1.1±5.6
	2071-2100	192.3±78.5	-11.9±6.4	85.6±31.0	-17.5±9.8	24.9±24.3	-21.3±26.5	145.8±83.9	-7.3±13.8	450.8±189.1	-11.9±7.0
	1961-1990	213.1±66.4		75.3±20.0		16.2±21.3		142.3±82.9		449.5±169.2	
CY	2021-2050	196.9±58.3	-6.8±7.5	67.7±24.5	-11.6±9.9	13.2±14.8	-13.5±24.9	146.6±73.6	7.8±16.0	426.9±158.4	-4.4±6.7
	2071-2100	172.5±62.9	-20.1±5.7	58.1±21.2	-24.3±10.9	9.7±11.9	-37.2±13.8	128.8±82.0	-11.0±14.6	371.4±166.3	-19±8.0
	1961-1990	288.8±97.9		111.4±32.9		17.2±17.1		164.5±105.5		585.3±230.6	
EA	2021-2050	266.9±91.3	-6.8±11.0	98.1±39.2	-13.2±11.5	15.1±15	-15.4±22.8	174.7±92.5	10.9±16.6	558.1±219.6	-4.2±7.7
	2071-2100	242.7±93.9	-16.8±6.2	88.4±33.2	-21.9±9.9	10.7±11.5	-38.9±12.1	146.6±91.7	-11.0±12.3	491.3±215.3	-17.1±6.0
	1961-1990	240.0±86.1		81.7±28.6		13.2±15.8		141.5±100.1		479.4±216.8	
D	2021-2050	216.9±78.9	-9.5±5.4	70.7±31.0	-14.8±10.3	10.2±11.0	-17.9±24.0	144.5±84.8	10.2±23.5	445.0±197.8	-6.4±7.9
	2071-2100	190.4±77.5	-21.7±5.7	61.4±28.4	-26.7±9.6	8.3±9.3	-35.4±20.0	122.5±90.3	-14.3±14.3	385.1±196.9	-21.2±7.3

Table 1. Continue.

Region	Time Period	Winter		Spring		Summer		Autumn		Annual	
		PRE	ΔPRE%	PRE	ΔPRE%	PRE	ΔPRE%	PRE	ΔPRE%	PRE	ΔPRE%
	1961-1990	274.3±110.7		111.5±43.9		22.7±31.4		156.1±67.2		567.8±224.3	
C	2021-2050	239.5±90.6	-11.8±4.2	92.5±36.9	-16.3±10.2	18.0±24.7	-21.8±13.3	151.9±49.2	2.4±16.5	504.7±183.3	-9.8±6.3
	2071-2100	197.9±75.5	-27.1±6.7	70.9±28.6	-35.5±12.2	11.8±16.0	-48.3±17.1	123.9±61.6	-21.7±12.5	407±164.4	-28.1±8.0
	1961-1990	181.1±20.3		128.9±29.4		54.9±56.5		141.0±47.6		507.4±111.8	
CEG	2021-2050	173.4±24.4	-4.4±5.6	115.9±27.7	-9.8±10.7	45.3±43.3	-20.1±17.6	144.3±36.4	4.4±9.9	480.5±97.9	-5.0±4.9
	2071-2100	158.3±23.2	-12.6±7.9	93.7±29.1	-28.1±10.4	36.5±35.9	-38.5±21.5	131.8±47.7	-6.9±12.5	421.8±102.4	-17.2±6.5
	1961-1990	195.7±23.9		107.6±21.3		33.8±36.5		140.6±43.5		479.6±81.1	
EP	2021-2050	180.7±27.4	-7.8±5.6	92.2±22.3	-14.8±7.5	26.9±27.9	-24.1±18.4	140.4±34.5	1.2±8.3	442.1±79.4	-7.9±4.6
	2071-2100	155.9±25.0	-20.3±8.1	72.9±20.6	-32.8±10.7	20.5±22.2	-46.4±18.8	120.8±39.9	-14.3±12.3	371.8±82.0	-23.0±7.0
	1961-1990	512.2±183.7		291.4±71.1		89.6±75.0		287.4±83.8		1185.4±302.9	
WG	2021-2050	475.2±197.9	-8.4±9.1	249.2±59.9	-14.2±6.2	72.0±58.8	-21.1±10.5	283.4±84.8	-1.5±3.4	1084.5±304.0	-9.0±4.3
	2071-2100	428.4±161.5	-16.7±9.7	204.8±52.4	-29.7±8.9	54.1±49.8	-44.4±15.0	240.5±72.2	-16.5±8.2	932.4±264.7	-21.8±5.8
	1961-1990	339.2±92.1		146.1±44.8		30.9±27.3		266.3±114.3		786.6±247.8	
I	2021-2050	310.3±93.7	-9±7.3	129.6±43.5	-11.5±9.3	25.3±22.1	-16.0±20.6	269.5±114.3	1.2±5.7	738.6±250.4	-6.6±5.3
	2071-2100	279.1±89.1	-18.4±6.2	109.0±35.1	-25.2±10.2	19.7±17.7	-36.0±16.3	240.6±123	-11.8±11.0	652.0±246.2	-18.2±6.8
	1961-1990	400.7±133.5		201±46.6		56.2±55.8		219.3±71.8		881.1±229.7	
WP	2021-2050	359.3±132.2	-10.8±10.5	164.6±43.7	-18.4±7.2	43±41.9	-27.1±14.3	215.7±63.9	-1.0±5.6	786.5±218.6	-10.9±5.7
	2071-2100	306.7±111.5	-23.8±7.2	138.2±39.2	-31.8±9.4	32.8±33.8	-47.6±16.3	174.2±62.3	-21.0±9.8	655.2±202.6	-26.2±6.0
	1961-1990	251.8±65.3		124.5±29.0		37.8±34.8		168.5±73.1		585.2±165	
GREECE	2021-2050	232.4±61.7	-7.6±5.8	109.3±30.3	-12.7±7.3	31.3±27.4	-18.5±12.1	171.3±64.3	3.4±8.6	546.9±154.2	-6.4±4.2
	2071-2100	208.3±60.6	-17.7±5.1	92.5±26.8	-26.4±8.8	24.5±21.7	-37.0±12.8	148.9±69.5	-12.8±10.6	476.5±155.3	-19.3±5.5

Analysis of the results show that the precipitation totals is likely to decrease in the wide area of Greece. The decrease in mean annual precipitation will be relatively low in the near future (2021-2050) and remarkable in the far future (2071-2100) (Fig. 2). More specifically, a decrease of approximately 5% appears during the period 2021-2050 compared to the period 1961-1990 (Table 1). This decrease is projected to be highest in Crete and Peloponnese, where it approaches 15%, while in other regions of Greece range from 5% to 10%, with an exception of the North Aegean (NA), where the mean annual precipitation tends to increase slightly. Similar results are extracted concerning the winter season. Higher decrease of mean precipitation is projected during spring, which is more remarkable in the southern regions of Greece (Crete, Peloponnese and Dodecanese), exceeding 15% against 10% at the rest regions. Concerning summer season, even higher decrease (>20%) is projected for the majority of Greece. In contrast to the other seasons of the year, the mean precipitation during autumn is expected to increase slightly by 3% for the whole Greece, in the near future. This increase is considered to be significant in the islands of the Aegean Sea (approximately 10%) and eastern continental Greece (approximately 5%). On the contrary, the mean precipitation at the western regions appears slight decrease, during autumn (Fig. 3).

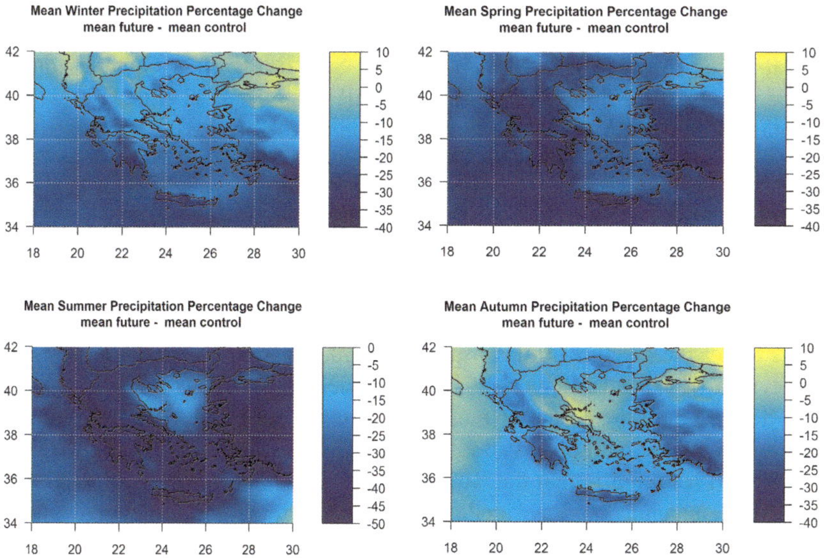

Fig. 4. Changes (%) in mean seasonal precipitation totals between the future period 2071–2100 (SRES A1B) and the control period 1961–1990.

As far as the end of the 21st century is concerned, future projections based on the ensemble mean simulations show that higher decrease in the precipitation totals is likely to appear. More specifically, the mean precipitation over the wide area of Greece is likely to decrease by 16% in winter, 26.5% in spring, 37% in summer, 12.5% in autumn and 19% in the whole year (Fig. 2, 4), during the period 2071-2100 relatively to the reference period 1961-1990. The decrease in mean annual precipitation is projected to be highest (25%) in Peloponnese and Crete, against 20% in other areas of Greece and less than 15% in North Aegean.

In winter season, during which highest precipitation totals are recorded over Greece, the highest decrease is expected in eastern continental regions, western Peloponnese, Cyclades, Crete and Dodecanese, exceeding 18%. On the contrary, in North Aegean the decrease appears to be less than 8% against 9 - 12% in other areas. During spring, concerning most areas of Greece, the mean precipitation is expected to decrease more than 20% with the highest decrease (approximately 30%) in West and Central-Eastern Greece, Peloponnese (>30%), and Crete (35%) while the decrease appears to be lower (<17%) in the Northern Aegean islands. During summer, the decrease is projected to reach or even exceed 40% over most parts of Greece. Even in the North Aegean Sea, a climate zone with the lower decrease in mean precipitation totals, the decrease appear to be more than 20%. However, with the exception of Northern Greece, the precipitation change in the rest areas remains low. During autumn, on one hand significant decrease in mean precipitation totals is expected to be approximately 20% over Crete and Western Peloponnese; while on the other hand, the decrease seems to be limited at 7% in Central-Eastern Greece and Northern Aegean Sea. In the other climatic zones of Greece, the decrease in mean precipitation will range from 11% to 13% with the exception of Western Greece and Eastern Peloponnese, where the decrease is projected to be higher, exceeding 15%.

4 Conclusions

In this study the changes in mean precipitation totals within the area of Greece were evaluated using 12 Regional Climate Models under SRES A1B. The precipitation derived from the ensemble mean simulations of the 12 RCMs was compared with the gridded station observations in a CRU like grid for the reference period (1961-1990) resulting in that the RCMS reproduce satisfactorily the spatial distribution of mean annual and seasonal precipitation in the region of Greece.

The findings of the performed analysis showed that the precipitation is likely to decrease within the 21st century. This decrease will be greater at the end of the 21st century reaching 20% for the whole area of Greece. Further research is needed in order to reveal the physical mechanisms, which are responsible for this future projected precipitation decrease.

References

Amanatidis GT, Paliatsos AG, Repapis CC, Bartzis JG (1993) Decreasing precipitation trend in the Marathon area, Greece. Int. J. Climatol. 13, 191–201

Dünkeloh A, Jacobeit J (2003) Circulation dynamics of Mediterranean precipitation variability 1948–98. Int. J. Climatol. 23, 1843–1866, doi: 10.1002/joc.973

Feidas H, Noulopoulou Ch, Makrogiannis T, Bora-Senta E (2007) Trend analysis of precipitation time series in Greece and their relationship with circulation using surface and satellite data: 1955-2001. Theor. Appl. Climatol. 87, 155-177, doi:101007/s00704-006-0200-5

Giorgi F (2006) Climate change hot-spots. Geophys. Res. Lett. 33, L08707, doi: 10.1029/2006GL025734

González-Rouco JF, Jimenez JL, Quesada V, Valero F (2001) Quality control and homogenization of monthly precipitation data in the southwest of Europe. J. Climate 14, 964–978

Hatzioannou L, Retalis D, Pasiardis S, Nikolakis D, Asimakopoulos DN, Lourantos N (1998) Study of the precipitation time series in SE Greece and Cyprus. Proc. 4th Greek Scientific Conference in Meteorology-Climatology-Atmospheric Physics, pp. 279-284, Athens

Hulme M, Barrow EM, Arnell NW, Harrison PA, Johns TC, Downing TE (1999) Relative impacts of human-induced climate change and natural variability. Nature 397, 688-691

IPCC (2007) Climate Change (2007): The Physical Science Basis. Contribution of Working Group I to the Fourth Assessment Report of the Intergovernmental Panel on Climate Change. Eds. Solomon, S., D. Qin, M. Manning, Z. Chen, M. Marquis, K.B. Averyt, M. Tignor and H.L. Miller. Cambridge University Press, Cambridge, United Kingdom and New York, NY, USA, 996 pp

Kandylis F, Repapis C, Kotini-Zambaka S (1989) Climatic changes during last 100 years in eastern Mediterranean. Proc. 1st Greek Conference in Geography, Athens

Metaxas DA, Philandras CM, Nastos PT and Repapis CC (1999) Variability of Precipitation pattern in Greece during the year. Fresen. Environ. Bull. 8, 1-6

Nakićenović N, Alcamo J, Davis G, de Vries B, Fenhann J, Gaffin S, Gregory K, Grübler A, Jung TY, Kram T, La Rovere EL, Michaelis L, Mori S, Morita T, Pepper W, Pitcher H, Price L, Raihi K, Roehrl A, Rogner H-H, Sankovski A, Schlesinger M, Shukla P, Smith S, Swart R, van Rooijen S, Victor N, Dadi Z (2000) IPCC Special Report on Emissions Scenarios, Cambridge University Press, Cambridge, UK

Nastos PT, Zerefos CS (2009) Spatial and temporal variability of consecutive dry and wet days in Greece. Atmos. Res. 94, 616–628

Pal JS, Giorgi F, Bi X (2004) Consistency of recent European summer precipitation trends and extremes with future regional climate projections. Geophys. Res. Lett. 31, L13202, doi:10.1029/2004GL019836

Piervitali E, Colasino M, Conte M (1998) Rainfall over the central-western Mediterranean basin in the period 1951-1995. Part I: precipitation trends. Geophysics and Space Physics 21C, 331–344

Tselioudis G, Zerefos C, Repapis C, Zanis P (2006) Climate Predictions for the Mediterranean Region, International Geosphere Biosphere Programme -Global Change Programme of the Academy pf Athens, Colloquium entitled "Climate Change", Athens, 41–50

Türkes M (1998) Influence of geopotential heights, cyclone frequency and southern oscillation on rainfall variations in Turkey. Int. J. Climatol. 18, 649–680

Xoplaki E (2002) Climate variability over the Mediterranean. PhD thesis, University of Bern, Switzerland ([http://sinus.unibe.ch/klimet/docs/phd_xoplaki.pdf])

Zanis P, Kapsomenakis I, Philandras C, Douvis K, Nikolakis D, Kanelopoulou E, Zerefos C, Repapis C (2009) Analysis of an ensemble of present day and future regional climate simulations for Greece. Int. J. Climatol. 29, 1614-1633, doi:10.1002/joc

Trends and variability of precipitation within the Mediterranean region, based on Global Precipitation Climatology Project (GPCP) and ground based datasets

P.T. Nastos

Laboratory of Climatology and Atmospheric Environment, Faculty of Geology and Geoenvironment, University of Athens, Athens, GR 15784, Greece. nastos@geol.uoa.gr

Abstract This study presents and analyzes the trends and variability of precipitation totals within the Mediterranean region, for the period 1980 - 2009, by using Global Precipitation Climatology Project (GPCP) gridded datasets and ground based observations from the Hellenic National Meteorological Service (HNMS) and the World Climate Data and Monitoring Programme (WCDMP) of the World Meteorological Organization. Besides, the trends of sea surface temperature (SST) and sea surface pressure (SLP) from NCEP/NCAR Reanalysis are investigated in order to explain the observed trends in precipitation. On the other hand, the influence of atmospheric circulation, by means of North Atlantic Oscillation Index (NAOI), on the precipitation variability is presented. The results showed that an anti-correlation, (statistically significant at 95% confidence level), exists between precipitation (mm/day), rain days and NAOI, within the rain season of the year (October-March), against a positive correlation within the dry season (April-September). Concerning the temporal distribution of precipitation, decreasing trends appear throughout the Mediterranean, especially in the eastern parts and within the rain season of the year, while these trends are positive within the dry period, but statistically insignificant (95% confidence level). Increasing trends of SST for both rain and dry season, and decreasing trends of SLP appear.

1 Introduction

Precipitation patterns and variability are of great scientific interest because they are involved in the energy balance of atmosphere, through the water cycle, having key role in climatic change. The sensitivity of precipitation in local and regional factors, such as land-sea interaction and topography, arouses the need of a denser network of rain gauges, which in most of the cases, is considered inadequate. This limitation in the rain gauges network comes to improve the Global Precipitation Climatology Project (GPCP), which is an element of the Global Energy and Water Cycle Experiment (GEWEX) of the World Climate Research program (WCRP).

Mediterranean appear to be a vulnerable region, concerning water scarcity and decreasing run off as adverse consequences of climatic change (IPCC 2007). Decreasing winter precipitation appears during the last few decades, mostly starting in the 1970s (Piervitali et al. 1997; Schonwiese and Rapp 1997). A decrease of cyclone frequency during winter in the western Mediterranean region is associated with a northward shift of the storm track and persistent high phase of NAOI (Hurrell and van Loon 1997). Cyclone activity presents large seasonal and spatial variability, with large differences from western to eastern Mediterranean and between cold and warm season (Lionello et al. 2006). The eastern Mediterranean, shows a tendency towards drier conditions (Kutiel et al. 1996, Turkes 1998) while the western and central areas, although showing negative trends in the number of wet days and/or the total rainfall amounts, indicate an increase in intense precipitation events over the period 1951-1996 (Brunetti et al. 2001, Alpert et al. 2002). Nastos and Zerefos (2009) concluded that the temporal variability of consecutive wet days shows statistically significant (confidence level of 95%) negative trends, mainly in the western region of Greece, characterized by large orographic precipitation amounts (Metaxas et al. 1999). Insignificant positive trends for consecutive dry days appear almost all over the country with emphasis in the southeastern region.

The goal of this study is on one hand, to investigate the trends of precipitation totals, precipitation rate and rain days (days with precipitation greater than 1 mm) from gridded (GPCP) and ground based observations (HNMS, WCDMP), during the period 1980-2009. On the other hand, possible impacts of atmospheric circulation, variability of SST and SLP are analyzed.

2 Data and Analysis

The precipitation datasets used in the analysis concern on one hand, gridded monthly values from the Global Precipitation Climatology Project (GPCP), which is an element of the Global Energy and Water Cycle Experiment (GEWEX) of the World Climate Research program (WCRP). Monthly mean precipitation estimates are being produced beginning January 1979 and going through nowadays. It was established in 1986 (Arkin and Xie 1994), and its primary goal is to provide a long time series of monthly mean precipitation data on 2.5° x 2.5° latitude–longitude grids. The GPCP satellite precipitation data (Huffman et al. 1997) are a combined product from microwave and infrared (IR) satellite-derived precipitation estimates and rain gauge data. However, other sources of precipitation data are also utilized by GPCP, such as the Global Historical Climatology Network (GHCN) - Climate Anomaly Monitoring System (CAMS) gauge analysis, Television and Infrared Observation Satellite (TIROS) Operational Vertical Sounder (TOVS)-based estimates, and the outgoing longwave radiation (OLR) precipitation index (Adler et al. 2003). Validation studies, starting with Adler et al. (2003) tend to show small biases, but regions of complex terrain have an increased risk of low bias due to deficiencies in the sitting of precipitation gauges, as well as deficiencies in the satellite algorithms in detecting orographic enhancement of precipitation. Sampling errors of the GPCP products are of the order of 10%–15% (Gruber and Levizzani 2006) and they are estimated using different sampling strategies and comparison with Tropical Rainfall Measuring Mission (TRMM) Microwave Imager (TMI) sampling.

On the other hand, monthly precipitation totals and annual number of rain days from 40 meteorological stations within the Mediterranean region were acquired from the Hellenic National Meteorological Service (HNMS) and the World Climate Data and Monitoring Programme (WCDMP) of the World Meteorological Organization, for the period 1980-2009. The influence of atmospheric circulation in the gridded and ground based precipitation variability is examined with respect to North Atlantic Oscillation Index. A dominant teleconnection pattern is the North Atlantic Oscillation (NAO), which is defined as a difference of standardized sea level pressure (SLP) time series from a station close to the centre of the Azores High (usually Lisbon, Azores or Gibraltar) and a station close to the centre of Islandic Low (Reykjavik) (Hurrell 1995).

3 Results and Discussion

In order to better understand the variability of GPCP gridded precipitation (mm) in the Mediterranean region, the spatial variation of GPCP trends are presented within the rain (October-March) and dry (April-September) season (Fig. 1).

During rainy season, increasing trends appear in the west Iberian region, while decreasing trends exist through the Mediterranean, more pronounced in the eastern Mediterranean (Fig. 1a). Concerning Greece, the precipitation trend is not uniform, appearing decreasing trends over southern region against slight increasing trends over central continental Greece. On the other hand, during dry season, increasing trends are observed in the majority of the Mediterranean, with emphasis over mountainous continental Greece (Fig. 1c), against decreasing trends over northern Italy peninsula. All the above mentioned trends are not statistically significant at 95% confidence level. These findings are in agreement with Xatzianastassiou et al. (2008), who studied the GPCP data in the broader Greek area for the 26-yr period 1979-2004, finding that the precipitation over the study area has not changed essentially during the 26-yr period, since a very small decrease in precipitation is computed, which is not statistically significant at 95% confidence level.

The annual trends present similar patterns with those of rain season (not shown). Nevertheless, statistically significant correlations appear between NAOI and precipitation totals; negative correlations within the rain season (-0.5 over Iberian, -0.3 over Greece) (Fig. 1b) against positive correlations within the dry season (Fig. 1d) in the west and central Mediterranean. Slight negative correlations appear in the east Mediterranean. In the recent years, a lot of studies have been carried out giving evidence that the frequency, persistence and severity of long period dry conditions in the Mediterranean countries have increased (Sahsamanoglou et al. 1992, Wigley 1992, Maheras et al. 1999, Houghton et al. 2001, Feidas et al. 2007, Nastos and Zerefos 2009).

In order to compare the observed trends of the seasonal gridded time series (GPCP) with the corresponding trends of the ground based time series (HNMS, WCDMP), the spatial average of annual precipitation totals from representative meteorological stations, for west, central and east Mediterranean were calculated, for the period 1980-2009.

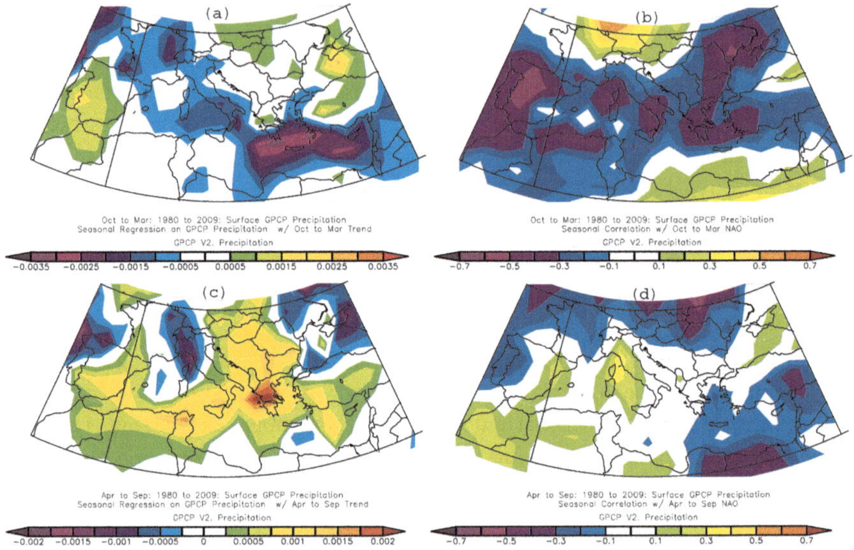

Fig. 1. Spatial variation of the trends for GPCP gridded precipitation (mm/day) within the rain season (a) and dry season (c) and the correlation coefficients between GPCP gridded precipitation and NAOI within the rain season (b) and dry season (d).

More specifically, the estimated time series of average annual precipitation totals within the west (Gibraltar: 5.12° W, 36.09° N; Oran: 0.36° E, 35.38° N; Barcelona: 2.09° E, 41.24° N; Ajjacio: 8.48° E, 41.55° N; Perpinian: 2.53° E, 42.44° N), central (Malta; 14.29° E, 35.51° N; Tunis: 10.14° E, 36.50° N) and east Mediterranean (Athens: 23.43° E, 37.58° N; Kalamata: 22.0° E, 37.04° N; Naxos: 25.37° E, 37.10° N; Heraklion: 25.11° E, 35.20° N; Nicosia: 33.36° E, 35.16° N; Larnaca: 33.38° E, 34.53° N; Antalya: 30.44° E, 36.52° N) are depicted in Figure 2 (left graphs). The annual number of rain days (days with precipitation greater than 1 mm) concerning west (Menorca: 4.14° E, 39.52° N; Alicante: 0.30° W, 38.22° N; Barcelona: 2.09° E, 41.24° N), central (Malta: 14.29° E, 35.51° N; Tunis: 10.14° E, 36.50° N) and east Mediterranean (Athens: 23.43° E, 37.58° N, Kalamata: 22.01° E, 37.04° N, Heraklion: 25.11° E, 35.20° N, Rhodes: 21.44° E, 36.24° N) are also depicted in Figure 2 (right graphs).

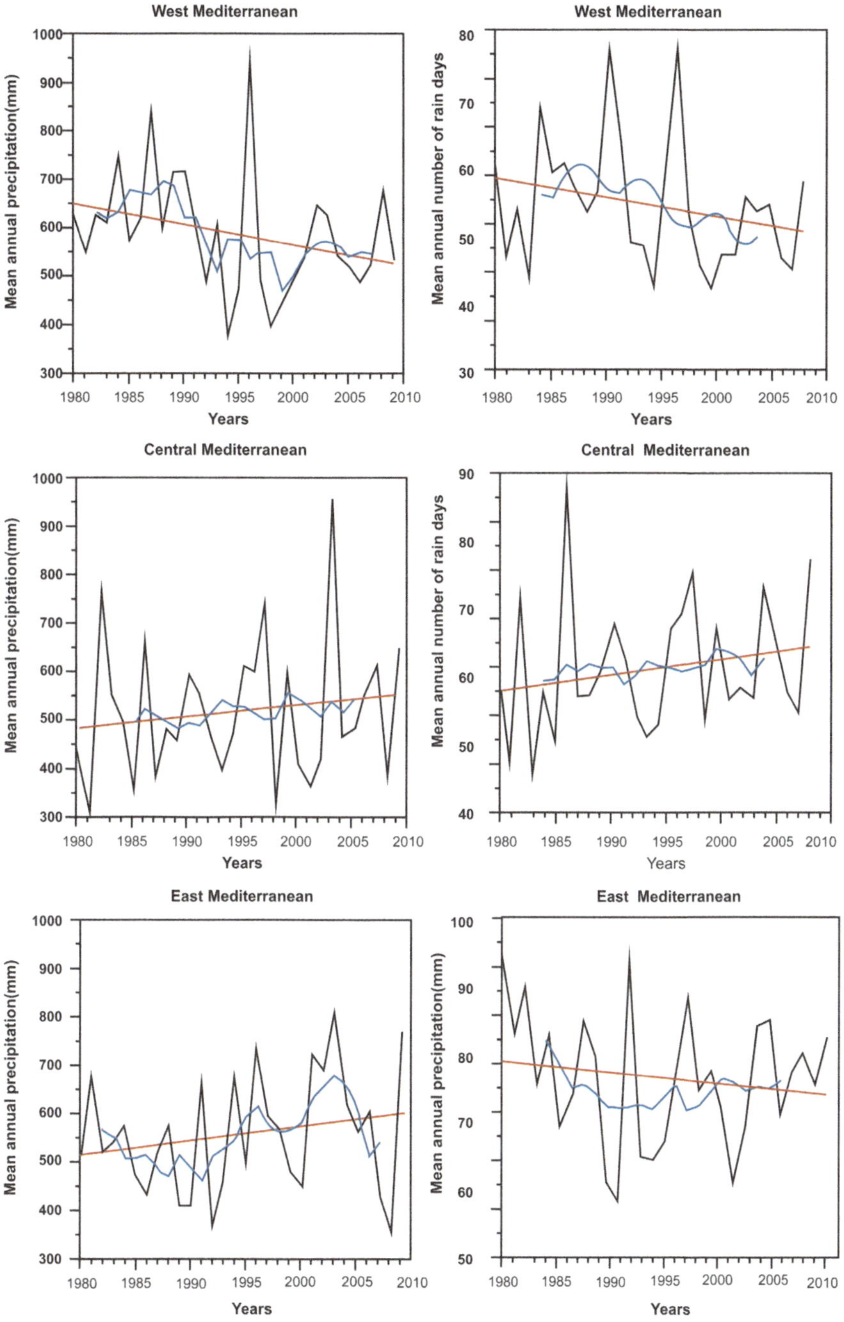

Fig. 2. Mean annual precipitation (left graphs) and mean annual number of rain days (right graphs) from the west (upper), central (middle) and east (lower) Mediterranean region, along with 9 years moving average (blue line) and linear trend (red line), for the period 1980-2009.

Insignificant increasing trends (at 95% confidence level) of mean annual precipitation for central and east Mediterranean against decreasing trends for west Mediterranean appear, while insignificant decreasing trends of annual rain days for west and east parts and increasing trends for central Mediterranean are observed (Fig. 2). These findings possibly explain the observed increasing trend of precipitation rate in the eastern Mediterranean (Fig. 1e, g). Besides, the decreasing trend in NAOI (not shown), although insignificant, is likely associated with the increasing trends mainly in the central and east Mediterranean, during the period 1980-2009. NAOI shows positive figures from 1980 to the middle of 1995, and since then negative ones dominate.

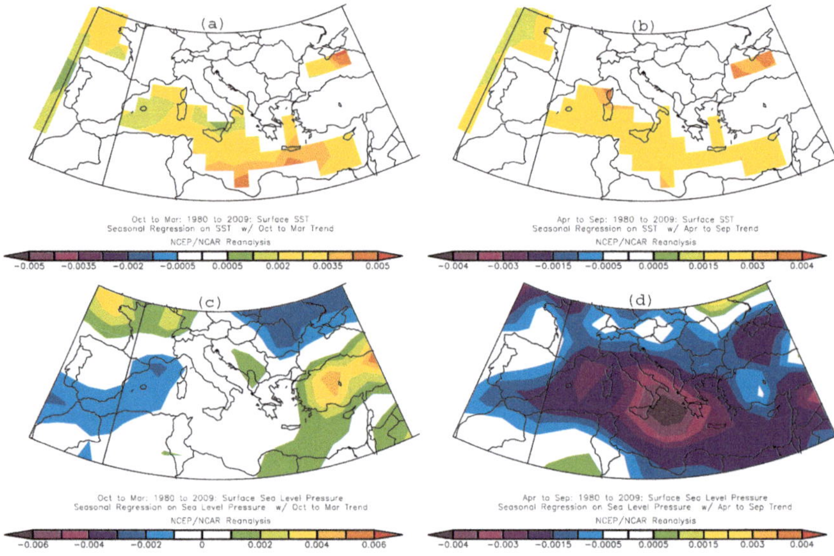

Fig. 3. Spatial variation of the trends for SST within the rain season (a) and dry season (b). Spatial variation of the trends for SLP within the rain season (c) and dry season (d).

Figure 3 presents the spatial variations of trends for sea surface temperature (SST) and sea level pressure (SLP), within the rain and dry season of the year, during the period 1980-2009. The observed increasing trends in SST and decreasing trends in SLP within the dry season (Fig. 3b, d) are likely associated with the increasing precipitation totals (Fig. 1c). Higher SST figures results in higher water content in the atmosphere, through the evaporation process. It seems that in dry season, when the barometric troughs move northwards, SST plays a key role in precipitation formation. On the other hand, increasing trends in SST (Fig. 3a) are associated with slight decreasing/increasing trends in SLP (3c) within the rain season, at west/east Mediterranean, against decreasing trends of precipitation totals (Fig. 1a). A decrease of cyclone frequency during winter in the western Mediterranean region is presumably associated with a northward shift of the storm track

(Hurrell and van Loon 1997), while in addition, such decline of cyclone frequency is suggested to continue as green house gas concentration increases, as shown by scenario simulations (Lionello et al. 2002).

4 Conclusions

In this study the spatial variation of precipitation trends using gridded monthly datasets from the Global Precipitation Climatology Project (GPCP) along with ground based observations from representative stations within the Mediterranean region are analyzed. The results showed that the trends of gridded GPCP data against those from stations are in a good agreement. During the examined period (1980-2009), insignificant decreasing/increasing trends for precipitation are observed within rain/dry season of the year, while a significant anti-correlation between NAOI and precipitation exists within the rain season against a significant correlation within the dry season. It seems that in dry season, SST plays a key role in precipitation formation while the decrease in cyclones frequency due to northward shift of the storm track within the rain season, is associated with decreasing trends of precipitation totals.

References

Adler RF, Huffman GJ, Chang A, Ferraro R, Xie P, Janowiak J, Rudolf B, Schneider U, Curtis S, Bolvin D, Gruber A, Susskind J, Arkin P, Nelkin EJ (2003) The Version 2.1 Global Precipitation Climatology Project (GPCP)

Alpert P, Ben-Gai T, Baharad A, Benjamini Y, Yekutieli D, Colacino M, Diodato L, Ramis C, Homar V, Romero R, Michaelides S, Manes A (2002) The paradoxical increase of Mediterranean extreme daily rainfall in spite of decrease in total values. Geophys. Res. Lett. 29, 311-314

Arkin P A, Xie P (1994) The Global Precipitation Climatolog Project: First Algorithm Intercomparison Project. Bull. Amer. Meteor. Soc. 75, 401-419

Brunetti M, Colacino M, Maugeri M, Nanni T (2001) Trends in the daily intensity of precipitation in Italy from 1951 to 1996. Int. J. Climatol. 21, 299-316

Feidas H, Noulopoulou N, Makrogiannis T, Bora-Senta E (2007) Trend analysis of precipitation time series in Greece and their relationship with circulation using surface and satellite data: 1955-2001. Theor. Appl. Climatol. 87, 155-177

Gruber A, Levizzani V (2006) Assessment of global precipitation: A project of the Global Energy and Water Cycle Experiment (GEWEX) Radiation Panel GEWEX, World Climate Resarch Programme, WMO. Assessment of Global Precipitation Climatology Project (GPCP), Executive Summary, 61 pp. [Available online at http://cics.umd.edu/_yin/GPCP/main.html.]

Hatzianastassiou N, Katsoulis B, Pnevmatikos J, Antakis V (2008) Spatial and temporal variation of precipitation in Greece and surrounding regions based on Global Precipitation Climatology Project data. J. Climate 21 (6), 1349-1370

Huffman G J, Adler P F, Arkin P A, Chang A, Ferraro R, Gruber A, Janowiak J, Joyce R J, McNab A, Rudolf B, Schneider U, Xie P (1997) The Global Precipitation Climatology Project (GPCP) combined precipitation dataset. Bull. Amer. Meteor. Soc. 78, 5-20
Houghton J T, Ding Y, Griggs DJ, Noguer M, van der Linden PJ, Dai X, Maskell K, Johnson CA, Eds. (2001) Climate Change 2001: The Scientific Basis. Cambridge University Press, 881 pp
Hurrell JW, Van Loon H (1997) Decadal variations in climate associated with the North Atlantic oscillation. Climatic Change, 36, 301-326
Hurrell JW (1995) Decadal trends in the North Atlantic Oscillation: Regional temperatures and precipitation. Science, 269, 676–679
IPCC (2007) The Physical Science Basis. Contribution of Working Group 1 to the Fourth IPCC Assessment Report, Chapter 11 Regional Climate Projections
Kutiel H, Maheras P, Guika S (1996) Circulation and extreme rainfall conditions in the eastern Mediterranean during the last century. Int J Climatol 16, 73–92
Lionello P et al (2006) Cyclones in the Mediterranean region: climatology and effects on the environment, In: P. Lionello, P. Malanotte-Rizzoli & R. Boscolo (Eds), Mediterranean Climate Variability, Amsterdam: Elsevier, pp. 324-372
Lionello P, Dalan F, Elvini E (2002) Cyclones in the Mediterranean Region: the present and the doubled CO2 climate scenarios. Clim. Res. 22,147-159
Maheras P, Xoplaki E, Kutiel H (1999), Wet and dry monthly anomalies across the Mediterranean basin and their relationship with circulation, 1860–1990. Theor. Appl. Climatol. 64, 189–199
Metaxas DA, Philandras CM, Nastos PT and Repapis CC (1999) Variability of Precipitation pattern in Greece during the year. Fresen. Environ. Bull. 8, 1-6
Nastos PT, Zerefos CS (2008) Decadal changes in extreme daily precipitation in Greece. Adv. Geosci. 16, 55–62
Nastos PT, Zerefos CS (2009), Spatial and temporal variability of consecutive dry and wet days in Greece. Atmos. Res. 94, 616–628
Piervitali E, Colasino M, Conte M (1997) Signals of climatic change in the Central-Western Mediterranean basin. Theor. Appl. Climatol. 58, 211–219
Sahsamanoglou HS, Makrogiannis T, Rossidis ZB (1992) Characteristics of rainfall in the greater region of the Mediterranean. Proc. First Panhellenic Conf. of Meteorology, Climatology and Physics of the Atmosphere, Thessaloniki, Greece, Hellenic Meteorological Society, 147-153
Schonwiese C, Rapp J (1997) Climate Trend Atlas of Europe based on observations 1891-1990. Kluwer Academic Publishers, Dordrecht, 224pp
Turkes M (1998) Influence of geopotential heights, cyclone frequency and Southern oscillation on rainfall variations in Turkey. Int. J. Climatol. 18, 649-680
Wigley TML (1992) Future climate of Mediterranean basin with particular emphasis on changes in precipitation. Climatic Change and the Mediterranean, L. Jeftic, J. D. Milliman, and G. Sestini, Eds., Edward Arnold, 15-44

Climatic influence on Lake Stymphalia during the last 15 000 years

I. Unkel[1], C. Heymann[1], O. Nelle[1], E. Zagana[2]

[1] Graduate School "Human Development in Landscapes", Institute for Ecosystem Research Kiel University, Olshausenstr. 40, D-24098 Kiel (Germany)
[2] Department of Geology, University of Patras, Rio 26500 Patras (Greece)
iunkel@ecology.uni-kiel.de

Abstract Known from the ancient myth of Heracles fighting the Stymphalian birds, the karst polje of Stymphalia (22°27'E, 37°51'N) is an ideal site to study the climate history of the area. Stymphalia is the only natural perennial lake on the Northern Peloponnesus, which provides a continuous sedimentary record of the entire Holocene and a large part of the Last Glacial. As a large and quite reliable water reservoir, Lake Stymphalia and its surrounding karst springs played an important role for the water supply of the region from ancient time until today. However, due to climate fluctuations, the water supply can change significantly, challenging the water management of the people living in the area. Here we present geochemical analyses of the uppermost part of a lake sediment core (STY-1), recording the changes in climate and water supply during the Holocene and the Late Glacial. The chronology is based on several ^{14}C dates combined to a Bayesian age-depth model. Using XRF elemental analysis, we compare the influx of terrestrial material (indicated by K and Rb) to the carbonate precipitation in the lake (indicated by Ca and Sr). The Rb/Sr ratio as a proxy for changes between dry/warm and wet/cold conditions indicate pronounced wet phases around 6800, 4000–3700–4000, 3500–3000 and 500–200 cal BP.

1 Introduction

Lake Stymphalia, or Stymphalos as it was called in ancient times, is the only larger natural lake of the Northern Peloponnesus, which is mostly perennially water-bearing through much of the Holocene period. As water is a rare resource on the Peloponnesus, it is a persistent and consistent factor in urban development and history (Crouch 1993). Settlement history in the karst polje of Stymphalia is known since at least the 5th century BC (Williams and Gourley 2005) and mentioned by Pausanias in his *"Description of Greece"* from the late 2nd century AD. During the time of the Roman emperor Hadrian (ca. 125 AD) an aqueduct was built channeling the waters of Stymphalia spring north-eastwards along an 85 km route to Acrocorinth (Lolos 1997). The Hadrianic water conduit possibly followed a dam-like structure parallel to the eastern shore of Lake Stymphalia through a Roman tunnel nowadays called Siouri to the neighboring polje of Skotini and continued further towards Acrocorinth.

The ancient activities in managing the water emphasize the importance of the rare resource in the region. If we believe the report of Pausanias, the lake was not perennial during the entire Holocene. He writes (Pausanias 8.22.3b) "In winter the spring makes a small lake in Stymphalos, and the river Stymphalos issues from the lake; in summer there is no lake, but the river comes straight from the spring." The water discharge from the springs providing Lake Stymphalia was very low in the early 1980s (Morfis and Zojer 1986) and the lake dried up completely for few years during the 1990s (Papastergiadou et al. 2007). All these reports show, that the lake level, which has a mean value of only 1.5 m in modern times, can fluctuate significantly due to climatic changes, leading to periods of entire desiccation.

Climatically, the region is located in the transition zone between the temperate mid-latitudes at the edge of the North Atlantic Oscillation and the tropical low-latitude region with its descending branch of the Hadley cell (Bout-Roumazeilles et al. 2007). Hence, a lake in such a climatically sensitive region is an excellent archive of climate change.

2 Geological and hydrogeological setting

The polje of Stymphalia (22°27'E, 37°51'N) is situated at about 600 m above sea level. The surrounding mountains reach up to about 2000 m. Mt. Ziria (2374 m) to the north of the polje marks the highest point of the area and is also the main catchment area of Stymphalia. The pre-Quaternary geology of the area is characterized by four units (Morfis and Zojer 1986; Röckel 1987): (1) Phyllite-Quarzite unit with quarzites and mica schists; (2) Tripolitza unit with phyllites (Tyros layers), limestones and Flysch; (3) Olonos-Pindos with limestones, radiolarites and marls; (4) Neogene deposits with lime-dominated conglomerates. The karstified limestones of the Tripolitza and Olonos-Pindos units represent prominent aquifers, while the almost impermeable phyllites and Flysch layers from aquicludes.

Annual precipitation ranges from about 800–1090 mm in the polje to about 1275 mm at the slopes of Mt. Ziria (Morfis and Zojer 1986). The modern water depth is 1.5 m on average and is directly linked to the discharge from several karst springs. The two most important springs are the villages of Kefalari (max. discharge 160 l/s) and Stymphalia (max. discharge 800 l/s) (Morfis and Zojer 1986). Depending on the lake level, the average surface area of the lake fluctuates between 3.5 and 5 km^2. There are two outflows of the lake on the southern margin of the polje. One is the natural sink hole (katavothre) in the southwest, which is not active any longer since the water management installations of the 20th century normally keep the lake level below the outflow level. Only for a tracer experiment in spring 1984, water was injected into the sink hole again. The experiment showed that underground discharge from lake Stymphalia re-emerges some 35 km to the South through the springs of Lerni and Kiveri, draining into the Gulf of Argos (Morfis and Zojer 1986). Astonishingly, this was already assumed by Pausanias (8.22.3c) “This river descends into a chasm in the earth, and reappearing once more in Argolis it changes its name, and is called Erasinos instead of Stymphalos.”

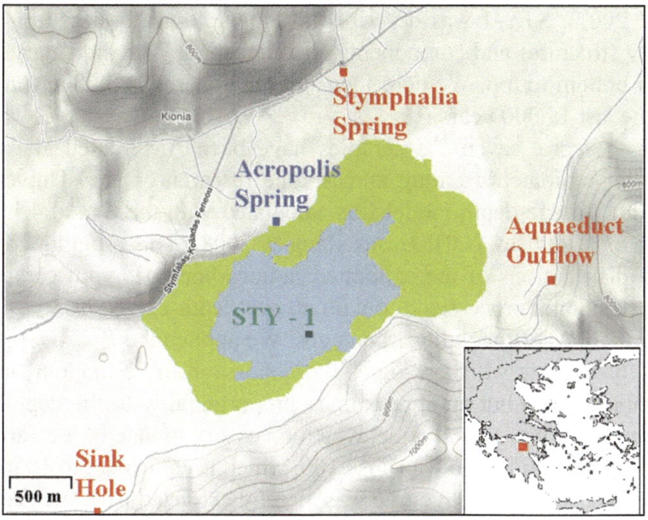

Fig. 1. Topographic map of the study area. Marked with red dots are the natural sink hole, the artificial outflow of the Hadrianic Aquaeduct and the position of the coring site STY-1. The blue area gives an approximate estimate of the open water surface, which changes seasonally and annually. Much of the lake area is covered by a dense reed (*Phragmites*) belt marked in green.

3 Material and methods

Water samples were collected in March 2010 in two polyethylene bottles. The first bottle of 0.5 L volume was filtered on-site through 0.45μm Millipore filters. It was then acidified to pH about 2 with 65% ultra pure HNO_3 and used for major cation and trace element analysis. The second non-acidified aliquot (1 l) was retained to determine non-metal ions. Temperature (T), pH and electrical conductivity (E.C.) were determined using an YSI 63 multi-parameter meter, while redox potential (Eh) using an ion/E.C. meter (Consort® C533). The major cations Ca^{2+}, Mg^{2+}, Na^+ and K^+ were determined by atomic absorption spectroscopy AAS (GBC Avanta). Trace element determinations were conducted using ICP-MS Perkin Elmer, ELAN 6100. Alkalinity and CO_2 were measured in situ by volumetric titration using bromocresol green-methyl red indicator by Hach® titration kits. Similarly, chlorides (Cl^-) were measured using the $AgNO_3$ method by Hach® titration kits, while SO_4^{2-}, NO_3^-, NO_2^-, PO_4^{3-} , F^- and NH_4^+ were determined using a spectrophotometer (Hach®, DR/4000). All analyses were conducted at the Laboratory of Hydrogeology, University of Patras (Greece), immediately after collection. The charge balance error for all chemical analyses is within a highly acceptance range (±5%). The results are presented in Figure 2 and Figure 3.

A 15.52 m long sediment core (STY-1) was recovered from Lake Stymphalia in March 2010 using an Usinger piston corer with 80 and 55 mm diameter (Mingram et al. 2007). STY-1 was described sedimentologically based on grain size, sedimentary structures and components, distinguishing 70 sedimentary units numbered from bottom to top. We here focus on the uppermost 3.50 m (units 52–70), spanning the last 15.000 years BP.

To date the core, seven ^{14}C samples have been AMS-dated at the Leibniz-laboratory for radiometric dating and isotope research of Kiel University (Germany) so far. An age-depth-model was calculated with the OxCal 4.1 calibration software (Bronk Ramsey 2001; Bronk Ramsey 2008). This program combines the probability distributions of the calibrated radiocarbon ages with certain assumptions on depositional processes to obtain an age-depth-curve with the 1σ (68.2%) probability margins of the radiocarbon ages. We applied the *P_Sequence* model of OxCal 4.1 to our data which assumes the deposition to be random (meaning no regular annual layers) but an approximate proportionality to the depth z is given (Bronk Ramsey 2008). As a *k*-value, which gives an estimate of the variation from a constant sedimentation rate we chose 100, which is equivalent to 10 mm calculation increments. An age-depth curve was directly extracted from the OxCal model, returning maximal and minimal ages for every 1 cm interval (Fig. 4). A curve based on the mean age values – $(age_{max}+age_{min})/2$ – was then used to plot the geochemical data of the core discussed hereafter.

X-ray fluorescence (XRF) analysis of the elemental composition was carried out at the Institute of Geosciences, Kiel University, with an Avaatech Core Scanner. The scanning resolution was 1 cm with varying tube voltage to record different elemental spectra discerning palaeoclimate variations and sedimentological patterns (for a detailed discussion of the method see Richter et al. (2006) and Tjallingii (2006)).

4 Results

We selected the elements Ca, K, Rb and Sr for reconstructing the climate-driven variability of the water budget at Stymphalia (Fig. 4). They have been normalized against Ti, which is assumed to be inert and not involved in geochemical or hydro chemical processes.

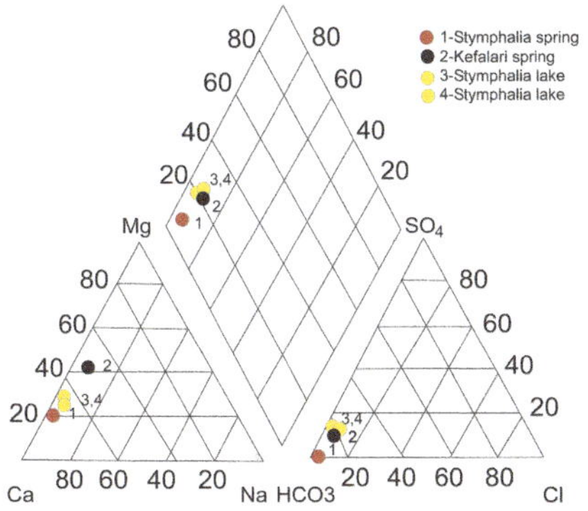

Fig. 2. Piper diagram based on the water samples taken in March 2010. Samples 1, 3 and 4 belong to the hydrochemical type of Ca-HCO_3, while sample 4 represents a Ca-Mg-HCO_3 type.

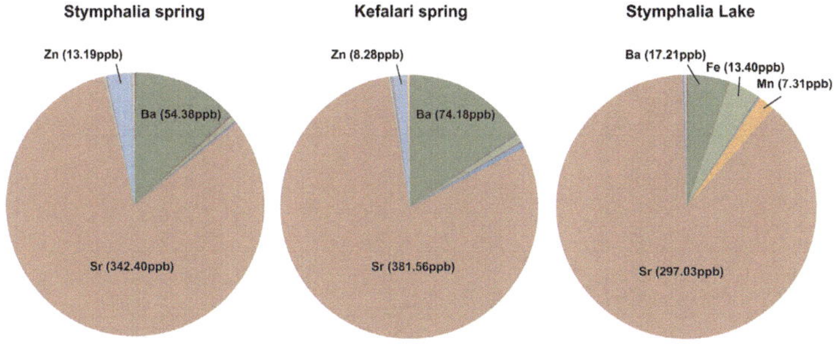

Fig. 3: The major trace elements in ppb. Sr is dominating, which is typical for carbonate rock.

Elements associated mainly with clay minerals (e.g., Rb, K, Al) and clastic sediments (e.g., Zr, Ti, Si) are correlated throughout the Glacial and Holocene. They yield information on the sediment input to the lake during the wet period of the year (winter/spring). Hence, high precipitation in the lake catchment is correlated with high values in K and Rb.

Ca and Sr are precipitated autochthonous in the lake as carbonates. As carbonate precipitation is the highest in the dry and warm seasons of the year (summer/autumn), high Ca and Sr values are interpreted as a proxy for especially dry/warm periods. Statistical correlation analyses following the XRF measurements of core STY-1 indicate, that the elements associated with clay minerals are anti-correlated with the elements associated with the carbonates during the Holocene but not during Glacial times (Heymann, in prep.).

The combination of these proxies is the Rb/Sr ratio, which is widely used as a relative measure for wet/cold or dry/warm conditions (Chen et al. 1999; Jin et al. 2001). In lake sediments, high Rb/Sr ratios indicate high Rb input and thus higher precipitation and higher weathering in the catchment area under wetter/colder conditions. In contrast, low Rb/Sr ratios indicate high Sr precipitation in carbonates due to warmer/drier conditions.

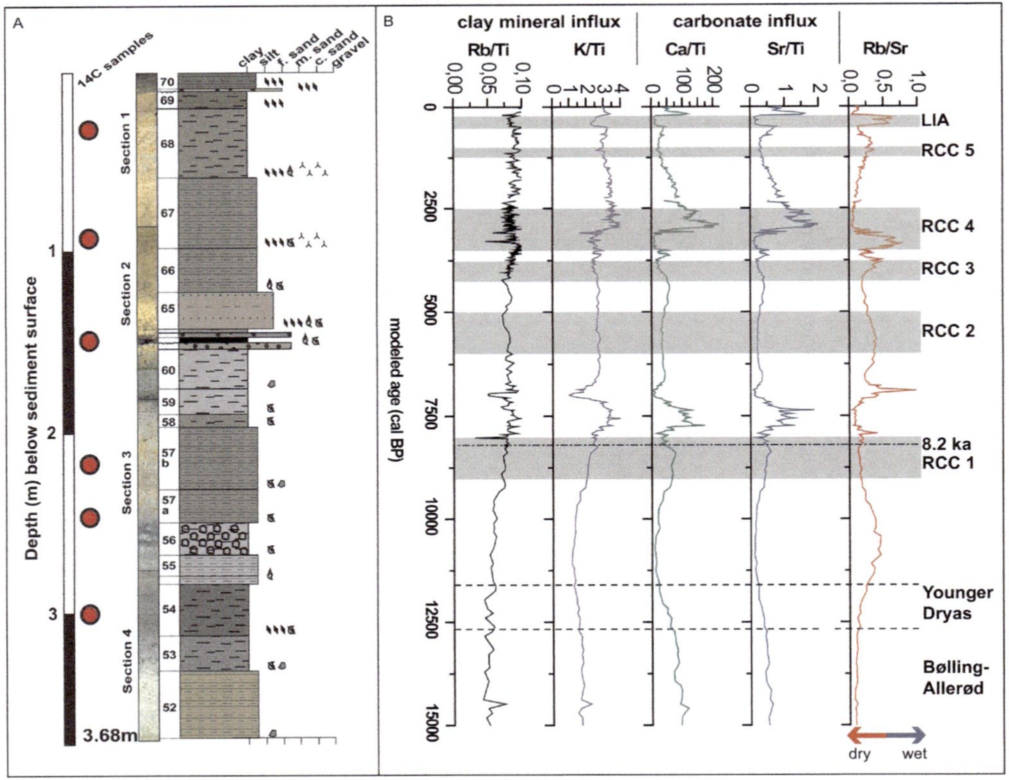

Fig. 4. a. lithological profile of core STY-1 with the position of the ^{14}C samples. **b**. The elements elements Rb, K, Ca and Sr normalized on Ti together with the Rb/Sr ratio plotted against the age derived from the ^{14}C age-depth-model.

A general trend to higher Rb and K values can be recognized from 15.000 cal BP to present (Fig. 4). K, Ca and Sr show a higher variability through time than Rb with pronounced peaks around 7500, 2900 and ca. 200 cal BP. The Rb/Sr ratio has the highest values around 6800, 3200 and 500–200 cal BP.

5 Discussion

The steadily increasing values of Rb over time can be interpreted as increasing weathering and erosion in the catchment area due to both natural and anthropogenic forcing. The eastern Mediterranean experienced an increase in winter precipitation during the early Holocene (Roberts et al. 2011). The oscillatory decline in precipitation after 6000 cal BP (Roberts et al. 2011) is not reflected in our Rb or K data, probably due to increasing erosion caused by anthropogenic activity.

Mayewski et al. (2004) describe six global events of rapid climate change (RCC) during the Holocene (grey bars in Fig. 4). As they point out, not all the events can be recognized in the same strength with the same effect in every region of the world. The youngest event dating approx. to 600-150 cal BP is known in northern Europe as Little Ice Age (LIA), associated with a period of severe flooding events in the Mediterrean (Grove 2001). This is in excellent agreement with our Rb/Sr ratio, indicating wetter conditions at Stymphalia. Wet periods inferred from the Rb/Sr ratio in a similar way appear around 3000–3500 (RCC 4), 3700-4000 (RCC 3) and around 6900 cal BP. The latter one has no equivalent in a global RCC and thus seems to reflect a local or regional signal.

6 Conclusions

Our geochemical analyses of the lake sediments from Lake Stymphalia have shown that the water supply in the region is not stable but fluctuates due to climate driven changes. As the comparison of our data with the Holocene RCC events as described by Mayewski et al. (2004) indicate, these changes can be either of global or regional scale. The geochemical data presented here is the starting point for a more detailed and comprehensive environmental reconstruction of the Stymphalia region and it needs to be compared to upcoming data from other proxies such as pollen and from other sites on the Northern Peloponnesus.

Acknowledgments We kindly thank Mathias Bahns, Kimon Christanis, Nikolaos Lambrakis, Görkim Oskay and Giorgos Siavalas for their invaluable support of our work. We thank the German research foundation (DFG) for funding the project through the GS "Human development in landscapes".

References

Bout-Roumazeilles V, Combourieu Nebout N, Peyron O, Cortijo E, Landais A, Masson-Delmotte V (2007) Connection between South Mediterranean climate and North African atmospheric circulation during the last 50,000 yr BP North Atlantic cold events. Quaternary Science Reviews 26(25-28):3197-3215

Bronk Ramsey C (2001) Development of the radicocarbon calibration program. Radiocarbon 43(2A):355-363

Bronk Ramsey C (2008) Deposition models for chronological records. Quaternary Science Reviews 27(1-2):42-60

Chen J, An Z, Head J (1999) Variation of Rb/Sr Ratios in the Loess-Paleosol Sequences of Central China during the Last 130,000 Years and Their Implications for Monsoon Paleoclimatology. Quat. Res. 51(3):215-219

Crouch DP (1993) Water management in ancient Greek cities. Oxford Univ. Press, New York, p 380

Grove AT (2001) The "Little Ice Age" and its Geomorphological Consequences in Mediterranean Europe. Climatic Change 48(1):121-136

Jin Z, Wang S, Shen J, Zhang E, Li F, Ji J, Lu X (2001) Chemical weathering since the Little Ice Age recorded in lake sediments: a high-resolution proxy of past climate. Earth Surface Processes and Landforms 26(7):775-782

Lolos YA (1997) The Hadrianic Aqueduct of Corinth (With an Appendix on the Roman Aqueducts in Greece). Hesperia 66(2):271-314, http://www.jstor.org/stable/148487

Mayewski PA, Rohling EE, Curt Stager J, Karlén W, Maasch KA, David Meeker L, Meyerson EA, Gasse F, van Kreveld S, Holmgren K, Lee-Thorp J, Rosqvist G, Rack F, Staubwasser M, Schneider RR, Steig EJ (2004) Holocene climate variability. Quat. Res. 62(3):243-255

Mingram J, Negendank J, Brauer A, Berger D, Hendrich A, Köhler M, Usinger H (2007) Long cores from small lakes-recovering up to 100 m-long lake sediment sequences with a high-precision rod-operated piston corer (Usinger-corer). Journal of Paleolimnology 37(4):517-528

Morfis A, Zojer H (1986) Karst Hydrogeology of the Central and Eastern Peloponnesus (Greece). In: Morfis A, Zojer H, Harum T, Hörmann M (eds) Steir. Beitr. Z. Hydrogeol.. Graz, pp 1-301

Papastergiadou E, Retalis A, Kalliris P, Georgiadis T (2007) Land use changes and associated environmental impacts on the Mediterranean shallow Lake Stymfalia, Greece. Hydrobiologia 584(1):361-372

Richter TO, van der Gaast S, Koster B, Vaars A, Gieles R, de Stigter HC, De Haas H, van Weering TCE (2006) The Avaatech XRF Core Scanner: technical description and applications to NE Atlantic sediments. Geological Society, London, Special Publications 267(1):39-50

Roberts N, Brayshaw D, Kuzucuoğlu C, Perez R, Sadori L (2011) The mid-Holocene climatic transition in the Mediterranean: Causes and consequences. The Holocene 21(1):3-13

Röckel T (1987) Geologie und Hydrogeologie des Poljes von Stymfalia / Peloponnes (Griechenland). In: Institut für Angewandte Geologie Karlsruhe (AGK). Univ. Karlsruhe, p 276

Tjallingii R (2006) Application and quality of X-Ray Fluorescence core scanning in reconstructing late Pleistocene NW African continental margin sedimentation patterns and paleoclimate variations. In: Department of Geosciences. University of Bremen, p 114

Williams H, Gourley B (2005) The fortifications of Stymphalos. Mouseion III(5):213-259

Climate change impact on the Almiros brackish karst spring at Heraklion Crete Greece

A.I. Maramathas[1], **I. Gialamas**[2]

[1]National Technical University of Athens, School of Chemical Engineering

thamar@chemeng.ntua.gr

[2]Hellenic Center of Marine Research

jgiala@hcmr.gr

Abstract The climate change due to global warming will undoubtedly affect the quantity and the quality of the available water resources. In this work the influence of the climate change in the coastal karst spring of Almiros at Heraklion of Crete in Greece was quantitatively investigated with the MODKARST deterministic mathematical model. The influence of the change of the parameters that affect the above-mentioned water resource (precipitation quantity and intensity, wet day number over year, mean sea level), based on the last predictions of the scientific community for the climate change at the end of 21th century, was examined separately, and in combination with two scenarios. The most important conclusions about the operation of the spring and the planning of the exploitation works under the predicted climate conditions are as follows: 1) the spring is mainly affected by the annual precipitation height and less, by the precipitation intensity, by the wet day number over the year and by the mean sea level rise, 2) it will not give fresh water during the whole year and sometimes, it will be dried in the summer, 3) due to fractal structure of the limestone porosity, the chloride concentration of the spring water will decrease during the depletion period, when the discharge becomes less than $2.80m^3/sec$, although there is not any precipitation during this period. From the MODKARST model simulation under the today's climate situation, it has been proved that an 18 meters upraising of the water outlet point of the spring, through the construction of a dam, will prevent seawater entering into the spring reservoir and the salinization of the spring water. Under the predicted climate changes, the necessary upraising must be 2 meters higher than the upraising under the today's climate situation that is it must be 20 meters.

1 Introduction

The majority of the relative scientists consider climate change; due to planet warming caused to natural and anthropogenic reasons, certain. There are a lot of published works that quantitatively predict the changes of the climatic indicators of the planet, in the next decades.

Various research teams predict for south-eastern Europe until the end of this century, a rainfall decrease from 30-45% (Giorgi et al 2004) until 70% (Raisanen et al 2004), an increase of rainfall intensity (Christiensen and Christiensen 2003), and an increase of dry periods duration and at the same time a decrease of the rainy-days number over the year (Polemio and Casarano 2004). As a result an important increase of the difference of the flowing water volume through South Europe rivers between the wet and the dry period of the year is predicted. Besides, some rivers will lose 80% of their water during summer (Santos et al. 2002). It is also predicted a decrease of the recharge of aquifers (Eitziger et al. 2003). The m.s.l.r (mean sea level rise) is expected between 0.5m and 1m (Melloul and Collin 2006).

The objective of this work is the presentation of the climate change influence on the operation of the periodically brackish karst spring of Almiros at Heraklion in the island of Crete in Greece, in a practical way so that one can get conclusions about its sustainable development. It has been emphasized on the exploitation works, which must take into account the environmental changes. Generally, the design of development projects for such springs must be based on the prediction of their behavior under the changing climate conditions, which is achieved by the use of mathematical models. In our case, the mathematical model MODKARST for brackish karst springs (Maramathas et al. 2003) has been used.

MODKARST model has already been used for the simulation of the Almiros spring from 1/9/1989 till 31/8/1997. The goal of the present work is to predict the change in the hydrograph (discharge versus time) and in the chloride concentration of the spring water versus time (chloride concentration curve) at the end of the 21th century, taking into account the climate changes as they have been described in the above-mentioned works. That is, change in rainfall height per year, change in the intensity and the distribution of the rainfall over the year (decrease of the rainy-days number per year and respective increase of the rainfall height per rainy day) and change in the m.s.l (mean sea level). The influence of each of the above-mentioned changes has been separately studied and in combination with two scenarios (Table 1).

Almiros spring lies 10 km west of Heraklion city in the island of Crete in Greece, at an elevation of 4m a.m.s.l (above mean sea level) and at a distance 1km from the sea. The discharge of the spring fluctuates between 4m^3/sec in the dry period (summer) and 70-80m3/sec in the wet period (winter) following heavy rainfall. At low discharge rates the spring water is brackish due to seawater entering through fractures and conduits connecting the sea with the spring reservoir (Maramathas et al. 2006). The water chloride concentration fluctuates between 35ppm at high discharge rates and 5000ppm at low discharge rates. During the time span of the above-mentioned MODKARST model simulation that is from 1989 till 1997 the spring gave fresh water for 357days (12.2%) among 2920 days in all. From MODKARST model simulation it has been proved that an artificial upraising of the spring outlet point, from 4m a.m.s.l where it is today, to 22m a.m.s.l through the construction of a dam, will prevent the sea-water entering due to freshwater pressure increase and the spring will give only freshwater over the year (Maramathas 2006).

Table 1. Climate changes and scenarios.

	Climate Changes
C_1	Annual rainfall height decrease by 30%
C_2	Annual rainfall height decrease by 45%
C_3	Mean sea level rise by 0.5m
C_4	Decrease of the annual rainy-days number by 28% and respective increase of the rainfall height per rainy day
C_5	Decrease of the annual rainy-days number by 41% and respective increase of the rainfall height per rainy day
	Scenarios
S_1	$C_1 + C_3 + C_4$ "optimistic"
S_2	$C_2 + C_3 + C_5$ "pessimistic"

2 Results

By the use of algorithms that have been created for this reason, the rainfall and the mean sea level (m.s.l) of the simulation period of the spring (1989-1997) have been adapted to the predictions of Table 1. Especially, in order to realize the changes C_1 and C_2, the daily rainfall of the simulation period was multiplied with the factors 0.70 and 0.55 respectively. In order to realize the changes C_4 and C_5, the contiguous rainy days of the simulation period were separated into groups of two and three adjacent days respectively and the total rainfall of each group was considered to have been realized in the day of the group with the higher rainfall. Finally, in order to realize the change C_3, the elevation of the spring outlet point was "lowered" by 0.5m. Then it was simulated, with MODKARST model, first the influence of the changes C_1, C_2, C_3, C_4, and C_5 separately and then the influence of the scenarios S_1 and S_2. The results are presented on Table 2. The results for the scenarios S_1 and S_2 are also presented in the Figures 1 and 2.

Table 2. Results.

Climate changes	Mean number of days per year that the spring gives freshwater			Mean annual freshwater volume ($10^6 m^3$)			Mean volume (m^3) of saltwater per cubic meter of freshwater		
	Today's situation	After climate change	Difference %	Today's situation	After climate change	Difference %	Today's situation	After climate change	Difference %
C_1	44.62	0	-100	183.12	109	-40.5	0.1103	0.1763	59.95
C_2	44.62	0	-100	183.12	78.37	-57.2	0.1103	0.1881	70.62
C_3	44.62	37.62	-15.68	183.12	182.25	-050	0.1103	0.1158	5.09
C_4	44.62	37.87	-15.12	183.12	176.5	-3.56	0.1103	0.1167	5.82
C_5	44.62	25	-43.97	183.12	169.62	-7.31	0.1103	0.1267	14.95
S_1	44.62	0	-100	183.12	107.5	-41.29	0.1103	0.1832	66.22
S_2	44.62	0	-100	183.12	77.75	-57.51	0.1103	0.1942	76.14

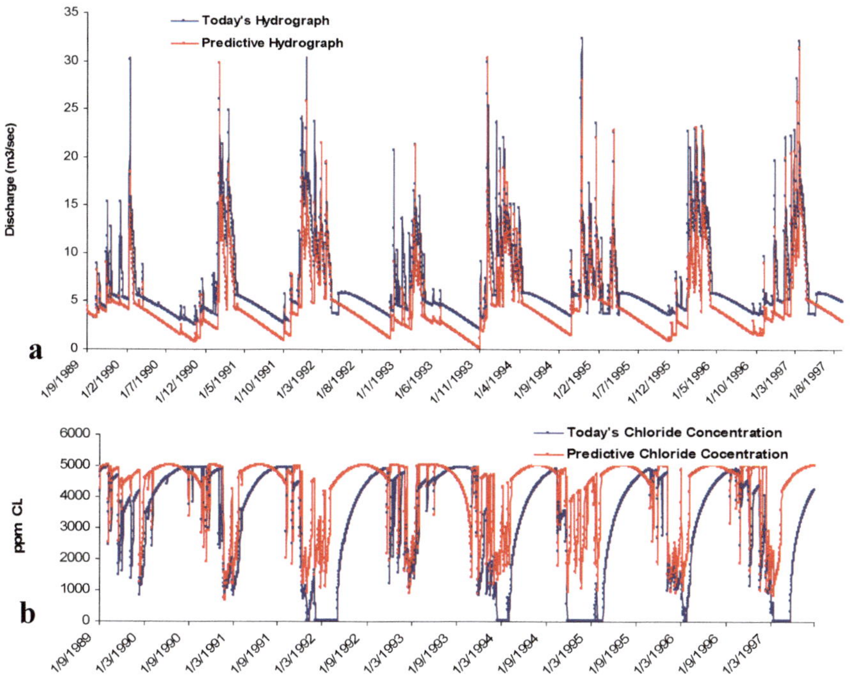

Fig. 1. a. Scenario S_1: Predicted spring hydrograph versus the spring hydrograph of today, **b.** Scenario S_1: Predicted chloride concentration curve versus the curve of today.

In Figures 1a and 2a the hydrograph for the scenario 1 and 2 respectively is compared with the today's hydrograph while in Figures 1b and 2b the chloride concentration curve for the scenario 1 and 2 respectively is compared with the chloride concentration curve of today. On Table 4 the necessary change of the height of the exploitation dam for each one of the two scenarios is presented.

Table 3. Critical Discharge for the reverse of the Cl^- curve gradient.

Year	Discharge for the reverse of the Cl^- curve gradient (m^3/sec)
1990	2.85
1991	2.93
1992	2.92
1993	1.87
1994	3.06
1995	2.92
1996	2.93
1997	2.93
Mean	2.80

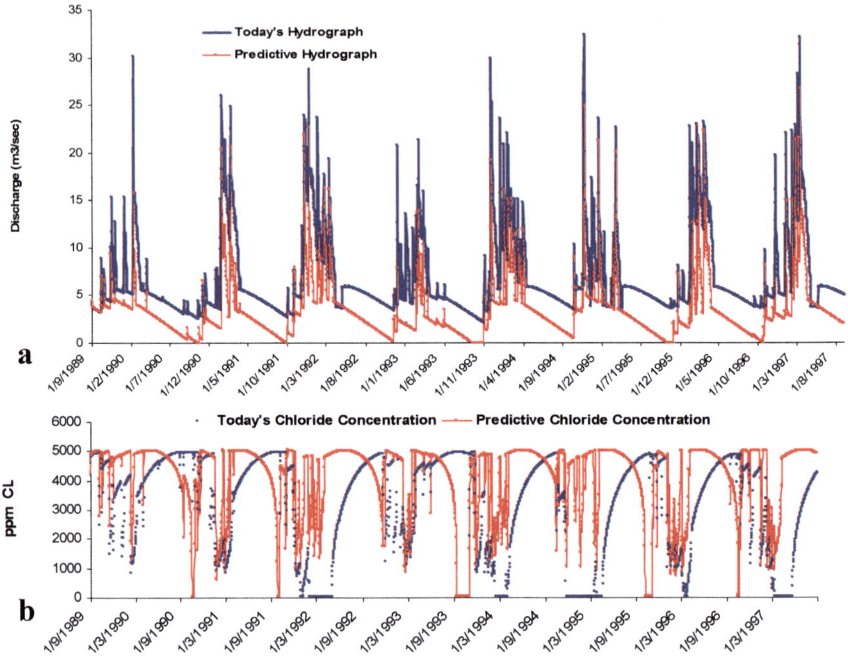

Fig. 2. a. Scenario S_2: Predicted spring hydrograph versus the spring hydrograph of today, **b.** Scenario S_2: Predicted chloride concentration curve versus the curve of today.

Table 4. Exploitation Dam Height variation.

	Dam Height (m)	Variation (m)
Today	22	-
Scenario S_1	24	2
Scenario S_2	24	2

3 Discussion and Conclusions

From Table 2 it is concluded that the spring discharge versus time as well as the chloride concentration of the spring water, mainly depends on annual rainfall height (C_1, C_2) and secondly on the rainfall intensity and distribution over the year (C_4, C_5). The m.s.l.r (C_3) affects a little the chloride concentration of the spring water and does not affect the spring hydrograph

In the scenario S_1 the spring gives only brackish water over the year while its discharge rate significantly decreases during the summer, reaching minimum values minimum at the end of October. Chloride concentration of the water is always over 700ppm and the higher values are shifted in the spring contrary to the chloride concentration of today, where the higher values are observed in the middle autumn (Fig. 1b). From the Figure 1a one can conclude that during the winter the discharge is not significantly affected contrary to chloride concentration of the spring water, which presents an important increase. An explanation for the above-mentioned phenomenon is that such karst aquifers present dual porosity; one is caused by fractures and the other by large conduits. Correspondingly, there are two kinds of discharge in the winter (recharge period of the spring); one from fractures that is similar to the base flow of a river and the other from conduits that is similar to the flood flow of a river. From the climate change, it is the discharge from fractures (base flow) that is mainly affected. Since the above-mentioned fracture discharge is related with the sea intrusion mechanism, it is fed by both the fresh water from the rainfall infiltration and the salt water from the sea. Consequently, as the fresh water decreases due to rainfall decrease, the fresh water pressure also decreases and the saltwater that comes from the sea increases, which implies an increase of the water chloride concentration. The whole spring discharge is not affected at the same degree as the water chloride concentration, since the discharge from conduits that dominates in the winter, remains significant.

In the scenario S_2 the hydrograph (Fig. 2a) as well as the chloride concentration curve (Fig. 2b) have the same qualitative features with the hydrograph and the chloride concentration curve of the scenario S_1. Nevertheless, there are some important quantitative differences, which are: a) The great decrease of the spring discharge or the drying of the spring in the end of autumn and b) the spring water, a month and a half for two months before the minimum discharge or the drying of the spring, is presented an increasing improvement of its water quality (decrease of the water chloride concentration) having for this period of year, lower chloride concentration than today (Fig. 2b). In the scenario S_1 there is a similar but not so clear tendency. This happens, because the chloride concentration curve in the end of spring or in the beginning of the summer, when the spring discharge is about 2.80m^3/sec (Table 3), presents a maximum value and consequently, during the next period (late summer and autumn) since the curve gradient is reversed, the chloride concentration of the spring water versus time, decreases. This phenomenon is unexpected since it is combined with a decrease of the spring discharge (no rainfall at this period), while normally it should be combined with an increase of the spring discharge. The above-mentioned behavior of the spring is caused as follows: As the water level in the karst system reservoir decreases, due to the absence of rainfall, the flow ceases in an increasing number of freshwater tubes. As a result, since the spring is elevated above the sea level the saltwater flow in the tubes, which comes from the sea and meets the above freshwater tubes, ceases too. Due to fractal structure of the spring reservoir, the sea intrusion points (points where the seawater tubes meet the freshwater ones) decrease, following the water level decrease, in a growing rate according to a power law (Maramathas and Boudouvis 2006). Consequently, the seawater, discharge to the spring reservoir at a larger decrease rate than the decrease rate of the fresh water discharge. This implies that the chloride concentration of the spring water continuously decreases. On Table 3 the critical discharge where the above-mentioned phenomenon starts for every year of the simulation period, is presented.

Climate change affects the required dam height in order to prevent the sea intrusion to the spring reservoir. In both scenarios the dam must be two meters higher than it is required today. That is, its height must be 24m instead of 22m, which is enough for today.

References

Christensen J H, Christensen O B (2003) Severe summertime flooding in Europe. Nature 421, 805-806

Eitzinger J, Stastna M, Zalud Z, Dubrovsky M (2003) A simulation study of the effect of soil water balance and water stress in winter wheat production under different climate change scenarios. Agric. Water Manage 61, 195-217

Giorgi F, Bi X, Pal J (2004) Mean, interannual variability and trend in a regional climate change experiment over Europe. II: Climate change scenarios 2071–2100. Climate Dynamics 23:7-8, 839-858

Maramathas A (2006) A new approach for the development and management of brackish karst springs. Hydrogeology Journal 14, 1360-1366

Maramathas A, Boudouvis A (2006) Manifestation and measurement of the fractal characteristics of karst Hydrogeological formations. Advances in Water Resources 29:1, 112-116

Maramathas A, Maroulis Z, Marinos-Kouris D (2003) A Brackish Karstic Springs Model. Application on Almiros Crete Greece. Ground Water 41(5), 608-620

Maramathas A (2003) Simulation of the brackish karst springs (in Greek) PhD Thesis. National Technical University of Athens, school of chemical Engineering section II

Melloul A, Collin M (2006) Hydrogeological changes in coastal aquifer due to sea level rise. Ocean and Coastal management, 29(5-6), 281-297

Polemio M, Casarano D (2004) Rainfall and Drought in Southern Italy (1821–2001). IAHS Publication 286, 217-227

Räisänen J, Hansson U, Ullerstieg A, Döscher R, Graham L P, Jones C, Meier H E M, Samuelson P, Willén U (2004) European climate in the late twenty- first century: regional simulations with two driving global models and two forcing scenarios. Climate Dynamics 22, 13–31

Santos F D, Forbes K, Moita R Eds (2002) Climate Change in Portugal: Scenarios, Impacts and Adaptation Measures. SIAM Project Report Gradiva Lisbon, 456

Climate Change Implications on Groundwater in Hellenic Region

G. Stournaras, G. Yoxas, Emm. Vassilakis, P.T. Nastos

Faculty of Geology and Geoenvironment, University of Athens, GR 157 84, Greece.
stournaras@geol.uoa.gr

Abstract There is a general consensus that climate change is an ongoing phenomenon. This will inevitably bring about numerous environmental problems, including alterations to the hydrological cycle, which is already heavily influenced by anthropogenic activity. The available climate scenarios indicate areas where rainfall may increase or diminish, but the final outcome with respect to man and environment will, generally, be detrimental. The research is focused on the climate change implications to the underground water tables of the mainland in Greece, excluding the island complex. It is based on regional climate models (RCMs) simulations under emission scenarios (SRES) A1B (3.5 °C), A2 (4.5 °C) and B2 (3.1 °C), for specific climatic zones of the Greek area during the period 2011-2100, which were estimated by the third report of IPCC.

1 Introduction

Climate change in combination with increased anthropogenic activities will affect water resources throughout the world. This paper gives information concerning the impacts of climate change on water resources, and particularly groundwater. Climate variability and change can affect the quantity and quality of various components in the global hydrologic cycle (Milly et al. 2005).

The Intergovernmental Panel on Climate Change (IPCC 2007) estimates that the global mean surface temperature has increased 0.6 ± 0.2 °C since 1861, and predicts an increase of 2 to 4 °C over the next 100 years. Global sea levels have risen between 10 and 25 cm since the late 19th century. As a direct consequence of warmer temperatures, the hydrologic cycle will undergo significant impact with accompanying changes in the rates of precipitation and evaporation. Predictions include higher incidences of severe weather events, a higher likelihood of flooding, and more droughts. The impact would be particularly severe in the tropical areas, which mainly consist of developing countries (Kumar 2005).

Although the most noticeable impacts of climate change could be fluctuations in surface water levels and quality, the greatest concern of water managers is the potential decrease and quality of groundwater supplies, as it is the main available potable water supply source for human consumption and irrigation of agriculture produce worldwide (Stournaras 2008). Because groundwater aquifers are recharged mainly by precipitation or through interaction with surface water bodies, the direct influence of climate change on precipitation and surface water ultimately affects groundwater systems.

Bearing the complexity of the linkage between climate change and groundwater in mind, it is necessary to adapt groundwater management accordingly and to consider the best options for developing and safeguarding groundwater resources. Main scope of this paper is to estimate the flux of groundwater's infiltration volume due to climate change.

2 Climate Scenarios

Fundamental cause forecasted at the future of anthropogenic changes of climate in world scale is the increase of greenhouse gases (GHG) due to human activities. For this aim, according to the third report of IPCC (2001) a big number of scenarios (in total 40) relative with the future development of emissions of greenhouse gases (Nakićenović et al. 2000) were made.

The total of scenarios covers the main uncertainties for the forecast of future emissions GHG based on the running knowledge and the initial "basic histories". Each one of the emission scenarios (SRES) constitutes simply a likely version of development of greenhouse gases. The simulations of the used regional climate models (RCMs) were carried out under SRES A1B, A2 and B2 respectively.

SRES A1B concerns economic growth that will be continued rapidly. The changes in the land use will not be enormous whilst the world population continues increasing rapidly up to year 2050 when it reaches about 9 billion and afterwards decreasing progressively. The main sources of energy that will be used are coming either from mining fuels or alternative sources which will have an impact of increase of CO_2 emissions within the 21st century, reaching the concentration of 720 ppm until the year of 2100.

SRES A2 refers in medium scale increase of income and particularly intense consumption of energy. The world population will continue increasing rapidly. The technological growth will be slow and partial, while medium to low changes in land use will take place. The increase of CO_2 emissions in the atmosphere will be rapid within the 21st century reaching the concentration 850 ppm, at the year 2100.

SRES B2 reports in continuation of world economic growth while the technological changes are lower compared with them of SRES A1. The increase of CO_2 concentration in the atmosphere will continue within the 21st century reaching 620 ppm until the year 2100.

3 Division of Greece in 12 climate zones

The climate of Greece shows high spatial variability due to intensive vertical partitioning and land/sea interaction. For this reason and taking into account that the spatial resolution of the RCMs data used in this study are quite satisfactory for describing the relief of Greece, climate change study was applied not only throughout the entire Greek peninsula as a whole but also for each one from the 12 climate zones that Greece has been divided according to Zanis et al. 2009. The 12 zones are the following (Fig.1): a) Mainland climate zones: Western Greece (WG), Central and Eastern Greece (CEG), Western and Central Macedonia (WCM), Eastern Macedonia-Thrace (EMT), Western Peloponnesus (WP) and Eastern Peloponnesus (EP), and b) Island complex climate zones: Crete (C), Dodecanese (D), Cyclades (CY) Eastern Aegean (EA), North Aegean (NA) and Ionian.

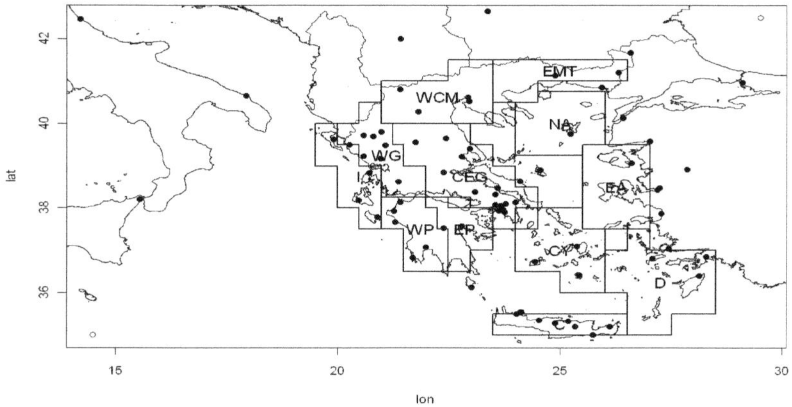

Fig. 1. The climate zones in which Greece was divided in. The black dots represent the 57 meteorological stations (from HNMS) used for the analysis (Zanis et al. 2009).

The above mentioned division was based on several climate and geographic criteria, the most important of which are the following: i) Pindos mountain range, which splits the mainland in the wet western area and the dry eastern area, ii) the frequency of either the tropical maritime or the polar continental air masses affect each area, which shows significant increase from south to north, especially during winter but also during the transitional seasons. Therefore the temperature shows high gradation along the north-south axis, iii) the high temperature variation between the island complex and the mainland areas, iv) the precipitation variation between the Eastern Aegean Islands and the Dodecanese where high values are recorded in contrast to extremely low precipitation values throughout the Cyclades Island complex (the climatic data concerning air temperature and precipitation totals from the 57 meteorological stations were acquired from Hellenic National Meteorological Service (HNMS).

4 Current Conditions

The status of Greece due to water reservation and management seems to be quite interesting, showing some peculiarities which indicate the level of true development and organization. From the water reservation point of view, Greece can be identified as a relatively rich territory concerning the rest Mediterranean frame (Stournaras 1998). The main reasons which are responsible for being identified as rich territory are related to the regime of the atmospheric precipitation. The precipitation totals in Greece shows an average value of 700 mm, which is rather interesting for the Mediterranean basin (Scoullos 2009). This is highly, related to the general factors which are responsible for the Greek climate and the weather phenomena, but also to the topography variation since the high relief near the coastline results in precipitation at mountainous areas. The most significant factor on this matter is the nearly north-south orientation of Pindos mountain range, the elevation of which in many cases exceeds 2,000 m. The wet air masses coming from the west (due to west atmospheric circulation) are blocked by this mountain range causing much more precipitation along the western than the eastern part of the country (Stournaras 2010). Additionally, due to the high relief contrast of the Greek peninsula, a dense drainage network pattern has been shaped leading the surface water to relatively large rivers which drain also the ground water in cases of additional surface discharge through springs. The cases of Transboundaries Rivers of northern Greece are referred as special cases as their entrance in Greece is being correlated with neighbor countries before their flow into north Aegean, e.g. Evros (Bulgaria, Turkey), Nestos, Strymon (Bulgaria), Axios (FYROM). The single case of reverse flow is represented by the river Aoos which starts from Greece and flows into Albania.

The underground water circulation is also a quite interesting matter since a large part of the Greek territory is consisted of permeable rocks, which accept the water through primary or secondary infiltration procedures and form the shape and the characters of the aquifers (Soulios 2010). Such cases can be identified in large continental basins or deltaic fans, the carbonate rocks of karst water tables (limestone, dolomites, marbles, and gypsum) and in extended igneous and metamorphic massifs (Stournaras et al. 2007).

Despite the fact that none of the 14 Water Districts (WD) of the country has not reported any problem for immediate water shortage according to Ministry of Development of Greece (2003), there are indications that the national water potential is decreasing, which is in agreement with the IPCC general findings for the climate change. Additionally, according to the results of the research project ENSEMBLES (http://www.ensembles-eu.org) the precipitation in central and northern Greece is gradually decreased during the last five decades. The reduction starts from 30 mm and regionally reaches 150 mm per decade. Also, the comparison between the river water flows for the periods 1900-1970 and 1971-1998 shows decreased water amount percentages between 5%-10% for the territory of

Greece in general, with the exception of Epirus District where the decrease percentages are not more than 2%-5% (Milly et al. 2005).

5 Observations and concessions for the estimation of water resources in Greece due to climate change

In order to estimate the country's expected water resources changes until the year 2100, several hydrological balance estimations were carried out for the periods 2021-2050 and 2071-2100. The main information came from the use of climate change modeling (precipitation and evapotranspiration in mm respectively). Specifically the simulations of the used RCMS were carried out for SRES A1B, B2 and A2.

Table 1. Modeling results for the main hydrologic balance parameters' change (R, I) of Greece's climate zones (CZ) and Water Districts (WD) for the periods 2021-2050 and 2071-2100 under different SRES scenarios.

A1B		2021-2050			2071-2100		
WD	CZ	R (hm^3)	I (hm^3)	Total (hm^3)	R (hm^3)	I (hm^3)	Total (hm^3)
1	WP	2455.32	1303.90	3759.22	2090.98	1208.15	3299.13
2	I	n/a	n/a	n/a	n/a	n/a	n/a
3	EP	-267.13	1214.01	946.89	-378.04	1091.76	713.72
4,5	WG	8247.09	6248.78	14495.86	7224.69	5936.71	13161.39
6,7,8	CEG	1875.80	2095.48	3971.28	2147.40	2064.74	4212.14
9,10	WCM	1018.33	1341.53	2359.86	1035.22	1330.11	2365.32
11,12	EMT	4239.73	1454.83	5694.56	3419.28	1335.75	4755.03
13	C	-142.94	471.34	328.41	-169.87	449.90	280.03
total				31556.08			28786.76

B2		2021-2050			2071-2100		
WD	CZ	R (hm^3)	I (hm^3)	Total (hm^3)	R (hm^3)	I (hm^3)	Total (hm^3)
1	WP	-528.02	436.98	-91.03	-256.60	420.48	163.88
2	I	n/a	n/a	n/a	n/a	n/a	n/a
3	EP	225.91	1543.55	1769.46	87.76	1345.88	1433.64
4,5	WG	2856.74	4315.04	7171.77	1625.28	3455.71	5080.99
6,7,8	CEG	3241.63	2224.22	5465.85	2567.20	1937.18	4504.38
9,10	WCM	72.06	1341.15	1413.21	34.37	1151.90	1186.27
11,12	EMT	2102.01	1260.46	3362.47	1144.64	1023.94	2168.57
13	C	-376.39	334.18	-42.21	-255.21	337.32	82.11
total				19049.52			14619.84

A2		2021-2050			2071-2100		
WD	CZ	R (hm^3)	I (hm^3)	Total	R (hm^3)	I (hm^3)	Total (hm^3)
1	WP	-528.02	436.99	-91.03	-254.82	388.75	133.93
2	I	n/a	n/a	n/a	n/a	n/a	n/a
3	EP	225.91	1543.55	1769.46	-41.55	1151.53	1109.98
4,5	WG	2856.74	4315.04	7171.77	2214.68	3487.28	5701.96
6,7,8	CEG	3241.63	2224.22	5465.85	1984.71	1655.22	3639.93
9,10	WCM	72.06	1341.15	1413.21	-70.43	1062.74	992.31
11,12	EMT	2102.01	1260.46	3362.47	1195.73	967.33	2163.06
13	C	-376.40	334.18	-42.21	-206.13	291.74	85.60
total				19049.52			13826.77

The interpretation of the results led to estimations of hydrologic balance parameters (Table 1) such as infiltration (I) and runoff (R) for the 2021-2050 and 2071-2100 period respectively.

It is obvious, according to the above estimations that the main parameters of hydrologic balance, which control the amount of groundwater, show a decrease correlated with current situation (Ministry of Development, 2003). Additionally, in some regions the runoff values are being described by a negative sign which means that they are showing a deficit character.

In order to describe the estimated change for every parameter of the hydrologic balance for the period of 2021-2100, a comparison of the results was made. The comparative results for all the described SRES scenarios led to the following conclusions (Table 2):

Table 2. Comparative results of the estimated change for every parameter of the hydrologic balance under different SRES scenarios (in %). (V: Precipitation, Etr: Evapotranspiration, I: Infiltration, R: Run off).

A1B				B2				A2			
V	Etr	I	R	V	Etr	I	R	V	Etr	I	R
- 5	- 2	- 5	- 12	- 15	- 12	- 16	- 35	- 22	- 20	- 21	- 37

According to the above resulted values at each SRES scenario, there is a clear decrease in groundwater's amount, due to climate change. The total precipitation amount is decreasing and subsequently the total infiltrating water is very likely to decrease from 5% (A1B) to 21% (A2) than today levels, and the total runoff will decrease from 12% (A1B) to 37% (A2) respectively throughout the mainland of Greece.

6 Estimations of groundwater decrease

In order to depict the correlation of rain volume V (hm^3) and volume of infiltrated water I (hm^3) cross plots diagrams were constructed (Fig 2 a, b, c), according to above resulted values at each scenario for the period 2011 - 2100. According to these diagrams it is obvious that due to precipitation's decrease the affection in ground water is instant.

Additionally, between 2011 and 2050 period the flux of infiltration water is rougher than between the period 2050 and 2100 respectively.

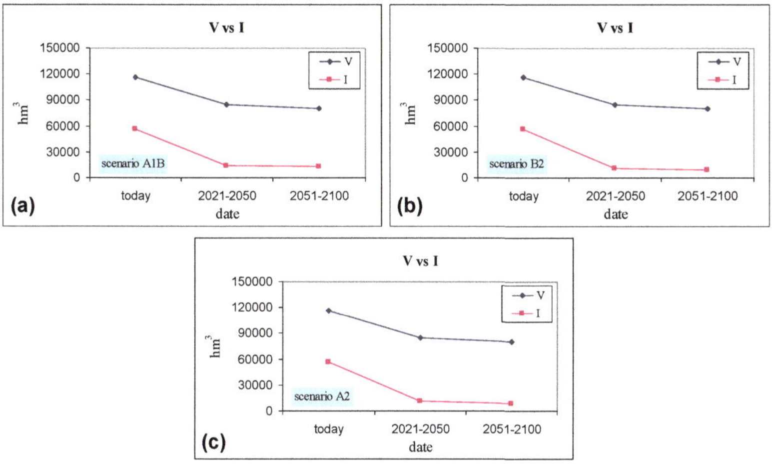

Fig. 2. Cross-plot diagrams showing the flux of precipitation's volume (V) versus volume of infiltrated water (I), (a) SRES A1B, (b) SRES B2, (c) SRES A2.

7 Conclusions

The implications of climate change on water systems in Greece and especially on groundwater systems can be summed up as following:

i. Decreased inflow and renewal of the aquifer water due to precipitation decrease and evapotranspiration increase.

ii. Increased salinization of the coastal and submarine water tables, especially the karstic ones and its advancing towards the mainland because of the land water decreasing potential due to less feeding and over pumping.

iii. Increased polluted load concentration in the coastal and sea water bodies due to less rarefaction.

iv. Intense disorganization of the deltaic fan areas as a result of either dam construction along the upstream zone (less flow, full deposition) and technical interference across the delta (leading the eroded material at a single mouth).

v. Pollution or exsiccation or the coastal wetlands.

vi. Aggravation of desertification phenomena due to water shortage and soil alterations (condensation etc.).

These conclusions should be combined with the fact that almost 75%-80% of the total water resources of the country are used for irrigation; it is more than obvious that these changes will have direct consequences on the type and extent of the cultivating areas along with the change of the agricultural methodologies.

Acknowledgments This paper is presented on behalf of the Bank of Greece Scientific Committee for the Study of the Climatic Change in Greece (EMEKA). Acknowledgment is addressed to the Bank of Greece for its initiative and to the Academy of Athens for its coordination of the discussed project.

References

IPCC (2001) Climate Change 2001: The Scientific Basis, Contributions of Working Group 1 to the Third Assessment Report of the Intergovernmental Panel on Climate Change. In: Houghton J. T, Ding Y, Griggs D J, Noguer M, Van der Linden P J, Dai X Maskell K, Johnson C.A. (eds.) Cambridge University Press, Cambridge, UK

IPCC (2007) Climate Change 2007: The Physical Science Basis. Contribution of Working Group I to the Fourth Assessment Report of the Intergovernmental Panel on Climate Change. In: Solomon S, Qin D Manning M, Chen Z, Marquis M, Averyt K B, Tignor M, Miller H L (eds.) Cambridge University Press, Cambridge, UK and New York, NY, USA, 966

Kumar K (2005) High-resolution climate change scenarios for India for the 21st century. Current Science 90, 334-345

Milly P C D, Dunne K A, Vecchia A V (2005) Global pattern of trends in streamflow and water availability in a changing climate, Nature 438, 347–350

Ministry Of Development (2003) Water Master Plan of Water Resources Management of Greece, Athens (in Greek)

Nakicenovic N, Alcamo J, Davis G, De Vries B, Fenhann J, Gaffin S, Gregory K, Grubler A, Yong Jung T, Kram T, Lebre La Rovere E, Michaelis L, Mori S, Morita T, Pepper W, Pitcher H, Price L, Riahi K, Roehrl A, Rogner H-H, Sankovski A, Schlesinger M, Shukla P, Smith S, Swart R, Van Rooijen S, Victor N and Dadi Z (2000) Emissions scenarios. Special Report of the Intergovernmental Panel on Climate Change, Cambridge University Press

Skoullos M (2009) Climate change in the Mediterranean region: Possible threats and responses in "Water Management", Water Security and Climate Change Adaptation: Early Impacts and Essential Responses By Sadoff Cn and Muller M, GWP, TEC background papers No 14

Soulios G (2010) General Hydrogeology. Volume 1, Unioversity Studio Press, Thessaloniki

Stournaras G (1998). Hydrology of Greece, Papyrus – Larousse Encyclopaedia, Volume: Greece

Stournaras G, Migiros G, Stamatis G, Evelpidou N, Botsialas C, Antoniou V, Vasilakis E (2007) The fractured rocks in Hellas, Groundwater Hydrology SV

Stournaras G (2008) Hydrogeology and vulnerability of limited extension fissured rocks islands, Ecohydrology & Hydrobiology 8, No 2-4, 391-399

Stournaras G (2010) The water in Mediterranean, in "Losing Paradise": The Water Crisis in the Mediterranean, Ashgate Publishing Ltd, UK

Zanis P, Kapsomenakis I, Philandras C, Douvis K, Nikolakis D, Kanelopoulou E, Zerefos C, Repapis C (2009) Analysis of an ensemble of present day and future regional climate simulations for Greece. Int. J. Climatol. 29, 1614-1633, doi:10.1002/joc

www.ensembles-eu.org. Research Project, EU

Climatic modelling and groundwater recharge affecting future water demands in Zakynthos Island, Ionian Sea, Greece

P. Megalovasilis[1], A. Kalimeris[2], D. Founda[3], C. Giannakopoulos[3]

1. Department of Geology, University of Patras, GR 26 504, Greece. pmegal@upatras.gr
2. Technological Educational Institute of the Ionian Islands, Zakynthos, Greece
3. Institute of Environmental Research & Sustainable Development, National Observatory of Athens, Greece

Abstract Zakynthos is a rapidly urbanized and touristic developing island situated in the Ionian Sea (central Mediterranean) where statistically significant negative precipitation trends prevail at least during the last four decades. During the last century precipitation has been decreased by 22%. Groundwater is the main water reservoir for domestic, agricultural and industrial use and thus is very important to sustain both its quality and quantity for present and future water demands. Climate simulations through an ensemble of Regional Climate Models, predict a dramatic decrease in rainfall by almost 25% for the period by the end of the current century (2071-2100) in the area. This fact indicates the risk of a pronounced future high water deficit in Zakynthos, causing even 39 hm^3 less groundwater natural recharge.

1 Introduction

Climatic modeling forms the basic framework together with future water demands estimations on making policy and taking the proper decisions that will assure future water demands. Groundwater is the main water reservoir in Zakynthos for domestic, agricultural and industrial use and its quality and quantity plays important role in economy and social life of island. The increase of agricultural water demands, the urbanization and tourist development has caused degradation of groundwater quality (Voudouris et al. 2006). Zakynthos (Zante) Island is located in the Ionian Sea, at the western part of the main orographic barriers of Greek peninsula, covering an area of 406 km^2 (Fig. 1). The population is highly variable along the year; while positive trends are recorded since 1980 (e.g. in the period 1991-2001 the rate of increment was equal to 19%). The permanent population amounts to 39,000 although emigration from non-EC countries inflates this value to 51,000. The population density (96 persons/km^2) is nearly equal to the mean value of the Greek islands, but almost 5 times less than the global mean for islands of similar size (such a population density would amount to a total population of about 250,000 residents). However during the last 140 yrs the population density due to permanent residents never exceeded 110 persons/km^2. In the period from April to October touristic arrivals (about 680,000 visitors per year) increases the population to a maximum value of 97,000 (σ = 19,700) during August (according to data of the last decade). In the period from 2003 to 2009 the touristic arrivals have increased by 16% and the overnight stays by 9.5%. The aforementioned population dynamics is accompanied by an intensive urbanization along the island. House construction was increasing at a mean rate of 2.6% yr^{-1} in the period 1991-2008. As a consequence, urbanization with 3% yr^{-1} increase rate and touristic development almost with 10% rate have a great impact on water consumption in the island. Groundwater is the primary source used to fulfill domestic and agricultural needs and on an annual basis, 4.9 10^6 m^3 of groundwater are drilled, with a maximum daily rate of 21,000 m^3 during the summer months. The increased water demands drive a series of environmental problems to aquifers of the island making groundwater not suitable for human consumption due to salinization and pollution from agricultural uses (Diamantopoulou and Voudouris 2008).

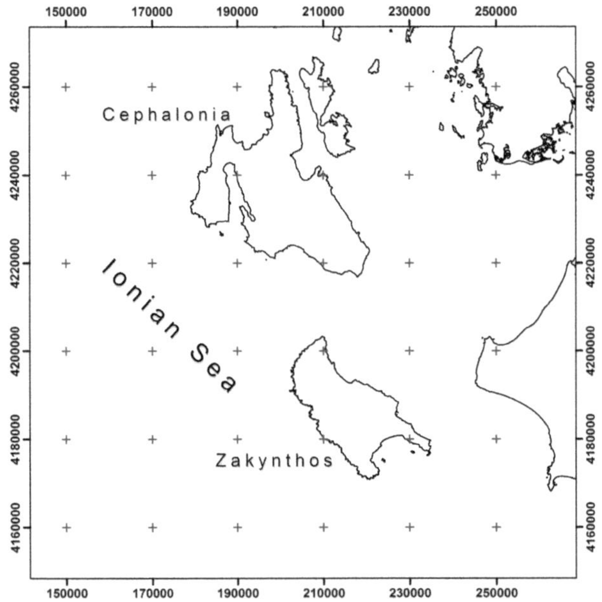

Fig. 1. Map of southern Ionian Sea. The computational grid boundaries (cells with horizontal resolution 25 × 25 *Km*) used in the regional climatic models are also indicated by the crosses.

2 Geological and Hydrogeological setting

Zakynthos Island is formed by two geotectonic zones the Pre-Apulian and the Ionian zone (Aubouin and Dercourt 1962). The Ionian zone is considered to be overthrusted to the Paxos zone. Sedimentation in Zakynthos island can be distinguished in carbonates Cretaceous to Miocene and clastic Plio - Quaternary. The Paxos zone in the island consists of Upper Cretaceous limestones that are overlain by Eocene limestones (Kati 1999). Limestones cover about 50% of the total area of the island (Vraxionas Mountain). An Oligocene zone to the East consisting of marly limestones and marls overlies on the Eocene deposits. The other part of the island is covered by Miocene and Pliocene rocks (gypsum deposits and marls with layers of marly sandstones respectively (Underhill 1988; 1989). But also some Pliocene sediments are also found (marls and sandstones). The younger deposits are of Pleistocene and Holocene age and their thickness is less than 10-15 m. They are developed to the eastern part and are consisting of alternating beds of gravels, clays, sands and silty clays.

Three main aquifers are developed in the geological formations of the Island: the karstic aquifer of carbonate formation (Paxos zone) in the west, the confined aquifer in Neogene deposits of Kypseli unit (mainly Pliocene deposits sandstones with marls and clays), in the east and the unconfined alluvial aquifer in the alluvial deposits in central-eastern part of the island. The hydrogeological and hydraulic characteristics of the island aquifers differ in many properties but their average thickness (AT) and their effective area (EA) are (e.g. Diamantopoulou and Voudouris, 2008): $Karstic_{AT}$= 100 m, $Karstic_{EA}$= 18 km^2, $Confined_{AT}$= 50 m, $Confined_{EA}$= 15 km^2 and $Aluvial_{AV}$= 6–10 m, $Aluvial_{EA}$= 59 km^2. The exploitable groundwater reserves were estimated by Diamantopoulou and Voudouris (2008) to be 3.3 10^6 m^3 yr^{-1}. The percentage of annual precipitation, which infiltrates through the carbonate rocks, has been estimated to be around 50%, while the percentages for confined and the shallow aquifer are 5-10% and 15% respectively of the total annual precipitation (Diamantopoulou 1999).

There is a lack of systematic discharge measurements in Zakynthos and values can be deduced only indirectly from the hydrological balance. Estimated values are different for any of the particular surface rock type: Limestones 2%, Neogene rocks 21-26% and 16% of the annual precipitation (Diamantopoulou 1999). Most of this water ends to the sea.

3 Climatic dynamics and precipitation regime

The main factors modulating the precipitation regime in the Ionian Sea can be classified in two categories: (***a***) the cyclonic activity ought to lee cyclones that develop in the south of the big mountain ridges of the northern Mediterranean coasts (in the so-called Mediterranean belt; Bartholy et al. 2009) but also at the Atlas mountains in Northern Africa and (***b***) the topography of the region. The moist air masses move from west towards the continental Greece. The humid air converges with a drier westerly flow, causing it to rise; further lifting is provided by islands and mountains.

During the rainy season, the Ionian Islands are strongly influenced by frontal depression systems originating in the Mediterranean belt. The great majority of these systems move towards the Ionian Sea (e.g. Alpert et al. 1990) and modulate peak values of the Mediterranean precipitation field, in the area between southern Italy and Greece (e.g. Fig.1 in Xoplaki et al. 2004). As a result, 90% of the rainy season precipitation in Greece is ought to Mediterranean cyclones activity (Maheras and Anagnostopoulou 2003).

Many studies have shown that along the 20th century (mostly after 1960s) a significant decline in cyclogenesis (amounting approximately to 15%) took place in the Mediterranean belt during the winter and spring epochs, while an increase in autumn was also detected (eg. Maheras et al. 2001; Lionello et al. 2006; Bartholy et al. 2009). It is remarkable that these negative trends prevailed particularly in the Italy-Greece region (eg. Norrant and Douguédroit 2005).

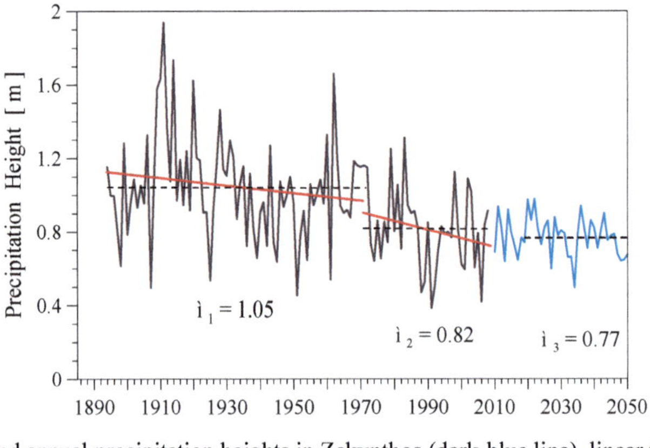

Fig. 2. Observed annual precipitation heights in Zakynthos (dark-blue line), linear trends (red lines) and mean levels μ_1, μ_2 (dashed lines) of the periods 1893-1971 and 1972-2008, respectively. The Ensembles-6 RCMs simulation for the period 2010-2050 is also shown (light-blue line) along with the predicted mean value (μ_3) for the period 2021-2050.

In accordance to the aforementioned climatic dynamics, the time-series of the total annual precipitation $H(t)$ in Zakynthos exhibits a strong negative trend, that enhanced after the mid 1970's (Fig.2). Associated with a southward gradient $\overrightarrow{\nabla H}$ in the Ionian Islands precipitation field, this negative trend intensifies in the southern Ionian (eg. see Fig.6 in Pnevmatikos and Katsoulis 2006) and reaches a value of -3.5 mm·yr^{-1} (p<0.0001) in Zakynthos. A climatic discontinuity detected through the sequential Mann-Kendall test in the precipitation series of the island about at 1971 (Kalimeris et al. 2010). Similar discontinuities have been also detected in most stations along the Greek peninsula between 1974 and 1984 (Feidas et al. 2007). Hence, two different periods are distinguished in the recent precipitation history (last 120 yrs) of Zakynthos: (*i*) the period before 1971, where the mean annual precipitation height was $\overline{H}_1 = 1.047$ m with a negative linear trend $\dot{H}_1 = -2.05$ mm·yr^{-1} (p=0.15), and (*ii*) the period after 1972, where the mean annual height dropped to $\overline{H}_2 = 0.816$ m (or by 22 %) with an enhanced negative trend $\dot{H}_2 = -4.8$ mm·yr^{-1} (p-value 0.15) which is one of the highest values in the Mediterranean. This sudden aggravation after 70's seems to be associated with the decline of the cyclogenesis in the Mediterranean belt and a persisting positive phase of the North Atlantic Oscillation and the East Atlantic-Western Russia pressure field anomalies (eg. Krichak at al. 2002; Lionello et al. 2006).

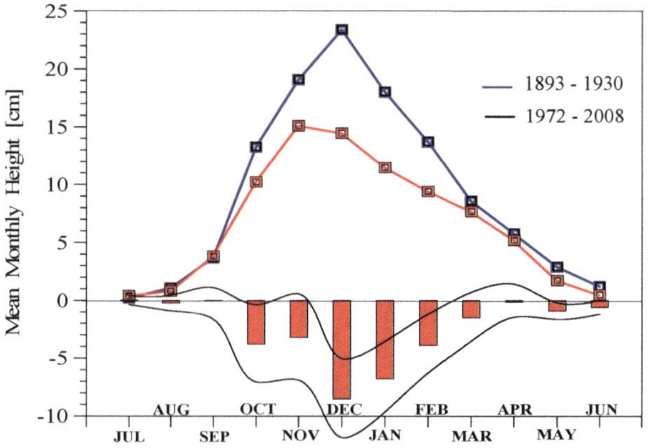

Fig. 3. Mean monthly precipitation heights of the periods 1893-1930 (blue line) and 1972-2008 (red line). The monthly variation ΔH_M between the two periods is also depicted (red bars) along with the 95% confidence limits for ΔH_M.

The examination of daily data (available for the periods 1893-1930 and 1955-2008) show that precipitation deficit in Zakynthos strongly affects the entire rainy season (October to March). For example, a comparison between the periods 1893-1930 and 1972-2008 show that the precipitation deficit is partitioned among autumn (by 24%) and winter (by 65%) months (Fig. 3). Furthermore, light and moderate rain events (H<15 mm/day) tend to increase after 1972 while medium and heavy (15<H<50) events tend to decrease. These changes were accompanied by an increase in the number of days of light rain and a decrease in all other classes.

Climatic models projections

To identify changes in the precipitation regime for the island of Zakynthos, regional climate model output has been used. More specifically, in an effort to reduce uncertainty associated with future climatic projections output from several regional climate models (RCMs) has been used (Christensen and Christensen, 2007). These models were developed at KNMI, CNRM, ETHZ, MPI, METO and METNO within the framework of the EU ENSEMBLES project [www. ensembles-eu.org]. They employ the A1B greenhouse gas emissions scenario and have a horizontal resolution of 25 km × 25 km.

The A1B scenario describes a future world of very rapid economic growth, global population that peaks in mid-century and declines thereafter, and the rapid introduction of new and more efficient technologies. The energy used to drive the economy is balanced between fossil fuel and other sources. By taking output from these 6 models, we produced an ensemble mean pattern (Ensemble-6) of the precipitation regime in the island of Zakynthos for the control (1961-1990) and the near future period (2021-2050). Simulations for the more distant future period 2071-2100 were also performed using 3 models (Ensemble-3). Given the current horizontal resolution of RCMs, numerical results correspond to the single land grid cell covering the island (Fig. 1). Furthermore, weighting factors were applied to these models according to the guidelines for the weighting produced within the ENSEMBLES project (see also Founda and Giannakopoulos 2009). Following this procedure, the KNMI model has a weight of 465 compared to 131 for CNRM, 81 for MPI, 86 for METO, 67 for ETHZ and 52 for the METNO model. The largest overall weight is associated with the KNMI model, mostly as a result of its successful representation of extremes, which shows the largest variability across models. Validation of the weighted ensemble model was performed by comparing output of simulations at the grid closest to the three studied stations with observations at the stations for the reference period, 1961-1990 (ENSEMBLES Deliverable D3.2.2).

Table 1. Observed and simulated mean annual precipitation height (in mm).

Period	Observations	Ensemble–6	Ensemble–3
1961 – 1990	837	807	775
2021 – 2050		770	756
2071 – 2100			600

Simulated and observed values of the mean annual precipitation for Zakynthos area are provided in Table 1 (and until 2050, in Fig.2). Observed and simulated values of the control period are in excellent agreement especially when the Ensemble of 6 models was used. A slight decrease of the annual rainfall amount with respect to the control period was simulated for the near future period 2021-2050 for both the Ensemble-3 and the Ensemble-6 simulations. It should be noted here that the Ensemble-3 simulation provides lower values of precipitation compared to observations and the Ensemble-6 simulation. A dramatic decrease by almost 25% was simulated for the period by the end of the current century (2071-2100) with the Ensemble-3 simulation, indicating the risk of a pronounced future high water deficit in Zakynthos, as well as, in the whole south Ionian region.

4 Estimation of water demands

Water services on the island are operated by DEYAZ (Public Water & Wastes Corporation of Zakynthos), and although the water quality has been deteriorated still manage to supply and to satisfy mainly domestic water demands. According to our analytic estimations, the local population water consumption is 3.7 10^6 m^3yr^{-1}, while the tourist water consumption loadings are 1.2 10^6 m^3 yr^{-1}. Unfortunately there are high water losses through the network of DEYAZ. An approximate estimation of water losses is close to 45% of the local population consumption. This means that another 1.7 10^6 m^3 yr^{-1} of groundwater is lost. Groundwater is also exploited from a great number of private boreholes and used for domestic, agricultural and commercial purposes. Our estimations show that 2.7 10^6 m^3 are drilled from April to October (where greater water demands are requested). Hence total annual water consumption amounts to: $3.7 + 1.2 + 1.7 + 2.7 = 9.310^6\ m^3$. This water volume is in accordance with other approaches of local authorities (e.g. of the Environmental Office) but is almost twofold than the estimation given by Diamantopoulou and Voudouris (2008) since population and boreholes loadings seems to be underestimated there.

Table 2. Water demands in Zakynthos according to various loading scenarios, in m^3.

Cases	Population	Tourists	Losses	Boreholes	Total
Current status (2010)	3.700.000	1.200.000	1.665.000	2.700.000	9.265.000
Scenario A	9.206.784	1.932.612	4.143.053	2.700.000	17.982.449
Scenario B	20.108.696	1.932.612	9.048.913	2.700.000	33.790.221

In an effort to predict future water demands at 2050 the following two scenarios of different cases were taken into account: (**A**) that of a steady development rate (equal to the observed during the last decade), and (**B**) that of a maximum estimated loading, being equal to the global population density of similar area islands. Results are given in Table 2. As can be seen almost 18 hm^3 of water will be needed in 2050 even if private borehole drilling remain constant (scenario A). According to the work hypothesis scenario B, the island would require 33.8 hm^3 of groundwater reserves. Moreover, adopting the infiltration percentages for each aquifer proposed by Diamantopoulou (1999), the effect of precipitation deficit according to the aforementioned climatic forcing, would decrease the groundwater almost by 8 and 39 hm^3yr^{-1} for the two climatic periods (2021-2050) and (2071-2100), respectively. This will be extremely critical because water demands due to development should be at least doubled in 2050.

5 Conclusions

Climatic Models are reflecting a dramatic precipitation deficit equal to -5% in the period 2021-2050 and -25.7% in 2071-2100, while the corresponding estimated groundwater renewal will decrease by 7.7 $hm^3\ yr^{-1}$ to 38.6 $hm^3\ yr^{-1}$, respectively. Declination of the groundwater reservoirs would turn to become critical and the Sustain Water Management inevitable. Future water demands estimations predict also that island aquifers will be extremely stressed in future due to development which will require almost double the water needs of present time if the current rate remain constant (scenario A). Even worst, according to aforementioned maximum loading scenario B, the future water demands would far exceed not only the exploitable groundwater reserves but also the total geological groundwater reserves.

References

Alpert P, Neeman BU, Shay-ElY (1990) Climatological analysis of Mediterranean cyclones using ECMWF data. *Tellus* 42A : 65-77

Aubouin J and Dercourt J (1962) Zone preapulienne, zone ioniene et zone du Gavrovo et Peloponnese occidentale. *Bull Soc Geol Fr* 4(6):785-794

Bartholy J, Pongrácz R, Pattantyús-Ábrahám M (2009) Analyzing the genesis, intensity and tracks of western Mediterranean cyclones. *Theor Appl Climatol* 96: 133-144

Christensen JH, Christensen OB (2007) A summary of the PRUDENCE model projections of changes in European climate by the end of this century. *Climatic Change* 81:7-30

Diamantopoulou P (1999) *Hydrogeological conditions of Zakynthos Island, Ionian Sea.* Ph.D., Department of Geology, University of Patras, Greece, pp. 160 (in Greek)

Diamantopoulou P and Voudouris K (2008) Optimization of water resources management using SWOT analysis : the case of Zakynthos Island, Ionian Sea, Greece. *Env Geol* 54:197-211

Feidas H, Noulopoulou Ch, Makrogiannis T, Bora-Senta E (2007) Trend analysis of precipitation time series in Greece and their relationship with circulation using surface and satellite data : 1955-2001. *Theor Appl Climatol* 87: 155-177

Founda D and Giannakopoulos C (2009) The exceptionally hot summer of 2007 in Athens, Greece – A typical summer in the future climate ? *Global and Planetary Change* 67: 227-36

Kalimeris A, Founda D, Giannakopoulos C, Pierros F, Filandras CM (2010) Long-term changes of precipitation in the Ionian Sea – Future projections. Proceedings of the 10th International Conference on Meteorology, Climatology and Atmospheric Physics, Patras, pp.737-743

Kati M (1999) *Deposit and diagenesis of Eocene formations of the Pre-Apulian Zone in Zakynthos Island.* Ph.D. thesis, Department of Geology, National University of Athens

Krichak SO, Kishcha P, Alpert P (2002) Decadal trends of main Eurasian oscillations and the Eastern Mediterranean precipitation, *Theor Appl Climatol* 72: 209-220

Lionello P, Bhend J, Buzzi A, Della-Marta PM, Krichak, S, Jansá A, Maheras P, Sanna A, Trigo IF, Trigo R (2006) Cyclones in the Mediterranean region : climatology and effects on the environment, in Mediterranean Climate Variability. In : *Developments in Earth and Environmental Sciences* 4, ed. Lionello P, Malanotte-Rizzoli P, Boscolo R. Elsevier, pp.324-372

Maheras P, Flocas H, Patrikas I, Anagnostopoulou C (2001) A 40 year objective climatology of surface cyclones in the Mediterranean region : spatial and temporal distribution. *Int J Climatol* 21: 109-130

Maheras P and Anagnostopoulou Ch (2003) Circulation types and their influence on the interannual variability and precipitation changes in Greece. In *Mediterranean climate variability and trends*, Springer-Verlag, pp.215-239
Norrant C and Douguédroit A (2005) Monthly and daily precipitation trends in the Mediterranean (1950-2000). *Theor Appl Climatol* 83: 89-106
Pnevmatikos JD and Katsoulis BD (2006) The changing rainfall regime in Greece and its impact on climatological means. *Met Appl* 13: 331-345
Underhill JR (1988) Triassic evaporates and Plio-Quaternary diapirism in western Greece. *J Geol Soc Lond* 145:269-282
Underhill JR (1989) Late Cenozoic deformation of the Hellenide foreland, western Greece. *Geol Soc Am Bull* 101:613-634
Voudouris K, Baltas E, Vasalakis A, Diamantopoulou P (2006) Pressures on water resources and their impacts on coastal aquifers: case studies from Greek islands. WSEAS (World Scientific and Engineering Academy and Society). *Trans Environ Dev* 2(11):1427-1434
Xoplaki E (2002) *Climate variability over the Mediterranean*, Ph.D. Thesis, university of Bern
Xoplaki E, Gonzalez-Rouco JF, Luterbacher J, Wanner H (2004) Wet season Mediterranean precipitation variability: Influence of large-scale dynamics and trends. *Clim Dyn* 23: 63-78

Hydrology

Using spectral analysis for missing values treatment in long-term, daily sampled rainfall time series

E. Fakiris, D. Zoura, K. Katsanou, P. Kriempardi, N. Lambrakis, G. Papatheodorou

Department of Geology, University of Patras, Rio-Patras, GR 26 504, Greece.
fakiris@upatras.gr

Abstract In the present paper a spectral time series reconstruction approach is demonstrated regarding its suitability for treating missing values in long term, daily sampled rainfall time series. The core of the proposed model consists of the CLEAN algorithm, which reconstructs time series after modifying their frequency spectrum to exclude artifacts originated by nonstationarity and missing values. Further filtering and normalization of the predicted values in places where periods of missing observations occur enhance the model's potentialities. For the validation of the model, eight daily sampled rainfall time series from rain stations located in the Louros basin, NW Greece, were used. Its ability to predict rainfall height for long periods (e.g. a month) of continuous missing observations is tested by introducing several artificial blanks of a month's duration each in the time series and analyzing the success of the model in order to correctly predict them. The model's accuracy indicator is built on the basis of a cross correlation function so that time lags of just a few days between observed and predicted values do not affect it. The model performed adequately, especially for the wet months where its accuracy reached levels greater than 75% implying that daily variability in rainfall height can be to some extent described in the frequency domain.

1 Introduction

In general, the problem of missing values is a common obstacle in environmental time series analysis and specifically in the context of precipitation–runoff process modeling where it is essential to have serially complete data. There may be various reasons for missing values, for instance equipment failure, errors in measurements or faults in data acquisition, as well as natural hazards. Whatever the reasons of their existence, missing values constitute a significant problem for water resources applications that require a continuous database (e.g. Ramirez et al. 2005). Time series analysis is a common tool in studying hydrological process (Lambrakis and Daskalaki 1999). Modeling rainfall data from a statistical perspective presents the challenge of dealing with a time series that is nonstationary, nonnegative and has a skewed distribution that includes a point mass at 0. These data considerations along with the complex environmental processes that control rainfall induce great limitations in predicting missing values and explain why no modeling technique has ever exhibited complete accuracy, especially when dealing with daily predictions. Presently there are a large number of statistical techniques available for dealing with missing values which can be divided into three categories (Little and Rubin 2002; Kalteh and Hjorth 2009): 1) Listwise/pairwise deletion, 2) Imputation-based procedures such as mean, regression and hot deck and 3) Model-based procedures which are an active field of research nowadays and include among others multiple imputations (e.g. Lo Presti et al. 2010), spatial modeling (e.g. Genton and Furrer 1998), Autoregressive (AR) modeling (e.g. Matalas 1967; Delleur et al. 1976) Self Organizing Maps (SOM), Artificial Neural Networks (ANN) and Expectation Maximization (EM) algorithm. Most of them require neighbor stations and search or assume good correlation between their observations to predict missing values.

In this study the rainfall data from a total number of 8 rain stations along Louros basin (NW Greece), with a daily sampling interval, were used (Fig. 1). The intense morphology of the study area and the scarcity of the stations severely affect the correlation of the observed precipitation even in nearby stations thus, leading to the need for using a non neighbor-dependent model to fill in the missing values. The proposed model utilizes spectral analysis and consists of: 1) Application of the CLEAN algorithm (Roberts et al 1987) to modify the raw (dirty) Discrete Fourier Transform (DFT) spectrum of the whole rainfall time series (including missing values), reconstruct the time series and predict missing values and 2) Normalization of the predicted values where missing ones occurred by using a suitable filter and statistical normalization of rainfall amplitudes. Although the application of spectral analysis techniques to treat missing values and nonstationarity in environmental time series is well documented (Schulz and Stattegger 1997; Heslop and Dekkers 2002; Dilmaghani et al. 2007; Fakiris et al. 2007), its applicability has not been demonstrated against rainfall time series. The reliability of the proposed method is validated by applying it to the raw time series after adding artificial blanks of a month's duration each in random places and comparing the observed values to the predicted ones. The overall accuracy of the proposed method was adequate. The wet days (peaks) were well predicted in most of the cases with correlations exceeding 75% and an average lag of less than 3 days between the real and predicted values. The relative success of this study implies that spectral analyses techniques applied to long term rainfall databases can be of value for predicting and forecasting even very short term variability in rain height, such as the daily one.

2 Study area – Data preparation

The study area is located in Epirus in NW Greece, and includes the drainage basin of Louros River, which occupies an area of 926 km^2 (Fig. 1). Evaporation and humidity are relatively high in the area throughout the year. Climate can be characterized as mild temperate, to continental, changing under the influence of geographical position and relief (Boltsis 1986). The hydrological regime of northern Epirus is characterized by uneven seasonal and regional distribution of precipitation. An average precipitation for the Louros watershed is 1150-1800 mm. Out of this 50% is evapotranspiration, 35% is recharge and 15% is discharge (Nikolaou 1991). Morphologically, the Louros karstic system is characterized by high elongated mountain ranges and narrow valleys, due to the tectonic (anticlinic structures) and geological conditions of the region and the lithologic alternation between limestones and flysch of the Ionian zone.

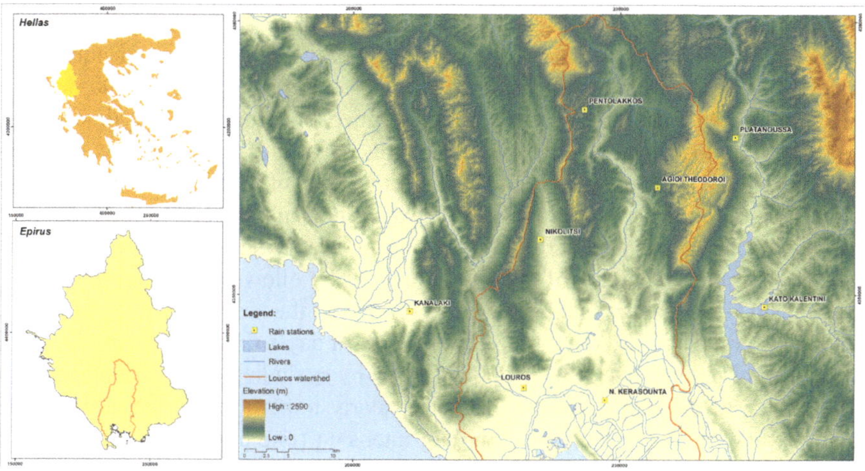

Fig. 1. Morphological - hydrological map of the study area including the rain stations used for the validation of the presented model for treating missing values.

The dataset used in the present study was provided by the Hellenic Ministry for the Environment and includes long term rainfall height time series from eight rain stations, sampled on a daily basis. The sampling period started in 1/1/1951 and lasted for 58 years until 31/12/2009. Each timeseries includes blanks ranging from 2 months to 3 years total duration, arranged in arbitrary single months through the dataset. The blanks that are considered in this study are in fact artificial blanks since considering account the true ones would lead to method validation inability. These were created by introducing twelve blanks of a month's duration each in every dataset so that each one of the twelve months of a year was missing in twelve random years. Table 1 includes the rain stations sorted by their altitudes and the years for which certain months have been blanked.

3 Missing values treatment - Methodology overview

3.1 The CLEAN algorithm

The CLEAN algorithm is an effective tool for predicting missing values in both stationary and non-stationary time series using spectral analysis and was introduced by Roberts et al. (1987). The main advantages of the algorithm are that it removes spectrum artifacts related to missing data; it provides clean stable peaks (Tiwari and Rao 2000) and does not require a formal statistical test (Negi et al. 1996). Furthermore, Vio et al. 1992 showed that the CLEAN algorithm is able to recover effectively most of the lost information even for significantly fewer number of data points. In the CLEAN algorithm, a raw frequency spectrum otherwise called dirty, is calculated by using DFT, which contains real peaks and side lobes. This dirty spectrum is then iteratively cleaned. The largest spectral peak is found and is subtracted with its side lobes from the original dirty spectrum. In the next iteration, the now largest peak is detected in the residual dirty spectrum and compensated for. The iterations are repeated until a defined noise level or number of iterations is reached. In this case the number of iterations was set to 2000. Inverse discrete Fourier transform (IDFT) is then applied to reconstruct the time-series, using a predefined time step interval; so that the new ones are stationary and their missing values are simulated. Usually the time step interval for the IDFT is defined as the minimum time difference between the samples. However, according to the Nyquist sampling criterion, the output time series should have at least double the sampling frequency of that of the raw ones. Following on from this, in the present study the spectral analysis was performed on the basis of half the sampling interval producing CLEANed time series of half a day's time step interval, which were then resampled back to one day's interval by excluding every second sample. The choice of the output time step is particularly important because a time step output equal or bigger than the sampling intension (one day in this case) would cause coarse time series and/or loss of information. To use the CLEAN algorithm with the available rainfall time series we used Baisch and Bokelman's (1999) computational implementation for MATLAB, modified to accept user specified time step interval.

3.2. Method accuracy and normalization of predicted values

After the CLEAN algorithm was applied to the data, the efficiency of the method was initially validated by comparing the non-blank parts of the raw time series against the corresponding parts of the reconstructed ones. The reconstructed time series proved to correlate almost perfectly (correlation coefficient > 98%) to the observed ones, indicating that the CLEANed spectrum contains all the information needed for successful rainfall time series reconstruction.

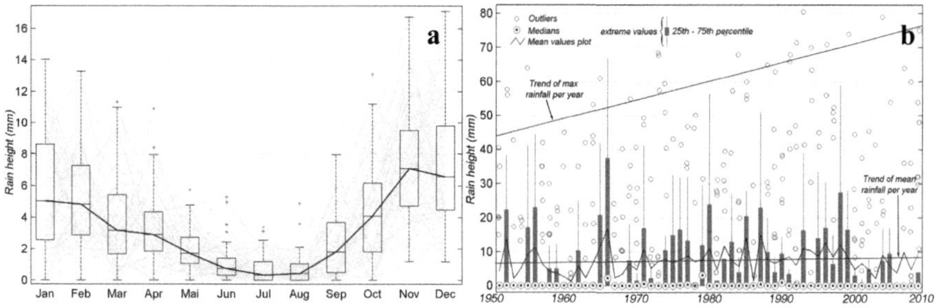

Fig. 2. Box-plots of: (a) the mean per month rainfall height and (b) the mean rainfall height of November, for all the available years of Kalentini station's dataset. Boxes correspond to 25th - 75th percentiles, linear extensions to extreme values and small crosses (or circles) to outliers. Black line segments cross the median values in (a) and the means in (b). Mean per month values for each year are plotted in grey in (a) and the trends of the mean and max values of November are presented in (b) using black thin lines.

However the correlation between the parts of the observed time series that have been blanked and the predicted ones was quite low. Visual comparison of the raw to the predicted values revealed that the majority of the high rain events (days where peaks occur) had accurately been predicted with a shift of only a couple of days in average. Even a lag of one day in the peaks' prediction in rainfall data can cause vertical reduction of the correlation coefficient. Consequently a cross-correlation function with a maximum lag of 7 days was used so that the method's accuracy estimator would allow small temporal misplacements of the predicted values in comparison to the true ones. An additional factor reducing the method's accuracy was that in many cases the predicted peaks were followed by a fade in – fade out effect so that a noisy increase precedes a high rainfall event and a noisy decrease follows it. To eliminate this effect we used an Infinite Impulse Response (IIR) notch filter to remove variations in the frequency domain that corresponded to periodicities greater than three days. After the previously described steps the correlation between the predicted and the observed values was significantly improved but the absolute values of high rainfall events had been transformed by the processing steps, raising the need for an amplitude standardization method. Using a statistical feature (such as mean) of the rest of the years for a particular month to estimate missing values is a common practice in environmental studies (e.g. Skoulikidis 2002). However, Fig. 2a shows that the mean rainfall height of a wet month is highly variable over the years implying that its use would be risky. Plotting the statistical features of the month under consideration against years (e.g. in Fig. 2b) can reveal important information about the environmental factors controlling the temporal evolution of precipitation. If a certain month (or other period of time) is blank in year x, its statistical features can be reliably predicted by performing linear interpolation between years x-1 and x+1. The statistical features that exhibit deterministic temporal patterns are standard deviation and extremes (Fig. 2b). Thus, we decided to standardize the processed predictions by forcing them to meet the interpolated standard deviation (std) of the missing period. This is feasible by first dividing each predicted value by the std of all the predicted values of the studied month and then multiplying it to the interpolated std as described previously. Using all the above steps to normalize the predicted daily values of the missing months lead to an average overall accuracy of the method greater than 75% with an average lag between true and predicted values of about 2 days (see Table 1). In Fig. 3 examples of comparisons between real daily rainfall height observations and predicted ones, in places where artificial blanks have been introduced, are presented. Furthermore, Table 1 shows in a centralized manner the results of the method validation analysis for the eight available rainfall time series. All the normalization and validation procedures described above in this paragraph have been computationally implemented in MATLAB through dedicated functions.

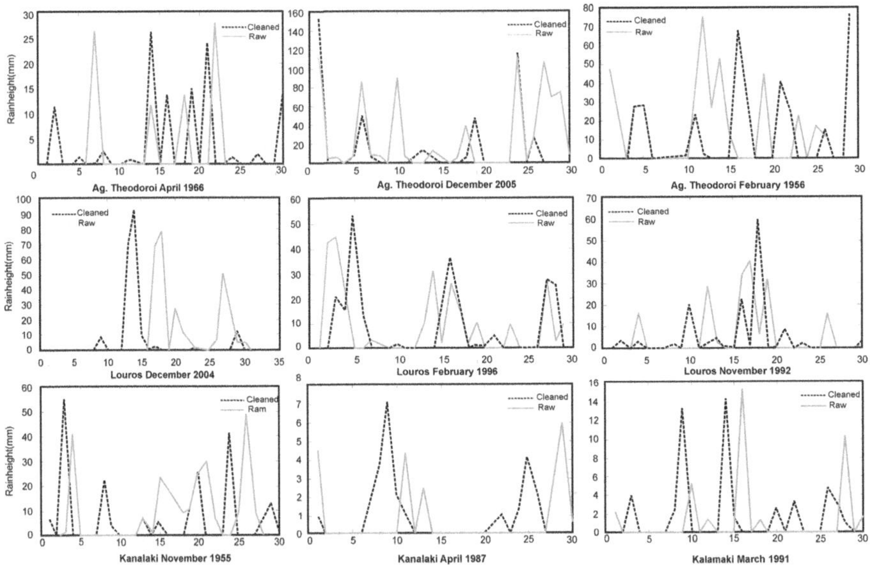

Fig. 3. Examples of comparisons between real daily rainfall height observation and predicted ones in places where artificial blanks have been introduced.

4 Discussion and Conclusions

In this paper, an attempt to use spectral analysis techniques in order to fill missing values in long term rainfall time series has been presented. The core of the implemented model consists of a well validated algorithm in environmental studies, called CLEAN.

Table 1. The model's validation results based on the treatment of artificial blanks of one month's duration each introduced to the eight available time series.

	Rain Station	Kanalaki	Louros	Kerasounta	Kalentini	Agioi Theodori	Nikolitsi	Platanousa	Pentolakos	M.Ac.(%)
	Alt. (m)	24	25	26	110	250	250	450	852	
JAN	Year	2006	1964	1973	1955	1951	1951	1952	1989	
	Ac.$^{(lag)}$	$0.95^{(0)}$	$0.93^{(7)}$	$0.65^{(0)}$	$0.74^{(-5)}$	$0.54^{(-2)}$	$0.62^{(0)}$	$0.73^{(0)}$	$0.82^{(0)}$	0.75
FEB	Year	1995	1996	1980	1961	1956	1957	1956	1987	
	Ac.$^{(lag)}$	$0.76^{(-5)}$	$0.91^{(0)}$	$0.89^{(0)}$	$1.00^{(0)}$	$0.77^{(-2)}$	$0.73^{(5)}$	$0.65^{(4)}$	$0.50^{(1)}$	0.78
MAR	Year	1991	1980	1970	1964	1961	1959	1960	1981	
	Ac.$^{(lag)}$	$0.61^{(2)}$	$0.88^{(2)}$	$0.84^{(1)}$	$0.79^{(-1)}$	$0.65^{(-2)}$	$0.61^{(3)}$	$0.71^{(-4)}$	$0.51^{(4)}$	0.70
APR	Year	1987	1976	1958	1968	1966	1960	1964	1975	
	Ac.$^{(lag)}$	$0.84^{(3)}$	$0.81^{(5)}$	$0.83^{(-4)}$	$0.52^{(3)}$	$0.71^{(-5)}$	$0.54^{(5)}$	$0.87^{(0)}$	$0.57^{(-2)}$	0.71
MAY	Year	1979	1988	1992	1972	2004	1967	1969	1974	
	Ac.$^{(lag)}$	$0.94^{(4)}$	$1.00^{(0)}$	$0.88^{(4)}$	$0.49^{(-2)}$	$0.70^{(-5)}$	$0.67^{(3)}$	$0.60^{(-5)}$	$0.73^{(4)}$	0.75
JUN	Year	1975	2007	1983	1976	1976	1970	1973	1972	
	Ac.$^{(lag)}$	$1.00^{(0)}$	$0.80^{(1)}$	$0.88^{(0)}$	$0.65^{(4)}$	$0.72^{(0)}$	$0.76^{(4)}$	$0.67^{(-6)}$	$0.72^{(-6)}$	0.77
JUL	Year	1971	1954	1970	1997	1981	1969	1977	1968	
	Ac.$^{(lag)}$	$0.85^{(-5)}$	$1.00^{(0)}$	$0.80^{(-4)}$	$0.68^{(-1)}$	$0.65^{(3)}$	$0.89^{(3)}$	$1.00^{(0)}$	$0.64^{(-4)}$	0.81
AUG	Year	1967	1958	1962	2001	1990	1978	1980	1962	
	Ac.$^{(lag)}$	$1.00^{(0)}$	$1.00^{(0)}$	$1.00^{(0)}$	$0.81^{(-4)}$	$0.68^{(-2)}$	$1.00^{(0)}$	$0.60^{(-6)}$	$0.42^{(5)}$	0.81
SEP	Year	1963	2001	1953	1980	1994	1987	1984	1953	
	Ac.$^{(lag)}$	$0.79^{(0)}$	$0.62^{(-4)}$	$0.79^{(-4)}$	$0.83^{(-1)}$	$0.88^{(1)}$	$0.87^{(0)}$	$0.77^{(4)}$	$0.55^{(5)}$	0.76
OCT	Year	1959	1970	1978	1984	1993	1995	1991	1955	
	Ac.$^{(lag)}$	$0.32^{(0)}$	$0.69^{(-3)}$	$0.79^{(-1)}$	$1.00^{(0)}$	$1.00^{(0)}$	$1.00^{(0)}$	$0.59^{(-5)}$	$0.66^{(-3)}$	0.76
NOV	Year	1955	1992	1988	1989	1996	2000	1993	1958	
	Ac.$^{(lag)}$	$0.44^{(2)}$	$0.82^{(-2)}$	$0.79^{(1)}$	$0.70^{(3)}$	$0.80^{(2)}$	$0.81^{(-1)}$	$0.88^{(-1)}$	$0.59^{(-2)}$	0.73
DEC	Year	1951	2004	1966	1993	2005	2006	1997	1960	
	Ac.$^{(lag)}$	$0.79^{(-4)}$	$0.84^{(6)}$	$0.86^{(-5)}$	$0.58^{(5)}$	$0.81^{(0)}$	$0.66^{(5)}$	$0.82^{(-1)}$	$0.73^{(-1)}$	0.76
M.Ac. (%)	0.77	0.86	0.83	0.73	0.74	0.76	0.74	0.62		

Ac.: Accuracy and Ac.

The CLEAN algorithm removes artifacts from the raw spectrum caused by non-stationarity and missing values and reconstructs the time series, predicting at the same time the missing values.

Application of the proposed model to rainfall time series with artificially blanked periods of one month duration each showed that the wet days (peaks) seemed to be well correlated between the predicted and the observed series, implying that daily variations can be indeed modeled in the frequency domain. Further filtering and normalization of the predicted values lead to a vast improvement in the proposed model's accuracy, reaching in some cases levels of up to 90%.

The accuracy indicator was built on the basis of a cross-correlation function so that it allows for a few days shift between real values and predictions. Deeper analysis of the model's validation results (Table 1) reveals the following information: 1) The model's accuracy is higher in wet months where it always exceeds 70% and the mean time lag between the predicted and observed peaks is around 2 days. This conclusion is drawn after ignoring the obvious absolute agreement that occurs for months with zero rainfall, mostly appearing in the dry (i.e. summer) months. 2) The accuracy of the model seems to decrease and the time lag between observed and predicted values to increase with the altitude of the rain stations. A probable cause for this deterioration of the model's performance with altitude is that the intense topography of mountainous areas leads to complex atmospheric processes with more random temporal expressions.

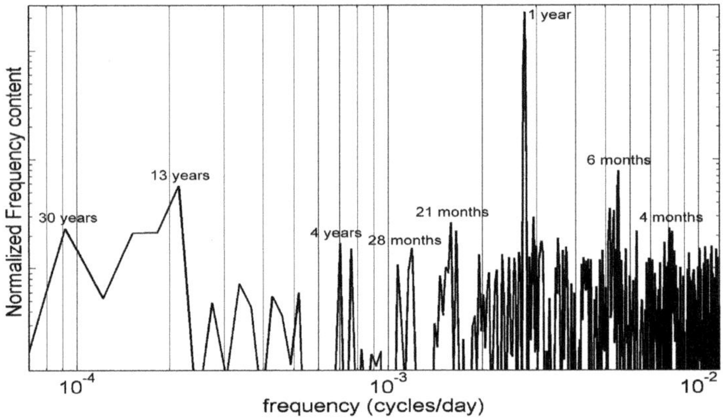

Fig. 4. The average CLEANed frequency spectrum of all the available rainfall time series in Louros basin. The major harmonic constituents detected are indicated along with their periodicity correspondence. Both axes are logarithmic.

Using the CLEAN algorithm with rainfall time series suffering from missing values and/or nonstationarity can also be of value for the production of a reliable frequency spectrum, suitable for analyzing their harmonic constituents and temporal periodicities. To illustrate a representative spectrum for Louros basin's available rainfall time series, we extracted the CLEANed spectrum for each one of the eight datasets and we averaged them (Fig. 4) over time. Apart from annual (1 year) and seasonal (3-4 months) periodicities there are others apparent in it, some of them possibly being harmonics of higher periodicities and others (e.g. 13 years) corresponding to the solar cycle and to the coupled ocean – atmosphere dynamics (Yadava and Ramesh 2007).

Despite methodological and computational improvements that the proposed model may need, it represents a fair demonstration of using spectral approaches towards predicting missing values in rainfall time series with daily sampling interval and by extension towards modeling and forecasting daily precipitation.

References

Baisch S, Bokelman GHR (1999) Spectral analysis with incomplete time series: an example from seismology. Computers & Geosciences. Res. 25, 739-750

Boltsis T (1986) Contribution to the study of the water equivalent of precipitation in Epirus region. PhD Study, University of Athens, Greece

Delleur JW, Tao PC, Kavvas ML (1976) An evaluation of the practicaly and complexity of some rainfall and runoff time series models. Water Resour. Res. 12, 953-979

Dilmaghani S, Henry IC, Soonthornnonda P, Christensen ER, Henry RC (2007) Harmonic analysis of environmental time series with missing data or irregular sample spacing. Sci. Technol. Res. 41, 7030-7038

Fakiris E, Papatheodorou G, Panagiotopoulos P (2007) An innovative data mining procedure using clean algorithm and factor analysis, for irregularly sampled temporal environmental data sets. Proceedings: Bulletin of the Geological Society of Greece, 11th G.S.G. International Congress, vol. XXXX:4, 1947-1958. Athens

Genton MG, Furrer R (1998) Analysis of Rainfall Data by Simple GoodSense: Is Spatial Statistics Worth the Trouble. Journal of Geographic information and Decision Analysis. Res. 2, 12-17

Heslop D, Dekkers MJ (2002) Spectral analysis of unevenly spaced climatic time series using CLEAN: signal recovery and derivation of significance levels using a Monte Carlo simulation. Physics of the Earth and Planetary Interiors. Res. 130, 103-116

Kalteh AM, Hjorth P (2009) Imputation of missing values in a precipitation–runoff process database. Hydrology Research. Res. 40:4, 420-432

Lambrakis N, Daskalaki P (1999) Environmental aspects and the exploitation of groundwater resources in Greece. Proc. of the XXIX Congress of IAH, Hydrogeology and Land Use Management, pp. 189-195. Bratislava

Little RJA, Rubin, D. B. 2002 Statistical Analysis with Missing Data. Wiley, New York

Lo Presti R, Barca E, Passarella G (2010) A methodology for treating missing data applied to daily rainfall data in the Candelaro River Basin (Italy). Environ. Monit Assess. Res. 160 (1-4) 1-22

Matalas NC (1967) Mathematical assessment of synthetic hydrology. Water Resour. Res. 3, 937-947

Negi JG, Tiwari RK, Rao KNN (1996) Clean periodicity in secular variations of dolomite abundance in deep marine sediments. Marine Geology Res. 133, 113-121

Nikolaou E (1991) Study of the regime of the groundwater systems of Epirus. Report Ins. for Geology and Mineral Exploration. p. 200. Preveza

Ramirez MCV, Velho HFDC, Ferreira NJ (2005) Artificial neural network technique for rainfall forecasting applied to the Sao Paulo region. J. Hydrol. 301, 146–162

Roberts DH, Lehar J, Dreher JW (1987) Time Series Analysis with Clean - Part One - Derivation of a Spectrum. Astronomical Journal. Res. 93 (4), 968

Schulz M and Stattegger K (1997) Spectrum: Spectral Analysis of Unevenly Spaced Paleoclimatic Time Series. Computers & Geosciences 23(9): 929-945

Skoulikidis NTh (2002) Hydrochemical character and spatiotemporal variations in a heavily modified river of western Greece. Environmental Geology Res. 43, 814-824

Tiwari RK, Rao KNN (2000) Solar and tidal reverberarions of deglaciation records from the tropical western Pacific a clean spectral approach, *Geojizika,* 16-17, 33-41

Vio R, Christiannis, Lossi O, Provenzale A (1992) Time series analysis in astronomy: An application to quasar variability studies. Astron. J. Res. 391, 518-530

Yadava MG, Ramesh R (2007) Significant longer-term periodicities in the proxy record of the Indian monsoon rainfall. New Astronomy 12, 544-555

Suitability of DSM derived from remote sensing data for hydrological analysis with reference to the topographic maps of 1/50000

K. Nikolakopoulos[1], **E. Gioti**[2]

[1] Institute of Geology and Mineral Exploration, 1 Sp. Louis Str. Olympic Village Entrance C, Acharnae, GR 13677, Greece. knikolakopoulos@igme.gr

[2] Department of Geography, Harokopio University of Athens 70, El. Venizelou Str, Athens 17671 Greece, linagioti@gmail.com

Abstract Hydrological indexes calculation requires the existence of spatial data such as the drainage network, the hydrological basin and the contours. For the Greek territory this data can be extracted from the topographic maps of the Hellenic Military Geographical Service (HMGS) or from remote sensing data. In this study the suitability and the accuracy of spatial information derived from remote sensing data are controlled with reference to the respective information from the topographic maps of 1/50.000. DSM from ALOS, CARTOSAT and airphoto stereopairs has been used for the extraction of the contours, the drainage network, the hydrological basin and the calculation of three hydrological indexes namely frequency, density and basin slope. Two areas with different physiographical and hydrogeological characteristics were selected. The first area is Antiparos Island and the second one is in Chalkidiki Peninsula.

1 Introduction

As remote sensing has the advantages of spatial, spectral and temporal availability of data covering large and inaccessible areas within short time is a very handy tool in hydrological studies. Every part of the spectrum can provide data suitable for different hydrologic topics. Microwave data can be used for surface soil moisture measuring and snow cover monitoring (Schmugge et al. 2002). Thermal data is very useful for under water springs detection (Karfakis and Nikolakopoulos 2002; Nikolakopoulos and Pavlopoulos 2003), for land surface temperature calculation, for near surface soil moisture (Moradkhani 2008) and evapotranspiration estimation (Lei et al. 2006). Landscape surface roughness can be estimated using lidar data. Visible and near infrared data can be used for water quality and snow covering monitoring. But most of all visible and near infrared provides quick and useful baseline information on the parameters controlling the occurrence and movement of groundwater like lithology, structural geology, geomorphology, soils, landuse, landcover, lineaments etc.

In this paper we will concentrate on the use of visible and near infrared data for the extraction of the spatial information such as the drainage network, the hydrological basin and the contours that is necessary in hydrological studies. DSM of high vertical and planimetric (horizontal) accuracy (Tsombos et al. 2008; Nikolakopoulos et al. 2009; Nikolakopoulos and Gioti 2010), created from two different satellite data sets with a spatial resolution of 2.5 meters (ALOS and Cartosat) and respective DSMs produced from airphoto stereopairs with a vertical accuracy of 3 meters are used. The suitability and the accuracy of spatial information derived from those DSMs are controlled with reference to the respective information from the topographic maps of 1/50.000 of the Hellenic Military Geographical Service (HMGS). Three hydrological indexes namely frequency, density and basin slope are used for the control.

Two areas with different physiographical and hydrogeological characteristics were selected for the evaluation. The first area is Antiparos Island and the second one is in Chalkidiki Peninsula (Fig. 1).

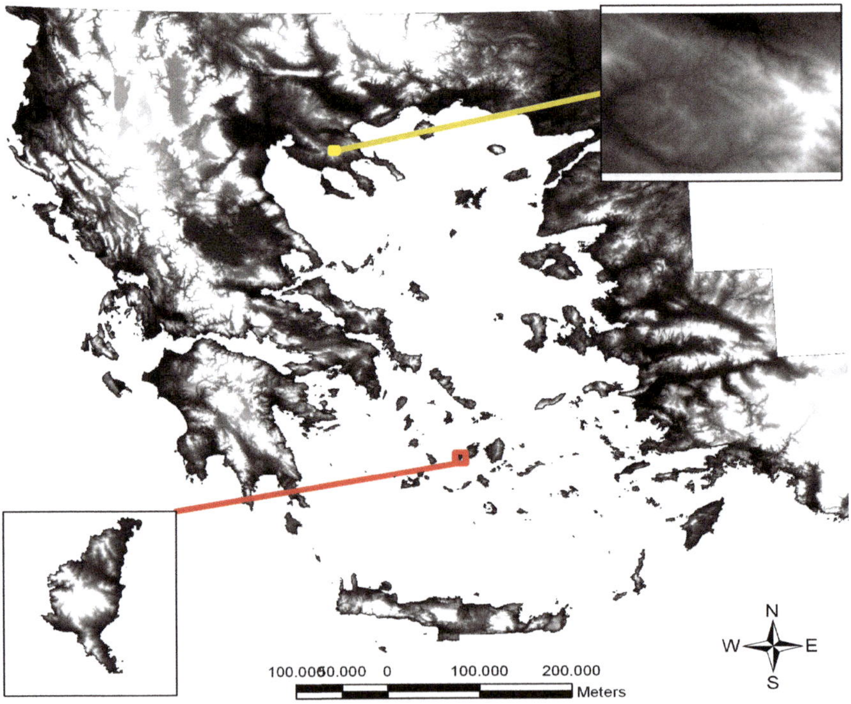

Fig. 1. Study areas.

2 Antiparos Island

The first study area is Antiparos Island. It is a semi-arid area with a quite extended drainage network. In order to present the results better, a central hydrological basin was selected. The extracted drainage networks from the ALOS, the Airphoto, the ASTER DSMs, the DTM from the digitized contours and the reference drainage network from the topographic maps of 1/50.000 of the HMGS are presented in Figure 2.

As it can be observed in Figure 2, the reference drainage from the topographic map has only three branches, which mean that the network is extremely limited. The drainage network from the HMGS DTM is obviously more extended in comparison with the reference even though its hydrological basin area seems to be quite small. Moreover, Aiphoto DSM presents a quite extended drainage network which is included in a wide basin area. ALOS DSM seems to give the most extended drainage network as it has many branches. Finally, ASTER DSM's central hydrological basin looks like ALOS's basin; nevertheless its drainage network is more limited.

Fig. 2. The drainage network of Antiparos Island. From left to right: Reference, DSM HMGS, Airphotos, ALOS, ASTER.

Apart from the visual comparison, a quantitative hydrological analysis has been accomplished. In particular, morphometric parameters that concern the drainage network such as the length, the basin area, the number of branches and the stream order, have been calculated and evaluated. Moreover basic hydrological indexes such as drainage density, frequency and basin slope have also been calculated in order to estimate the hydrological texture of the drainage network. All the results are presented in Table 1. As it can be observed ALOS DSM drainage network length is 34km while those of ASTER DSM, HMGS DTM, Airphoto DSM and the reference are 7.23km, 6.93km, 20.8km, 3.7km respectively. This means that ALOS and Airphoto drainage networks have the greatest values. Moreover, it is quite interested to be mentioned that ALOS DSM drainage network seems to have many differences to the others as it is composed by 163 branches and it has the greatest stream order. Furthermore, it is important that the indexes of frequency and density are extremely high for this drainage network (291 and 61 respectively). Airphoto DSM drainage network gives also high values in the same indexes. Its drainage frequency value is 7.3 and its drainage density value is 3. The reference drainage network, as it was expected, gives quite small values for these two indexes. Except of the ASTER basin all the others present similar slope values that range between 18 and 19%. ASTER's basin slope is quite smaller, only 10%.

Table 1. Basic hydrological parameters and indexes of Antiparos DSM's drainage network.

	Basin area (km^2)	Drainage length (km)	Number of branches (Strahler)	Stream order - Strahler	Stream order- Shreve
DSM ALOS	5.593	34.467	163	5	131
DSM ASTER	5.488	7.233	34	3	26
DTM HMGS	4.213	6.939	29	4	25
DSM AIRPHOTOS	6.903	20.820	51	4	43
REFERENCE TOPO-MAP	4.213	3.780	3	2	2
	Drainage frequency	Drainage density	Contour length (km)	Contour interval (km)	Basin slope (%)
DSM ALOS	29.143	6.162	50.920	0.02	18.208
DSM ASTER	6.195	1.318	27.969	0.02	10.192
DTM HMGS	6.884	1.647	40.332	0.02	19.147
DSM AIRPHOTOS	7.388	3.016	62.207	0.02	18.022
REFERENCE TOPO-MAP	0.712	0.897	41.950	0.02	19.916

As it was observed from the visual and the quantitative analysis, ALOS DSM gives an extended drainage network with the highest density and frequency indexes while Airphoto DTM gives a more limited drainage length and smaller values for the same indexes. Although, the basin area of the Airphoto DSM is wider than ALOS's and the drainage network's form differs from each other. The above results can lead to the assumption that there is a spatial error. To detect this error, a Quickbird orthophoto satellite image has been used. The image interpretation has shown that ALOS DSM drainage is the correct. That happens because part of the Airphotos drainage network seems to have a direction parallel to the road network and as a result the basin area is getting wider (Fig. 3).

3 Chalkidiki peninsula

The second study area is in Chalkidiki peninsula. It is an area with dense vegetation cover. In order to present the results better, a central hydrological basin was also selected in this case. The extracted drainage networks from the CARTOSAT, the Airphoto, the ASTER GLOBAL DSMs, the DTM from the digitized contours and the reference drainage network from the topographic maps of HMGS, are presented in Figure 4.

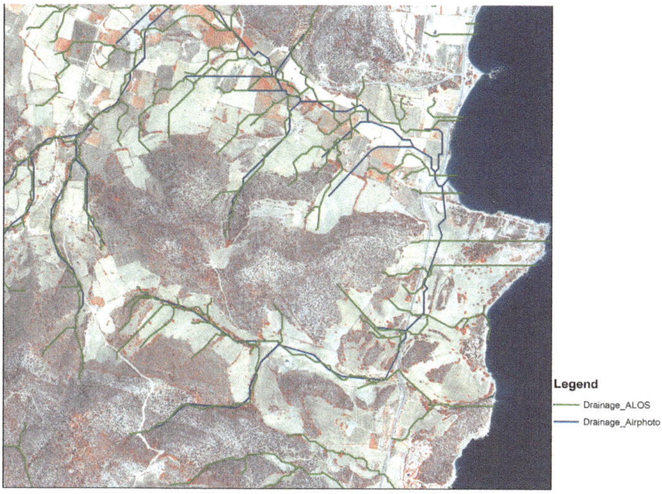

Fig. 3. Visual comparison of ALOS and Airphotos drainage network for error detection. Orthorectified Quickbirt satellite image.

As it can be observed in Figure 4, the reference drainage network is quite extended and well-shaped. Almost the same visual results can be observed to CARTOSAT and Airphoto drainage networks as well as to their respective hydrological basins. The drainage network derived from the ASTER GLOBAL DSM is limited and it has fewer branches than the others. Finally, the drainage network extracted from HMGS's DTM seems to be incorrect, because the central hydrological basin which was selected for the study is separated in three hydrological basins.

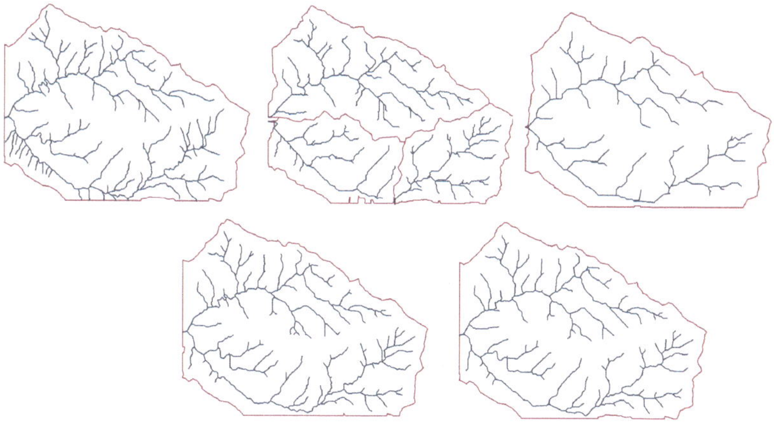

Fig. 4. The drainage network of Chalkidiki Peninsula. From left to right: Reference, DSM HMGS, ASTER GDEM, CARTOSAT, Airphotos.

The above visual results were also controlled by a quantitative analysis. Specifically, Table 2 presents all the basic morphometric parameters for the five drainage networks of Chalkkidiki peninsula as well as some basic hydrological indexes. As it can be observed the reference drainage network has the greatest length and number of branches. Moreover its basic indexes such as frequency, density and slope, present the highest values which are 4, 2.4 and 64 respectively. In addition, drainage networks extracted from CARTOSAT and Airphotos present also great values in all parameters and indexes. In particular the values of the basin area, the number of branches, the stream order, the drainage frequency and density are almost the same in those three drainage networks. The only difference was found in the drainage length and basin slope where the values of CARTOSAT and Airphotos networks are smaller in comparison to the reference drainage network.

Table 2. Basic hydrological parameters and indexes of Chalkidiki peninsula DSM's drainage network.

	Basin area (km^2)	Drainage length (km)	Number of branches (Strahler)	Stream order - Strahler	Stream order-Shreve
DSM ALOS	28.796	64.733	115	5	85
DSM ASTER	28.399	41.211	66	4	41
DTM HMGS	28.051	53.401	112	4	43
DSM AIRPHOTOS	28.803	60.123	116	5	80
REFERENCE TOPO-MAP	28.803	70.992	116	5	84
	Drainage frequency	**Drainage density**	**Contour length (km)**	**Contour interval (km)**	**Basin slope (%)**
DSM ALOS	3.994	2.248	387.455	0.02	26.91
DSM ASTER	2.324	1.451	274.858	0.02	19.35
DTM HMGS	3.993	1.904	563.987	0.02	40.21
DSM AIRPHOTOS	4.027	2.087	360.256	0.02	25.01
REFERENCE TOPO-MAP	4.027	2.465	924.677	0.02	64.20

The drainage network from the HMGS DTM, as already mentioned, is incorrect because it is separated in three different hydrological basins. Nevertheless, its hydrological indexes and especially basin slope contain high values. Finally, the drainage network extracted from ASTER GLOBAL DSM present the smallest values in all parameters and indexes after the visual comparison. It is interesting to observe that the number of branches is 66 while all the other networks have a number of branches between 112 and 116.

4 Discussion

What's worth discussing is that in the case of Antiparos Island, the reference drainage network which comes from the topographic maps of HMGS in a scale of 1:50.000 seems to be incorrect. This specific network is insufficient and apparently inappropriate for the evaluation of the other drainage networks, which were derived from the DSMs. Moreover the drainage network of ALOS DSM, is truly extended compared to the other drainage networks of this particular study area. It is impressive that the values of all the morphometric parameters and hydrological indexes of this drainage network are higher than those of the other drainage networks.

As far as the study area of Chalkidiki peninsula is concerned, interesting topics were traced. To begin with, the reference drainage network from the digitized contours has a great quality. This can be proved from the visual results as well as the quantitative analysis. Furthermore, it is of great interest the fact that, the drainage networks from CARTOSAT and Airphoto DSM look very much alike. Another important observation derived from this study is the limited expansion of the drainage network from ASTER GLOBAL DSM. This network is characterized as limited because it's composed of a really small number of branches. To conclude the discussion on this particular study area it has to be mentioned that the HMGS' DTM drainage network present some ambiguities. The drainage network was separated into three independent branches and as a result the hydrological basin was also separated into three parts.

5 Conclusions

The suitability of DSM derived from remote sensing data for hydrological analysis was controlled in the present paper. The shape and the accuracy of spatial data derived from DSMs created from satellite data was examined with reference to the respective information extracted from the 1/50000 topographic maps.

It was proved that the DSMs created from medium resolution satellite stereo pairs such as Cartosat or ALOS are suitable for hydrological analysis. The respective DSM from the ASTER data is less accurate and inappropriate for such analyses. This is due to the spatial resolution difference between the ALOS-Cartosat and ASTER data. The first have a spatial resolution of 2.5m while the latter has a spatial resolution of fifteen meters.

It is also remarkable that ALOS and Cartosat can provide very accurate spatial data. The comparison of different hydrological indexes proved that the spatial data derived from the satellite data are very similar to the respective data extracted from airphoto stereopairs.

Finally, it has to be pointed out that in some cases the hydrological data extracted from the topographic maps of 1/50000 of the HMGS are inappropriate for hydrological analysis.

Acknowledgments ALOS data was provided by the European Space Agency.

References

Karfakis I and Nikolakopoulos K, (2002) Use of Landsat TM Images for the Detection of Water Outflows in the Coastal Area of South Attiki Peninsula, Greece, Proc. of SPIE, Vol. 4881, p. 692-700

Lei Y, Shu Y, Li H and Zheng L, (2006) Integrated remote sensing and hydrological models for water balance in mountain watersheds, Proc. of SPIE Vol. 6359 p. 63590X-1

Moradkhani H, (2008) Hydrologic Remote Sensing and Land Surface Data Assimilation, Sensors Vol. 8, p. 2986-3004

Nikolakopoulos K, Tsombos P and Lathourakis G, (2009) Comparison of different along the track satellite stereo-pair for DEM Extraction, Imagin [e,g] Europe, I. Manakos and C. Kalaitzidis (Eds), IOS Press 2010, p. 303-310

Nikolakopoulos K and Gioti E (2010) Accuracy control of ALOS DSM, '30th EARSeL Symposium: Remote Sensing for Science, Education and Culture', Rainer Reuter (Editor), EARSeL 2010, p.507-514

Nikolakopoulos K and Pavlopoulos K, (2003) Joint Use of Multitemporal Satellite Data & Hydrogeological Studies for the Detection and Surveillance of Under Water Springs in the Gulf of Argolis, Greece, GIS AND RS: ENVIRONMENTAL APPLICATIONS Proceedings of the International Symposium, organized by The European Commission COST 719 & The University of Thessaly, p. 239-248

Schmugge T, Kustas W, Ritchie J, Jackson T. and Rango A, (2002) Remote sensing in hydrology Advances in Water Resources 25, p.1367-1385

Tsombos P, Nikolakopoulos K and Lathourakis, (2008) DEM creation from Cartosat data and comparison to DEM from other sources, Proc. of SPIE, Vol. 7106 p. 71061C1-12

A GIS method for rapid flood hazard assessment in ungauded basins using the ArcHydro model and the Time-Area method

M. Diakakis

Department of Geology and Geoenvironment, National & Kapodistrian University of Athens, Greece. diakakism@uoa.geol.gr

Abstract Geomorphology plays an important role in hydrological processes, as it largely controls a catchment's hydrologic response. Study of morphometric characteristics becomes even more significant in ungauged basins, especially when systematic hydrological data are scarce. A very well suited tool for accurate quantification of these characteristics is GIS software and especially ArcHydro model. In this work, four catchments in Greece, with rich flooding history are studied in terms of flood hazard spatial distribution. ArcHydro is used for rapid computation of catchment and sub-catchment characteristics. Based on these computations and with the aid of the time-area method, instantaneous unit hydrographs are developed in numerous locations across each catchment. Subsequently, peak discharge values extracted from these hydrographs, are interpolated across each catchment to visualize the runoff pattern, highlighting the areas where there is an inherent tendency for higher peak flow rates. In this way, the method allows the detection of locations where higher peak discharge creates a higher flood hazard potential. Results of the method are compared with historical floods damages for verification purposes, showing good correlation with the resulting flood hazard zonation.

1 Introduction

In Greece, flooding occurs mostly in catchments drained by ephemeral torrents with little or no water at all for most of the year. In this context and given the scarcity of instrumental hydrological data, the exploitation of geomorphology becomes essential for flood hazard assessment across a basin. Previous works showed that understanding a basin's response to extreme rainfall based on geomorphological indices can be valuable when studying flood hazard in ungauged catchments (Muzik 1996a; Martini and Loat 2007; Lastra et al. 2008).

One of the most widely used tools to assess this rainfall-generated runoff is the instantaneous unit hydrograph. Several methodologies have been developed for instantaneous unit hydrographs derivation based on morphometric parameters. One of these techniques is the time-area method, the basic idea of which is the compilation of a "distance-area" histogram, which illustrates how different parts of a catchment contribute to the runoff at its outlet. This histogram can be translated into a unit hydrograph indicating the distribution of runoff from partial watershed areas as a function of time (Clark 1945; Maidment et al. 1996; Muzik 1996b). Muzik (1996b) and Maidment et al. (1996) used "time-area" diagrams and constructed unit hydrographs in order to simulate peak discharges and recommended that in spatial hydrology, it would be desirable to produce hydrographs for more than one part of the watershed. Pattison et al. (2008) acknowledges peak flow magnitude as a primary flood hazard indicator. Fountoulis et al. (2007), Saghafian et al. (2008) and Boglis et al. (2009) ranked sub-catchments' outlets within a basin according to a series of calculated morphometric characteristics in order to locate areas with elevated flood hazard.

In this work, instantaneous unit hydrographs based on the time-area method are compiled with the aid of GIS software and ArcHydro model (Maidment 2002), in multiple locations along the drainage networks of four catchments in Greece. Calculated peak flow rates reflect runoff conditions across four drainage basins. In this way, areas where peak flow rates are elevated are identified, highlighting higher flood hazard conditions.

2 Study areas

The method was applied in four catchments in Greece with rich flooding history and recent, well-documented flash flood events. The first one drained by Charadros torrent is situated in Attica, approximately 25 km NE of Athens. The basin consists of a hilly area on the north and west draining into a low lying part. The main torrent reaches the sea, passing through a small valley where the village of Marathon and Bei settlement are situated. The 2nd basin, drained by Rapentosa torrent is situated also in the NE of Athens. The torrent network drains the north side of Penteli Mt. and reaches the sea through the famous for its battle Marathon plain, passing through Vranas and agricultural land. The two catchments share a drainage divide for a certain length (Fig. 1a). The 3rd catchment studied is near Corinth, approximately 70 km west of Athens in northeastern Peloponnese and is drained by the Xerias river. The torrent network drains the north side of Trapezona Mt. and reaches the sea in the north, passing through the town of Corinth (Fig. 1b). The 4th basin is situated in south Peloponnese drained by Evrotas River. The river network drains the east side of Taygetos Mt and the west side of Parnon Mt. The main tributaries converge near the town of Sparti, passing through the floodplain moving southwards, towards the Laconic Gulf.

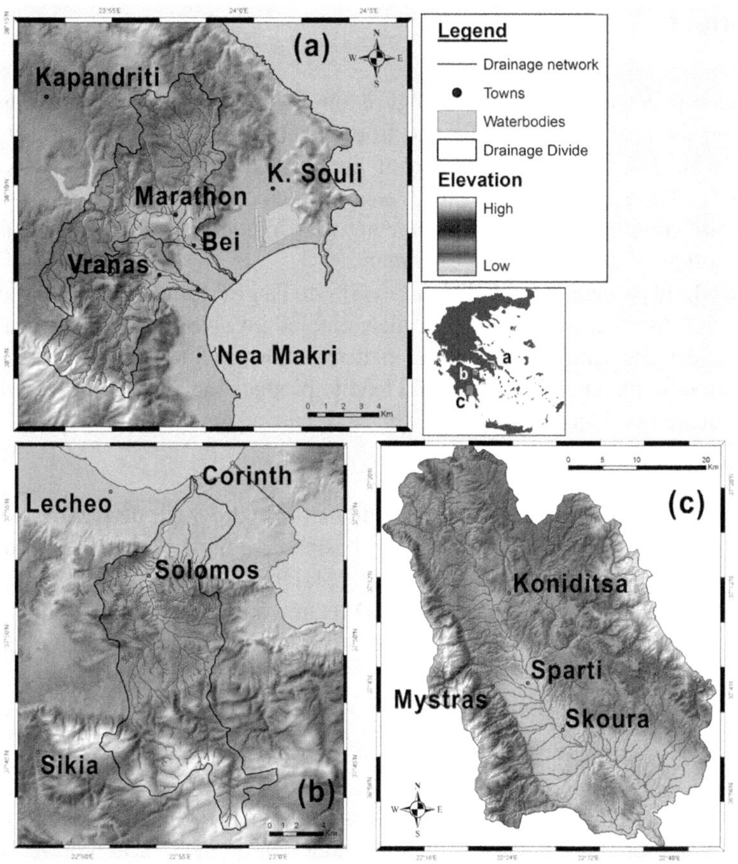

Fig. 1. a, b, c. Location map illustrating the four catchments: Charadros and Rapentosa (a), Xerias (b) and Evrotas (c).

In this work, a portion of the basin (last 4km of the river) is excluded due to substantial lack of topographic data. All four rivers have flooded their banks numerous times in the past. The most notable flood events are illustrated in Table 1.

Table 1. Catchment area and notable events in Charadros, Rapentosa, Xerias and Evrotas Rivers.

	Charadros	Rapentosa	Xerias	Evrotas
Area	60.2 km^2	37.5 km^2	162.9 km^2	1653.3 km^2
Notable events	1987 Nov. 12th [a]	1980 Oct. 27th [a]	1972 [b]	1928 Nov. 28th [c]
	2001 Jan. 14th [a]	1998 Nov. 20th [a]	1990 Aug. [b]	1999 Nov. 7th [c]
	2003 Jan. 26th [a]	2001 Nov. 4th [a]	1997 Jan. 13th [b]	2000 Dec. 6th [c]
	2005 Nov. 25th [a]	2005 Nov. 25th [a]		2005 Nov. 24th [c]

[a] Diakakis (2010), [b] Lekkas et al (1997), [c] Fountoulis et al (2007)

3 Method

The basic concept of the methodology is the compilation of a flood hazard map based on the peak flow rates derived from instantaneous unit hydrographs across each basin. The method makes use of three basic parts (modules): "ArcHydro module", "Time-area module" and "Velocity calculation module" (Fig. 2). Hydrographs are compiled along the drainage network in numerous locations assumed to be the outlets of theoretical sub-catchments.

The whole procedure is carried out in GIS and a grid format is used to represent the spatial distribution of values of all variables involved. Calculated peak flow rates reflect the catchments' runoff pattern, showing locations where intrinsic morphometric parameters of the basin tend to produce higher peak flows, which in turn indicate flood hazard potential. The steps carried out are as follows:

a) ArcHydro Module

- Assignment of sub-catchments outlets for hydrograph derivation along the drainage networks of the four basins. Automatic delineation of sub-catchments using ArcHydro Model (Maidment 2002).
- Automatic development of Flow Direction and Slope grids based on the initial DEMs, with the use of ArcHydro Model for each sub-catchment.
- Development of ''Flow-distance'' grids for each sub-catchment using ''Flow Length'' tool of ArcGIS 9.3, based on data produced in step 3.

b) Velocity calculation Module

- Assignment of Manning roughness coefficient values based on land use maps for every basin ("n" grid). Production of an "i" grid representing excess rainfall intensity using unitary rainfall. Production of a "Flow Distance" grid, representing distances from the drainage divide ("x" grid).
- Development of an overland flow "Velocity grid" based on the combination of flow at equilibrium equation (as given by Overton and Meadows 1976) with Manning's equation. A combination which gives the following equation (Kilgore 1997 and Christofidis 2008):

$$V_0 = \frac{(i_e * x)^{0,4} * S_0^{\ 0,3}}{n^{0,6}}$$

- where: i_e the rainfall excess intensity (m/s), n the Manning roughness coefficient, x is the distance along the flow path (m), S_0 the slope (m/m) and V_0 flow velocity of overland flow. Calculations were based on simple mathematical functions between grids representing the parameters involved (step 5). "Velocity grid" was used, as a weight grid that indicated the time needed for water to pass through each cell of the grid.

- Combination of "Velocity" grid with "Flow Distance" grids with the use of "Flow Length" tool of ArcGIS 9.3 to produce a "Time-Area" grid for every sub-catchment. "Time-Area" grids represented travel times across each catchment as a function of velocity and travel distance.

c) Time-Area Module

- Production of "Time-Area" histograms for every sub-catchment based on their respective "Time-Area" grids. Time of concentration used for each sub-basin was computed with Kiprich formula (Kiprich 1940) using automatically calculated needed parameters with ArcHydro.
- Development of hydrographs based on histograms of step 8, by converting Area to Discharge (by applying a 1 mm uniform rainfall and dividing by the associated time segment).
- Computation of hydrograph peak discharge for every sub-catchment outlet and interpolation (in GIS) of their values across their respective basin.
- Ranking catchments in 5 classes of hazard according to peak flow values.

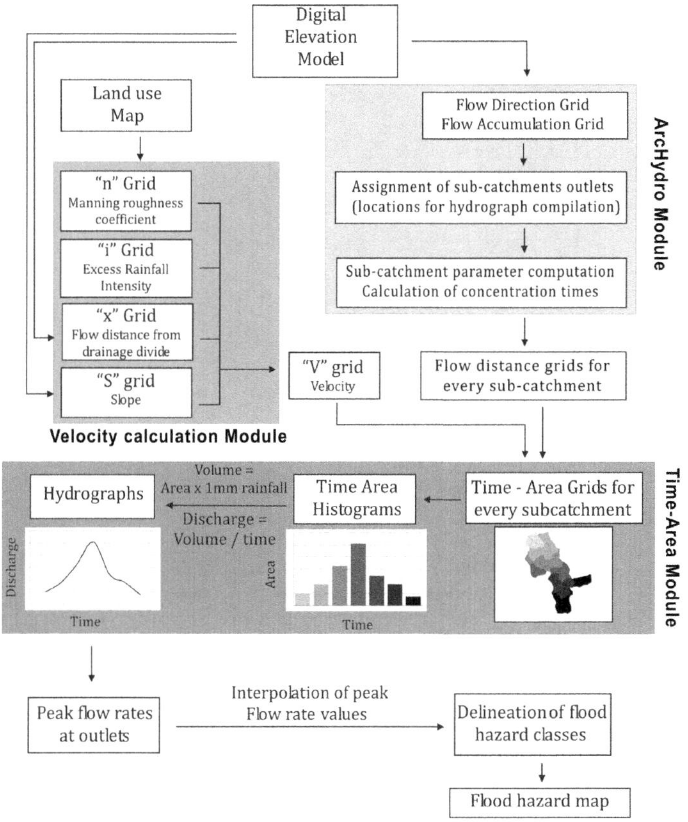

Fig. 2. Main steps of the methodology followed for every catchment.

The steps followed in the methodology are summarized in Figure 2. Land use data from CORINE land cover program (EEA 2000) were used to assign Manning roughness coefficient values.

For verification purposes, past flood damage locations for each catchment were compiled in a list and subsequently plotted against flood hazard maps. Damages included agricultural land, household appliances and a limited amount of structural problems on buildings.

4 Results and Conclusions

Results were produced in the form of maps categorizing the catchments in 5 classes according to the spatial distribution of peak flow rate values. Red color was assigned to areas with the highest peak flow rates, whereas green color to the lowest ones. Past flood damages as recorded by Diakakis 2010, Lekkas et al (1997) and Fountoulis et al (2007) were plotted on the maps.

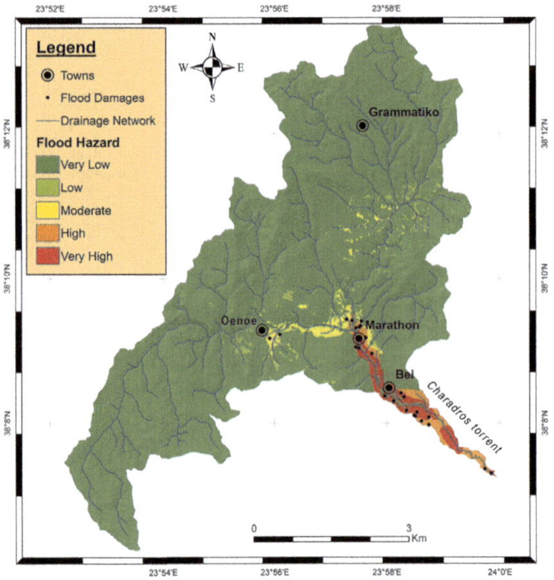

Fig. 3. Flood hazard map of Charadros catchment. Hazard zonation is projected against flood damages recorded by Diakakis (2010).

The majority of damage locations coincided with ''Very high'', ''High'' and ''Moderate'' hazard zones in their vast majority. The process presented takes advantage of computer-assisted computation of morphometric characteristics (like flow paths length, catchment area and basin shape) to compile hydrographs and subsequently use them to predict areas where discharge potential is higher.

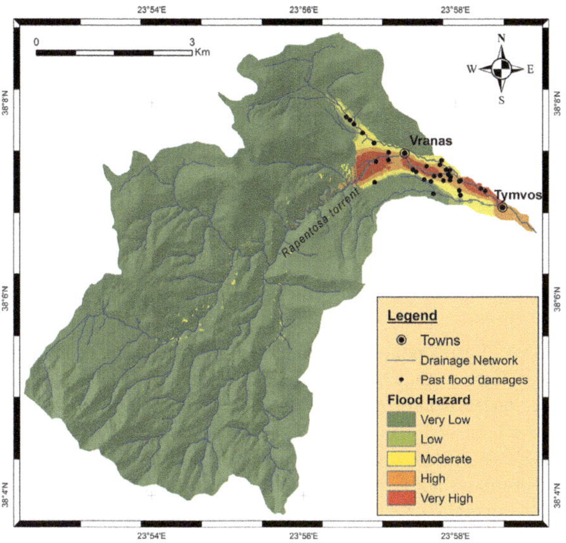

Fig. 4. Flood hazard map of Rapentosa catchment. Hazard zonation is projected against flood damages recorded by Diakakis (2010).

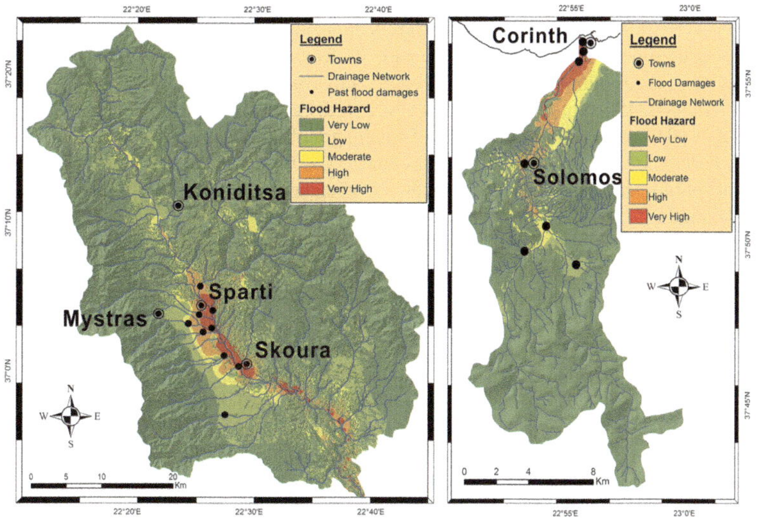

Fig. 5. Flood hazard map of Evrotas (left) and Xerias (right) catchment. Hazard zones are projected against flood damages recorded by Fountoulis et al (2007) and Lekkas et al (1997) respectively.

The results are considered important as they demonstrate that areas with higher peak discharges show good correlation with past flood damage locations, proving the ability of the method proposed to reflect the runoff pattern and identify the higher flood hazard areas. One of the limitations is that human induced flooding problems cannot be reliably predicted with this process. However, the method visualizes the areas where there is natural tendency for high discharge potential.

A significant advantage of the process is considered to be its ability to produce results using only good quality topographical and land use data. In this way, the methodology is able to produce results in ungauged catchments, which is the case for the majority of the watercourses in Greece.

References

Boglis A, Evelpidou N, Vassilopoulos A., Lekkas DF, Gournellos T, Fountoulis I (2009) Urban flood modeling in Karlovassi area-Samos Island, Greece. In: Proceedings of the 11th international conference on environmental science and technology, vol 1. Chania, Crete, 3–5 September 2009, 83–91

Christofidis A (2008) Development of a GIS-based rainfall-runoff model. PhD Thesis. Department of Water Resources and Environmental Engineering, Faculty of Civil Engineering, National Technical University of Athens

Clark CO (1945) Storage and the unit hydrograph. Trans ASCE 110:1419–1446

Diakakis M (2010) Flood history analysis and its contribution to flood hazard assessment. The case of Marathonas, Greece. Bull Geol Soc Greece 43:1323–1334

EEA (2000) Corine land cover. European Environment Agency, Commision of the EC, http://www.eea.europa.eu/publications/COR0-landcover Accessed: 15 April 2010

Fountoulis I, Mariolakos I, Andreadakis Emm, Sambaziotis E, Karagiozi E (2007) Strategic planning of anti-flood protection for Laconia Prefecture. In: Nikolaides et al. (ed) Antiflood protection master plan for Laconia prefecture, Environment, LIFE, EnviFriendly, LIFE05 ENV/GR/000245. Available at: http://www. evrotas.gr/archive.php Accessed 25 July 2010

Kirpich ZP (1940) Time of concentration of small agricultural watersheds. Civil Eng 10:362–368

Lastra J, Fernandez E, Diez-Herrero A, Marquinez J (2008) Flood hazard delineation combining geomorphological and hydrological methods: an example in the Northern Iberian Peninsula. Nat Hazards 45(2):277–293

Lekkas E, Lozios S, Skourtsos E, Kranis H (1997) Floods, geodynamic environment and human intervention. The case of Corinth (Greece), In: C.A.Brebbia, J.L.Rubio, J.L.Uso (eds) Risk Analysis, Wit Press, Computational Mechanics Publications, 2, 135-144

Maidment DR (2002) ArcHydro: GIS for water resources. ESRI Press, Redlands, California

Maidment DR, Olivera F, Calver A, Eatherall A, Fraczek W (1996) Unit hydrograph derived from a spatially distributed velocity field. Hydrol Process 10:831–844

Martini F, Loat R (2007) Handbook on good practices for flood mapping in Europe. European exchange circle on flood mapping (EXCIMAP), Paris/Bern. http://ec.europa.eu/environment/water/flood_risk/ flood_atlas Accessed on 13 April 2010

Muzik I (1996a) Flood modelling with GIS-derived distributed unit hydrograph. Hydrol Process 10:1401–1409

Muzik I (1996b) A GIS-derived distributed unit hydrograph. Proceeding of HydroGIS 96: application of geographic information systems in Hydrology and Water Resources Management, April 1996, Vienna, IAHS Publ. no. 235

Overton DE, Meadows ME (1976) Stormwater modeling. New York: Academic press

Pattison I, Lane SN, Hardy RJ, Reaney S (2008) Sub-catchment peak flow magnitude and timing effects on downstream flood risk. 10th Nat Hydrol Symposium, 15–17 Sept 2008 Exeter. http://www.hydrology.org.uk/Publications/exeter/44.pdf Accessed on 15 April 2010

Saghafian B, Farazjoo H, Bozorgy B, Yazdandoost F (2008) Flood intensification due to changes in land use. Water Resour Manag 22:1051–1067

Flood hazard evaluation in small catchments based on quantitative geomorphology and GIS modeling: The case of Diakoniaris torrent (W. Peloponnese, Greece)

E. Karymbalis, Ch. Chalkias, M. Ferentinou, A. Maistrali

Department of Geography, Harokopio University of Athens, 70, El. Venizelou Str, GR 176 71, Greece. karymbalis@hua.gr

Abstract This study deals with the evaluation of flash flood hazard in the ungauged Diakoniaris torrent drainage basin which is a fifth order stream located in northwestern Peloponnese. Diakoniaris drains an area of 27.83 km^2 and discharges into the Gulf of Patras. Its lower reaches flow through the city of Patras and has suffered from intense human interferences due to increased urbanization during the last decades. For the purposes of the study the quantitative geomorphological characteristics of the basin were estimated and studied focusing on the investigation of the hierarchical drainage by stream order. The longitudinal profiles as well as the stream power diagrams of the main stream channel and its major tributaries were constructed and analyzed. A routing model based on runoff travel time was created within GIS environment in order to estimate the discharge at the outlet of the basin for an assumed extreme rainfall event.

1 Introduction

Floods are a global common geo-environmental hazard with consequences on the social and economic structure, on local and regional scale. Several regions in Greece suffer from frequent and hazardous extreme flash floods. Flood phenomena in Greece are usually caused by intense rainstorms. Many drainage basins are relatively small with steep slopes, configured by torrents with braided main channel morphology. These systems are usually dry for most of the year but become particularly active during extreme flash flood events of low frequency but high magnitude. Such extreme discharge may cause significant damage and life loss because the incorporation of flood prevention as a parameter of urban planning context is relatively ignored by planners and public (Despiniadou and Athanasopoulou 2006).

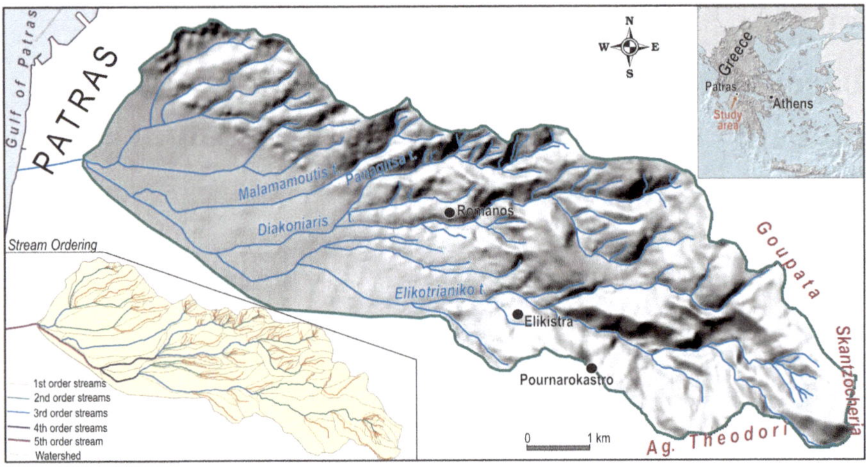

Fig. 1. The Digital Elevation Model of the Diakoniaris torrent drainage basin and ordering of the drainage network according to Strahler (1957).

The aim of this study is to highlight the natural factors responsible for flash flooding in the ungauged drainage basin of Diakoniaris torrent, located in northwestern Peloponnese, based on quantitative geomorphology and GIS hydrological analysis.

The Diakoniaris torrent is an ephemeral flow and drains an area of 27.83 km^2 (Fig. 1). The drainage basin is elongated along an almost E-W trending axis and reaches an elevation of 1,837 m at its southeastern boundary (Panachaikon Mt.). The main channel follows an E-W flow direction for about 10 km and discharges into the Gulf of Patras. At the mouth of the torrent a fan has developed where the western suburbs of the city of Patras are located. The Diakoniaris drainage network belongs to the parallel pattern with tributary streams tending to stretch out in a parallel-like fashion following the slope of the surface. The catchment can be morphologically divided into two parts: the eastern mountainous area of rough relief and the highest altitudes (Goupata (1,328 m), Skantzocheria (1,837), Ag. Theodori (1,301 m) and the western area of lower elevations and gentle slopes. Geologically the drainage basin is composed of Alpine and post-Alpine sediments (Fig. 2). The Alpine formations belong to the geotectonic unit of Olonos-Pindos, occupy the eastern part of the catchment and consist mainly of thin to medium platy pelagic limestones of Coniacian-Maestrichtian age and semi-permeable radiolarites of Jurassic-Lower Cretaceous age. A very small part of the eastern basin is composed of Maestrichtian-Paleocene flysch formations (which include sandstones and marls alternating with thin platy limestones) (Tsoflias 1984). The Pliocene formation includes marine, brackish and lacustrine deposits (marls, clays, sandstones and conglomerates) are developed at the central part of the basin and their hydrogeological behavior varies due to lateral transitions and alternations. At the western part of the basin, recent torrential deposits occur along the lower reaches of the torrent and its tributaries. The city of Patras, which is located at the low-lying area of the alluvial deposits of the torrent, has often suffered extensive damage during extreme rainfall events. The torrent experienced extreme discharge events in December 17th of 1962 and October 25th of 1997 while the most severe flood happened on December 16th of 2001 and caused 2 people to lose their lives and extensively damaged houses and cultivations regionally (Stamatopoulos 2005).

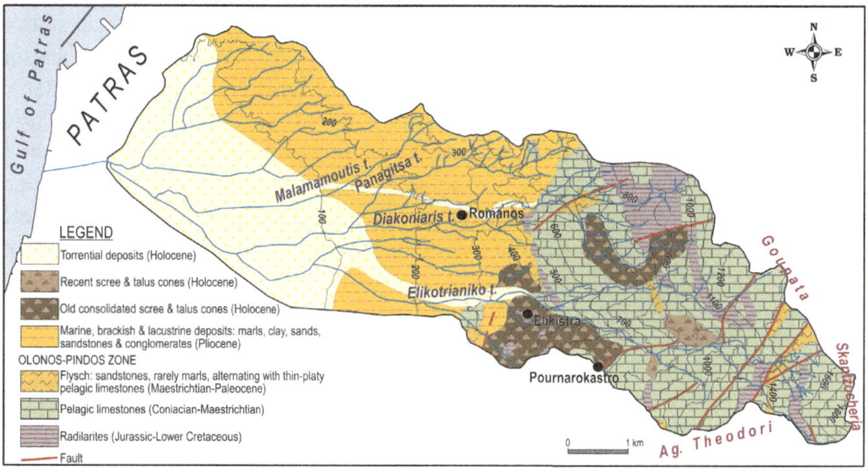

Fig. 2. Geological map of the Diakoniaris torrent drainage basin (based on the 1:50,000 scale geological map of Institute of Geology and Mineral Exploration, Tsoflias 1984).

2 Methods and data

For the purposes of this study a spatial database was designed and implemented consisting of primary GIS layers (contour lines, elevation points, streams, geological formations) derived from analogue maps. The post processing of these layers within a GIS environment produced secondary thematic layers (Digital Elevation Model and its products like hillshade, flow direction, flow accumulation, flow length, stream ordering, and drainage basins) which are the outputs of standard GIS functions. An accurate 20m x 20m resolution DEM was constructed with the use of the available topographic maps of the Diakoniaris basin and the ANUDEM interpolation algorithm (Hutchinson 2003) within ArcGIS software package. This interpolation algorithm produces hydrologically consistent outputs and maintains landscape integrity (Gallow et al. 2007). The quantitative geomorphological and morphometric characteristics of the Diakoniaris torrent drainage basin were estimated and studied. The drainage network derived from the Digital Elevation Model was quantified by ordering the channels according to Strahler (1957) and quantitatively analyzed while bifurcation ratios and the hierarchical drainage by stream order was also investigated in order to detect irregularities of the drainage network organization (Garbrecht et al. 1988). The constructed spatial database and GIS technology were utilized for the quantitative measurements. In order to estimate direct runoff at the outlet of the catchment produced by an assumed spatially homogeneously distributed extreme rainfall event (90 mm/12 h) a routing model (Olivera and Maidment 1999, Malesse and Graham 2004) was created within GIS environment. This extreme event represents the maximum 12hour rainfall of the meteorological station of Araxos for the last 12 years. This meteorological station is the nearest to the basin with continuous rainfall data since the Patras station does not provide continuous measurements for this time period. The model is based on the travel time. Thus, overland flow travel time as well as channel flow time estimations were implemented (Du et al. 2009). For the estimation of travel time, flow velocities for both overland and channel flows are required. The overland flow travel velocity was estimated for each cell out of the channel network with the combination of kinematic wave approximation with Manning's equation according to the following formula:

$$V = \frac{(i_e x)^{0.4} S_o^{\ 0.3}}{n^{0.6}}$$

In this formula V is the overland flow travel velocity, i_e is the net incoming flux (m/s), x is the upwards flow length (m), S_o is the slope of the cell (m/m), and n is Manning coefficient estimated from the literature. Moreover, channel flow travel velocity was estimated for the pixels with high values of flow accumulation according to the combination of Manning's equation with the contiguity equation by using the following final formula:

$$V = S_o^{\ 0.3} Q^{0.25} n^{-0.75}$$

V is the channel flow travel velocity, S_o is the slope of the cell (m/m), Q is the cumulative discharge (m^3/ sec) and n is Manning coefficient.
Since V was calculated for each cell of the basin under investigation by following the aforementioned procedure, the travel time for each cell was computed from cell velocity and the travel distance (flow length, FL) as

$$T_c = FL / V$$

The classification of T_c values is the final step in order to locate isochrones (lines of equal travel time to the outlet) within the basin and create the time-area diagram of the study area. The direct runoff at time $t=n \cdot \Delta t$ is assumed as the aggregation of the runoff pulses from each of the time zones properly lagged in time:

$$Q_n = \sum_{i=1}^{n} \frac{P_{ij} \cdot A_i}{\Delta t}$$

where P_{ij} is the total excess rainfall for all cells in the isochrone zone during time interval j and A_i is the area of each isochrone zone. The general concept is to incorporate GIS-based hydrological analysis functions and background spatial data for the construction of the flow time layer. Next, this layer is integrated with a uniformly distributed rainfall for the creation of the direct runoff by convolution. Additionally, the total stream power profiles along main branches of the study area were constructed. The longitudinal profiles of the main stream and four of its major tributaries were also drawn and analyzed. The stream power diagrams for the assumed rainfall event were used in combination with the corresponding longitudinal profiles in order to interpret the energy potential for each part of the channels under investigation. The calculation of the total stream power diagrams for each main stream of the study area was performed using the following equation (Knighton 1999):

$$\Omega = \gamma \cdot Q \cdot S$$

where γ is the specific weight of the water (9807 N/m^3), Q is the discharge (m^3/sec) and S is the energy slope derived from the DEM (m/m). Since there are no available discharge recordings, the Q parameter was based on flow accumulation area (from the analysis of the DEM) and the assumed rainfall event of 90 mm for 12 h.

3 Results and discussion

The quantitative geomorphological analysis of the Diakoniaris drainage network showed relatively high mean bifurcation (3.10), cumulative channel length (3.79) and basin area (4.10) ratios.

Table 1. Relation between the numbers of streams, mean cumulative channel length (in km) and mean drainage basin area (in km^2) per order for the Diakoniaris torrent drainage network.

Stream order	Streams Number	Bifurcation Ratio	Mean ratio	Channel length	Length ratio	Mean length ratio	Basin area	Area ratio	Mean area ratio
1	73			0.54			0.15		
2	16	4.6		3.48	6.4		1.11	7.40	
3	6	2.7	3.1	12.69	3.65	3.79	4.25	3.83	4.10
4	2	3.0		28.84	2.27		9.63	2.27	
5	1	2.0		81.45	2.82		27.83	2.89	

Particularly high values are observed for the ratios between the 1st and 2nd order streams (Table 1) with the basin area ratio reaching 7.40. This means that there are numerous 1st order channels that drain directly into the 2nd order streams (without producing them).

Table 2. Number of streams, channel length and basin area for each order that drain directly into streams of higher order for the drainage network of Diakoniaris torrent.

Stream order	Number of streams	%	Channel length	%	Basin area	%
1st to 2nd	33	45.2	20.05 km	51.0	5.29 km^2	50.0
1st to 3rd	8	11.0	5.56 km	14.1	1.46 km^2	13.8
2nd to 3rd	4	25.0	10.53 km	18.9	3.20 km^2	18.1
3rd to 5th	2	33.3	21.50 km	28.2	7.82 km^2	30.7

These junctions are located at the northern and eastern part of the network at the upper reaches of the Diakoniaris and its tributaries Panagitsa and Elikotrianiko streams (Fig. 1). Thus, during extreme rainfall events, the surface runoff of an area of 5.29 km^2 is directly added to the main 2nd order channels discharge. Table 2 includes the number of streams, channel length and basin area for each order, which drain directly into higher order streams.

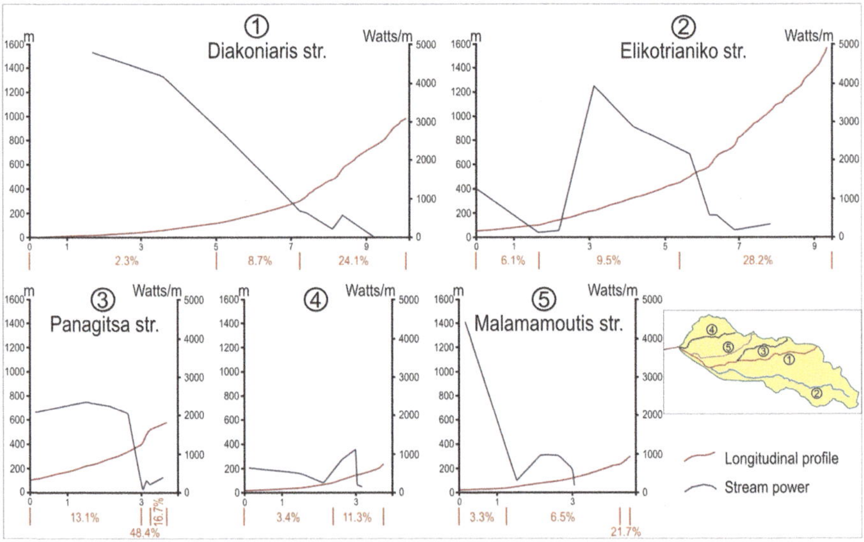

Fig. 3. Longitudinal profiles and stream power diagrams of Diakoniaris torrent and four of its major tributaries. Insert map shows the location of each stream in the catchment.

The most important irregularities concern the 1st order streams that drain directly into the 2nd order channels as well as the 3rd order streams that drain directly into the 5th order main channel of the Diakoniaris. For the 1st order streams, 33 channels which correspond to the 45.2% of their total number flow directly into 2nd order streams. The total length of these streams corresponds to the 51.0% of the total 1st order channel length and drain 50.0% of the 1st order drainage basin area. The two 3rd order streams that drain directly into the 5th order Diakoniaris stream join the main torrent channel at its lower reaches (0.96 and 1.77 km upstream from the torrent mouth) within the city of Patras, increasing its discharge during extreme precipitation as they carry the surface runoff of an area of 7.82 km^2. The banks of the Diakoniaris from the estuary up to 1,100 m upstream are now covered and constitute one of the main transportation axes (El. Venizelos Av.) of the city of Patras.

Longitudinal profiles (Fig. 3) revealed that Diakoniaris and Elikotrianiko streams contain three segments. An upper mountainous region with significantly high mean channel slopes (24.1 and 28.2%), a middle part with medium slopes (8.7 and 9.5%) and a downstream region with lower values (2.3 and 6.1%). The Diakoniaris part of gentle channel slope corresponds to the lower 5 km before the river mouth while for the Elikotrianiko stream this section corresponds to its flow for 1.7 km upstream of its confluence with the main torrent channel. The channel length of the other three tributaries is significantly smaller. The upper high slope (21.7%) section of tributary 5 is of very small length while the middle section of tributary 3 has a mean slope of 48.4% and corresponds to a knick-point formed due to the lithologic contrast between flysch and limestones. Intense and abrupt variations in the relief result in the decrease of water velocity in low gradient areas and the consequent accumulation over a short time period of a vast volume of water, which the main stream channel cannot accommodate. The most interesting stream power diagrams along the main channels of the drainage network, for the assumed extreme 12 hour rainfall event, are the ones that correspond to the main channel of the Diakoniaris, Elikotrianiko and Malamamoutis streams (numbers 1, 2 and 5 in Fig. 3 respectively). These streams release a large amount of power, reaching up to almost 4,500 watts/m through high gradient sites, as is shown by the combined longitudinal profile-stream power diagrams (Fig. 3). High power values are also observed along the two elongated streams at the confluence of major 3rd and 4th order tributaries with the main channel (especially at the lower reaches of the Diakoniaris) due to their high contribution of surface runoff to the main channel.

The application of the GIS based runoff model lead to the isochrones map within the catchment (Fig. 4a) as well as to the time travel-cumulative area diagram of the studied basin (Fig. 4b) for the assumed 90mm/12 h extreme rainfall event. The diagram shows that the runoff of almost the entire basin requires about 4 hours to reach the outlet. Thus the response of the basin to intense rainfall is relatively immediate. This approach estimated that the maximum direct discharge for the simulated event at the outlet of the watershed is 65 m^3/sec. The implementation of the same modeling approach for the outlet of Elikotrianiko torrent estimated 45 m^3/sec as the maximum discharge at this point. Although validation of this maximum discharge is impossible, since there are no gauging records, it is in agreement with the results of previous approaches to this basin (Daniil et al. 2004).

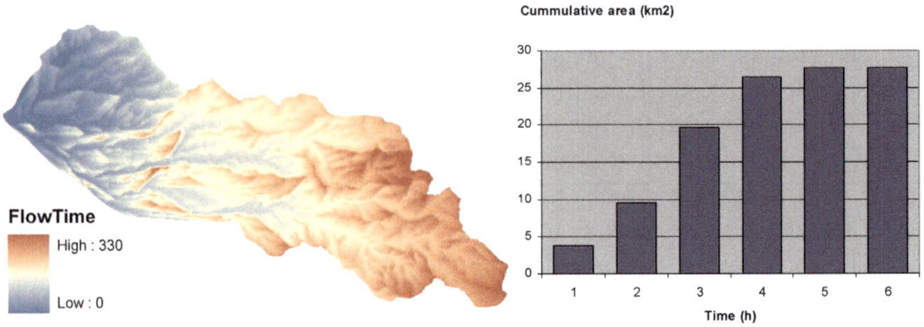

Fig. 4. a, b Isochrones map within the catchment and time travel time-area diagram.

4 Conclusions

Apart from intense rainfall, among the most important natural causes that are responsible for flash flood events at the lower reaches of Diakoniaris torrent are geomorphic features of the drainage network. The quantitative geomorphological analysis showed that the pattern and characteristics of the drainage network enhance flash floods. These features include irregularities in the hierarchical drainage by stream order and especially the junction of 3rd order streams with the 5th order main channel, the relatively high value of the bifurcation ratio (4.6), the extremely high basin area ratio (7.40) between the 1st and 2nd order streams as well as high channel gradients in the upper reaches of the two main stream channels. During extreme rainfall events, the main streams of Diakoniaris, Elikotrianiko and Malamamoutis release a large amount of power reaching up to almost 4,500 watts/m through high gradient sites and at the confluence of major tributaries. Abrupt changes in stream power and runoff characteristics occur at channel junctions where the surface runoff experiences discrete increase as two tributaries merge. According to the flood susceptibility assessment the response of the basin to intense rainfall is relatively immediate. The direct runoff for the hypothetical extreme event at the outlet of the watershed is 65 m^3/sec. Human interference in the main channel of the torrent, particularly with respect to its path through the city of Patras, seems to be detrimental during extreme flood events.

References

Callow JN, Van Niel KP, Boggs GS (2007) How does modifying a DEM to reflect known hydrology affect subsequent terrain analysis? Journal of Hydrology. 332(1-2), 30-39

Daniil EI, Bouclis GD, Michas SN, Lazaridou PL, Lazarides LS (2004) Hydrologic issues in flood management and control in an urban environment in Greece. Protection and restoration of the environment VII Water Resource Management – Environmental Fluid Mechanics Intl. Conf. of Protection and Restoration of the Environment VII, 28 June-1 July, Mykonos, Greece, (CD-ROM)

Despiniadou V, Athanasopoulou E (2006) Flood prevention and sustainable spatial planning. The case of the River Diakoniaris in Patras. Proceedings if the 46th Congress of the European Regional Science Association (ERSA), Volos

Du J, Xie H, Hu Y, Xu Y, Xu CY (2009) Development and testing of a new storm runoff routing approach based on time variant spatially distributed travel time method. Journal of Hydrology 369 (1-2), 44-54

Garbrecht J, Shen HW (1988) The physical framework of the dependence between channel flow hydrographs and drainage network morphometry. Hydrological Processes. 2, 337-355

Hutchinson MF (2003) ANUDEM Version 4.6.3. Australian National University. Centre for Environmental Studies, Canberra

Knighton D (1999) Downstream variation in stream power. Geomorphology 29, 293-306

Melesse AM, Graham WD (2004) Storm runoff prediction based on a spatially distributed travel time method utilizing remote sensing and GIS. Journal of the American Water Resources Association 40 (4), 863-879

Olivera F, Maidment D (1999) Geographic information systems (GIS)-based spatially distributed model for runoff routing. Polygraph International 1, 1155-1164

Stamatopoulos L (2005) The flood risk from Diakoniaris River in urban area of Patras, NW Peloponnese, Greece. Bulletin of the Geological Society of Greece. XXXVIII, 69-76

Strahler A (1957) Quantitative analysis of watershed Geomorphology - Am. Geophys. Union Trans. 38(6), 913-920

Tsoflias P (1984) Geological Map of Greece, scale 1:50,000, Chalandritsa Sheet. Inst Geol. And Miner. Explor., Athens

Preliminary flood hazard and risk assessment in Western Athens metropolitan area

M. Diakakis[1], M. Foumelis[2], L. Gouliotis[1], E. Lekkas[1]

[1] Department of Dynamics Tectonics and Applied Geology, National and Kapodistrian University of Athens, Panepistimioupoli-Zografou, GR 157 84, Greece, diakakism@geol.uoa.gr, legouliot@geol.uoa.gr, elekkas@geol.uoa.gr

[2] Department of Geography, Harokopio University of Athens, Athens, GR 176 71, Greece, mfoumelis@hua.gr

Abstract The increase in urban population and the continuous pressure for cities' expansion along with the increase in urban flooding phenomena in Greece and worldwide, stress the need for enhancement of flood risk mitigation efforts. West Athens urban area, in Greece, experienced a significant population clustering since the 1950s leading, in some occasions, to a poorly-planned development, even in areas with imminent flood risk. An issue becomes apparent, taking into account the rich flooding record, the extended damages in property and infrastructure and the 76 flood victims during the last century in the area. In this work, flood hazard is assessed in 10 municipalities of West Athens, with the application of a GIS-based methodology that exploits catchment morphometric characteristics to delineate flood hazard zones. Historical flood events are reconstructed to provide better understanding of the flooding problem in the area. Finally flood hazard was studied in conjunction with vulnerability to estimate flood risk spatial distribution. The results showed that areas around Fleva and Eschatia torrent, especially Mpournazi, parts of Ilion and Kamatero and some parts of Peristeri presented the highest flood hazard and risk values. Additionally, moderate flood risk appeared in several mountain torrents in west parts of Petroupoli and Peristeri.

1 Introduction

The augmentation of cities' population has contributed to the increase of urban flood risk during the last decades worldwide. During the last years the efforts towards flood risk mitigation have been enhanced in EU and in Greece with numerous research projects, civil protection agencies' initiatives and improved legal framework. One of the most substantial legal binding instruments is the EU Directive 2007/60 which creates the obligation for completion of preliminary flood hazard assessments in river basins in all member states.

West part of Athens metropolitan area in Greece is not exception to this flood-prone regime, presenting a rich flooding record during the last century. Flood history data were collected based on numerous sources such as local and central government organizations, the General Secretariat of Civil Protection, and the press. Based on these data 14 flood events between 1900 and 2010 were identified (Table 1) inducing significant damages and killing 76 individuals. Historical events were reconstructed in spatial terms (Fig. 1) based on spatial information provided by the same sources, in order to identify areas with increased historical flood frequency.

In this work, preliminary flood hazard and risk assessments are carried out in West Athens metropolitan area, with the aid of a GIS-based methodology that uses morphometric characteristics of the catchment to predict the areas with high intrinsic flood hazard. The study area is an administrational entity in West Athens comprising of 10 municipalities. It consists of a significant portion of urban areas, situated in lowland, and a hilly part with ephemeral mountain torrents. Most of the torrents drain into Kifissos River which is the main drainage route in Athens basin, just outside the eastern limit of the study area. Kifissos River was not included in this study as the major parts of the main torrent are outside of the east border of the study area. The dense population and the augmented social, industrial and commercial activity in the area, lead to increased vulnerability in natural disasters.

Table 1. Historical floods in the study area between 1900 and 2010. Information is based on the GSCP (2007), local organization archives and press archives of the National Library's Digital Newspapers Collection (2010).

Flood data	Location	Torrent	Victims
26/10/1930	Sepolia	Kifissos	2
06/11/1961	Bournazi, Ilion, Ag. Fanourios, Mykoniatika	Fleva	40
Jan 1972	Western Athens	–	–
02/11/1977	Peristeri (mainly), Ilion, Antoupolis, Petroupolis	Fleva, Giorgiza, Vathi	25
28/10/1978	Aghioi Anargyroi	Kanapitseri	–
07/10/1980	Ilion, Petroupolis, Peristeri , Aghioi Anargyroi, Antoupolis	Giorgiza, Fleva, Vathi	–
27/10/1980	Ilion, Kamatero	Fleva, Euripidon	–
27/10/1986	Ilion, Peristeri	Fleva	–
12/11/1987	Petroupolis, Peristeri, Ano Liosia	Giorgiza	–
25/02/1988	Petroupolis, Ilion, Mpournazi	Fleva, Giorgiza, Chaidarorrema	–
09/12/1989	Aghioi Anargyroi	Kanapitseri	–
21/11/1994	Aghioi Anargyroi, Aigaleo	Kifissos, Podoniftis, Chaidarorrema	9
27/01/1996	Western Athens	–	–
08/07/2002	Aigaleo, Koridallos	Kifissos	–

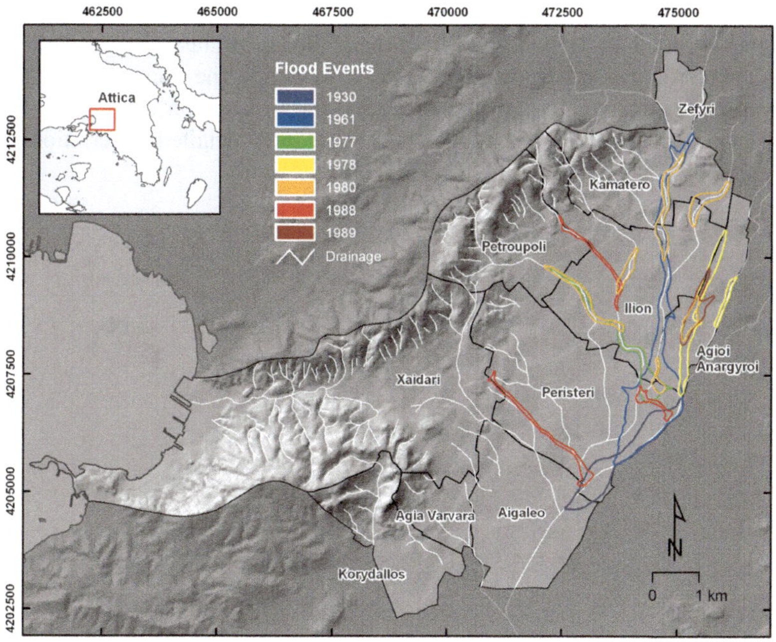

Fig. 1. Location map of the study area. Municipalities of West Athens under examination, drainage network and extent of past flooding phenomena.

2 Data and methods

A GIS-based methodology was used for hazard assessment, exploiting morhometric characteristics of the catchments under study. The basic concept of the methodology was the compilation of a flood hazard map based on the peak discharge values derived from multiple instantaneous unit hydrographs across the catchments under study. Unitary hydrographs were compiled along the drainage network in numerous locations assumed to be the outlets of theoretical sub-catchments. Hydrographs were compiled with the use of ArcHydro Model (Maidment 2002) and the Time-Area method (Clark 1945). Time-Area histograms for every sub-catchment were converted to hydrographs by applying a 1 mm uniform rainfall. Finally, calculated peak flow rate values of these hydrographs highlighted locations where intrinsic morphometric parameters of the basin tend to produce higher peak flows, which in turn indicate flood hazard potential. The whole procedure was carried out in GIS and a grid format is used to represent the spatial distribution of values of all variables involved. Geological data used were based on geological maps (Gaitanakis 1982, Katsikatsos et al. 1986, Papanikolaou et al. 2002) and additional detailed field mapping, locally in 1:5.000 scale

(Gouliotis 2002). Land use data were based on CORINE maps (EEA 2000) and topographic maps of 1:5000 scale from the Hellenic Military Geographic Service (HMGS). Geology and land use were used to define the concentration time in every sub-catchment, a parameter necessary for hydrograph derivation. A summary of the methodology is shown in Figure 2. A detailed description of the methodology is available from Diakakis (2010).

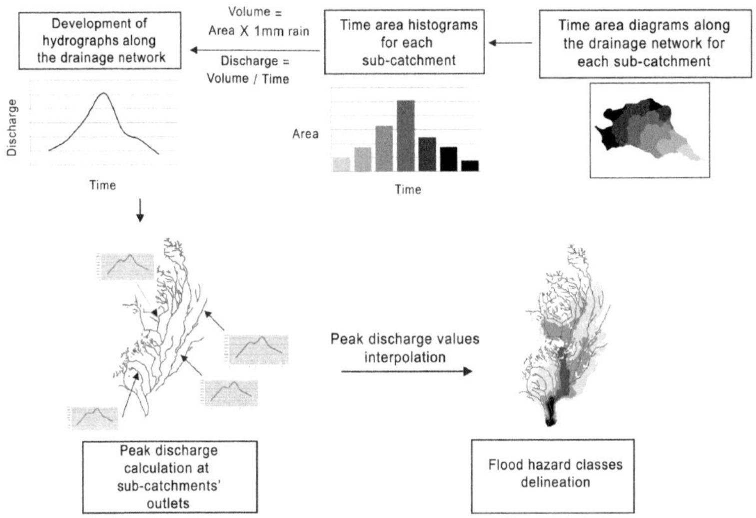

Fig. 2. Basic steps of the methodology.

Subsequently, vulnerability was determined, by categorizing the study area into four classes according to the significance of elements present, and taking into consideration the impact from possible flood damage. Data on infrastructure, housing, road network, public facilities and critical services were provided by the Development Association of West Athens (ASDA). Detailed description of this classification is illustrated in Table 2, while a sample of vulnerability distribution within a selected part of the study area is shown in Figure 3.

Table 2. Categorizing the study area into four classes according to the significance in terms of vulnerability of elements and infrastructure at risk.

Vulnerability class	Elements distributed to each class
Very high	Town halls, Police & Fire brigade stations, Hospitals / Clinics, Schools
High	Churches, Public Services, Ministries
Moderate	Houses, Exhibition centers , Theatres & Cinemas, Camps, Banks
Low	Parks, Sports facilities, Cemeteries, Parking lots

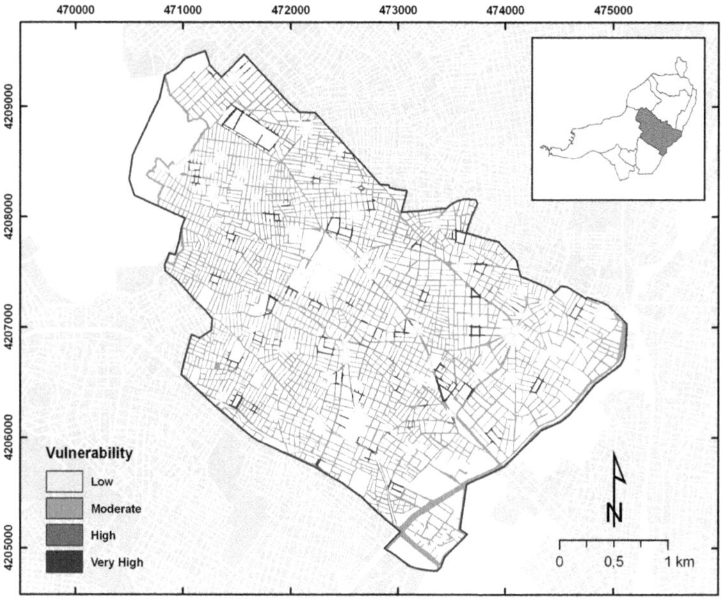

Fig. 3. Sample map of a part of the study area, showing different values of vulnerability.

Hazard and vulnerability maps were transformed in a grid format that allowed simple mathematical functions between them in a GIS environment. In this context, "Vulnerability grid" was multiplied on a cell to cell basis with "Hazard grid". Risk was determined as a function of hazard and vulnerability with these two parameters operating as amplifying factors to the risk level, according to the following equation:

$$R = H \times V$$

Where R is risk, H is hazard and V is vulnerability. The final risk map was the result of this multiplication in a grid format and categorized in five classes (Very high, High, Moderate, Low and Very Low).

3 Results and discussion

Reconstruction of historical events highlighted the areas where there were significant flooding problems in the past. As far as the hazard distribution is concerned the analysis showed that higher values are mainly observed around Fleva torrent in the areas of Ilion, Mpournazi and parts of Ilion and Peristeri. In addition, high hazard values appear along Eschatia torrent mainly across Kamatero (Fig. 4). Comparison of hazard distribution with flood reconstruction based on historical evidence presented a good correlation.

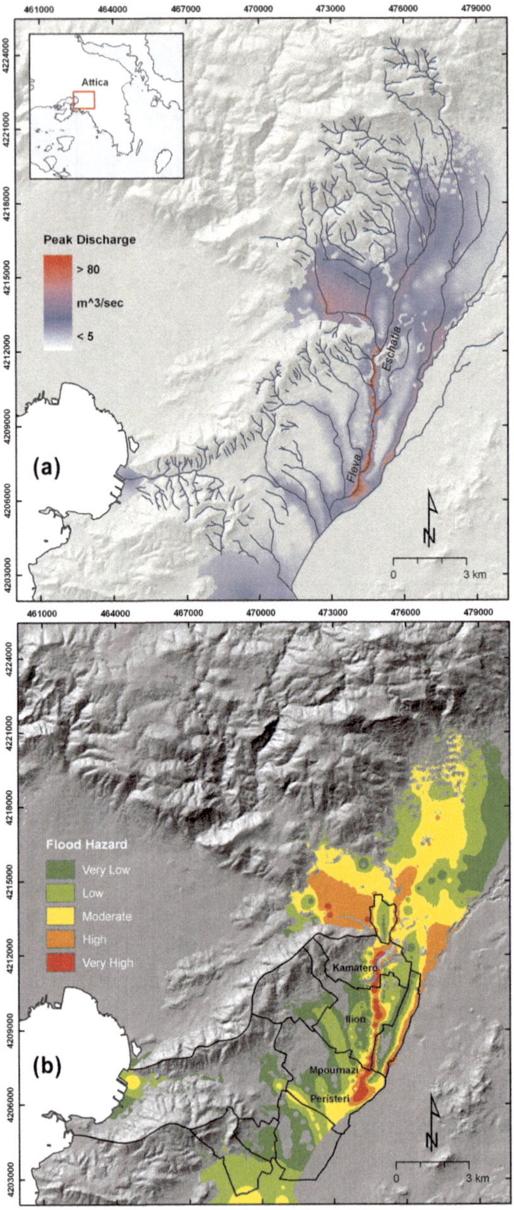

Fig. 4. **(a)** Flood hazard map of the study area in absolute peak discharge measures (m^3/sec) and **(b)** in relative terms as delineated hazard zones.

Subsequently, hazard was multiplied with vulnerability resulting in a preliminary risk map (Fig. 5).

Based on risk distribution maps one can identify the locations around Fleva torrent especially Mpournazi, Ilion, Kamatero and parts of Peristeri and Agioi Anargyroi as the ones with the major flooding issues. Moderate flood risk appears at the border between Peristeri and Aigaleo while lower risk is presented at some parts of Petroupoli and Ilion along ephemeral torrents draining from the Aigaleo hill.

Although this approach does not give any quantitative expression hazard and risk, it certainly highlights and prioritizes the locations where actions have to be taken to mitigate risk. One of the advantages of the approach is its ability to produce results in short time and with low data and cost requirements, as it can function with commonly available inputs such as land use, geology and topography. In this sense the method fulfils the requirements of a preliminary flood risk assessment in the context of the EU and Greek legal framework and identifies the areas where further study (e.g. 1-D modelling) is required.

Moreover, in this study, it is shown that the use of historical data combined with the use of a method based in geomorphology could be a useful and reliable tool that contributes to a better understanding of the flooding issues.

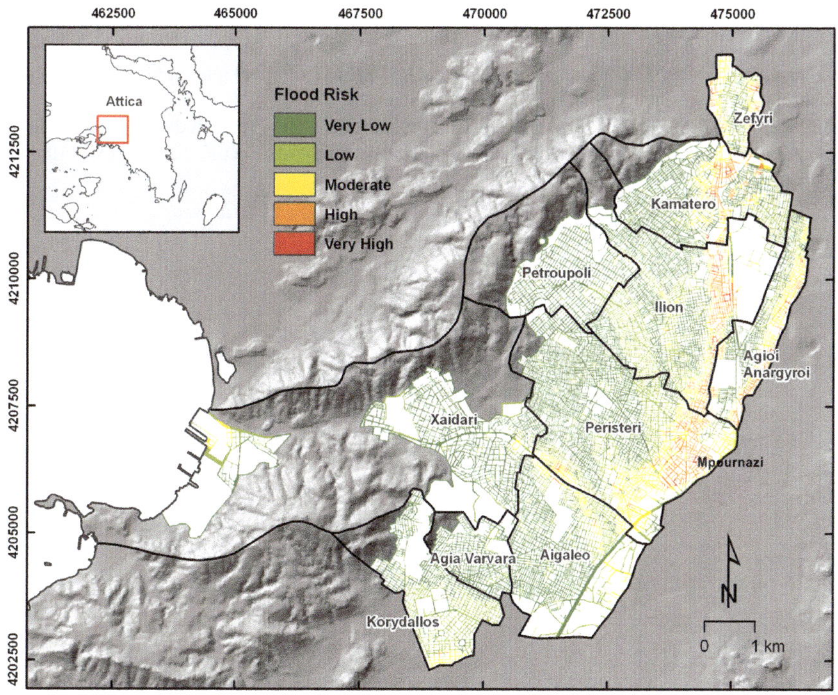

Fig. 5. Flood risk map of the study area. Higher risk values appear in Mpournazi, Ilion, Peristeri and Kamatero mainly in areas around Fleva torrent.

Acknowledgments This work was carried out within the framework of the Project "Emergency Operational Planning in Development Association of West Athens (ASDA) for Civil Protection and Natural and Environmental Hazards Defence" The authors wish to acknowledge ASDA for providing data necessary for the completion of this study.

References

Clark CO (1945) Storage and the unit hydrograph. Trans ASCE 110:1419–1446

Diakakis M (2010) Flood history analysis and its contribution to flood hazard assessment. The case of Marathonas, Greece. Bull Geol Soc Greece 43:1323–1334

Digital Newspapers Collection (2010) E-efimeris: Digital Newspapers Collection of the Greek National Library. Available at: http://www.nlg.gr/digitalnewspapers/ns/main.html Accessed 10 January 2011

EEA (2000) CORINE Land Cover Technical Guide – Addendum 2000. European Environment Agency

Gaitanakis P (1982) Geological map of Greece, "Athinai-Piraieus" sheet, scale 1:50.000, I.G.M.E. Publications

Gouliotis L (2002) Geological structure of Mt. Poikilo/Geological mapping in 1:5.000 scale. University of Athens, Department of Geology and Geoenviroment, Athens

GSCP (2007) Flooding in Attica. General Secretariat of Civil Protection, Ministry of Interior, Greece

Katsikatsos G, Mettos A, Vidakis M & Dounas A (1986) Geological map of Greece, "Athinai-Elefsis" sheet, scale 1:50.000, I.G.M.E. Publications

Maidment DR (2002) ArcHydro: GIS for water resources. ESRI Press, Redlands, California

Papanikolaou D, et al. (2002) Geological-Geotechnical study of Athens' Plain, OASP, Athens (in Greek)

Effects on flood hazard in Marathon plain from the 2009 wildfire in Attica, Greece

M. Diakakis

Department of Geology and Geoenvironment, National & Kapodistrian University of Athens, Greece. diakakism@geol.uoa.gr

Abstract In August 2009 a wildfire in northeast Attica, in Greece, burned approximately 187 square km of forest and agricultural land, inducing significant damages in properties and infrastructure. The fire extended in 19 hydrological basins affecting substantially the hydrological processes in the area and raising significant questions about changes in their flood risk regime. In this work, possible changes in peak discharge and the total runoff volume after intense rainfall, due to the wildfire effects in Charadros and Rapentosa catchments in Marathonas, Greece, are investigated. The results showed that peak discharge resulting from the 50-year return period design storm will be increased between 13.8% and 19.8% in Rapentosa and Charadros. Peak flow rate resulting from the 100-year design storm will increase between 11.4% and 16.7% in the two basins. Total runoff volume presents an increase of 9.5% to 16.7% in the two catchments depending on the storm scenario. Furthermore, it is shown that a new probability regime has been introduced, as discharges connected with storms with annual probability of 1%, before the fire, will now require storms of 1.6% annual probability of occurring in Charadros and 1.3% in Rapentosa.

1 Introduction

In 22nd of August 2009, a wildfire that started near Gramatiko village in northeast Attica moved southwards for 5 days burning forest and agricultural land, extending to approximately 187 sq km (Papanikolaou et al. in press). Apart from extended areas of invaluable forest, the fire damaged properties, infrastructure and agricultural production after by migrating into urban-wildland interface zones. In total, 19 hydrological basins were affected by the disaster, all with significant portions of natural vegetation burned, a fact that raised questions about the extent that their natural hydrological processes were affected.

Wildfire role in fluvial processes is well established. Shakesby and Doerr (2006) suggest that runoff portion of a storm increases in a wildfire-hit area due to changes in the chemical properties of surficial material. Hevley (1980) showed that concentration times are reduced in a basin after a wildfire, a conclusion supported also by Keith and Ward (1998) and Shakesby and Doerr (2006) who suggested that peak discharges also tend to increase. Cerda and Doerr (2005) notice an increase in total runoff volume immediately after a wildfire, but suggest that this trend might alter after several years due to land use changes and possible vegetation recovery. Moreover, vegetation removal reduces evaporotranpiration and ability of the vegetation canopy to withhold a portion of rainfall.

One of the methods that have been used successfully in quantifying these changes after a wildfire is the SCS-CN (SCS 1972; McLin et al 2001; Candela et al 2005).

In this work, we investigate the hydrological regime in two of these catchments, Rapentosa and Charadros, after the wildfire of 2009 burned 82% and 87% of their total area respectively. Their hydrology is investigated in terms of:

a) Possible changes in peak discharge values after intense rainfall events.
b) Possible changes in total runoff volume
c) Possible changes in annual flooding probability regime

For the purposes of this study 50- and 100-year return period storm scenarios have been used as rainfall inputs.

2 Study area

The two catchments studied are relatively small, drained by ephemeral torrents with little or no water at all for most of the year. The first one drained by Charadros torrent (60.2 km^2) is situated in Attica, approximately 25 km north east of Athens. In terms of geomorphology, the basin consists of a hilly area on the north and west draining into a low lying part. The main torrent reaches the sea, passing through a small valley where the village of Marathon and Bei settlement are situated and at the final part of its course through Marathon plain and Paralia coastal settlement. The second basin, drained by Rapentosa torrent (37.5 km^2), is situated also in the north east of Athens. The torrent network drains the north side of Penteli Mountain and reaches the sea through the famous for its battle Marathon plain, passing through small settlements (Vranas and Tymvos) and agricultural land. However, due to the construction of a dam in 2004 along the main torrent of Rapentosa, only the lower 11.23 km^2 of the basin are studied in this work, as the part upstream of the dam does not contribute to runoff. The two catchments, as shown in figure 1, share a drainage divide for a certain length.

In terms of geology, the area consists of a series of Mesozoic formations, accompanied by postalpine Cenozoic sediments. Upper Miocene brecciocongomerates and fluviolacustrine formations can be found on top of the Mesozoic series. Marathon plain is dominated by Holocene alluvial deposits accompanied by some talus cones (IGME 2002). Both Charadros and Rapentosa have flooded their banks numerous times in the past (Diakakis 2010).

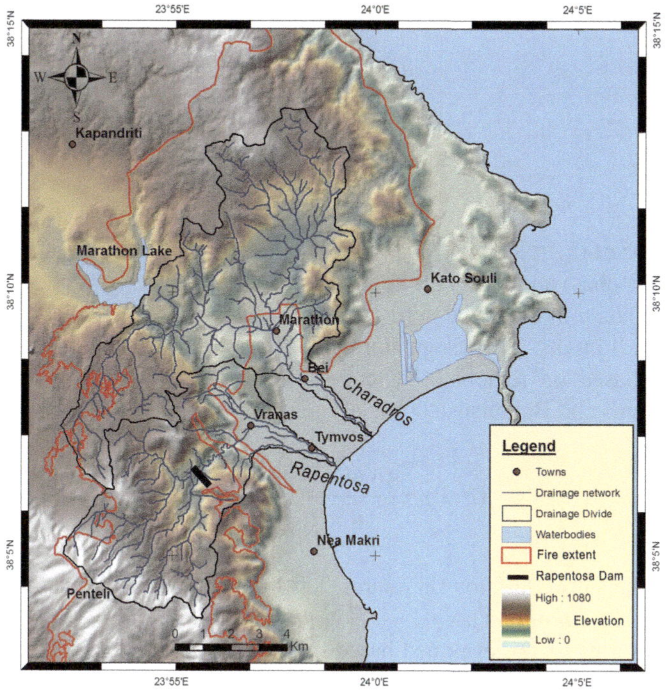

Fig. 1. Map of the north east part of Attica, showing the two catchments and the main towns and villages. Charadros is the northernmost basin. The red line denotes the 2009 fire extent.

3 Data and methods

The curve number (SCS-CN) method (SCS 1972, NRCS 2004) was used to define retention capabilities of different parts of the two catchments before and after the 2009 fire. Geologic formation data (IGME 2002) together with land use and vegetation data before (EEA 2000) and after the wildfire (Papanikolaou et al. in press) were used to define CN parameter in each basin before and after 2009. In the case of Charadros the average CN was found to be 65 and 69 before and after the fire. In the case of Rapentosa average CN was calculated at 65 and 68 before and after the fire.

Special retention (S) was calculated based on CN parameter distribution with the use of equation [1] (NRCS 2004). Subsequently, based on equation [2] of the SCS-CN method (NRCS 2004) runoff was calculated for every increment of the input rainfall in each basin.

$$CN = \frac{1000}{(10 + S)} \quad [1]$$

$$Q = \frac{(P - 0.2S)^2}{(P + 0.8S)} \quad [2]$$

where Q is runoff (in inches), P is rainfall (in inches), S is a special retention parameter (in inches) and CN is the "curve number" parameter, assigned to show infiltration capacity of every combination of soil and land use.

As rainfall inputs, 24h design storms of 50- and 100-year recurrence intervals were used based on regional IDF curves expressed by the following equation (Koutsoyiannis and Baloutsos 2000):

$$i(d,T) = \frac{40.6(T^{0.185} - 0.45)}{(d + 0.189)^{0.795}} \quad [3]$$

where d is the storm duration (in hours), T is the recurrence interval (in years) and I is the intensity (in mm/hr).

Design storms were developed based on the alternate blocks technique (Chow et al. 1988). Subsequently, in order to estimate peak discharge for each scenario, synthetic hydrographs were developed for both catchments with the use of:

a) SCS dimensionless hydrograph (SCS 1972)
b) Geomorphologic Unit Hydrograph (GUH) based on the time-area method (Clark 1945).

Additionally, in order to investigate possible changes, the total volume of runoff for each scenario before and after the fire, was calculated with the use of SCS-CN method (equation 2).

Finally, it was examined whether discharges connected previously with 50-and 100-year storms, will now be triggered by different recurrence interval storms.

4 Results

According to the analysis, both methods (SCS and GUH) of discharge estimation showed small differences in peak discharge values and anticipated change. (Tables 1 and 2). In the case of Charadros a mean peak discharge of 284 m^3/s was calculated for the pre-fire period and a mean peak discharge of 340.8 m^3/s for after the fire, indicating an average change of 19.8% for the 50-year storm scenario. The 100-year storm scenario, showed an average increase of 16.7% (Table 1).

Table 1. Peak discharge estimation based on SCS and GUH methods for 50- and 100-year design storms in Charadros catchment, before and after the 2009 fire.

Catchment	Return period	Estimation method	Peak Discharge (m^3/s)		Change (%)
			Before fire	After fire	
Charadros	50-years	SCS	280	332.7	+18.8 %
		GUH	289	348.9	+20.7 %
		Mean	284.5	340.8	+19.8 %
	100-years	SCS	369.3	428.3	+16.0 %
		GUH	366.3	429.9	+17.4 %
		Mean	367.8	429.1	+16.7 %

In the case of Rapentosa, in the 50-year storm scenario, mean change in peak discharge was estimated at 13.8% while in the 100-year storm, mean change was estimated at 11.4% (Table 2). Figure 2 presents differences in synthetic hydrographs before and after the fire for both basins and for both 50- and 100-year storm scenarios.

Table 2. Peak discharge estimation based on SCS and GUH methods for 50- and 100-year design storms in Rapentosa catchment, before and after the 2009 fire.

Catchment	Return period	Estimation method	Peak Discharge (m^3/s)		Change (%)
			Before fire	After fire	
Rapentosa	50-years	SCS	106.5	121.8	+14.4 %
		GUH	107.8	122.1	+13.3 %
		Mean	107.1	121.95	+13.8 %
	100-years	SCS	141.6	158.4	+11.9%
		GUH	142.1	157.6	+10.9 %
		Mean	141.85	158	+11.4 %

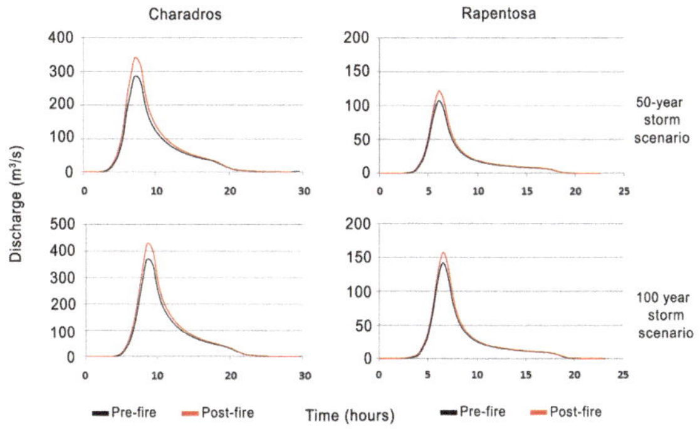

Fig. 2. Change in shape and peak discharge between before- and after-fire simulated hydrographs in Charadros (a) and Rapentosa (b) for 50- and 100-year floods respectively. Discharge values used for the compilation of these hydrographs come from averaging SCS and GUH hydrographs' respective discharge values.

As far as the total volume of runoff is concerned, assuming a uniformly distributed storm of 124 mm in 24 hours (50-year return period) and one of 148 mm in 24 hours (100-year return period) we can calculate the total volume of rainfall over a basin by multiplying with catchment area (Table 3). Taking into account the runoff volume estimates based on the SCS method different portions of rainfall are expected to runoff in each scenario indicating that infiltration capacity of both catchments is reduced after the 2009 fire (Table 3). Based on calculations of the SCS method, Charadros presented an increase of 12.9% in runoff volume induced by 50-years return period storm and a 9.5% after a 100-years storm. Rapentosa presented a 16.7% increase in runoff volume after a 50-years storm and an increase of 12.5% in 100-years storm (Fig. 3, Table 3).

Table 3. Calculations of total rainfall and runoff volumes in both catchments, for both 50- and 100-year storm scenarios, before and after the 2009 fire.

Catchment	Scenario	Basin Area	Total Volume ($\times 10^6$ m^3)	Runoff before fire ($\times 10^6$ m^3)	Runoff after fire ($\times 10^6$ m^3)
Charadros	124 mm	60.2 km^2	7.5	3.1	3.5
	148 mm		8.9	4.2	4.6
Rapentosa	124 mm	11.23 km^2	1.4	0.6	0.7
	148 mm		1.7	0.8	0.9

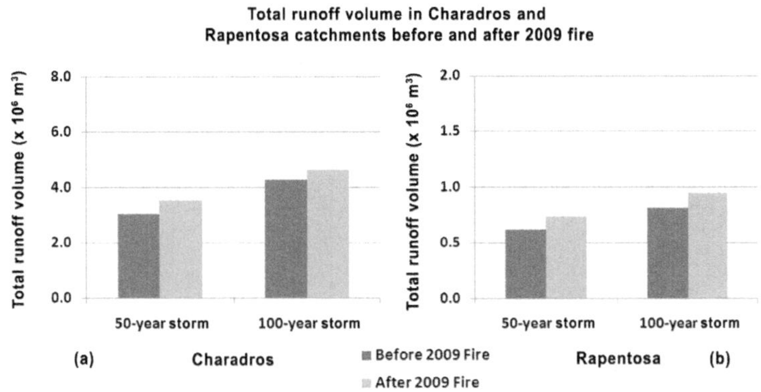

Fig. 3 a, b Total runoff volume change in Charadros (a) and Rapentosa (b) catchments before and after the wildfire of 2009, produced by 50- and 100-year recurrence interval storms.

As far as probability regime is concerned, the analysis showed substantial changes in this field too. In the case of Charadros, a peak discharge of 284.5 m^3/s would be the result of a storm of 2% annual probability (124 mm) before 2009. After the 2009 forest fire, on the basis of different CN parameters this peak discharge requires a storm of 111 mm which according to the regional IDF curves (equation [3]) has an annual probability of 3.2%. Similarly, a peak discharge of 367.8 m^3/s would be the result of a storm of 1% annual probability (148 mm) before 2009. After the 2009 fire, this peak discharge requires a storm of 132 mm which has an annual probability of 1.6%.

The same pattern develops in Rapentosa catchment as peak discharges connected with storms with annual probability of 2% and 1% before 2009, after the fire can be triggered by storms of 2.6% and 1.3% annual probability respectively (Fig. 4).

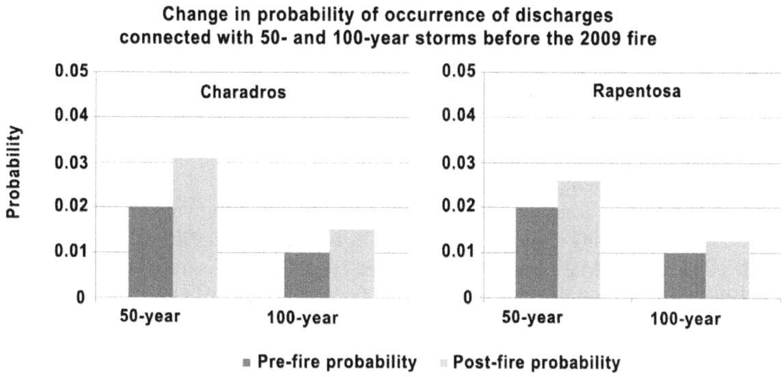

Fig. 4. Change in probability of occurrence of discharges connected previously with the 50- and 100-years storms. The darker columns denote the probability of storms connected with the above-mentioned discharges. The light-colored ones show the new probability (of the same discharge) after the 2009 fire.

5 Conclusions

The results showed that the 2009 wildfire in northeast Attica had a substantial impact on flood regime of the Marathon plain, affecting peak discharge, total runoff volume and probability regime in both Charadros and Rapentosa catchments.

In this work, by developing synthetic hydrographs with the use of SCS and geomorphologic unit hydrograph method, it was shown that peak discharges of 50 and 100-years return period storms will increase between 10 and 20% in both basins. In terms of total runoff volume, it was shown that an increase between 9.5% and 16.7% is expected.

In terms of probability, discharges caused by storms of 2% annual probability before 2009 will now experience an increase to 2.6% - 3.2%. Respectively, discharges connected with storms with 1% annual probability before 2009 will now experience an increase to 1.3-1.6%.

Generally, increases in 50-year discharges presented more extended changes than the 100-year ones, indicating that in more intense storms, changes in retention capability of a catchment due to a wildfire are less significant than in lower intensity ones.

Estimated measures of peak discharges, total runoff volumes and probability cannot be considered static as catchment hydrology is a dynamic natural system especially after a wildfire. Therefore, vegetation and man-made changes could gradually lead to modifications of retention capabilities of the two catchments. In addition, predictions of changes calculated in this paper are based on the assumption of a not changing climatic regime.

References

Candela A, Aronica G, Santono M (2005) Effects of forest fires on flood frequency curves in a Mediterranean catchment. Hydrol Sci J 50:193-206

Cerdà A, Doerr SH (2005) Long-term soil erosion changes under simulated rainfall for different vegetation types following a wildfire in eastern Spain Int J Wildland Fire 14:423- 437

Chow VT, Maidment DR, Mays LW (1988) Applied Hydrology. McGraw-Hill

Clark CO (1945) Storage and the unit hydrograph. Trans ASCE 110:1419-1446

Diakakis M (2010) Flood history analysis and its contribution to flood hazard assessment. The case of Marathonas in Greece. Bull Geol Soc of Greece 43(3):1323-1334

EEA (2000) Corine land cover. European Environment Agency, Commission of the European Communities, http://www.eea.europa.eu/publications/COR0-landcover Accessed: 15 April 2010

Helvey JD (1980) Effects of a north central Washington wildfire on runoff and sediment production. Water Resour Bull 16:627–634

IGME (2002) Kifissia sheet. Geological map of Greece series. Sheet 208, 1:50000, Athens, Institute of Geology and Mineral Exploration

Keith S, Ward R (1998) Floods: physical processes and human impacts. Chichester: J. Wiley & Sons

Koutsoyiannis D, Baloutsos G (2000) Analysis of a long record of annual maximum rainfall in Athens, Greece and design rainfall inferences. Nat Haz 29:29-48

McLin SG, Eeckhout MEV, Springer EP, Lane LJ (2001) A characterization methodology for post-wildfire flood hazard assessment. International Congress on Modelling and simulation MODSIM 2001, 10-13 December 2001

NRCS (2004) National Engineering Handbook, Section 4 Hydrology, Soil Conservation Service Washington US

Papanikolaou D, Diakakis M, Aggelopoulos A (in press) Geoenvironmental impact after the 2009 wildfires in northeastern Attica, Greece. Proc of Forest Fires in East Attica, Policies and Remediation Measures Meeting, Geotechnical Chamber of Greece, December 8, 2009, Athens, Greece

SCS (1972) National Engineering Handbook, Section 4 Hydrology, Soil Conservation Service Washington US

Shakesby R, Doerr SH (2006) Wildfire as a hydrological and geomorphological agent. Earth-Sci Rev 74:269- 307

Flash flood event of Potamoula, Greece: Hydrology, geomorphic effects and damage characteristics

M. Diakakis, E. Andreadakis, I. Fountoulis

Department of Geology and Geoenvironment, National & Kapodistrian University of Athens, Greece. diakakism@uoa.geol.gr

Abstract The geoenvironmental setting is a determinative factor in catchments' response during heavy precipitation events. In this work, the flash flood of 2008 in Potamoula (Aetoloakarnania, Greece) is investigated in terms of hydrologic and geomorphologic features. The study area is a relatively small, partly forested, rural catchment situated in western Greece with steep mountainous topography and with flysch being the prevailing lithology. In 2nd of October 2008 a storm of high intensity (measured up to 280 mm in 24 hours) produced noteworthy quantities of runoff which inundated the lower parts of the small valley, inflicting damage to a significant number of structures and killing two people. Field investigations were carried out to record the extent and characteristics of damage, the physical attributes of flooding such as the peak discharge, the geomorphic effects and the geological factors affecting the local hydrology. The results were investigated in comparison with geomorphological and geological evidence showing that this event, although extreme, corresponded very well to the geological record of the area. Finally, runoff response of the catchment was assessed with respect to the geology of the basin.

1 Introduction

Platanorema is a relatively small catchment (28 km^2), and a part of the drainage network of Acheloos River. It is located in Aetoloakarnania in western Greece, 216 km northwest of Athens and 15 km north of Agrinio. It is inhabited by approximately 830 people with Potamoula being the largest village (Fig. 1). Platanorema stream network drains a hydrological basin of the 5th order. The drainage network has an asymmetric dendritic form. High incision rates are obvious along the streams, along both the bedrock and the alluvial deposits. Along the main riverbed, there are well developed alluvial terraces, mainly in the southwest part of the basin. Due to the steep forested slopes, towns and villages in this and numerous other catchments are settled near the main rivers, in flat areas dominated by river deposits.

Platanorema has suffered floods in the past. On the basis of eyewitness accounts and web sources, one can identify at least three significant events in 1960, in 1961 and in the autumn of 1999. However, the 2008 flood is considered to be the most damaging.

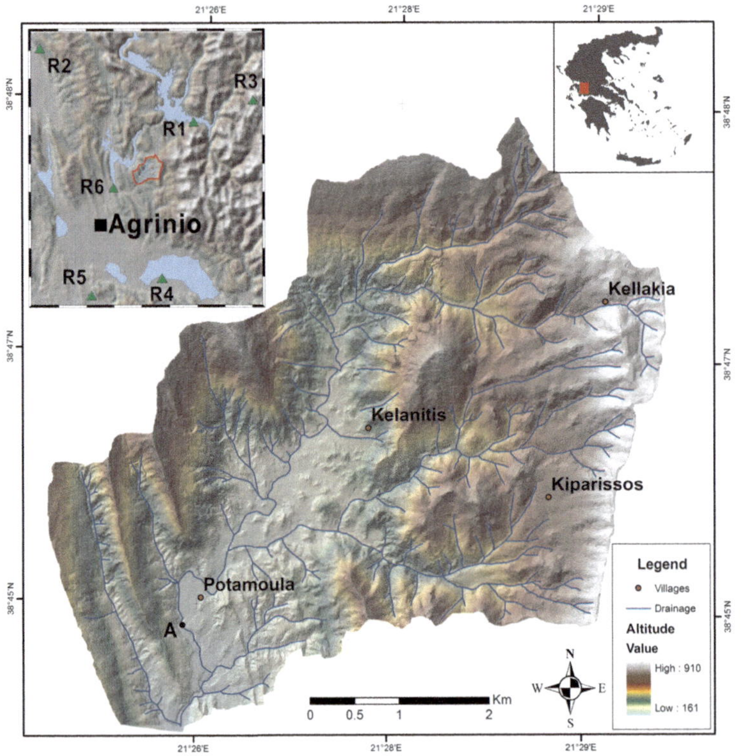

Fig. 1. Map of the catchment under study. In the upper left corner, there is a location map illustrating the rain gauge network in the area. Location A represents the peak discharge reconstruction location.

The impressive intensity of the event posed a series of questions, considering the nature of the phenomena, and the role of the contributing factors. Goal of this work is to investigate the following issues:

- the peak discharge during the event
- the damage characteristics and distribution
- the geomorphic effects

In terms of geology, the upper Eocene - Oligocene flysch is the prevailing geological basement in the area. According to IGME (1998), the dominant lithologies of the catchment are:

- Sandstones: thin to medium bedded, alternating with fine to coarse-grained sandstones and thin layers of shales, with occasional occurrence of conglomerates.
- Shales: grey to greenish, bearing alternating layers of pelites, marly silts, silty sandstones as well as fine-grained sandstones in bands of 0.20-0.50m.

Most of the catchment area consists of the shales series (~70%), while the main sandstone occurrences outcrop along the drainage divide or near it. The tectonic deformation of the area is rather intense; with medium to large scale folds with axes trending NW-SE to N-S and plunging south, together with extensive strike-slip or oblique-slip right-lateral faults trending NE-SW.

2 Methodology

A post flood field investigation in the area was carried out to measure the channel geometric characteristics (Fig. 2) and examine the peak water level indicators. Based on these measurements, peak discharge was computed at location A (Fig 1) with the use of Manning's formula (Manning 1891, Flamant 1891). Location A was chosen as one of the few locations that cross section and water level was clearly measurable.

$$Q = \frac{A}{n} R^{2/3} S^{1/2}$$

where: n is the roughness coefficient, R=A/P is the hydraulic radius, P is the wetted perimeter, A is the cross-sectional area , S is the channel slope at the location. Maximum water surface elevation was measured in the field using vegetation debris and driftwood. For calculation purposes water flow in this particular occasion was assumed to be uniform.

In order to calculate Manning roughness coefficient necessary for the peak discharge computation, vegetation and flow type were also investigated along the main river in accordance with Cowan's methodology (Cowan 1956).

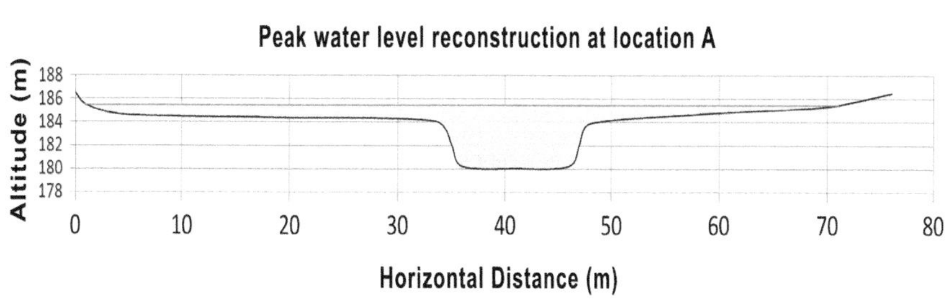

Fig. 2. Cross sectional area at location A, where peak flow discharge was calculated.

According to Cowan (1956) the mean value of "n" can be calculated through the following equation:

$$n = (n_b + n_1 + n_2 + n_3 + n_4) \times m$$

where: n_b, is a base value of n for a straight, uniform, smooth channel in natural materials, n_1 is a correction factor for the effect of surface irregularities, n_2 is a value for variations in shape and size of the channel cross section, n_3 is a value for obstructions, n_4 is a value for vegetation and flow conditions, m is a correction factor for meandering of the channel. In this study for the specific cross section (fig. 2) the following values where calculated: $n_b = 0.04$, $n_1=0.003$ $n_2=0$, $n_3=0.032$, $n_4= 0.04$, $m = 1.5$. Based on these values and Cowan equation Manning coefficient value is calculated at 0.17 (Table 1).

Moreover, rainfall data from 6 gauges (Fig. 1) were examined in order to study the spatial characteristics of the storm that caused the event. This examination showed a total rainfall of 201mm across the small basin (although one of the rain-gauges recorded 280mm in 24 hours). Based on the storm record, calculated peak flow rate was also cross-checked with the application of SCS-CN rainfall-runoff method (SCS 1972) to assure consistency of the result. Subsequently peak discharge in this catchment was compared to peak flow rates of other catchments in Greece during flood phenomena, in order to assess the role of flysch formation and its hydrological response. In addition, the geomorphic effects of the flood were documented along with geological observations on the older river flood deposits in the area. For this purpose, measurements were carried out concerning the grain size of the 2008 and older flood events' sediments.

Furthermore, a field survey was carried out to record the type of damage in structures and their spatial distribution. Finally, damages were projected against older flood deposits to assess their spatial correlation. This step was carried out to examine whether the 2008 event was predictable on the basis of geomorphology in terms of extent and damage pattern.

3 Results

Peak discharge was computed at location A at 94.5 m^3/s with the use of Manning's formula as shown in Table 1.

Table 1. Calculation of peak flow rate at location A, based on Manning's formula.

Parameter	A - Cross sectional area (m^2)	P - Wetted Perimeter (m)	S – Slope (m/m)	n – Manning coefficient	Q - Peak flow (m^3/sec)
Value	95	72	0.0198	0.17	94.5

Based on the application of SCS-CN method, peak flow rate was computed at 87.8 m^3/s (Fig. 3) showing good correlation with the Manning formula results. The application of the SCS-CN method showed 115mm of runoff out of 201mm storm over the basin. The two results give a mean value for peak discharge of approximately 91m^3/s which results in peak discharge per unit area of 3.65 $m^3/s/km^2$. This value was found similar compared with peak discharge calculated in other catchments in the region during notable storms (e.g. 201mm storm in Xerias catchment in 1997 flood event, Baloutsos et al. 2000, Gaume et al. 2009).

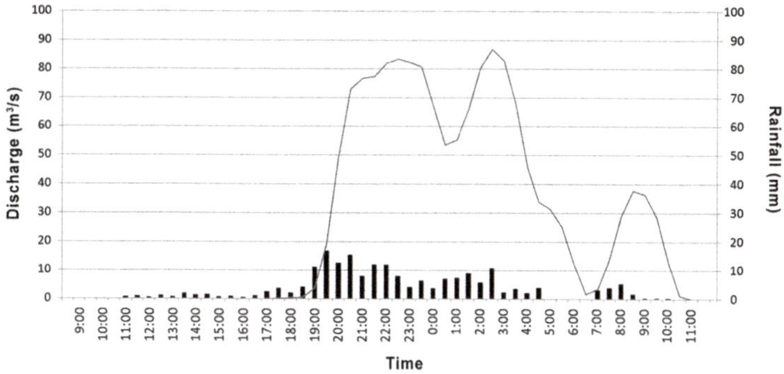

Fig. 3. Storm rainfall across the catchment during the 2nd and the 3rd of October 2008 and associated flow rate at location A, calculated through the SCS-CN method.

Concerning spatial distribution of damages when projected against older flood deposits they showed good correlation especially with the limits of the active floodplain (Fig. 4). As far as the damage characteristics are concerned it was found that 72 structures were partly or completely damaged. Their vast majority was situated on the active floodplain or older alluvial terraces. The following table illustrates the portion of the structures situated on terraces (Table 2).

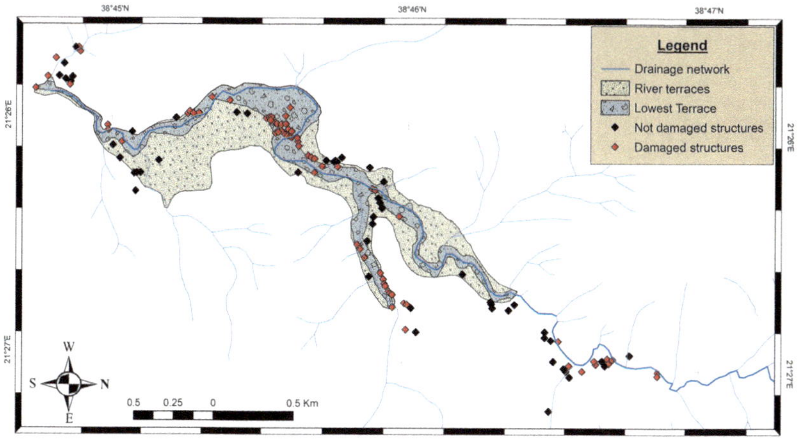

Fig. 4. Damages spatial distribution in comparison with the extent of river terraces and specifically the lower terrace (active floodplain).

Table 2. Percentage of damaged structures against the respective geologic units.

Area / Zone under consideration	Total structures in considered area/zone	Damaged structures in considered area/zone	Percentage
Alluvial terraces	76	57	75 %
Lowest terrace	54	50	92.6%
Whole Catchment	617	72	11.7 %

As far as the type of damage is concerned, a very critical factor on the structural performance of the building was found to be the type of foundation. Buildings without deep foundations were more severely damaged because flood water eroded the ground material under their basal part (Fig. 5a, b). Buildings with deep foundations presented a better structural performance, resulting in damages only in ground floor masonry and utensils.

With respect to the geomorphic effects of the event alluvial sediments were deposited on the lower terrace mainly in the form of cobbles and boulders in places with high water velocity (Fig. 5c). Measurements that were carried out showed

Fig. 5. Structure with superficial foundations destroyed after erosion of the ground material below their basal part (a). Illustration of the erosion mechanism (b). Cobbles deposited on the lower terrace during the flood in areas with high water velocity (c) and boulders deposited in the lower terrace during older flood episodes (d).

that most of the cobbles were lower than 30cm but a portion of them measured up to 60cm on the terrace at Potamoula settlement, indicating water velocities between 0.8 – 2.8 m/s according to the Hjulström curve (Novak 1973). These values show good correlation with the value extracted from the flow rate calculations in location A, further downstream, which showed a mean velocity of about *1m/s*. In addition, measurements of the grain size of older flood deposits presented boulder sizes of approximately 1m, showing even higher energy deposition environment. On the contrary, in areas with lower flow speed fine grain sediments were deposited in various thicknesses (up to 30 cm in places) (Fig. 6a). During the event erosion processes were intensified leading to enhancement of transportation of sediment material and minor failures on the river sides (Fig. 6b).

Fig. 6. a Fine grain sediments deposited in areas with low flow velocity (Potamoula terrace), during the flood event of October 2008. **(b)** One of the locations where the event enhanced erosion and processes and sediment was supplied in the river.

4 Discussion and conclusions

Results showed that the portion of runoff during the storm reached 57% of the total rainfall (115mm out of 202mm total). This value is similar to the 56,9% of runoff discharged during a 201mm storm in Xerias river in Corinth in 1997 (Baloutsos et al 2000). However, it is important that the two catchments have substantial differences in geology and landuse, a fact that indicates that in high intensity storms permeability is not a dominant factor in runoff.

Damages presented a very good correlation with the spatial extent of the lower terrace proving that the study of geomorphology is a useful tool when studying flood hazard. The examination of the type of damage showed that structural standards are very important in buildings situated in high hazard areas and that based on this geomorphic criterion a long term policy of removal of infrastructure from flood zones has to be adopted.

Geologic material eroded from specific locations was deposited (in various grain sizes) continuing the process of terrace-building. Study of the geomorphic effects in general showed that this event although extreme, followed the pattern of fluvial geomorphic processes imprinted in the geology and the geomorphology of the catchment, proving that it was nothing but a normal part of this processes in acceleration.

References

Baloutsos G, Koutsoyiannis D, Economou A, Kalliris P (2000) Investigation of the hydrologic reponse of the Xerias torrent watershed to the rainstorm of January 1997 using the SCS method. Geotechnical Scientific Issues 11:77-90

Cowan WL (1956) Estimating hydraulic roughness coefficients. Agricultural Engineering, 37:473-475

Flamant AA (1891) Mécanique appliquée – Hydraulique. Baudry et Cie, coll. Encyclopédie des travaux publics, Paris, (réimpr. 1900), 686 p

Gaume E, et al (2009) A compilation of data on European flash floods. Journal of Hydrology, 367 (1–2): 70–78

IGME (1998) Geological Map of Greece 1:50.000, Alevras Sheet. Institute of Geology and Mineral Exploration

Manning R (1891). On the flow of water in open channels and pipes. Inst. Civ. Eng. Ireland Trans. 20, pp. 161–207

Novak ID (1973) Predicting coarse sediment transport: The Hjulström curve revisited. In "Fluvial Geomorphology" (M. Morisawa, ed.), Publ. Geomorphol., pp. 13-25. State University of New York, Binghamton

SCS (1972) National Engineering Handbook, Section 4 Hydrology, Soil Conservation Service Washington US

Hydrograph analysis of Inountas River Basin (Lakonia, Greece)

C. Gamvroudis[1], N. Karalemas[1], V. Papadoulakis[2], O. Tzoraki[1], N.P. Nikolaidis[1]

[1]Technical University of Crete (TUC), Department of Environmental Engineering, Chania, Greece, christos.gamvroudis@gmail.com [2] v.papadoulakis@lakonia.gr

[2] Department of Environment and Hydrology, Region of Peloponnesus, Regional Unit of Lakonia, Sparta, Greece

Abstract Three approaches have been used to analyze the hydrograph of Inountas River basin in two reaches. Flow in each reach was estimated by combining the water level records with monthly field flow measurements. Base-flow separation has been accomplished by using the SWAT (Soil and Water Assessment Tool) baseflow filter program. The fraction of water yield contributed by baseflow was estimated between 0.76-0.62 for Kladas reach and 0.79-0.70 for Vassaras reach respectively. Flow duration curve for each reach was constructed from daily flow data. The flow-probability relationship, Q_{50} for Vassaras reach had a value of 0.164 and for Kladas 0.198. The ratio Q_{90}/Q_{50} for Vassaras reach was 0.689 indicating that 68.9% of Vassaras stream flow was contributed from groundwater storage. The calculated recession factors were for Kladas reach 0.024 (10/04/2010-04/06/2010) and for Vassaras reach 0.009 (20/03/2010-27/05/2010) and 0.005 (09/07/2010-11/10/2010).

1 Introduction

The hydrologic basin of Inountas River (or Kelefina River), a tributary of Evrotas River is located at the south-eastern portion of Peloponnesus, Greece (Fig. 1). Inountas River has a watershed area of 343.5 km^2 and is a complex hydrologic and hydrogeological system consisting of temporary tributaries, high mountainous areas and springs which are the main contributors to base-flow.

River hydrograph analysis, specifically separating base-flow (long-term discharge from natural storages) and quick flow (direct response to rainfall event) is important to estimate the groundwater discharge. Many techniques have been used to analyze base-flow from stream hydrographs since the last decades. In general, these methods can be categorized to three approaches, baseflow separation, frequency analysis and recession analysis.

Baseflow separation can be accomplished either using filtering procedures or graphical techniques. Data processing or filtering techniques provide an easily automated method indicating the baseflow response of the watershed (Brodie and Hostetler 2005).

Frequency analysis indicates the relationship between the magnitude and frequency of stream flow discharges from the hydrographic record. Flow duration curve (FDC) is one of the most informative methods of displaying the complete range of river discharges from low flows to flood events and it is a relationship between any given discharge value and the percentage of time that this discharge is equaled or exceeded (Smakhtin 2001).

During the dry weather periods, water is gradually removed from the catchment by evapotranspiration and groundwater/soil water discharge into a stream. Stream flow discharge depletion during these periods is known as «recession», and is reflected in stream flow hydrograph by a recession curve. Recession curve is the specific part of the hydrograph after the crest (and the rainfall event) where stream flow diminishes. Recession period lasts until stream flow begins to increase again due to rainfall (Smakhtin 2001). Recession segments can be analyzed to define the discharge processes that make up baseflow. The equation mostly used considering linear storage is (Maillet 1905) $Q_t=Q_o \times e^{-\alpha t}$ where Q_o is the initial flow at the start of the recession segment, Qt is the flow at time t and α is a constant known as the cut-off frequency. The term $e^{-\alpha}$ can be replaced by k, called the recession constant which is commonly used as an indicator of the extent of baseflow.

2 Study area

The study area includes the hydrological basin of Inountas River one of the largest tributaries of Evrotas River (Fig 1a). The geologic formation of the basin is comprised mainly of alpine formations of four geotectonic units. The Limestones of Mani (or Plattenkalk) Unit represent the indigenous unit of the area of study. At the base of the stratigraphic column of Mani Unit, is the Phyllitic Basement Rocks. Over the phyllitic unit lies the carbonate unit, consisting of marble and crystalline limestone with alternations of siliceous schist. The column ends with a transition to the slightly metamorphosed flysch. In the hydrologic basin of Inountas River, only parts of the carbonate formations of the unit are found, at its southern and eastern part. Phyllite - Quartzite (or Arna) Unit, which overlies Mani Unit, consists of metamorphic rocks, characterized by strong deformation. The rocks of the unit consist of schist, micaceous schist and quartzite, meta-conglomerates, mafic and ultramafic rocks. An extended area of the basin is covered by this unit. The main branch of Inountas River flows mostly on the geological formations of Phyllites - Quartzite. Tripoli Unit lies over the Phyllite - Quartzite Unit. The stratigraphic column starts with a slightly metamorphic formation, known as Tyros Beds, which occupy a small surface in the eastern part of the basin.

Over Tyros Beds lays a carbonate sequence, the main feature of which is the continuous neritic sedimentation with limestone and dolomites. This part of the unit covers a significant area of the basin; however, in most places it is characterized by small thickness. Above the carbonate sequence, lays the flysch of the unit, which has limited appearances in the central and southern sections of the basin. Pindos Unit overlies all the aforementioned units. In the area of study, the upper part of this unit, which consists of platty limestone with clay alternations, is found as remaining nappes. Above the alpine formations post-alpine formations of negligible extent are found at several places. All the above lead to the conclusion that the major water tables in the basin are three and they are hosted within the karstic aquifers. Water tables located within the limestone of Pindos Unit are not particularly important, due to the limited extent of the formations, and the fact that they do not extend out of the hydrologic basin of Inountas River. The other two aquifers are capable of altering the hydrological balance of the basin, because they cover a wide area, and extend outside the basin.

The impermeable bedrock of the marbles of Mani Unit, do not participate in the area of study, so it is almost certain that the amount of water infiltrating in these formations will flow outside the basin. This view is supported by the fact that there are no springs that discharge these rocks. In contrast, the impermeable bedrock of the limestone of Tripoli Unit is Phyllite - Quartzite Unit. As a result, part of this limestone is discharged by springs within the basin (Springs Polydroso, Vamvakou, etc.). There is also a large part of the limestone of Tripoli Unit that is not discharged by springs. The most extended limestone area, where this phenomenon is observed, is located in the southern part of the basin, between the villages Vresthena, Chrisapha and Agriani, despite the fact that it is surrounded mostly by Phyllite - Quartzite. The southern part of this limestone area, north-east of Chrysapha, is an exception as limestone and a carbon mass that lies further to south are joined. Map data and fieldwork showed that the contact of limestone with their bedrock takes place at a high altitude, which prevents the flow of karst water to the south. However, this does not occur north-west of Agriani. In this case, even though the limestone of Tripoli Unit is in contact with the marbles that lay towards the East at high altitudes, the Phyllite - Quartzite Unit between the limestone and the marbles has disappeared. Therefore, there are great chances that the same phenomenon is happening beneath the mass of limestone of Tripoli Unit, resulting in the drainage of the karst water within them, to the underlying marbles, through which it exits the hydrological basin (Fig 1b). The most important parameter is that the water infiltrating in the carbon mass is not only water that comes directly from precipitation. The largest tributary of Inountas River, which is called Sophronis River, flows above these rocks. Sophronis River subbasin covers 141.5 km^2 of the entire hydrological basin that extends to an area of 343.5 km^2. Although Sofronis River drains a large area, it has no flow during the longest part of the year. Flow takes place only after heavy rainfall and only for extremely limited time during wet seasons. Moreover, Sofronis River is fed by several springs, the most important being those in Polydroso. Therefore, the largest amount of water in the hydrographic network of Sofronis River, due either to springs or to runoff, infiltrates into the carbonate rocks or is driven outside the hydrologic basin.

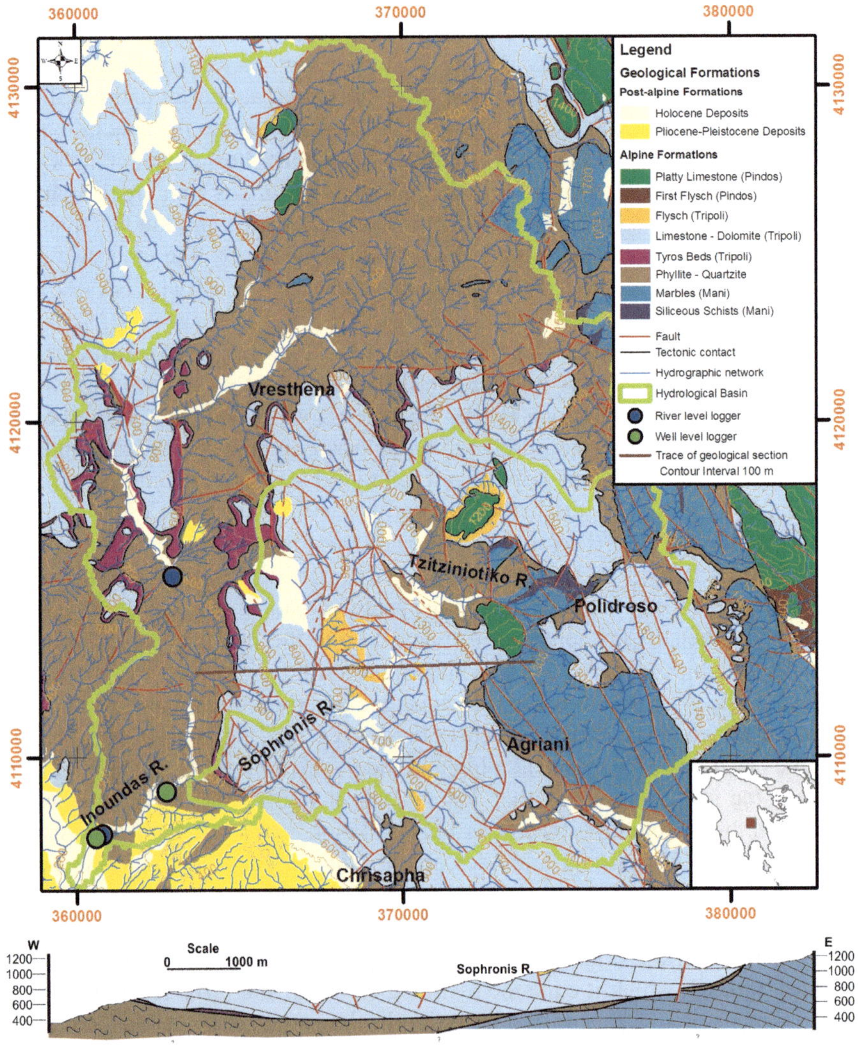

Fig. 1 a. Geological map of the study area (Psonis 1990, Special Environmental study of Parnonas Mountain – Moustos wetland 1999), modified, **b.** Geological cross-section.

3 Methodology

For the purposes of this study, one level logger was installed in the upstream (Vassaras reach) and one in the downstream part (Kladas reach) of Inountas River measuring the level of water table every 15 min. Flow was measured on monthly basis using a flow current meter (OSS PC1).

Flow in each reach was estimated (Fig. 2 a,b) by combining the water level records with the monthly field flow measurements. Rating curves were created using the open-channel flow equation, in which the flow and channel parameters are constant with respect to distance and time, and the slope of the water surface is equal to the energy gradient line. The semi-empirical Manning flow equation for one dimensional uniform steady flow is used for the simulation of water flow. The most common relationship between stage and flow is (USGS 1982; Maidment 1993) $Q=a\times(H-H_o)^b$ where Q is the flow (m^3/s), H is the stream stage (m), a and b are constants, and H_0 is the stage at which the discharge is zero. This type of relationship is established by fitting a curve to the measured data.

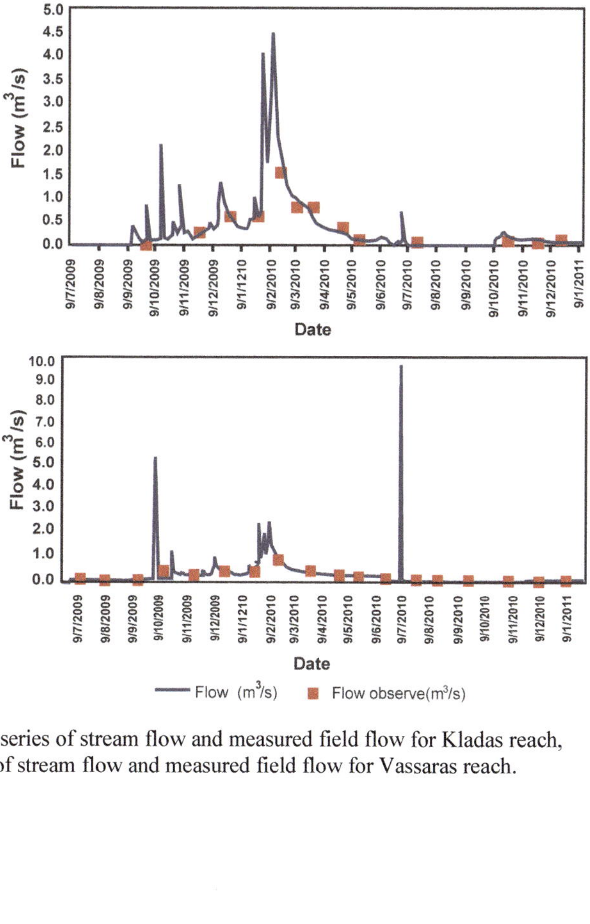

Fig. 2. a. Time series of stream flow and measured field flow for Kladas reach, **b.** Time series of stream flow and measured field flow for Vassaras reach.

The base-flow has been separated from daily stream flow time-series using the SWAT (Soil and Water Assessment Tool) baseflow filter program according to the methodology outlined by Arnold and Allen (1999) and Arnold et al. (1995) which is a modification of the recession curve displacement method. Results are presented in Figure 3 a, b.

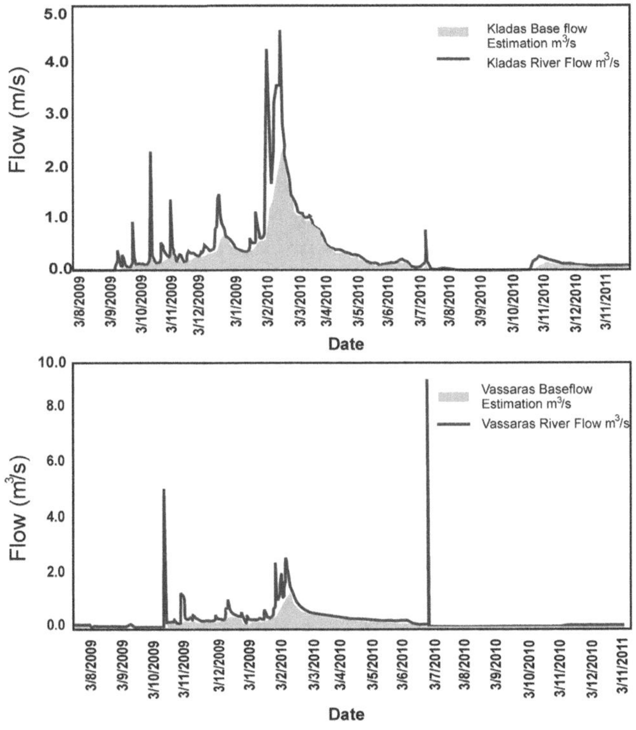

Fig 3. a. Time series of stream flow and estimated baseflow for Kladas reach, **b.** Time series of stream flow and estimated baseflow for Vassaras reach.

A flow duration curve for each reach is constructed from daily flow data. Probability P that a given flow will be equaled or exceeded is defined by $P=100\times(m/n+1)$ where m is a unique ranking number to each flow starting from 1 for the maximum flow to n for the minimum flow and n is the number of flow measurements. The flow-probability relationship is presented as a log-normal plot in Figure 4. The median flow (Q_{50}) which is the flow which is equal or exceeds 50% of the time has been calculated. Base-flow is significant if this part of the curve has a low slope, as this reflects continuous discharge to the stream. Also the ratio of the flow which is equal or exceeds 90% of the time, to that of 50% of the time (Q_{90}/Q_{50}) has been calculated indicating the proportion of stream flow contributed from groundwater storage (Nathan and McMahon 1990).

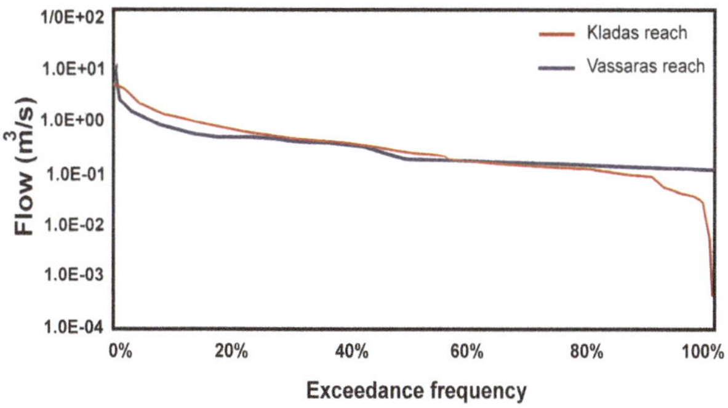

Fig. 4. Flow duration curve for Vassaras and Kladas reaches.

Recession segments for each reach were selected from the hydrograph and were analyzed to gain an understanding of the flow processes that make up base-flow. Hydrograph for this segment was then plotted on semi logarithmic scale.

4 Results

Rating curves have been calculated for two periods for each reach (due to change of location of Kladas recorder in June 2009 and repair of Vassaras recorder in June 2010). The rating curve was calculated as $Q=2.04\times H^{1.29}$, $R^2=0.95$ before June 2009 and $Q=2.04\times H^{1.29}$, $R^2=0.95$ after that period for Kladas reach. For Vassaras reach was $Q=6.36\times H^{5.50}$, $R^2=0.69$ before June 2010 and $Q=3.79\times H^{3.86}$, $R^2=0.97$ after that period. Kladas (downstream) water quantity was 16.2×10^6 m^3 for 04/08/2009-11/01/2011 and Vassaras (upstream) quantity 15.8×10^6 m^3 for the same period.

The fraction of water yield contributed by base-flow fall between 0.76-0.62 for Kladas reach and 0.79-0.70 respectively for Vassaras reach.

For Vassaras reach, the Q_{50} was 0.164 and for Kladas 0.198. The part of the curve with flows below the median flow represents low-flow conditions. The ratio Q_{90}/Q_{50} for Vassaras reach is 0.689 indicating that 68.9% of Vassaras streamflow is contributed from groundwater storage. On the other hand Q_{90}/Q_{50} is 0.363 for Kladas reach. In general, the steep flow for low-flows suggests relatively small contributions from natural storages like groundwater (Brodie and Hostetler 2005). In our case this steep flow is result of human activities such as direct river abstractions for agricultural purposes.

From the hydrograph of Kladas reach, can be observed that the recession curve is from 10/04/2010 until 04/06/2010 then is interrupted with rainfall an then the river dries. In Vassaras reach the recession curve is from 20/03/2010 until 27/05/2010 and then appears again from 09/07/2010 until 11/10/2010. The recession curve equation is $Q_t=0.389\times e^{-0.024t}$, $R^2=0.85$ for Kladas and $Q_t=0.535\times e^{-0.009t}$, $R^2=0.95$ and $Q_t=0.153\times e^{-0.005t}$, $R^2=0.86$ for Vassaras. The cut-off frequency α is

0.024 for Kladas (10/04/2010-04/06/2010) and 0.009 (20/03/2010-27/05/2010) and 0.005 (09/07/2010-11/10/2010) for Vassaras.

5 Conclusion

This work presented an analysis of Inountas River hydrograph. The Sofronis sub-basin comprises 40% of the area and according to the hydrogeological conditions a large fraction of Sofronis River flow infiltrates into the carbonate rocks and outside the hydrologic basin. Sofronis River contributes only to the downstream part (Kladas) of the Inountas basin as confirmed from the baseflow separation analysis.

The fraction of water yield contributed by base-flow to Kladas reach is 69% and to Vassaras reach 75%. The ratio Q_{90}/Q_{50} for Vassaras is 68.9% (groundwater storage from karstic springs) while this ratio for Kladas is 36.3% due to human activities such as direct river abstractions.

Acknowledgments Funding for this work was provided by the European Community's Seventh Framework Programme (FP7/2007-2011) under grant agreement 211732 (MIRAGE project).

References

Arnold JG, Allen PM (1999) Automated methods for estimating baseflow and ground water recharge from streamflow records. Journal of the American Water Resources Association 35(2): 411-424

Arnold JG, Allen PM, Muttiah R, Bernhardt G (1995) Automated baseflow separation and recession analysis techniques. Ground Water 33(6): 1010-1018

Brodie RS, Hostetler S (2005) A review of techniques for analysing baseflow from stream hydrographs. Managing Connected Water Resources Project, Bureau of Rural Sciences, ABARE, the Australian National University

Maidment DR (1993) Handbook of Hydrology. McGraw-Hill, New York

Maillet E (1905) Essais d'hydraulique souterraine et fluviale. Librairie Sci. Hermann Paris

Nathan RJ, McMahon TA (1990) Evaluation of Automated Techniques for Base Flow and Recession Analyses. Water Resources Research Vol. 26, No. 7, pp 1465-1473, July 1990

Psonis C (1990) Geological map of Greece, scale 1:50000, Sparti Sheet. Inst. Geol. and Miner. Explor., Athens

Smakhtin VY (2001) Low flow hydrology: a review. Journal of Hydrology 240:147-186

Special Environmental study of Parnonas Mountain – Moustos wetland (1999), Ministry for the Environment, Planning and Public Works, Athens

USGS (1982) Measurement and computation of streamflow: volume 2. Computation of discharge. US Geological Survey Water-Supply Paper 2175. US Government Printing Office

Hydrologic modelling of a complex hydrogeologic basin: Evrotas River Basin

O. Tzoraki[1], V. Papadoulakis[2], A. Christodoulou[1], E. Vozinaki[1], N. Karalemas[1], C. Gamvroudis[1], N.P. Nikolaidis[1]

[1]Technical University of Crete (TUC), Department of Environmental Engineering, 73100, Chania, Greece
[2]Department of Environment and Hydrology, Region of Peloponnesus, Regional Unit of Lakonia, Sparta, Greece

[1] ourania.tzoraki@enveng.tuc.gr, [2]v.papadoulakis@lakonia.gr

Abstract Climate change is expected to affect mostly water resources in arid and semi-arid areas, imposing in this way significant constraints in satisfying water demands in these regions. Under these conditions, water resources management should focus in increasing the efficiency of water uses by minimizing water losses and improving on the water balance accounting of the watersheds. Evrotas River Basin is a typical case of a water resource that has been under intense human pressure due to extensive water abstractions. Pumping from numerous wells (approximately 3000 private wells) and surface water abstractions impose significant uncertainty in water balance estimation and a new approach was developed to constrain the system and improve on the water balance assessment. The ETD (Enhanced Trickle Down model), a physically-based watershed model was coupled to a karstic model and were used to simulate the hydrologic response of the watershed. Ten years of data were used to simulate the hydrologic regime of the watershed. Losses from the karst due to evapotranspiration and infiltration to the deeper aquifer were estimated to be 45% of its annual volume. Karst is recharged annually by a water volume of 306Mm^3 and only 24% (75Mm^3) recharges the river as karst baseflow. The annual rainfall volume (for the non-karstic area of the watershed) was estimated to be 629Mm^3, the evapotranspiration was 655Mm^3, stream discharge 152Mm^3 and stream abstraction and irrigation were 77Mm^3. On the average, the amount of irrigation (77Mm^3) corresponded to 900 mm (per year for agricultural areas) which is almost twice the recommended irrigation levels. The modelling approach presented in this work can be used to constrain the uncertainty in the hydrologic budget of basins with complex hydrogeomorphology.

1 Introduction

The effects of climate change are expected to be more profound in arid and semi-arid regions of the world like the Mediterranean according to the IPCC scenarios. Rigorous watershed management plans are required to be developed in order to address the decline in water supply as well as comply with relevant legislation such as the Water Framework Directive. The geomorphology of the Mediterranean is characterized by steep slope mountains, flashy hydrographs, temporary rivers as well as ground water reservoirs found in karstic areas and alluvial deposits which make the hydrologic characterization very difficult.

Limited modelling tools have been developed for arid and semi-arid hydrology (Tzoraki et al. 2009) due to difficulty to model flush floods (Vivoni et al. 2006) and low flow regime (Nikolopoulos et al. 2011), the fluctuation in floodplain inundation (Tockner et al. 2000) and spring baseflow of the hydrograph (Fiorillo F. 2009, Rimmer and Sallingar 2006).

The objective of this study was to develop a methodology to estimate the hydrologic mass balance of Evrotas River Basin. Karst was simulated using the KARSTIC MODEL (Kourgialas et al. 2010, Tzoraki et al. 2009) and the non-karst area using the ETD MODEL (Nikolaidis et al. 1994). The limited available meteorological and flow data, in addition to the complicated hydrogeology and geomorphology of the basin, imposed significant constraints in the modelling exercise.

2 Site description and data

Evrotas River is located in south-eastern Peloponnese in the Prefectures of Laconia and Arcadia covering an area of 2,050km^2, and discharges into the Laconic Gulf after crossing 90 km of semi-mountainous areas and floodplains. The Evrotas River Basin expands between the mountain complexes of Taygetos and Parnonas, where numerous ephemeral and intermittent streams discharge into the main course. The main tributaries are Oinountas, Magoulitsa, Gerakaris, Kakaris, and Rasina (intermittent flow), Mariorema, Xerias, (episodic flow). The drainage density of the basin is 2.12 km/km^2 and the river network is classified as 5th order according to Strahler system. The basin has a typical Mediterranean climate with mild and cold winters and prolonged hot and dry summers with an average annual temperature of 16°C. Monthly mean temperatures are typically 4-11°C in the winter and 22-29°C in the summer. The majority of rainfall occurs from October to March. The mean annual precipitation is 802mm and evaporation 1754mm.

Geomorphologically, 41.1% of the basin area has an elevation higher than 600m, 46.2% of the area ranges between 150-600m, and just 12.7% has an elevation from 0 to 150m. The watershed is covered 61.2% by forest, 37.9% by agricultural areas, and 0.7% by urban areas. The inhabitants are 66000 and the biggest town is Sparta with 18000 inhabitants. The main activities in the watershed are agriculture, livestock and small food industries.

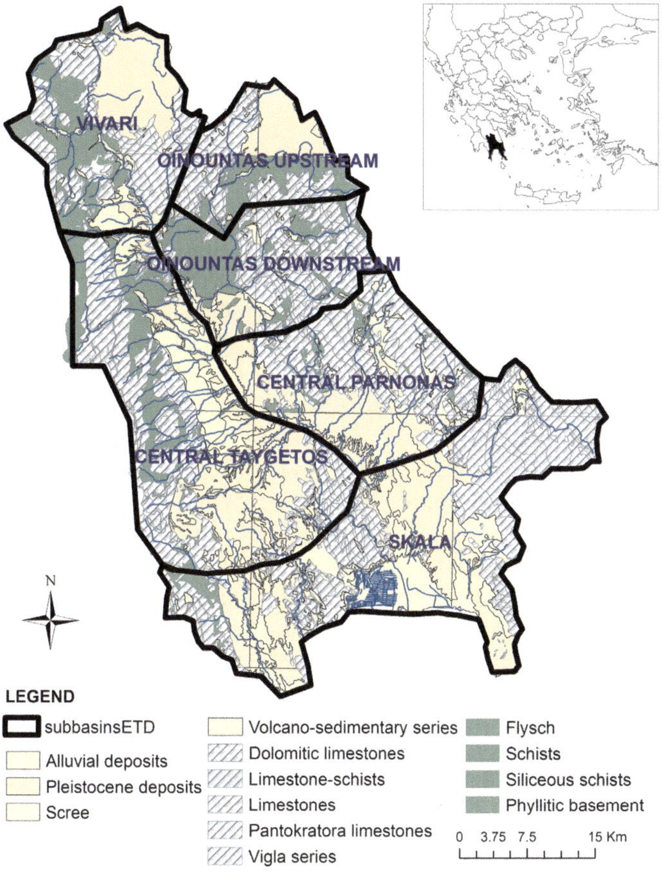

Fig. 1. Geology, subbasins and Hydrographic network of Evrotas River Basin.

The Taygetos mountain created by intense uplift and folding during the alpine orogeny straddles two geological zones: the first zone is the autochthonous Plattenkalk series comprising intensely folded phyllitic basement rocks of Permo-Triassic age overlain by middle Triassic to upper Eocene dolomitic limestones and crystalline limestones; and the second the allochthonous phyllitic quartzite series which consists of folded phyllites, schists and quartzites of Permian age (Pope and Millington 2000). A series of alluvial fans, restricted to piedmont zone, adjacent to the east edge of Taygetos Mountain, represents several depositional events and comprises significant water storage. Taygetos karst appears high transmissivity (10^{-3}-$10^{-4}$$m^2/s$) and recharges the north - west part of Sparta shallow alluvial aquifer through several springs. Sparta unsaturated zone has variable thickness 20-30m and the saturated hydraulic conductivity varies between $8 \cdot 10^{-5}$-$7 \cdot 10^{-7}$m/s (Antonakos and Lambrakis 2000). Sparta aquifer appears low transmissivity values ($3 \cdot 10^{-3}$-$14 \cdot 10^{-3}$), is recharged by surface water and springs and through numerous wells is used for the irrigation of the plain.

3 Methodology

Precipitation and stream flow have been measured since the 1970s by the Land Reclamation Service at six stations: Elos (4 m), Riviotissa (163.5m), Vrontamas (280m, it operates since 1953), Perivolia (490m), Sellasia (590m) and Vasaras (646m). In addition, there is a network of six automatic level loggers (Onset Computers, HOBO pressure transducer U20-001-04) that record the water level continuously and monthly measurements of flow are taken to construct the rating curves. Three weather stations in the catchment provide daily records of precipitation, temperature and evaporation. The watershed was subdivided into 6 subcatchments (Vivari, Oinountas upstream and downstream, central Parnonas, Central Taygetos and Skala) (Figure 1) following hydrologic, geologic and climate delineation criteria. Vivari subbasin is characterized by high altitude, flysch and limestone geologic series and numerous springs. Oinountas upstream subbasin is extended mainly in Phyllite - Quartzite series that consists of schists, metaconglomerates, quartzites and metabasalts. Oinountas downstream is characterized by a dense drainage area and high flows. Central Parnonas subbasin covers the south part of Parnonas Mountain and its streams experience mainly episodic or ephemeral flow in contrast to Central Taygetos subbasin that includes the most important related to flows streams and the main river corridor. Finally the Skala subbasin includes the south part of the basin and even though in its north part the river is totally infiltrated into the karst, in its south part the river flow is originated by an extended springs system and ends in a heavily modified deltaic system.

The KARSTIC MODEL (that assumes the existence of an upper (faster) and a lower (slower) reservoir that represent different karstic formations of the studied area) was coupled with a watershed model, the Enhanced Trickle Down model (ETD) a physically-based model in order to simulate the hydrology of Evrotas river watershed. In the KARSTIC MODEL the recession flow constants of upper (ku) and lower reservoir (kl) relate to the rate of discharge of the karstic springs. The inverse of these coefficients is the hydraulic retention time of the discharged water for each reservoir. Three snow-melt related coefficients (n, k, and c) define the snow melt process. Two coefficients, a1 and a2 relate to geologic setting and hydraulics of the karst. Finally, the product e*Akarst is the equivalent surface area of the karst in km^2. The three precipitation and snow melt coefficients control the slope of the hydrograph during the melting season. Finally, e*Akarst adjusts the peaks of the hydrograph. The area of the karst that contributes to the spring flow is not known.

ETD model is a compartmentalized model treating each of the five compartments ((1) atmosphere, (2) snow, (3) soil, (4) unsaturated zone, (5) surface water and (6) groundwater)) as a continuous stirred reactor. The model is a combination of four submodels. The hydrologic submodel, used in this study, is the component that generates flows in and out of the compartments. It simulates snowmelt, interflow, overland flow, groundwater flow, frost-driven processes, seepage and evapotranspiration. It solves hydrologic equation in steady state and needs lumped parameters. The terrestrial portion of the watershed is divided into three components (soil, unsaturated zone and groundwater). The ETD model was modified in order to incorporate the upstream flow (QUPSTREAM), the karstic flow (QKARST) that is the contribution of the karstic springs, the irrigation water (QIRR) that is diverted from the groundwater to the agricultural soils and the surface water abstraction (QWITHDR) for irrigation. A schematic representation of the ETD hydrologic model is shown in Figure 2. The first number in parenthesis indicates the direction of flow and the second the source of flow, for example Q(2,1) means that flow from compartment 1 (atmosphere) is input to compartment 2 (snow). QOUT is the outflow of the basin and QOVERLAND FLOW is the basin surface runoff.

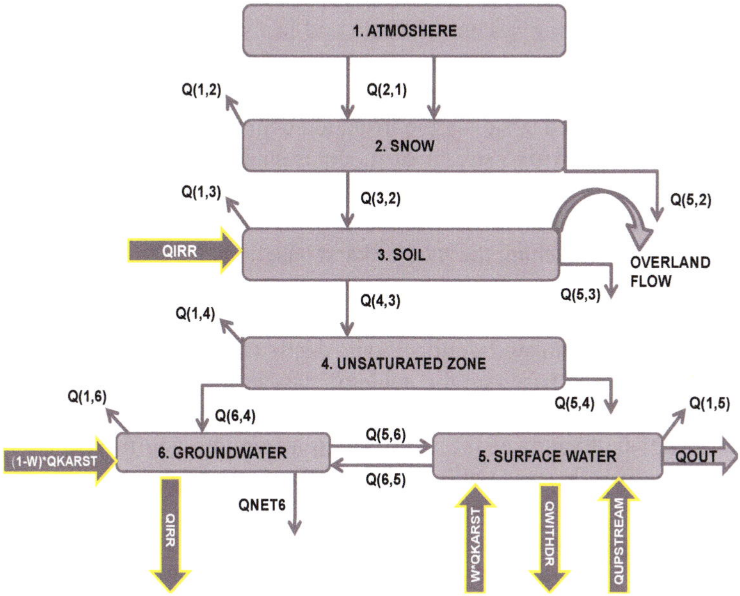

Fig. 2. Schematic representation of the revised version of the ETD model.

4 Modeling Results

The discharges of the six main karstic formations (limestones, dolomitic limestones) of Taygetos, Skortsinos, Vivari, Karyes, Parnonas and Skala were estimated by the KARSTIC MODEL. The runoff data for the period of 2007-2009 (daily values) were used in order to calibrate the karstic flow model and for the period 1999-2007 (monthly values) in order to validate the model. The area of each karst was estimated in GIS by the examination of the geologic maps. Detailed analysis of rainfall-elevation relationship extrapolated new time series of precipitation of the karst and non karst areas. The temperature lapse rate was estimated for Vivari, Skortsinos and Karyes to be -4.2°C/km and for Taygetos and Parnonas -5.1°C/km. The Karst system was simulated using one reservoir with a flushy response to rainfall events. The snow melt exponent was set equal to 0.25 and the snow melt factor was calibrated between 0.315-0.5. The reservoir hydraulic time indicates that the springs have a characteristic response time of 2-3 months (retention time 63-106 days). Losses from the karst due to evapotranspiration were estimated to be 45% of its annual precipitation.

Despite the important areal extend of Parnonas karst (434km^2, 18% of the basin), streams in that part of the basin experience only ephemeral flow. Parnonas karst is discharged partly into the river and is infiltrated in deeper reservoir, lost or contributes to the spring system of Skala and sea springs near the Laconic gulf. The springs of Vivari (mean annual outflow 390mm/y, 110km^2), Skortsinos (390mm/y, 16km^2) and Taygetos (480mm/y, 271km^2) are important drinking water sources and are used to satisfy the irrigation demands of the basin. Karyes karst contributes to Arachovitiko stream of Inountas subbasin with an average flow of 0.47 m^3/s, since Vivari karst contributes to Evrota with an average flow of 1.36 m^3/s (Figure 3a). The annual average recharge for Evrotas Karst was 306Mm^3 and 24% (75Mm^3) of it reaching the river as karst baseflow.

Karstic model outputs (daily time step) were used as inputs to the ETD model. The ETD model was used to simulate the hydrologic part of the non-karst parts of the basin. The simulated aquifer zone depth (depth of bedrock), DEPBER, (2-50m) was selected based on the geologic data in the watershed. The hydraulic conductivity of soils, KPERC3, was selected to allow a wide range of possible behaviors (0.001-0.008m/d) of the soils. The transpiration coefficient, Kb, was allowed to vary over the range of 0-17.78 μm/day, accordingly to the water demand variation of the plants in the basin.

The hydrologic simulation results of ETD for central Taygetos subbasins are presented in (Vrontama station) Figure 3b. The ETD model was able to capture the seasonal and interannual variability of the flow very well. The mean annual flow of Vivari and Central Taygetos subbasins estimated 1.6m^3/s and 4.76m^3/s and the Root Mean Square Error equal to 0.05m^3/s and 0.12m^3/s accordingly. The annual average hydrologic balance of Evrotas River basin up to Vrontamas station was estimated as follows on an annual basis: the precipitation (for the non-karstic

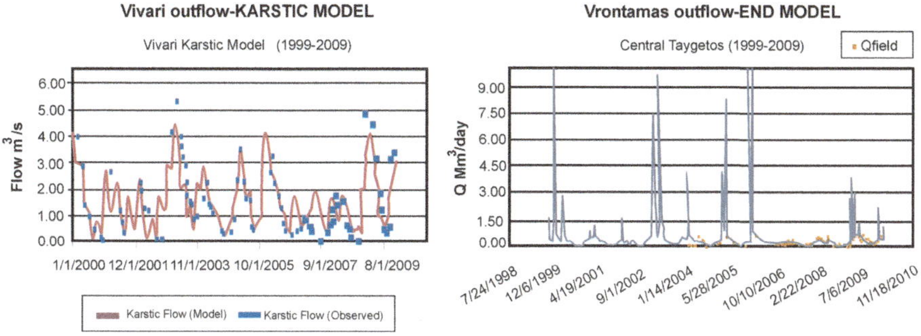

Fig. 3. Comparison of simulated and observed flow data for Vivari and Central Taygetos Subbasins (Vrontama outflow).

area of the watershed) was 629Mm^3, the karstic recharge was 306Mm^3, the evapotranspiration was 655Mm^3, stream discharge was 152Mm^3, stream abstraction (withdrawals) and irrigation were 77Mm^3. The amount of irrigation was estimated using data from the electric power company for agricultural electricity use. The electrical power generator consumption P (kW) and the volume of pumping water are related according the equation: P (kW) = Q*H*1.15/222, where Q in m^3 the volume of pumping water and H in meter the pumping depth. The power P that consumed in 2008 was equal to 13.67 GW for Sparti region and 38.62 GW for Gytheio area. Taking in consideration the different pumping depths (10-250m) of the irrigated land (84761acre) the annually water volume consumption estimated equal to 72Mm^3/year up to Vrontamas station. The irrigation water rate was 0.9m/m^2 compared to the recommended value of 0.5m/m^2 annually. The direct river abstraction was estimated to be 5Mm^3/year.

5 Conclusions

The Karstic Model was coupled with the watershed model ETD to simulate successfully the hydrology of Evrotas River Basin and assess its hydrologic balance. The 10 year annual average precipitation of the basin was estimated to be 1285 Mm^3/yr and almost the half volume (629 Mm^3/yr) is the precipitation of the karst areas. Karst areas springs guide 306 Mm^3/yr into the river and the groundwater of the non karst areas. The average annual evapotranspiration was 805 Mm^3/yr (63% of the annual precipitation) and the river runoff was 152 Mm^3/yr (12% of the annual precipitation. The annual irrigation and water withdrawals were estimated to be 77 Mm^3/yr or 6% of the precipitation. The amount of irrigation used in the area is approximately double the recommended irrigation for the products cultivated in the area. Significant losses exist from Parnonas karst to the Lakonic gulf and the Skala area. The modelling framework used in this work simulated successfully both wet and dry hydrologic years and it can be used for the management of watershed in semi-arid areas.

Acknowledgments This work was funded by the European Union projects LIFE05-EnviFriendly (www.EnviFriendly.tuc.gr) and FP7-MIRAGE. We would like to thanks Anoussis Kostas and Katsimalis George that provided the hydrologic data.

References

Antonakos A and Lambrakis N (2000) Hydrodynamic characteristics and nitrate propagation in Sparta aquifer, Wat. Res. Vol. 34, 16: 3977-3986

Nikolaidis NP, Hu H, Ecsedy C, (1994) Effects of Climatic Variability on Freshwater Watersheds, Case Studies, Aquatic Sciences, 56(2):161-178

Kourgialas N, Karatzas G, Nikolaidis NP (2010) An integrated framework for the hydrologic simulation of complex geomorphological river basin. Journal of Hydrology 381:308-321

Pope RJ and Millington AC (2000) Unravelling the patterns of alluvial fan develpement using mineral magnetic analysis: examples from the Sparta basin, Lakonia, Southern Greece. Earth Surface Processes and Landforms 25, 601-615

Tzoraki O, Nikolaidis N, Trancoso R, Neves R and Braunschweig F (2009) Modeling of instream biogeochemical processes of temporary rivers, Hydrological processes, 23(2), 272-283

Tockner K, Malard F, Ward JV. (2000) An extension of the flood pulse concept. Hydrological processes 14:2861-2883

Nikolopoulos EI and EN Anagnostou (2011) Sensitivity of a mountain basin flash flood to initial wetness condition and rainfall variability. Journal of Hydrology In Press,

Vivoni ER, Bowman RS, Wyckoff RL, Jakubowski RT, Richards KE (2006) Analysis of a monsoon flood event in an ephemeral tributary and its downstream hydrologic effects. Water Resources Research, 42 (3)

Rimmer A and Salingar Y (2006) Modelling precipitation-streamflow processes in karst basin: The case of the Jordan River sources, Israel. Journal of Hydrology, 331:524- 542

Evolution tendency of the coastline of Almyros basin (Eastern Thessaly, Greece)

G. Chouliaras, A. Pavlopoulos

Department of Sciences, Agricultural University of Athens, Iera Odos 75, GR-11855 Athens, apvlo@aua.gr

Abstract The Almyros basin, (Eastern Thessaly, Greece) formed during the Pliocene-Pleistocene N-S extensional phase that affected the entire Aegean area. Historical evidences show a prograding character of the Almyros coastline, developed along the Pagasitikos gulf. This tendency is enhanced by the actual form of the riverbeds witnessing a vertical (in depth) erosion of the riverbeds. This evolution is probably due to the continuing tectonic action and not to deposition of alluvial sediments as no important sediment load is observed. The fault system that could favor the progradation of the coastline should be the NW-SE and NNW-SSE slip-normal faults.

1 Introduction

The Almyros basin (Fig. 1), is based at SW part of the Pagasitikos Gulf (Eastern Thessaly, Greece), was formed during the Pliocene-Pleistocene as a result of the two main normal fault systems that affected the area. Their directions are NW-SE to NNW-SSE and ENE-WSW to E-W (Doutsos 1980; Caputo and Pavlides 1993; Galanakis 1997; Galanakis et al. 1998; Zovoili et al. 2004). The recent E-W normal faults and NNW-SSE oblique slip faults gave to the Almyros basin its actual form. Plio-Pleistocene formations (Fig. 1), constitute the main formations of the Almyros basin composed mainly of argillaceous beds, sandstones and river and coastal conglomerates (Marinos et al. 1962; Katsikatsos et al. 1978).The Holocene formations, situated at the eastern part of the basin, are restricted close to the sea front and include continental deposits and coastal sands and conglomerates. The main and most important drainage branches are the rivers of Platanorema, Xirias and Holorema (Fig. 1). The water flow in these rivers is intermittent and during the summer, the rivers get dry.

Quaternary scattered volcanic centers are also located, In the region of Almyros and are associated to the post-alpine volcanic activity of the Central Aegean area. Therefore, we can find imprints on the basin's landscape formed by the short living and weak activity of Microthebes, Achilion and Zerelia volcanoes.

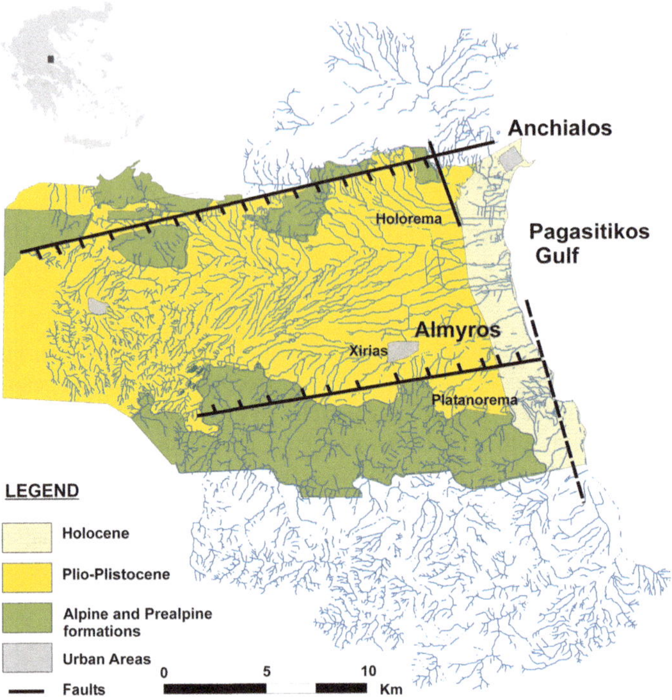

Fig. 1. Simplified geological map and drainage network of the studied area.

The present study is a preliminary report of an extensive research carrying out in Almyros basin. Its main purpose is to examine the coastline changes during the Holocene, based on physical and anthropogenic impacts.

2 Historical facts

In Greek language, "Almyros" means *salty*. The ruins of the ancient city of Alos are located approximately 10km to the south and 4km inland. Alos was a well known port in the ancient years and mostly important in the Persian wars (3rd century BC). The above information could be an indication that during historic times the coastline was situated west of the recent one.

3 Coastline shift

Evaluating the historical facts and the changes of the coastline, it is obvious that there is a prograding tendency. However the question is whether this progradation is due to alluvial fill or to tectonic activity. Based on the first field observations, the following remarks have to be made:

- The riverbeds, traversing the alluvial deposits, exhibit vertical (in depth) erosion forming abrupt lateral sides (Fig. 2).
- When the rivers cross the Plio-Pleistocene formations two groups of terraces are distinguished (Fig. 2B). The mean depth from the higher terrace down to the river bottom is 9-10m. The gradient of the surface of these formations is about 2% and the in depth erosion evident.

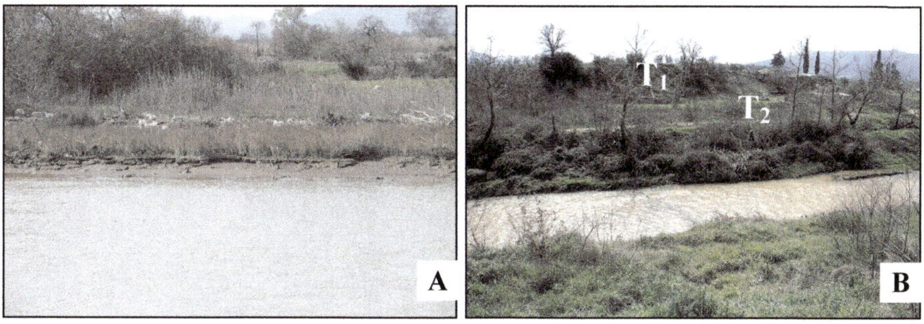

Fig. 2. Vertical (in depth) erosion of the riverbeds developed on Holocene (A) and Plio-Pleistocene (B) formations. T1 and T2: river terraces.

4 Discussion-conclusions

The vertical erosion of the riverbeds observed in Plio-Pleistocene and alluvial formations of the Almyros basin enforce the arguement of an uplift occurring during the Quaternary. As the basin was a result of normal faulting, this uplift could be a contradiction to the general morphotectonic evolution of the area. A mechanism responsible for such an evolution could be relative to the causes:

1. The sediment load of the drainage system is important during the wet and flooding periods so the accumulated sedimentary volume is dissected during the normal (low flow) periods.
2. The ratio uplift/immersion due to the tectonic activity is >1.

The first cause isn't entirely convincing as the size of the rivers, their interrupted flow and their sediment load are not capable enough to produce serious transformation on the coastline. The anthropogenic impact during the ancient years and up to the recent times could not be also a significant reason for coastline progradation (Fig. 3), as it is shown on the land-use map.
The most important human activities are restricted at the SE coastal part where mining processing plants were installed.

The second point related to the tectonic activity seems more valid. Zovoili et al. (2004) refer that the offshore segment of the Nea Anchialos fault causes an uplift of the north part of Almyros basin. Our observations indicate that this uplift is generalized at the whole area of the Almyros basin. The main faults that contribute to these movements, according to our observations, should be dip-slip faults oriented NW-SE to NNW-SSE.

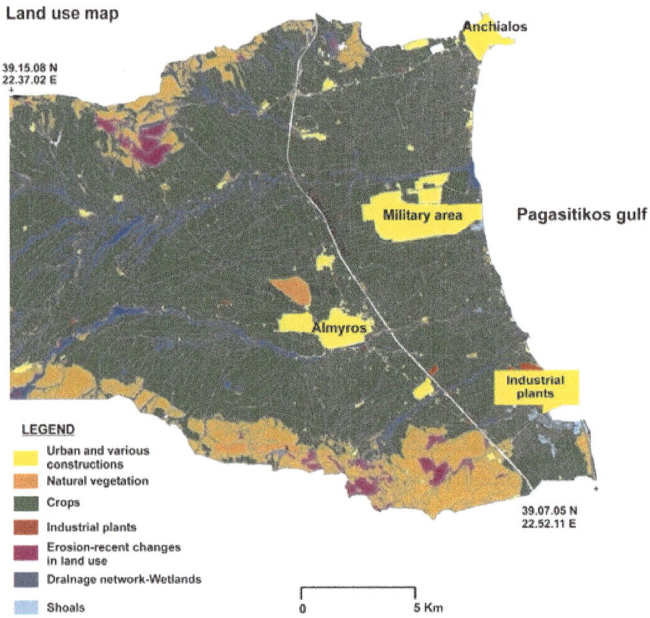

Fig. 3. Land use of Almyros basin

References

Caputo R, Pavlides S (1993) Late Cenozoic geodynamic evolution of Thessaly and surroundings (central-northern Greece). Tectonophysics, 223, 339-362

Galanakis D (1997) Neotectonic structure and stratigraphy of Neogene-Quaternarysediments of the Almyros-Pagasitikos, Pilion, Oreon-Trikeri and Maliakos basins. Unpublished PhD thesis, Thessaloniki

Galanakis D, Pavlides S, Mountrakis D (1998) Recent brittle tectonic in Almyros-Pagasitikos, Maliakos, N. Euboea and Pilio. Bull. SGG, XXXIV, 263-273 (in Greek)

Katsikatsos G, Mylonakis J, Vidakis M, Hecht J, Papadeas, G (1978) Geological map of Greece 1/50.000 sheet "Volos". IGME

Marinos G, Anastopoulos I, Maratos G, Melidonis N, Andronopoulos B (1962). Geological map of Greece 1/50.000 sheet "Almyros". IGME

Marinos G (1962) Sur deux volcans embryonaires du type maare, près d'Almyros-Thessalie. Bull. SGG V/1, 108-114 (in Greek)

Zovoili E, Konstantinidi E, Koukouvelas I (2004) Tectonic Geomopphology of escarpments: The cases of Kompotades and Nea Anchialos faults. Bull. SGG XXXVI, 1716-1725

An insight to the fluvial characteristics of the Mediterranean and Black Sea watersheds

S.E. Poulos

Department of Geography & Climatology, Faculty of Geology & Geoenvironment, National & Kapodistrian University of Athens, Panepistimioupolis, Zografou 15784, Attiki, Greece. poulos@geol.uoa.gr

Abstract The drainage basin of the Mediterranean Sea, accounting for some 4,184 10^3 km^2 (including the R. Nile), is larger than that of the Black Sea (2,320 10^3 km^2). Estimates of the annual water load for the Mediterranean and Black Sea rivers are 550 km^3 and 395 km^3, respectively. According to these figures, the corresponding values of annual water yield 130 m^3/km^2 (Med) and 170 m^3/km^2 (Black) show that Black Sea Rivers have a larger water capacity. In the case of the Black Sea, the large river systems (>10,000 km^2) drain 87% of its catchment supplying the 60% of its freshwater load. In contrast, the medium and small (<10,000 km^2) Mediterranean Sea rivers, although they drain 16% of its watershed, supply almost 70% of its water load. The impact of river damming has already reduced the freshwater fluxes by 40% for the Mediterranean and <10% for the Black Sea.

1 Introduction

River systems play a major role in the formation and evolution of the coastal zone, as the effect of their freshwater and sediment fluxes could be extended many kilometers from their mouth area influencing, therefore, a large part of the inner continental shelf and the associated coastal waters. Within this framework, the 30th CIESM Workshop had investigated the role of the rivers in the formation and evolution of the coastal zone of the Mediterranean and Black Seas (CIESM 2006), incorporating also aspects of human interference. Furthermore, over the last century the human impact has played also an important role i.e. "greenhouse effect", dam construction and other management schemes. Nowadays, the anticipated climatic change is expected inevitably to affect precipitation levels (Solomon et al. 2007) that consequently will change freshwater discharges.

The Mediterranean and Black Seas are two basins interconnected by the Sea of Marmara, through which they exchange water masses. The drainage basin of the Mediterranean Sea accounts for some 4,135 10^3 km^2, while along its coastline drain more than 200 rivers of various size. The Black Sea watershed is 2,350 10^3 km^2, including more than 130 rivers.

The present contribution provides comparative information regarding the freshwater fluxes of the Mediterranean and Black Seas drainage basins, with respect to their variable size of the incorporated river watersheds and the effect of river flow regulation (e.g. dam construction).

2 Data collection – methodology

For the calculation of the water load of the Mediterranean rivers, its catchments have been divided into 5 physiographic regions (Fig. 1), on the basis of their geographical locations and climatological conditions (mainly upon prevailing precipitation levels; Poulos and Collins 2002). Subsequently, the water yields (m^3/km^2) for each of the 5 regions was estimated. This estimation is based upon the calculation of the weighted-average of the corresponding water yields, from field-measurements of the water fluxes of more than 70 rivers discharging along the Mediterranean coastline (Poulos and Collins 2002; UNEP/MAP 2003; MEDHYCOS 2001; CIESM 2006). Subsequently, the mean (weighted) value for each Region was calculated, using the known water yields and the weighted area of the watersheds, corresponding to the measured water fluxes.

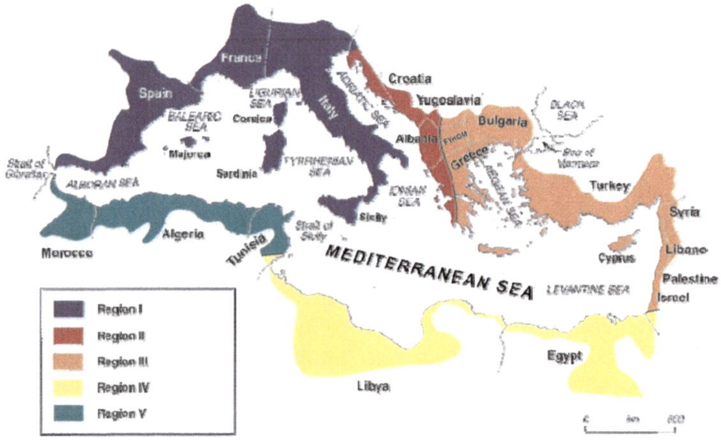

Fig. 1. The watershed of the Mediterranean Sea, including its division into five physico-geographical regions (I-V) (originally, from Poulos and Collins (2002).

Similarly, the drainage basin of the Black Sea was divided into 3 physiographic regions (Fig. 2). The water discharge of this region was calculated on the basis of existing data (Jaoshvilli 2002; Dimitrov et al. 2003) that covers almost the total area of Black Sea drainage basin. For river flow regulation, data and information have been abstracted from ICOLD (1998), Vörösmarty et al. (1998), Jaoshvilli (2002), UNEP/MAP (2003) and Poulos and Collins (2002).

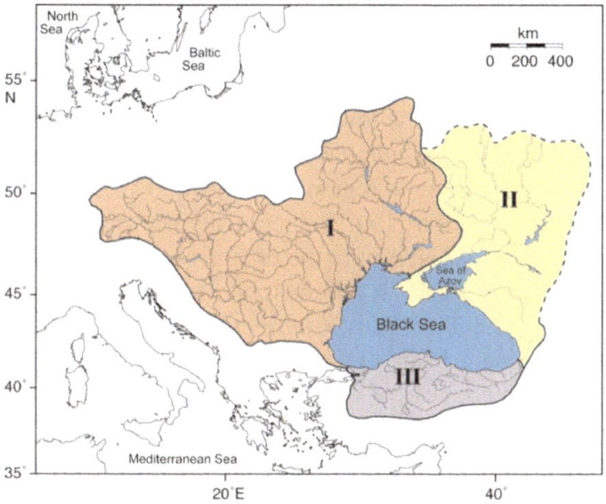

Fig. 2. Schematic presentation of the three regions of the watershed. The watershed of the Black Sea, including its division into three physicogeographical regions (I-III).

3 Mediterranean and Black Sea freshwater inputs

The total estimated volume of freshwater discharge for the total area of the Mediterranean Sea watershed (some 4,184,200 km^2) reaches annually the 550 km^3 (Table 1); this value corresponds to a mean water yield of some 130 $10^3 m^3/km^2$.

Table 1. Hydrological characteristics of the five regions of Mediterranean's watershed.

Reg.	CA (10^3 km^2)	CA/TA (%)	WL (km^3)	WL/TWL (%)	Y ($10^3 m^3/km^2$)	No. Rivers
I	614.6	14.7 (46.8)[2]	256.8	46.7 (55.1)	417.9	>90
II	107.0	2.6 (8.1)	90.1	16.4 (19.3)	842.2	>35
III	337.3	8.1 (25.7)	98.6	17.9 (21.1)	292.2	>42
IV	2,9284.0	70.0 (4.4)	86.7	15.8 (0.7)	29.6	>7
V	196.9	4.7 (15.0)	17.8	3.2 (3.8)	90.3	>18
Sum	4,184.2		550.0		131.4	>200
S(-Ni)[1]	1,314.2		466.5		355.0	

Key. (1): Total area excluding that of R. Nile; (2): In parentheses the statistical values are given for the total area, excluding the catchment of R. Nile (for region locations see Fig. 1); CA: catchment area; TA: total area; WL: water load; TWL: total water load; Y: water yield.

Higher freshwater discharges are associated with the European part (Regions I, II & III) of the Mediterranean watershed in comparison to its African part (regions IV & V), despite the presence of the extraordinary in size catchment of the R. Nile; this is related obviously the lower levels of precipitation in the region of the North Africa. Analogous is the number of the rivers draining the 5 regions, whilst the highest water yields belong to region III (some 840 $10^3 m^3/km^2$), when the least productive is the Nile' region (IV). It should be also pointed out the relatively small contribution of the R. Nile in terms of freshwater input (90 km^3), as it provides approximately only the 16.4% of the total freshwater volume (~280 km^3), although its watershed (2,870,000 km^2) represents almost 70% of the total area of the Mediterranean drainage basin (3,442,000 km^2).

The fluvial characteristics of the smaller in size Black Sea watershed (2,319,000 km^2) are given in Table 2. Its total freshwater discharge capacity reaches annually the 395 km^3. This means that the rivers out-flowing into the Black Sea have freshwater yields (on an average some 170 10^3 m^3/km^2) larger than those of the Mediterranean drainage basin (some 130 10^3 m^3/km^2). Such a difference is related to the fact that the Black Sea rivers drain a larger hinterland area with mean levels of precipitation >750 mm/yr, whilst the Mediterranean's catchment is characterized by a mean precipitation of <350 mm/yr (Poulos and Collins 2002). The latter figure remains smaller (some 650 mm/yr) than that of the Black Sea, even when the R. Nile is excluded from the analysis.

Table 2. Schematic presentation of the three regions (I, II & III) of the Black Sea watershed.

Reg.	CA (km^2)	CA/TA (%)	WL (km^3)	WL/TWL (%)	Y ($10^3 m^3/km^2$)	No. of Rivers
I	1,545.8	66.7	265.1	67.2	171.5	18
II	443.7	19.1	32.9	8.3	74.1	11
III	329.5	14.2	96.6	24.5	293.2	>107
Sum	2,319.0		394.6		170.1	>136

Key: CA: catchment area; TA: total area; WL: water load; TWL: total water load; Y: water yield.

4 Size-based analysis of the Mediterranean and Black rivers

For the needs of the present investigation, river watersheds have categorized into six classes with respect to their size; these classes are presented in Table 3 together with the number of the Mediterranean and Black Seas rivers. Subsequently, on the basis of the above size classification, the fluvial characteristics of the rivers discharging along the coastlines of the Mediterranean and Black Seas are presented in Table 4.

Table 3. River classification, according to the size of their watersheds.

	Size of catchment area (in km^2)					
No. rivers	Very small ($<10^2$)	Small (10^2-10^3)	Medium (10^3-10^4)	Large (10^4-10^5)	Very large (10^5-10^6)	Mega ($>10^6$)
Med/nean	>62	33	76	28	0	1
Black Sea	>38	63	23	9	3	0

Table 4. Hydrological characteristics of the Mediterranean and Black Seas rivers, according to their size classification (presented in Table 3).

River Size	CA (10^3 km^2)	CA/TA (%)	WL (km^3)	WL/TW (%)	Y ($10^3 m^3/km^2$)	Ra (10^{-6})
	Mediterranean Sea Rivers					
Very small & Small	375.2	9.0	137.9	25.1	367.6	>22.7
Medium	237.4	5.7	104.6	19.0	440.7	18.2
Large	701.6	16.8	224.0	40.7	319.2	6.7
Very large	0.0	0.0	0.0	0.0	0.0	0.0
Mega	2,870.0	68.6	83.5	15.2	29.1	0.2
Total (excl. Nile)	1,314.2	31.4	466.5	84.8	355.0	>47.6
Total (T)	4,184.2	100.0	550.0	100.0	131.4	>47.8
			Black Sea Rivers			
Very small & Small	69.1	3.9	34.0	12.5	492.3	>43.6
Medium	61.1	2.6	26.6	6.7	435.0	9.9
Large	419.1	18.1	75.9	19.2	181.1	3.9
Very large	1,769.6	76.3	271.8	68.9	153.6	1.3
Mega	0.0	0.0	0.0	0.0	0.0	0.0
Total (T)	2,319.0	100.0	394.6	100.0	170.1	>58.6

Key: CA: catchment area; TA: total area; WL: water load; TWL: total water load; Y: water yield; and Ra: river's density ratio (number of rivers (see Table 3) normalized to the total catchment area.

The Mediterranean watershed is characterized by the presence of the mega-sized Nile River and the absence of very large (10^5-10^6 km^2) rives (see also Table 5). The reverse situation applies to the Black Sea, where the very large rivers represent the ¾ of its total watershed area, whilst a mega river does not exist. The large rivers is more important in the case of the Mediterranean, draining >50% of its catchment and providing the 40% of its freshwater flux, than those of the Black Sea; the latter accounts for only 18% of its catchment and 19% of its freshwater water volume. Another difference of the two watersheds is given by the role played by the small and medium sized rivers; the latter representing about the 47% of Mediterranean's catchment provide almost the 50% of its freshwater volume, while for the Black Sea draining less than 7% of its watershed provide, approximately, the 19% of its freshwater. Another interesting observation is the substantially increased yields (>400 $10^3 m^3/km^2$) of the medium and small rivers with respect to either their corresponding average values and/or to the values of large rivers (see Table 4).

Table 5. The major river systems (>10.000 km^2) of the Mediterranean and Black Seas.

River (associated country)	CA (10^3 km^2)	CA/TA (%)	WL (km^3)	WL/TWL (%)
	Mediterranean Sea			
Segura (Spain)	15.0	0.36	1.0	0.18
Ebro (Spain)	85.8	2.05	17.0	3.09
Rhone (France)	96.0	2.29	54.0	9.82
Tevere (Italy)	17.0	0.41	7.4	1.35
Po (Italy)	54.3	1.30	48.9	8.89
Neretva (Croatia)	13.0	0.31	12.0	2.18
Drin (Albania)	19.6	0.47	11.8	2.15
Pinios (Greece)	10.8	0.26	3.8	0.69
Axios (Greece)	23.7	0.57	5.3	0.96
Strimonas(Greece)	16.6	0.40	5.2	0.95
Evros (Greece-Turkey)	27.5	0.66	3.2	0.58
Cediz-Nehri (Turkey)	18.0	0.43	2.3	0.42
Büyükmenderes (Turkey)	25.0	0.60	3.1	0.56
Seyhan (Turkey)	14.0	0.33	4.8	0.87
Ceyhan (Turkey)	19.1	0.46	6.8	1.24
Moulouya (Marocco)	51.0	1.22	1.6	0.29
Chellif (Algeria)	44.0	1.05	1.3	0.24
Mejerda (Tunisia)	22.0	0.53	0.9	0.16
Nile (Egypt)	2.870	68.59	90.0	16.36
Total:	3,442.3	82.27	280.4	50.98
	Black Sea			
Danube (Rom-Ukr)	817.0	35.23	200.0	50.68
Dnieper (Ukr)	503.0	21.69	53.0	13.43
Dniester (Ukr)	72.1	3.11	10.2	2.58
Don (Russia)	442.0	19.06	28.0	7.10
Kuban (Russia)	63.5	2.74	12.8	3.24
Rioni (Geo)	13.4	0.58	9.6	2.43
Chorokhi(Geo)	22.1	0.95	8.7	2.20
Kizil Irmak (Tur)	78.6	3.39	5.9	1.50
Sakarya (Tur)	56.5	2.44	5.6	1.42
Yesir Irmak (Tur)	36.1	1.56	5.3	1.34
Total	2,104.3	90.74	339.1	85.94

Key: CA: catchment area; TA: total area; WL: water load; TWL: total water load.

The major anthropogenic influence affecting the water and sediment fluxes, over the last century, is the construction of river dams for hydroelectric power generation, irrigation, public water supply and flood control. Obviously, this retaining of freshwater within the reservoirs leads not only to the strong smoothing of the river's hydrograph seasonality, but also to the reduction of freshwater fluxes due to direct evaporation from the reservoir and/or the use of the river water further downstream to dam position. Characteristic examples for freshwater reduction mainly after dam construction are provided in Table 6. UNEP/MAP (2003) report has estimated that at the end of the 20th century the freshwater capacity has been reduced to 333 km^3; this figure corresponds to a reduction of approximately 40%. In contrast, in the case of the Black Sea, it has been estimated that after damming freshwater fluxes (data from Jaoshvilli 2002; Dimitrov et al. 2003) have been reduced by only 8% (CIESM 2006). The later is explained by the physiography of the catchment area, which incorporates very large rivers with very low relief ratio that does not favour the construction of dams.

Table 6. Characteristic examples of reduced freshwater fluxes following river damming.

River/Country	Reduced water flow (%)	River/Country	Reduced water flow (%)
Ebro (Spain) [1]	-53.8	Nile (Egypt) [2]	-93.0
Tiber (Italy) [1]	-31.2	Medjerdah (Tunisia) [3]	-55.6
Ofanto (Italy) [1]	-42.8	Moulouya (Morocco) [4]	-76.1
Adige (Italy) [1]	-35.8	Dniester (Ukrania) [5]	-10.0
Cetina (Croatia) [1]	-88.5	Dnieper - (Ukrania) [5]	-17.9
Shkumbini (Albania)[1]	-33.6	Don (Russia) [5]	-12.5
Semani (Albania) [1]	-29.4	Rioni (Georgia) [5]	-74.4
Axios (Greece) [1]	-47.1	Inguri (Georgia) [5]	-39.2
Nestos (Greece) [1]	-54.8	Kamchea (Bulgaria) [5]	-30.7

Key. (1): UNEP/MAP (2003); (2): Abu el Ella (1993); (3): Zahar & Albergel (1999); (4): Snoussi et al. (2002); (5): Jaoshvilli (2002).

5 Conclusions

The drainage basin of the Mediterranean Sea, accounting for some 4,135 10^3 km^2 (including the R. Nile), is larger than that of the Black Sea (2,350 10^3 km^2). Estimates of the potential (not regulated) annual water load for the Mediterranean rivers are 550 km^3 and, for the Black Sea, 395 km^3. Black Sea rivers present, conclusively, a larger freshwater capacity than those of the Mediterranean rivers as shown by their corresponding yields of ~170 10^3 m^3/km^2 and ~130 10^3 m^3/km^2 and the relatively higher density ratio (Ra) of Black Sea rivers. Another difference is the fact that the large Black Sea rivers (>10,000 km^2) drain 87% of its catchment, providing ~60% of its freshwater input, while the large Mediterranean Sea rivers (excluding the R. Nile) represent only the 18% of its catchment, supplying ~27% of its freshwater flux. This is also related to the low freshwater inputs by the Nile river (some 86.5 km^3 before damming) despite its mega-size (2,870 10^3 km^2). A reverse situation applies to medium and small rivers (<10,000 km^2), whose contribution is greater in the case of the Mediterranean than in the case of the Black Sea. The main human interference is related to the regulation of rivers' flow through the construction of hundreds of dams. The impact of river flow regulation, mainly due to dam operation has already reduced Mediterranean freshwater flux by 40%, which is much greater than what applies to the Black Sea (<10%).

References

Abu El Ella EM (1993) Preliminary studies on the geochemistry of the Nile river basin, Egypt. Mitt. Geol.-Paläont. Inst. Univ. Hamburg, 74, p. 115-135

CIESM (2006) Fluxes of small and medium size Mediterranean Rivers: impact on coastal areas. Ciesm Workshop Monographs no 30, Monaco www.ciesm.org/online/monographs/ trogir06.pdf

Dimitrov P, Solakov D, Peychwev V Dimitrov D (2003) The source Provinces in the Black Sea. Institute of Oceanology. Academy of Sciences, Varna, v. 4 pp.29-35

ICOLD (1998) World Register of Dams. Published by the International Commission on large Dams, Paris

Jaoshvilli S (2002) The rivers of the Black Sea. EEA Technical report no 71, 58pp

MEDHYCOS (2001) The Mediterranean hydrological cycle observing system. MEDHYCOS phase II, period 2002-2005, report no. 17, 36 pp <http://medhycos.mpl.ird.fr/>

Poulos SE Collins MB (2002) A quantitative evaluation of riverine water/sediment fluxes to the Mediterranean basin: natural flows, coastal zone evolution and the role the dam construction. (In:) Jones SJ & Frostick LE (eds) Sediment Flux to Basins: Causes, Controls and Consequences. Geological Society, London, Special Publications, 191, 227-245

Solomon S, Qin D, Manning M, Chen Z, Marquis M, Averyt K B, Tignor M, Miller H L (eds) (2007) Contribution of Working Group I to the Fourth Assessment Report of the Intergovernmental Panel on Climate Change, Cambridge University Press, Cambridge, United Kingdom

Snoussi M, Haida S, Imassi S (2002) Effects of the construction of dams on the Moulouya and the Sebou rivers (Morocco). Regional Environmental Change, 3, 5-12

UNEP /MAP, (2003. Riverine transport of water, sediments and pollutants to the Mediterraneran Sea. Mediterranean Action plan Technical Reports Series No. 141, UNEP/MAP, Athens, 2003, 128 pp

Vörösmarty C J, Fekete BM Tucker BA (1998) Global River Discharge. Database (RivDIS) V. 1.1. Available on-line [http://www.daac.ornl.gov] from Oak Ridge National Laboratory Distributed Active Archive Center, Oak Ridge, TN, U.S.A

Zahar Y, Albergel J (1999) Hydrodynamique fluviale de l'oued Medjerdah à l'aval du barrage Sidi Salem. Evolution récente. Paper presented at Hydrological and Geochemical Processes in Large River Basins. Manaus Conference, Brazil

Flooding in Peloponnese, Greece: a contribution to flood hazard assessment

M. Diakakis, G. Deligiannakis, S. Mavroulis

Department of Geology and Geoenvironment, National & Kapodistrian University of Athens, Greece. diakakism@geol.uoa.gr, gdeligian@geol.uoa.gr, smavroulis@yahoo.gr

Abstract Flooding is one of the most important types of disasters in southern Greece with many victims and extended damages over the last century. The increase in population together with the augmented pressure for urban expansion has increased flood risk during the last decades considerably across southern Europe. Peloponnese is not an exception in this regime, recording a significant amount of flood events and casualties. In this work, a catalogue of flooding phenomena during the period 1896-2010 in Peloponnese has been compiled based on numerous sources. Based on this record the temporal and spatial distribution of flood events and victims was examined. In total, 102 events were identified, causing 57 human casualties and inflicting extensive damage across the study area. Results showed seasonality patterns with more events clustering in November. They also showed that urban and coastal environments tend to present higher flood recurrence rates than mountainous and rural areas. An increasing trend in reported flood event numbers and casualties during the last decades was discovered. Moreover, spatial patterns were identified highlighting areas with higher flood recurrence rates across the region studied.

1 Introduction

Flooding problem becomes an increasingly significant issue in the Mediterranean region as population expands to river deltas and coastal areas that are subject to inundation mostly from small rivers and ephemeral mountain torrents. Greece is not an exception in this regime having a very rich flooding record since the historical times.

During the last decade efforts to mitigate flood risk have been enhanced with new practices in Greece and across Europe in scientific as well as in civil protection terms. Moreover, the legal framework has been enriched with new legal binding instruments (e.g. European Commission Directive 2007/60). In Greece, regular recording of flood events by civil protection agencies started relatively recently, limiting the systematic official records to the last two decades. On the other hand, regional authorities, damage compensation organizations and the press documented disasters in official reports or in anecdotal form, maintaining archives with an extensive amount of data that were not systematically evaluated as a whole until now.

Previous works suggested and demonstrated that examination of flood history is an important part in flood hazard assessment (Potter 1978, Bayliss and Reed, Benito, Salvatti 2009, Diakakis 2010). In this context and given the scarcity of instrumental and systematic hydrological data, flood history examination has been considered, a useful tool to assess preliminarily flood hazard spatial distribution in Peloponnese.

Goal of this paper is to exploit the possible benefits of the examination of a historical flood record in Peloponnese, in terms of civil protection, strategy and operations.

2 Hydrological setting

Greece has evolved through a complicated geotectonic process during the alpine orogenic cycle, leading to the development of a variety of physiographic elements. Peloponnese is one of the characteristic examples of this process (Fig. 1).

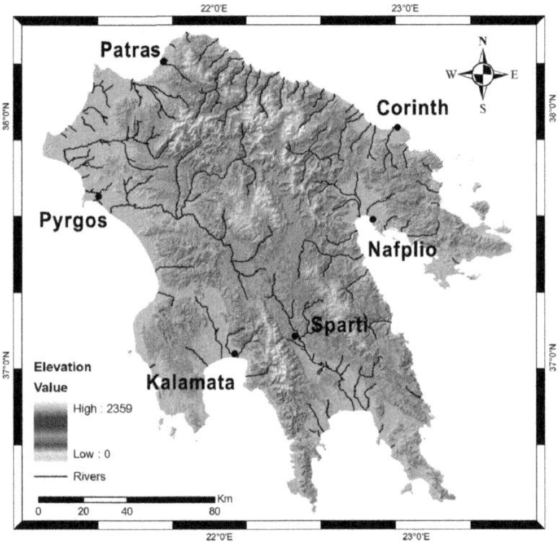

Fig. 1. Main physiographic elements of Peloponnese.

A large part of Peloponnese is mountainous, so most urban development takes place in the lowlands. Most of the population is clustering in the coastal regions (ex. the 3rd largest city of Greece, Patras), while during the summer there is a noteworthy population increase in smaller towns or villages along the coastline, contributing to augmented flood vulnerability.

A large part of the drainage network of Peloponnese consists of ephemeral mountain torrents and small to medium size drainage basins with limited amount of discharge for most of the year. However, there is a number of larger river basins situated mostly in the southern (Evrotas basin) and western (Alfeios and Pinios basins) parts of Peloponnese. Hydrological data are generally scarce as most of the basins are ungauged, making historical records valuable in flood studies.

3 Data and methods

A large number of reports and documents were collected from several sources, such as:

- Official flood damage delineation declarations from government organizations related to disaster recording, such as regional authorities, the Earthquake Rehabilitation Center (2010) and the General Secretariat of Civil Protection (GSCP 2007).
- The press, with the following databases:
  - the Digital Newspapers Collection of the Greek National Library (Digital Newspapers Collection 2010)
  - the Greek National Newspapers Archive (2010) of the Library of the Hellenic Parliament (microfilm database)

A systematic record of the above data was essential, so a database was developed with detailed data on each flood event. Information was standardized to match the database's requirements. Then data were stored with information on location, flooded river, flooded villages, towns or settlements, and casualties for each event. Finally, 102 entries in total were stored in this database. Simple checks such as cross-referencing between different sources of information were carried out to assure quality control of the record and standardization of information.

The development of a database improved our capabilities for data analysis and made them easier to retrieve and process. It also highlighted possible lack of information and presented the record in a more systematic form in a way that it could be evaluated for Peloponnese as a whole and for specific prefectures at the same time.

Temporal and spatial distribution of events and casualties were derived by using simple statistical methods for quantitative analysis of the data. Interesting conclusions were also drawn by plotting flood locations on a map in GIS environment, after connecting the developed catalogue with GIS databases. This process was essential in order to extract the spatial density of events across Peloponnese. Further analysis was performed in Peloponnese as a whole, in order to examine possible differences per month (Table 1).

Table 1. List of flood events in Peloponnese between 1896 and 2010.

Date	Location	Casualties	River	Date	Location	Casualties	River	Date	Location	Casualties	River
15/12/1896	Amaliada	1	-	28/11/1966	Pyrgos	0	-	21/10/1994	Argolida	0	-
4/11/1924	Kalamata	15	Nedon	1972	Corinth	4	-	21/10/1994	Isthmos	0	-
25/12/1927	Patra	0	Glafkos	19/11/1979	Pyrgos	1	-	9/11/1994	Lakonia	0	Evrotas
28/11/1928	Aigio	0	Selinountas	19/11/1979	Kalamata	1	-	12/1/1997	Argos	0	Inachos
28/11/1928	Gastouni	0	Pineios	16/11/1980	Argos	0	-	12/1/1997	Aigion	1	Vouraikos
28/11/1928	Sparti	10	Evrotas	18/11/1983	Messini	0	Riaka	12/1/1997	Corinth	6	Xirias
26/10/1930	Patra	0	-	24/11/1985	Pyrgos	0	-	25/11/1997	Argos	0	Xirias
9/10/1949	Ilia	0	-	24/11/1985	Prasidaki	0	Neda	25/11/1997	Isthmia	0	-
11/11/1957	Messini	0	-	4/11/1986	Vartholomio	0	-	20/6/1998	Argolida	0	-
19/12/1960	Alysos	0	Pineios	5/11/1986	Amaliada	0	-	23/11/1998	Argolida	0	-
19/1/1960	Manolada	0	Selinountas	24/8/1990	Patras	0	-	24/11/1998	Nafplio	0	-
18/12/1962	Patra	0	Meilichos	29/10/1990	Vartholomio	1	-	1/1/1999	Argolida	0	-
27/12/1962	Patra	0	Glafkos	15/11/1990	Laconia	0	Evrotas	7/11/1999	Sparta	1	Evrotas
3/2/1963	Patra	0	Diakoniaris	16/11/1990	Argos	0	Inachos	19/11/2000	Corinth	1	-
21/1/1963	Patra	0	Diakoniaris	7/11/1991	Kalamata	0	-	6/12/2000	Laconia	0	Evrotas
1/12/1966	Pyrgos	1	Alfeios	Ιαν-92	Astros	0	Tanos	5/11/2001	Lechaina	2	-
28/4/1966	Pyrgos	0	-	21/11/1993	Lefktro	0	-	16/12/2001	Patras	2	Diakoniaris

Table 1. Continue.

Date	Location	Casualties	River	Date	Location	Casualties	River	Date	Location	Casualties	River
7/11/2002	Megalopolis	0	-	24/11/2005	Vlachiotis	0	Vorvas	8/10/2006	Nafplio	1	-
7/11/2002	N.Manolada	0	-	24/11/2005	Vlachiotis	0	Lakas	20/10/2006	Akrogiali	0	-
1/1/2003	Megalopolis	0	-	24/11/2005	Klada	0	Kelefina	23/11/2006	Ilia	0	-
26/1/2003	Therapnes	0	Evrotas	24/11/2005	Agia Eirini	0	Magoulitsa	11/2/2007	Corinth	0	-
26/1/2003	Klada	0	Kelefina	24/11/2005	Mystras	0	Skatias	26/5/2007	Dimitsana	8	Lousios
26/1/2003	Vlachiotis	0	Lakas	24/11/2005	Polydentro	0	Paroritis	20/9/2007	Corinth	0	-
26/1/2003	Vlachiotis	0	Vorvas	24/11/2005	Riviotissa	0	Xerias	28/9/2007	Ilia	0	-
5/2/2003	Patras	0	-	24/11/2005	Retsa	0	Sochiotiko	17/11/2007	Argolida	0	-
5/2/2003	Zacharo	0	Alfeios	24/11/2005	Rasina	0	Rasina	17/11/2007	Dafni	0	Alfeios
17/3/2003	M. Spilia	0	Vlamandreas	24/11/2005	Kampos	0	Sminos	17/11/2007	Corinth	0	-
17/3/2003	Nom.Trochalia	0	-	24/11/2005	Selinitsa	0	Ardelis	17/11/2007	Messinia	0	-
17/3/2003	Krokeai	0	Lefkatiano	24/11/2005	Aigies	0	Fournolagado	11/12/2008	Corinth	0	-
26/3/2003	Iria	0	-	24/11/2005	Molaoi	0	Chilorema	11/12/2008	Sparta	0	Evrotas
15/10/2004	Ilia	0	-	24/11/2005	Vrontamas	0	Xirias	6/9/2010	Argos	0	-
6/3/2005	Prasidaki	0	Neda	24/11/2005	M. Spilia	0	Vlamandreas	23/9/2010	Corinth	0	-
15/9/2005	Nafplio	0	-	24/11/2005	Nom.Trochalia	0	-	26/9/2010	Pyrgos	0	-
1/10/2005	Pelopio	0	-	24/11/2005	Krokeai	0	Lefkatiano	27/10/2010	Vartholomio	1	-

4 Results

Statistical processing of the data showed 57 casualties resulting from 102 flood events in total during the period 1896-2010. The most important events in terms of casualties are shown in Table 1.

As far as monthly distribution is concerned the analysis showed that November is the month with the richest flood record (50% of total events) followed by January (12%), October and December (10% each) with flood casualties following approximately the same distribution monthly distribution (Fig. 2).

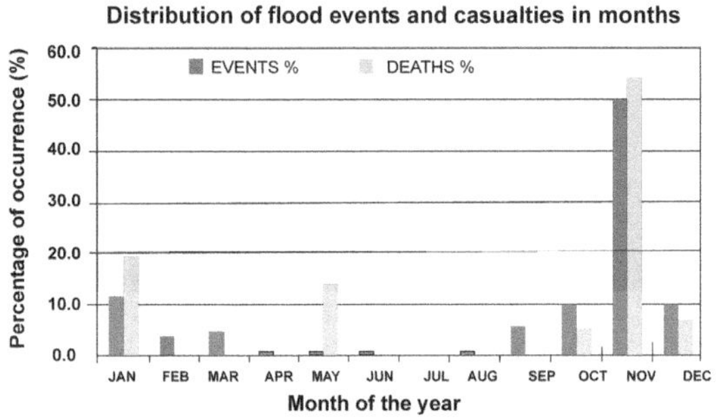

Fig. 2. Distribution of flood events and casualties in months expressed as percentage of total 102 events. November and January present the highest proportion with 50% and 12% of the total events. October and December follow with 10% each. In terms of casualties, November and January record the most victims with 54.4% and 19.3% respectively.

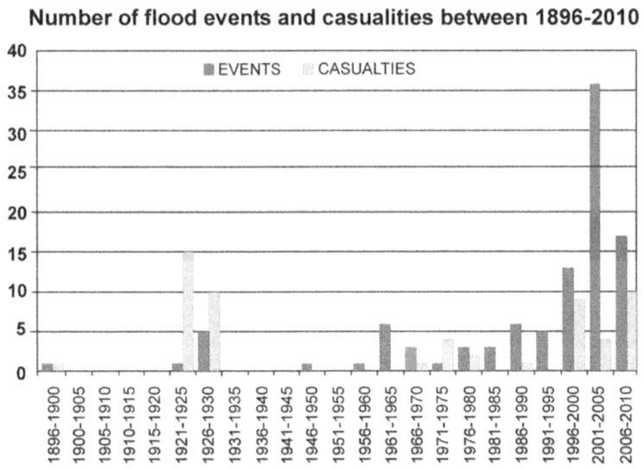

Fig. 3. Temporal distribution of flood events and casualties in 5-year segments between 1896 and 2010.

As far as the temporal distribution of flooding is concerned, events and casualties were grouped in five-year periods between 1896 and 2010, in order to examine their 5-year occurrence variations (Fig. 3). The analysis shows an increase in reported flood events and deaths since 1960 and especially in the last 25 years.

This increasing trend can be attributed mainly to improvements of means of reporting and increase in population density since the first decades of the century in the area. Natural trends cannot be established based on these data due to possible lack of information in the early century especially during wartime.

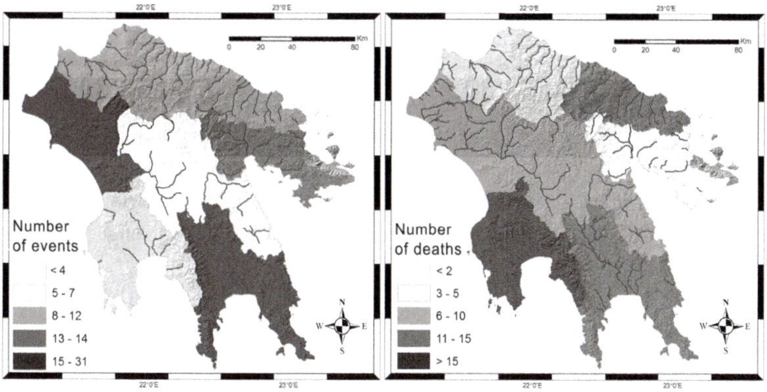

Fig. 4. Number of flood events (left) and flood victims (right) per administrational unit in Peloponnese during the period 1896-2010.

The spatial distribution of flood events was examined primarily on the basis of second level administrational entities ("Prefectures") and expressed as the number of events per unit (Fig. 4). Laconia and Ileia presented the highest concentration of events while Arkadia the lowest. In terms of flood casualties Messinia, followed by Korinthia and Laconia recorded the most victims (Fig. 4). Finally, flood events were plotted on the map to illustrate spatial patterns within the administrational units and the study area as a whole. Based on this projection spatial density was calculated across Peloponnese (Fig. 5).

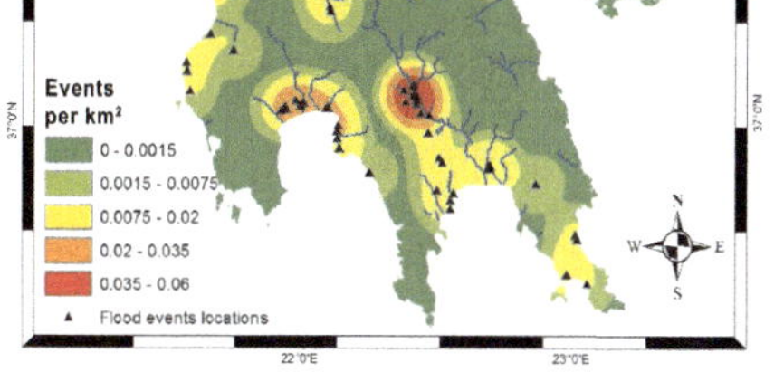

Fig. 5. Locations of 102 flood events and spatial density of in events per square kilometer across Peloponnese between 1896 and 2010.

5 Conclusions

The analysis showed that Peloponnese has suffered numerous floods during the last century that induced significant damages to properties and infrastructure and took 57 lives. In terms of seasonality the month with the richest record is November (50%) followed by January (12%). It is important to note that there is an increasing number of floods and victims reported during the last two decades that can be attributed mainly to population increase and improvement of reporting means and not in a existence of natural trend. In terms of spatial distribution Sparti, Patra, Corinth, Argos, Kalamata and Pyrgos are the locations with the richest flood record as shown by spatial density analysis. Floods appear to occur mostly in coastal and low altitude areas. In terms of administrational entities, the prefecture of Laconia presents the most flood events.

Analysis of the historical record can be used to preliminarily identify higher risk areas and can be useful for operational decision-making and civil protection.

References

Bayliss AC, Reed DW (2001) The use of historical data in flood frequency estimation. Report to MAFF, Centre of Ecology and Hydrology, Wallinford, 87p. Available at: http://nora.nerc.ac.uk/8060/01/BaylissRepN008060CR.pdf Accessed 9 January 2011

Benito G, Lang M, Barriendos M, Llasat MC, Francés F, Ouarda TBMJ, Thorndycraft VR, Enzel Y, Bárdossy A, Coeur D, Bobée B (2004) Use of Systematic, Palaeoflood and Historical Data for the Improvement of Flood Risk Estimation, Review of Scientific Methods, Nat. Hazards 31(3):623–643

Diakakis M (2010) Flood history analysis and its contribution to flood hazard assessment: The case of Marathonas, Greece. Bull Geol Soc Greece 43:1323-1334 Available at: http://www.geosociety.gr/GSG_XLIII_3.pdf Accessed 9 January 2011

Digital Newspapers Collection (2010) E-efimeris: Digital Newspapers Collection of the Greek National Library. Available at: http://www.nlg.gr/digitalnewspapers/ns/main.html Accessed 10 January 2011

Earthquake Rehabilitation Center (2010) Disaster delimitations, Earthquake Rehabilitation Center (YAS), Ministry of Public Works Available at: http://www.yas.gr Accessed 10 October 2010

ELSTAT (2001) Population Census 2001: De facto population, Area and population by range, density and elevation zones. Hellenic Statistical Authority. Available at: http://www.statistics.gr/portal/page/portal/ESYE Accessed 10 January 2011

GSCP (2007) Flood-prone areas of Attica. General Secretariat of Civil Protection. Ministry of Interior, Greece

Greek National Newspapers Archive (2010) National Newspapers Microfilm Database. of the Library of the Hellenic Parliament, Hellenic Parliament, Greece

Potter HR (1978) The use of historic records for the augmentation of hydrological data, Institute of Hydrology Report No.46 Wallingford 59p. Available at: http://nora.nerc.ac.uk/5773/01/IH_046.pdf Accessed 5 January 2011

Salvatti P (2009) Societal landslide and flood risk in Italy. Nat Hazards Earth Syst Sci, 10:465-483

Estimation of sedimentation to the torrential sedimentation fan of the Dadia stream with the use of the TopRunDF and the GIS models

A. Vasiliou, F. Maris, G. Varsami

Lab of Mountainous Watersheds and Risk Management, Department of Forestry and Management of the Environment and Natural Resources, Democritus University of Thrace, mail to: fmaris@fmenr.duth.gr

Abstract The present paper examines the maximum flow rate and soil erosion of the watershed of the Dadia stream. The use of empiric forms took place, in order to calculate the maximum flow rate, while the volume of soil erosion was estimated by the Universal Soil Loss Equation (USLE) and the modified Gavrilovič method. The methods above classified the sub-basins and the TopRunDF Model was applied to the most important watershed of all. The TopRunDF is a simulation model based on GIS, developed by Scheild and Rickenmann (2009) and predicts the repository and the amount of sediment deposition on the torrential fan. An extensive use of GIS took also place, for the simulation parameters to be defined and for the model to be wholly depicted.

1 Introduction

The stream of Dadia (local name Diavolorema) flows across the National Park of Dadia - Lefkimi - Soufli, and is part of the forest complex Dadia - Lefkimi - Soufli. The area under study is on the east part of Thrace, in the prefecture of Evros, southeast of the town of Soufli, near the village of Dadia and can be found at latitude 40 59' - 41 15' and longitude 2 19' - 2 36' east of Athens. The area possesses a high ecological value, due to the presence of a variety of raptorial birds, many of which are rare in Europe. As a result of its position, the area is an important habitat for many reptiles and birds. It is also a crossroads for migrating birds and hosts ideal avian nesting habitats. This complex is a separate area of the network "Natura 2000" named "Evros Mountains". In 2006 under a joint ministry decision the area was characterized as National Park with the name "National Park of Dadia - Lefkimi - Soufli". The area under study appears to have a highly multifarious terrain and a dense hydrographic network. The stream of Dadia ends up in Evros River along with all its debris. The area of the basin is 163.92 km^2 (Maris and Vasileiou 2010). The purpose of this study is to examine, quantify and determine the spatial soil erosion of the study area. In order to have a better and closer investigation of the drainage network, the basin was divided into 29 sub-basins and following that, all those which were less than 1 km^2 were removed, having the total number reduced to 20 sub-basins (Fig. 1).

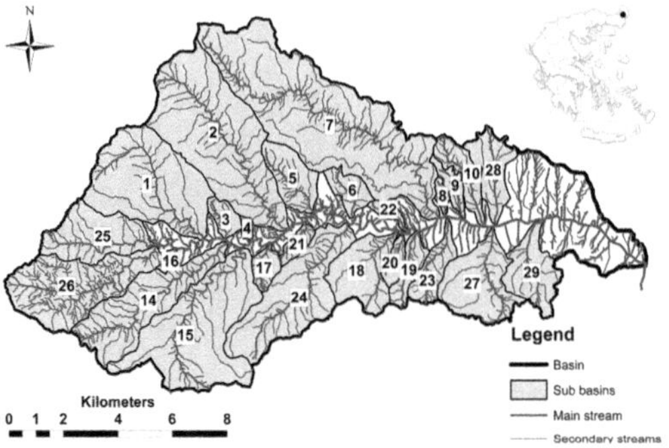

Fig. 1. a, b: Study area and sub-basins of the stream of Dadia.

2 Research Method

The TopRunDF model is a two dimensional runout simulation tool for debris flows to predict the amount of sediment deposition on the torrential sedimentation cone. It was created in 2009 by Scheild Rickenmann at the Institute of Mountain Risk Engineering, University of Natural Resources and Applied Life Sciences in Vienna (Scheidl and Rickenmann 2010). The TopRunDF is written in Visual Basic 6.0 and operates as a single executable program using ARC-Objects. Based on the topography of the torrential fan the model combines the simple flow routing algorithm described in Hürlimann et al. (2008) with the area-volume relation. The input parameters of TopRunDF are the debris-flow volume V, the mobility coefficient is a dimensionless parameter which contains some information on the flow properties during the depositional phase, the starting point of the deposition (fan apex) and the digital terrain model of the area. The output results predict the deposition areas as well as deposition height of a debris-flow event on the fan. The main advantage of this model is that it doesn't use very demanding and time-consuming data, while the accuracy of the results is remarkable.

2.1 Volume of the debris-flow event

The estimation of soil erosion was conducted with empirical models and in particular according to the Universal Soil Loss Equation (Universal Soil Loss Equation, U.S.L.E.) and the Gavrilovič method (Gavrilovič 2006).

2.1.1 Universal Soil Loss Equation

The general equation of soil degradation is based on multi-annual measurements of the intensity of erosion of standard surfaces, conducted by Wischmeier and Smith in various areas of the U.S. during the period 1930-1952 (Wischmeier and Smith 1965). In the present study in order to apply the method in areas with different land use and with a larger inclination, we modified the parameters we used. The USLE is expressed with the following equation: $A = R \cdot K \cdot LS \cdot P \cdot C$, where A represents the potential long term average annual soil loss, R is the rainfall and runoff factor, K is the soil erodibility factor, LS is the slope length-gradient factor, C is the crop/vegetation and management factor, P is the support practice factor. The torrential streams bank erosion, which is empirically estimated at 20% of the surface erosion, is also added to the total basin erosion (Maris et al 2006).

R Factor

In order to calculate the R factor, in this study, we used the following formula, which was developed in Germany and has been applied to Greek conditions (Chrysanthou and Pyliotis 1995):

$$R = 0.83 \cdot N\text{-}17.7,$$

where R mean annual amount of precipitation is the rainfall and runoff factor and N (in mm) (mm/year) For the calculation of the N factor, we used the average annual rainfall, calculated by the Soufli Meteorological Station. For the conversion of the point rainfall to areal rainfall we used the Kriging method (Matzarakis 2006).

K Factor

To calculate the K-factor we used the WRBFU.kml file from the European Soil Data Center.

Table 1. K factor according to soil classification (Da Silva 2005).

Soil Type	K Factor	Soil Type	K Factor
Dystric leptosols	0.0254	Eutric Cambisol	0.0508
Chromic luvisols	0.041	Chromic Vertisol	0.0367

The format of the file is compatible with the program Google Earth, where the satellite image of the basin we study, was taken from. The file had been georeferenced with the Arcmap program, with which we digitized the soil.

LS Factor

To calculate the LS factor we used the Arcmap 9.3 program and the following formula (Ma Jianguo 2001)

$$LS = (\text{Flow Accumulation Cell Size} / 22.13)^{0.4} (\text{Sin Slope} / 0.0896)^{1.3}$$

P Factor

The P factor was given a value of 1 for all the sub-basins, as there was no adjustment work found in any of them.

C Factor

To calculate the C factor, we used the land use database of the European Environmental Agency Corine 2000, from the Arcmap program, and the land uses with their codes as they are presented in Table 3.

Table 2. Land use factor (Zarris et al. 2002).

Land use	C Fact.	Land use	C Fact.	Land use	C Fact.	C Fact.	C Fact.
112	0.001	312	0.001	241	0.18	324	0.02
211	0.3	313	0.001	242	0.18	331	0.6
213	0.15	321	0.3	243	0.1	332	0.45
221	0.2	322	0.45	244	0.05	333	0.45
222	0.2	323	0.03	311	0.001	512	0.000

2.1.2 The Gavrilovič method

The Gavrilović Method is a parametric distributed model for qualifying the erosion severity and estimating the total annual sediment yield of a catchment. The equation has the following format: $W = T \cdot h \cdot \pi \cdot (\sqrt{Z^3}) \cdot F$ (m^3/year) W is the average annual production of sediments (m^3/year), π is the mathematical constant 3.14159, F is the area of the catchment, h Mean annual amount of precipitation (mm/year), Z: Coefficient of erosion, T : Temperature coefficient.

The T factor is calculated using the following formula:

$$T=(to/10+0.1)1/2$$

Where to is the mean annual temperature for the average altitude of the mountain basin. In order to calculate the h factor, we used the same methodology we used for calculating the N factor, the USLE method. The Z is the coefficient of erosion calculated as: $Z = x \cdot y \cdot (\varphi + \sqrt{J})$, where J is the average slope of the surface of the basin as a tangent angle, x the factor expressing the geological medium resistance reduction during erosion, depending on the status, the cultivation area and the presence of vegetation, Table 3 (Globevnik et al. 2003).

Table 3. Values of x for different land uses according to the Corine 2000.

Use	Price x	Use	Price x	Use	Price x	Use	Price x
311	0.05	313	0.05	313	0.05	323	0.05
324	0.5	243	0.4	243	0.4		

The erosion coefficient of rock and soil resistance y, which is depended on the petrological and edafological composition of the basins, has its values according to the corresponding bibliographical tables (Kotoulas 2001). In the study area, the values used for the y factor is 0.3 for the Crystalline igneous, 0.9 for the schist and 1.2 for Neogene deposit. The f coefficient is the value for the observed erosion processes. The general table where all the values were taken from is found in the literature (Kotoulas 2001). For each basin there was a different value after considering all these factors.

2.2 Mobility coefficient

The mobility coefficient Kb, is a dimensionless parameter defined by the user. For back calculation, it is recommended to estimate K_{Bobs} using the empirical relation: $K_{Bobs} = B_{obs} \cdot V_{obs}^{-2/3}$, with B_{obs} representing the observed planimetric deposition area [m2], and V_{obs} the observed volume [m3] of the debris-flow event to be simulated. Due to the fact that TopRunDF, in this case, was used to predict potential deposition areas, kBpred, the mobility coefficient was calculated based on the average slope of the channel Sc as well as on the average slope of the fan Sf, as show below:

$$K_{Bred} = 5.075 Sf^{-0.1} \cdot Sc^{-1.68}.$$

In the cases where the model is used for prediction, it is necessary to take into account the factor of uncertainty. In this case, this factor is equal to 2. Using the extension 3D analyst of ArcGis 9.3 we calculated the average slope of the streams of that area and the slope of the cone sedimentation.

2.3 Start point of the simulation

The model is given the simulation starting point, with coordinates X (east) and Y (north). Those coordinates must lay within the applied digital terrain model and have to be defined in the same projection. Starting point can be a distinct change within the longitudinal flow-profile or obstacles forcing the debris flow to deposit. The user is asked to complete a lot of simulations in order to be able to achieve reasonable results. Therefore, we conducted several tests for determining the co-ordinates that lead to more plausible results.

2.4 Digital elevation model-DEM

The TopRunDF model was created to perform high-resolution digital elevation models derived from LIDAR on a grid with size 2.5 m x 2.5 m. The digital elevation model must be provided in an ascII form to ensure its independence from any GIS commercial program, (Scheild et al. 2008). In the present paper, due to the lack of measurement instruments, we used the classical digitalization of contour lines of 4 meters, which originate from georeferenced maps of the Hellenic Military Geographical Service. Following that, they were condensed by 1 meter, they were transformed to a cloud of points and finally we created a digital terrain model in ascII grid 2.5m x 2.5m, so as to be compatible with the requirements of the TopRunDF model.

2.5 Monte Carlo Recurrence Factor (MCI number)

The model uses the Monte Carlo iteration number to simulate lateral spreading of the flow area. The MCI number indicates the recurrences in the estimation of the deposit areas. It was experimentally estimated that the model behaves better when the repetition rate equals to 50.

2.6 Water Discharge

To calculate the water discharge, due to lack of data, we resulted in using empirical models. The most important are those of Friedrich $Q_{max}=24.12 \cdot F^{0.516}$, of Valentini $Q_{max}=(30/F^{1/2}) \cdot F$, of Wundt $Q_{max}=13.8 \cdot F^{0.6}$ and finally those of Klement-Wunderlich $Q_{max}=5.5 \cdot F^{5/6}$, where F is the area of basin.

3 Results

According to the USLE method the highest values of the sediments come to 37722.95 m^3/yr (basin 14), 31737.7 m^3/yr (basin 7), 29208.23 m^3/yr (basin 15), 22515.4 m^3/yr (basin 2), 17720.45 m^3/yr (basin 1). The corresponding results of the modified method Gavrilovič represent 31467.37 m^3/yr (basin 14), 28503.85 m^3/yr (basin 7), 25874.25 m^3/yr (basin 15), 22364. m^3/yr (basin 2), 19621.42 m^3/yr (basin 1). Therefore both methods are identical both quantitatively and in terms of classification. The most significant referring to debris flow is basin 14 while at the same time appears notable discharge which the mean value is 8.24 m^3/sec.

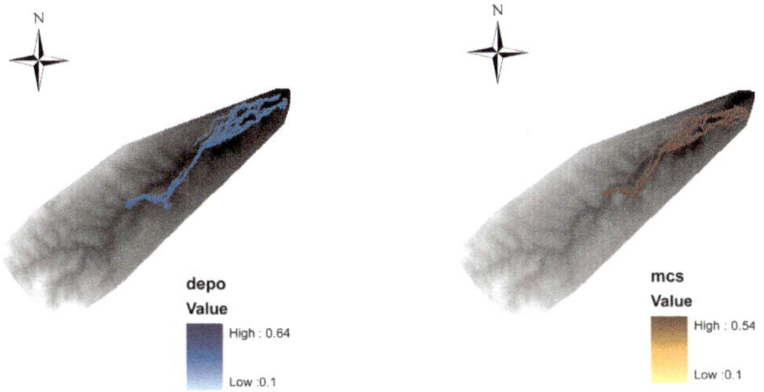

Fig. 2. a. maximum possible overflow of the sub-basin 14, **b.** amount of sediment deposition of the 14 sub-basin of the stream of Dadia.

Because of its importance, therefore, chosen be applied to the TopRunDF. Using GIS the Mobility coefficient is calculated to 64. The repetition number of Monte Carlo simulation is defined to 50. The total volume traded through the streams is about 40.000 m^3. The TopRunDF estimated 58.487 possible deposition areas from which 15600 it simulated. The maximum deposition height is 0.64 cm. The spatial representation of the results obtained in 2 separate files. The msc file shows the inundated simulation area combined with the overflow possibility of each related cell. The depo.asc file shows the deposition area and the deposition height of each cell (Fig. 2).

4 Conclusions

The stream is characterized by an important amount of Dadia torrential phenomena and its morphological evolution affects drastically the protected area. The upland watersheds of the area are characterized by the domination of crystalline-igneous formations, their large forest cover and the intense action of the climatic factors upon them. Specifically, the important precipitation and the intensive temperature fluctuations, especially in the areas upstream, are the factors which crystallize the influence of the torrential phenomena. Thus, the fragmentation and the foliation of the material is favored. The whole procedure results in the production of sediments. It is therefore necessary to apply the new model TopRunDF in order to be able to determine precisely the deposition areas of the debris-flow and their sedimentation, as well as the latters' implementation rhythm. In conclusion, TopRunDF enables early planning and exact location of the technical and agriculture projects that are required for the area's protection.

References

Chrysanthou V, Pyliotis A, (1995) Assessment of sediment influx into a reservoir under construction. 6° Pan-Hellenic Conference HHA, Thessaloniki pp.355-362

Da Silva AM, Alvares CA (2005) Levantamento de informacoes e estruturacao de um banco dados sobre a erodibilidade de classes de solos no estado de Sao Paulo. UNESP, Geociencias, 24:1, 33-41

Gavrilovic, Z, Stefanovic, M, Milocevic, M, Cotric, J (2006) ''Erosion Potential Method'' An Important Support for Integrated Water Resource Management. Balwois International Conference on Water Observation and Information System for Decision Support. Ohrid 23-26/5

Globevnik L, Holjevic D, Petkovsek G, Rubinic J (2003) Applicability of the Gavrilovič method in erosion calculation using spatial data manipulation techniques, Proceeding of the international symposium XXIII General Assembly of the international union of geodesy and geophysics,Sapoporo,Japan,8-9 July, pp.224-232

Hürlimann, M, Rickenmann D, Medina V and Bateman A (2008). Evaluation of approaches to calculate debris-flow parameters for hazard assessment. Engineering Geology 102, 152-163

Kotoulas D (2001) Arrangements of torrential streams. Part I. Publications of the Aristotle University of Thessaloniki, Thessaloniki (In Greek)

Ma Jianguo (2001) Combining the USLE and GIS/Arcview for Soil Erosion Estimation in Fall Creek Watershed in Ithaca, New York

Maris F, Karagiorgos K, Anastasiadis S, Vassiliou A, Karagiannis I (2006) Soil loss evaluation in the Polifitou Lake basin using Geographical Information System, International Conference on Sustainable Management and Development of Mountainous and Island Areas, Naxos, Greece,29 September-1 October, pp.302-313

Maris F, Vasileiou A (2010) Hydrology and the torrential environment In The Dadia-Lefkimi-Soufli forest national park, Greece: Biodiversity, management and conservation. WWF publication ISBN 978-960-7506-10-8

Matzarakis A (2006) Die Klimadaten in dieser Thesis stammen aus dem Buch Das Klima von Evros p 28

Scheidl, C and Rickenmann D (2010) Empirical prediction of debris-flow mobility and deposition on fans. Earth Surface Processes and Landforms, 5, 157-173, doi: 10.1002/esp.1897

Wischmeier, W. H. and Smith, D. 1965 Predicting Rainfall Erosion Losses from Cropland East of the Rocky Mountains. Agric. Handbook 282, U.S. Gov. Print. Office. Washington D.C

Zarris D, Lykoudi E, Koutsogiannis D, (2002) Sediment Yield Estimation from a Hydrographic survey: A case study for the Kremasta reservoir basin. At 5th International Conference, Water Resources Management in the Era of Transision. Athens 4-8/9, pp. 338 - 345

Continuous media Hydrogeology

Modeling of groundwater level fluctuations in agricultural monitoring sites

V. Vircavs, V. Jansons, A. Veinbergs, K. Abramenko, Z. Dimanta, I. Anisimova, D. Lauva, A. Liepa

Department of Environmental Engineering and Water Management, Latvia University of Agriculture, Akadēmijas 19, Jelgava, LV 3001, Latvia, valdis.vircavs@llu.lv

Abstract The interaction of surface water and groundwater was studied using modeling with regard to climate change. Groundwater level was simulated for reference climate period (1961–1990) and future climate period (2071–2100). Groundwater level difference between minimal and maximal values in both periods as well as seasonal tendencies of water level fluctuations in two monitoring sites Bērze and Mellupīte and impact on drainage runoff was determined. The objective of presented study is to investigate the overall status of fluctuations of groundwater level and to evaluate the possibility to use improved METUL model for the groundwater level modeling. However, it is important to consider that climate change could have a significant impact on many aspects of the hydrological processes, especially on groundwater level fluctuations. Relationship between meteorological conditions and groundwater are determined as seasonal variations of water level. Climate change directly influenced water cycle elements especially shallow groundwater as a significant part of water resources. The results show that groundwater level variations might be higher in the future than currently expected.

1 Introduction

Both groundwater regime and balance has a significant role in the agricultural land as an integral part of water cycle - surface and subsurface water interaction. Most of the groundwater resources, especially shallow aquifers below soil surface are sensitive to the contamination and other anthropogenic factors (Okkonen et al. 2009).

Numerous studies on groundwater level and flow modeling have been carried out elsewhere using the model HBV by Bergström (1992) and the model MODFLOW by McDonald and Harbaugh (1988). Recent studies on the groundwater level modeling show observed calculated and forecasted changes as mentioned by Almedeij and Al-Ruwaih (2006). Impacts of the future climate on hydrological cycle, especially on water level fluctuations are discussed by Dragoni

(1998) and Dragoni and Sukhija (2008). Moreover, in order to simulate impact of the future climate on water level, 11 regional climate model (RCM) data (Sennikovs and Bethers 2009) in this study were used.

The 11 future models were selected from different RCM models, which covered and provided climate data for modeling period (2071–2100). Climate change impact, especially on groundwater level in agricultural areas has not been studied in Latvia. In agricultural areas where intensive farming activity take place, high groundwater level is typical that makes groundwater sensitive and increases the risk to be highly influenced by climate change. The Latvia University of Agriculture (LUA) started groundwater monitoring in 7 observation wells in two agricultural run-off monitoring sites Bērze and Mellupīte in 2005. In Latvia, the amount of precipitation is higher than evaporation and surface runoff. Therefore, in the most of the agricultural lands in Latvia subsurface drainage systems for water level control have been constructed. The conceptual groundwater model METUL for Latvia was prepared by Krams and Ziverts (1993). This model is site oriented two dimensional mathematical model based on daily meteorological data. Previous studies of water level modeling in Latvia were used for design purposes of agricultural land drainage.

The main goal was to determine water regime and to control water level in agricultural areas for better farming conditions in spring and autumn seasons. In this study METUL model was calibrated using measurements of the daily water level data for the period (2006–2009). Groundwater level modeling in future climate conditions with METUL model RCM meteorological data (Goderniaux et al., 2009) was modified by Sennikovs and Bethers (2009). Reference climate period (1961–1991) was compared with future climate period (2071–2100) with regard of impact on groundwater level changes. The aim of the present study is to calibrate and adapt parameters of the METUL conceptual mathematical model using different regional climate model data and to verify the modeled groundwater level fluctuations at two agricultural monitoring sites during period 2006–2009.

2 Materials and methods

Description of the agricultural run-off monitoring sites

Agricultural run-off monitoring sites Bērze and Mellupīte are located in different parts of Latvia and represent regions with different climate, soil, slopes, crops and farming intensity (Fig. 1). The Bērze agricultural monitoring site is located in the central part of Latvia in Zemgale plain lowlands. The thickness of quaternary deposits is 10 – 20 m. Calcic Cambisol according to FAO classification (2006) is the dominating soil type in Bērze monitoring site. The composition of the top soil is the following: 10.5% clay, 54.96% silt and 34.54% sand. In monitoring site four wells (BG-1, BG-2, BG-3 and BG-4) have been constructed. BG-1 is 23.0 m deep and location of screen within the profile is 13.0 – 17.0 m., BG-2 and BG-4 is 5.70 m deep and location of screen is 2.0 – 6.0 m. BG-3 is 8.0 m deep and location of screen is 4.0 m - 7.70 m.

Fig. 1. The LUA groundwater monitoring sites in Latvia.

The Mellupīte agricultural monitoring site is situated in the western part of Latvia. The thickness of quaternary deposits is 10 – 20 m. According to the granulometric composition dominated soil is clayey loam. Most of the agricultural land is drained with tile drains (depth 1.1 – 1.3 m, spacing between drains 10 – 32 m) Vagstad et al. (2001). The composition of the top soil is 11.9% clay, 40.95% silt and 47.15% sand. In monitoring site three wells (MG-1, MG-2, and MG-3) have been constructed. MG-1 is 7.0 m deep and location of screen within the profile is 2.0 - 6.70 m. MG-2 is 4.5 m deep and location of screen is 2.0 m - 4.20 m. MG-3 is 11.0 m deep and location of screen is 6.0 m - 10.70 m.

Description of METUL model

The METUL model is a two dimensional conceptual mathematical groundwater level model based on the observed daily meteorological data. The data for the groundwater level modeling has two main data sets: meteorological data (precipitation, air temperature, relative humidity) for model input and groundwater level observation data (borehole data) for the model calibration (Fig. 2).

Calculation of water balance consists of three parts: the estimation of snow cover, estimation of the water balance of the vadose zone and estimation of groundwater levels (Krams and Ziverts 1993).

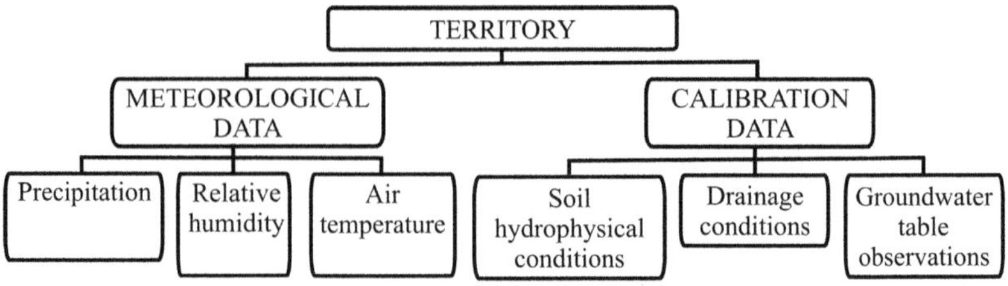

Fig. 2. Scheme of data necessary for groundwater level modeling with METUL model.

Data sets for modeling

Meteorological data were measured in the nearest meteorological stations of Latvian Environment, Geology and Meteorology Centre (LEGMC) in Saldus and Dobele. In the mentioned agricultural monitoring sites Bērze and Mellupīte groundwater level and temperature data of groundwater wells were measured using data logger "Geolog micro" with accuracy of 1 mm and frequency of measurements 1 h. In the Ensembles project (Hewitt and Griggs 2004) 11 RCM were developed for future climate prediction. For calculation of mean meteorological parameters RCM data sets modified by Sennikovs and Bethers (2009) were used.

Statistical methods

Coefficient of determination (R^2) was used to compare the observed and modeled data for the METUL model calibration. Characteristics of groundwater regime involve the arithmetical mean, minimal and maximal values of parameters. Estimation of the differences between water levels in reference with the climate period and the future climate period was carried out using the t-test and F-test.

3 Results and Discussion

The meteorological factors, such as, precipitation, air temperature and relative humidity, has impact to both daily water level fluctuations and seasonal changes. Water balance components are closely related to each other (Jan et al. 2007). The impact of one component on another during different seasons and intensity was discovered. The observed and modeled water level fluctuations in calibration period (2006–2009) of the Bērze and Mellupīte agricultural monitoring sites are analyzed and shown in Figure 3.

Groundwater recharge mainly depends on percolation from the vadose zone to groundwater as mentioned by Goderniaux et al. (2009). The observed data show that water table depth varies from 0.90 m to 0.70 m below soil surface in October, November and December in Bērze and Mellupīte monitoring sites. In January and February, sometimes in March, water level decreases approximately by 0.5 m, but in spring period (April, May) - increases again up to the 0.70 m below soil surface. The precipitation increases, while the relative humidity decrease and that relation of the increasing air temperature and relative humidity was observed in summer. In the studied monitoring sites Bērze and Mellupīte water level fluctuates insignificantly approximately 1.90 m below soil surface during summer months (June, July, August, and September) as shown in Figure 3.

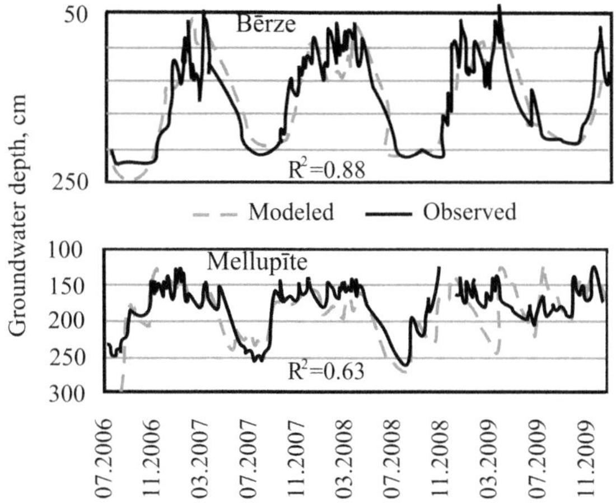

Fig. 3. Observed and modeled groundwater level in Bērze and Mellupīte monitoring sites.

The coefficient of determination (R^2) gives the proportion of the fluctuation of the predicted one variable in comparison with other variable set (Steel and Torrie, 1960). The calculated determination coefficient testifies the following values: R^2=0.88 in Bērze, R^2=0.63 in Mellupīte monitoring site (Fig. 3). The 11 RCM were used for predictions of future groundwater level. The obtained data show the increase of annual precipitation rates from 1.7 mm to 2.1 mm per day and mean temperatures from 6.2°C to 9.7°C. In the future climate period minimal water level depth will be increased in Bērze and Mellupīte monitoring sites, while the maximum groundwater level depth will decrease at the same time. The obtained and modeled data of minimum and maximum water levels are shown in Table 1.

Comparisons of seasonal changes of groundwater level fluctuations for both modeling periods are shown in Figure 4b. The pattern of water table fluctuations is similar in Bērze and Mellupīte monitoring sites within winter and summer seasons. However the most significant changes could be expected in winter season. Over 30 year period trends of water level fluctuations for reference and future period show the similar trends for Bērze monitoring site. In the future groundwater level could increase in winter and spring seasons but could decrease in autumn seasons (Fig. 4b).

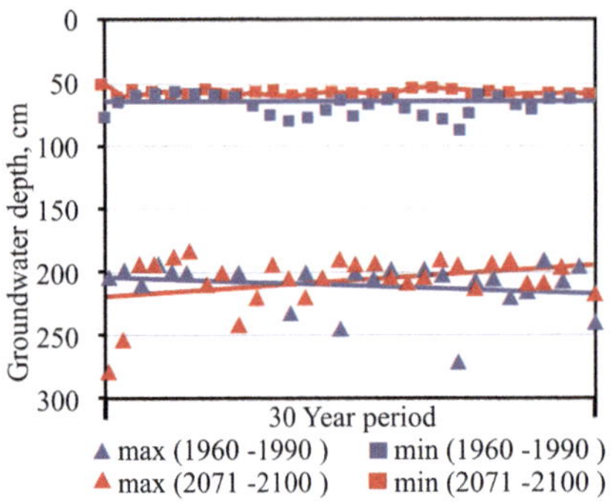

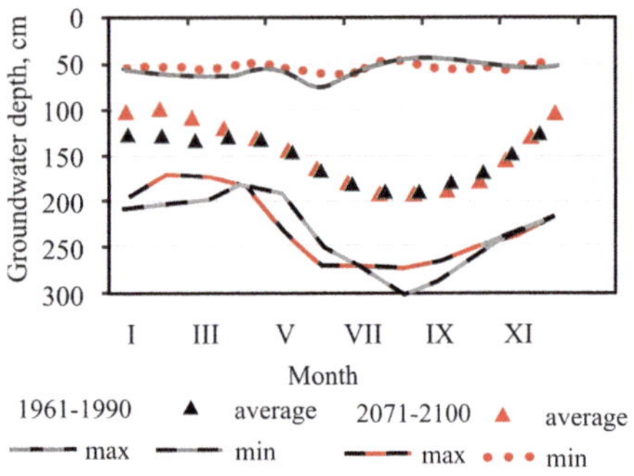

Fig. 4. a. Seasonal water level fluctuations in Bērze monitoring site. **b.** Modeled groundwater level fluctuations in Bērze site.

The modeled data show that due to climate changes water levels reach the levels closest to soil surface (minimum depth) with a positive trend and the deepest water levels with expressed negative trend (maximum depth). However, the main changes could be forecasted for minimum water levels (Fig. 4b, Table 1).

Table 1. Trend changes of water levels in LUA groundwater monitoring sites.

Period	Bērze		Mellupīte	
	Min	Max	Min	Max
1961-1990	0%	-7.5 %	-4.1 %	-5.9 %
2071-2100	-5.5 %	+10.1%	-7.4 %	+1.8%

Between water level and subsurface drainage runoff in Bērze monitoring site correlation coefficient is (R^2=0.75) (Fig. 5a). Runoff of subsurface drainage was observed if water level reached 2.0 m below soil surface.

According to climate changes the calculated probability shows that groundwater depth above 1.0 m and the intensive drainage runoff could be increased from 16% (58 days/year) in reference period to 28% (102 days/year) in the future climate period in Bērze monitoring site (Fig. 5b). Drainage runoff starts in water level ~ 2.0 m below soil surface (Fig. 5b).

Between water level and subsurface drainage runoff in Mellupīte monitoring site correlation is (R^2=0.67). If water level is ~ 2.20 m below soil surface in drainage area, runoff of subsurface drainage is starting. The similar calculations set increase of groundwater level that could prolong drainage runoff by 9% (33 days/year) and intensive runoff by 14% (51 days/year). The possible changes of water level fluctuations are considered in conditions of climate change. Results show that despite of the increase of annual precipitation rates in modeled future climate period (2071-2100) compared to reference climate period (1961-1990) water levels could be decreased. During summer and autumn seasons water level fluctuation amplitudes may increase in future period, but the average groundwater level could decrease. On the other hand expected higher water level in winter and spring seasons could prolong subsurface drainage runoff in the future climate period.

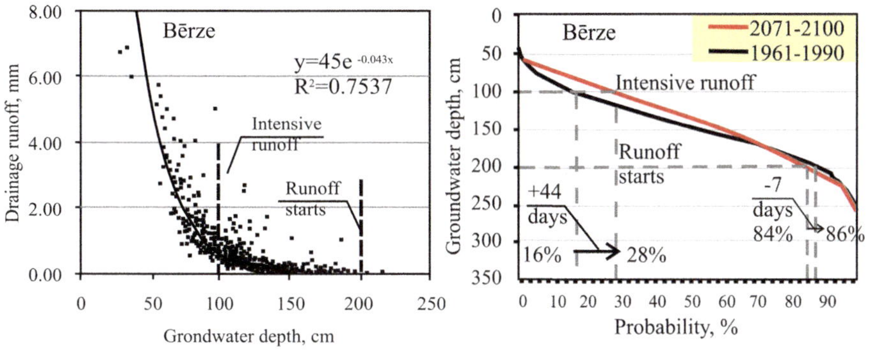

Fig. 5. a. Impact of water level on drainage runoff in Bērze site, b. Probability curves of groundwater level in Bērze site.

4 Conclusions

This study recommended that application of METUL model for prediction of water level could be used for reference period and future climate period. After calibration and verification process METUL model for groundwater level modeling could be used for water table simulation in agricultural run-off monitoring sites Bērze and Mellupīte. Climate change impact on groundwater regime fluctuations in agricultural areas of Latvia firstly has been occurred. Significant groundwater level changes in future (2071–2100) in winter and spring seasons are expected. It suggests that water level could increase in these seasons, while in summer and autumn seasons water level could decrease in comparison with the referred climate period (1961–1990). Groundwater aquifer shows its vulnerability to seasonal variations of water level fluctuations and can be attributed to climate impact in time period 2071–2100. It must be taken into consideration that the rise of groundwater regime amplitudes within seasons might influence water resource availability in the future. It is concluded that climate change impact with regard to agricultural activities should be restricted. Further research could be based on upscaling method of another territory with different soil type, infiltration conditions and other hydrophysical circumstances. Moreover the autocalibration will be implemented on METUL. The algorithm of autocalibration Nelder-Mead optimization method allows finding parameters which give the best correlation between modeled and observed groundwater levels. This work has supported by the ESF project "Establishment of interdisciplinary scientists group and modeling system for groundwater research".

References

Almedeij J, Al-Ruwaih F (2006) Periodic behavior of groundwater level fluctuations in residential areas. J.Hydrol. 328, 677– 684

Bergström S (1992) The HBV model – its structure and applications. SMHI Report Hydrol. 4, Norrköping

Dragoni W, Sukhija BS (2008) Climate change and groundwater: a short review. In: The Geological Society of London (eds.) Climate change and groundwater. Special Publications pp. 1–12. London

Dragoni W (1998) Some considerations on climatic changes, water resources and water needs in the Italian region south of the 43°N. In: Issar, A. S., Brown, N., (eds.) Water, Environment and Society in Times of Climatic Change, pp. 241–271. Kluwer, Dordrecht

FAO (2006) Guidelines for soil description. 4th edition. Report of the Food and Agriculture Organization of the United Nations (FAO), Rome

Goderniaux P, Brouyere S, Fowler HJ, Blenkisop S, Therrien R, Orban P, Dassargues A (2009) Large scale surface-subsurface hydrological model to assess climate change impacts on groundwater reserves. J. Hydrol. 373, 122–138

Hewitt CD, Griggs DJ (2004) Ensembles-based Predictions of Climate Changes and their Impacts. Eos, 85

Jan CD, Chen TH, Lo WC (2007) Effect of rainfall intensity and distribution on groundwater level fluctuations. J. Hydrol. 332, 348–360

Krams M, Ziverts A (1993) Experiments of conceptual mathematical groundwater dynamics and runoff modelling in Latvia. Nordic Hydrol. 24, 243–262

McDonald MG, Harbaugh AW (1988) A modular three dimensional finite difference groundwater flow model. U.S. Geological Survey Techniques of Water Resources Investigations. Book 6, A1, Washington

Okkonen J, Jukama M, Kløve B (2009) A conceptual approach for assessing the impact of climate change on groundwater and related surface waters in cold regions (Finland). Hydrogeol. J. 18, 429–439

Sennikovs J, Bethers U (2009) Statistical downscaling method of regional climate model results for hydrological modelling. 18th World IMACS/MODSIM Congress, 3962–3968

Steel RGD, Torrie JH (1960) Principles and procedures of statistics. McGraw-Hill, New York

Vagstad N, Deelstra J, Loigu E, Iital A, Jansons V, Sileika S,S (2001) Monitoring and assessment of nonpoint source pollution - Baltic Sea Regional project. Jordforsk - Norwegian Centre for Soil and Environmental Research Report 49/01

Groundwater level monitoring and modelling in Glafkos coastal aquifer

A. Ziogas, V. Kaleris

Department of Civil Engineering, University of Patras, Rio-Patras, 26500, Greece.
alzio@upatras.gr, kaleris@upatras.gr

Abstract Glafkos coastal aquifer is located at north Peloponnesus (Greece), south of the city of Patras. It is the most important source of fresh water for irrigation, industrial use and water supply of the city of Patras. However, a competent groundwater management policy for the protection of groundwater resources against pollution due to human activities and the impacts of a possible change in climate has not been established. A calibrated groundwater model is the basis for the implementation of such a management plan. Despite the fact that significant information about the geology and the hydrogeology of the aquifer has been presented in previous studies there exist considerable uncertainties concerning the recharge mechanisms of the aquifer. In order to get insight into the dynamic behaviour of the aquifer a groundwater level monitoring network consisting of seven boreholes has been constructed during the years 2007-2008 in cooperation with the Municipal Enterprise of Water Supply and Sewage of Patras (DEYAP). From the electrical resistivity profiles registered in the seven boreholes, hydraulic conductivity profiles have been derived. From the analysis of the time series of the groundwater level, differences in the behaviour of the seven observation wells have been identified, which are important for understanding and modeling the dynamics of the aquifer and for the delineation of the capture zone of the water supply wells.

1 Introduction

The plain basin of Glafkos River is located at north Peloponnesus (Greece), south of the city of Patras (Fig. 1). It is formed mainly by Quaternary deposits which host the Glafkos coastal aquifer (Lambrakis et al. 1997). The eastern, landwards part of the aquifer consists of coarse-grained sands and pebbles and is characterized as an unconfined aquifer while the coastal part consists of fine-grained sediments where the succession of multiple low permeability layers of clays and high permeability layers form the confined coastal part of the aquifer (Mandilaras et al. 2008).

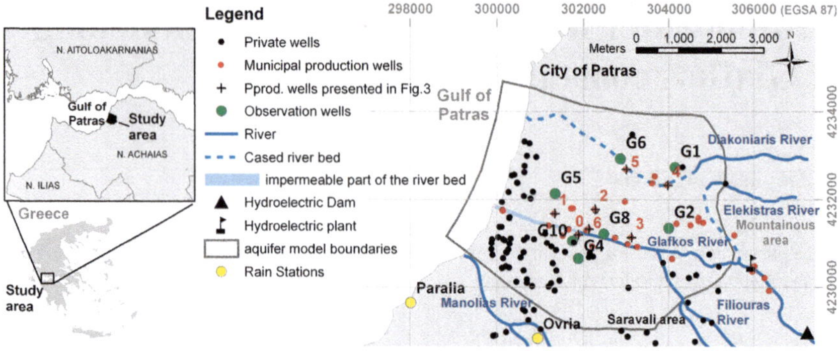

Fig. 1. Glafkos coastal aquifer.

A large number of pumping wells (Fig. 1) has been constructed in the coastal plain of Glafkos River during the past four decades, to account for water demand for irrigation, industrial use and water supply of the city of Patras. Due to saltwater intrusion problems which arose in the early 90's, wells located in the coastal zone at a distance less than 1 km were set inactive and the Municipal Enterprise of Water Supply and Sewage of Patras (DEYAP) rearranged the pumping schedule of municipal wells by operating them only during the dry period of the year, i.e. approximately from May to October. For the remaining period of the year surface water from Glafkos River has been used. The majority of the industrial wells are inactive since the mid of 90's, following the reallocation of local industries, which occurred at that time. The use of the groundwater of Glafkos aquifer has been reduced but the groundwater resources are still of significant importance both for the water supply during the dry period of the year as well as for the irrigation. Therefore, for the development of a groundwater management policy for the protection and sustainable use of the groundwater, a calibrated groundwater model is required. Although information on the geology and hydrogeology of Glafkos aquifer has been presented in detail in previous studies (Voudouris 1995, Mandilaras 2005), the recharge mechanisms still remain unclear and the available data are not sufficient for the calibration of a groundwater model (Ziogas 2006). For this purpose, a groundwater level monitoring network was designed and constructed in cooperation with DEYAP during 2007-2008 (Fig. 1). This work presents information obtained from the construction and operation of this network and the use of this information in order to appropriately simulate the recharge of the aquifer.

2 Groundwater level monitoring network – hydrogeological parameters

The network of the seven monitoring wells (Fig. 1) was constructed in cooperation with DEYAP in the frame of an INTERREG project. The depth of the wells ranges from 90 to 120m. The diameter of the boreholes is $11^{7/8}$in and the diameter of the screens is 6in. The space between the borehole wall and the screens is filled with gravel. Before the installation of the screens electrical resistivity logs were recorded.

Investigations concerning the hydraulic conductivity and the porosity of the aquifer have been presented by Lambrakis et al. (1997), Voudouris (1995) and Mandilaras (2005). The presented values were estimated by means of pumping tests and therefore they are depth integrated. The analysis of the electrical resistivity profiles recorded in the seven monitoring wells during their construction led to the estimation of the depth distribution of the hydraulic conductivity (Fig. 2a) assuming isotropic conditions. The analysis is based on the formation factor F, which is defined as the ratio of the resistivity Ro of the porous material saturated with an electrolyte, to the bulk resistivity Rw of the same electrolyte (Archie 1942, cited in Bear 1972). Ro can be measured in the field using a two electrode arrangement, called the normal log (Driscoll 1986). In fresh water aquifers, the electrolyte is the aquifer water. Thus, Rw results from measurements of the water electrical conductivity. From the known values of Ro and Rw the porosity profile of the formation is estimated using an appropriate relationship between F and the porosity. Then, from the porosity profile, the hydraulic conductivity profile is estimated using an appropriate hydraulic conductivity-porosity relationship. The analysis of the resistivity profiles of the seven observation wells as well as the evaluation of the assumptions, on which the method is based, is presented in Kaleris and Ziogas (2011).

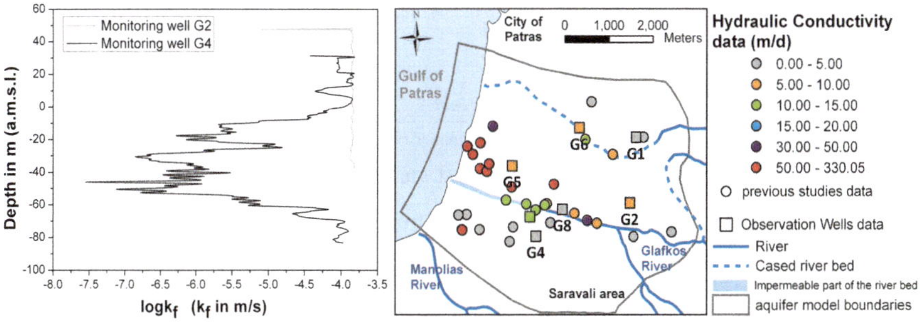

Fig. 2 a. Hydraulic conductivity profiles for two monitoring wells obtained from the analysis of electrical resistivity profiles, **b.** Locations, at which investigations to hydraulic conductivity have been performed - Depth integrated values.

The estimated hydraulic conductivity values from previous studies and the depth integrated values from the hydraulic conductivity profiles resulted from the discussed previously analysis are generally in good agreement (Fig. 2b). Although the above estimates cover spatially a significant part of Glafkos aquifer, the spatial distribution of the observation points in the aquifer does not allow the estimation of a reliable variogram for a kriging interpolation. Therefore, a regionalization of the hydraulic conductivity values has been performed empirically.

3 Groundwater level data

Before the installation of the groundwater level monitoring network mentioned above, groundwater level measurements were conducted by DEYAP at 17 municipal pumping wells. Such measurements are available for the period 1982-2009 and have been performed on a monthly basis. The time series of some of the wells exhibit large data gaps. However, the most significant problem with these data is that many of the measurements were conducted when the pumping wells were in operation. These values are strongly influenced by pumping and are not appropriate for the investigation of the recharge mechanisms of the aquifer. The monitoring network of the seven observation wells is operating since July 2008. In the period July 2008 to March 2009 groundwater levels were registered manually in monthly intervals. Since April 2009 groundwater levels are automatically registered on a daily basis. As shown in Fig. 3 and as it is expected, the behaviour of the groundwater level observed in the monitoring wells is totally different from the behaviour observed in the pumping wells.

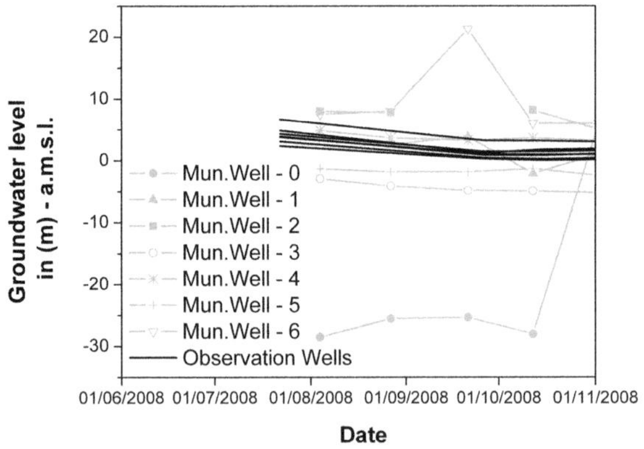

Fig. 3. Comparison between groundwater level data from the observation wells and the municipal pumping wells (see Fig. 1) operated by DEYAP.

The lowest recorded value of water level is 0.1m and the largest one is 17.0m. The pattern of the groundwater level variation in time is similar in all of the observation wells, as it is shown in Fig. 4a. Further, groundwater levels are very well correlated to each other with correlation coefficients between 0.93 and 1.00.

For the investigation of the recharge mechanisms a harmonic analysis of the groundwater level time series of all the wells has been performed for the two complete hydrological years contained in the time series, i.e. from 01/10/2008 to 30/09/2010. The time series of the daily values of each well contains two significant harmonics, one with a period of two years and one with a period of one year (annual cycle). The first one corresponds to an intermediate term variation of the groundwater level due to the corresponding variation of the hydrological regime and should be considered as a non-linear trend. The 2-year and the 1-year harmonic explain together more than 95% of the variance of the time series of each observation well. This implies that the time series of the groundwater level do not contain any significant high frequency harmonics (short term fluctuations).

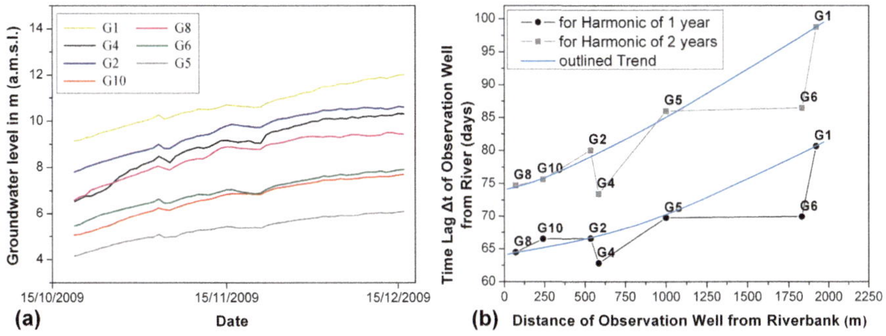

Fig. 4. a. Groundwater level time series obtained from the seven observation wells, **b**. Comparison of the time lag of the water level in each well for its 2-year and 1-year harmonic in refer to the corresponding harmonics of the river discharge.

Groundwater dynamics in the Glafkos aquifer seems to be strongly influenced by the river (see also Lambrakis et al. 1997; Voudouris 1995; Mandilaras 2005). In order to investigate the effect of the river on the groundwater level we estimated the harmonics of the time series of river discharge and compared them with the harmonics of the groundwater level measured in the observation wells. River discharge measurements are conducted on a daily basis at the hydroelectric dam of Glafkos River (Fig. 1) and have been available by the Public Power Corporation (DEH). The river discharge contains two significant harmonics with similar periods as these of the groundwater level, i.e. with a period of one and two years. However, these harmonics explain only 50% of the variance of the time series. This means that the discharge time series contain a large number of high frequency components (short term fluctuations) each of which explains a small portion of the variance. These short term fluctuations do not influence the groundwater level. Between the corresponding harmonics of the river discharge and the groundwater level there is a time lag which is shown in Fig. 4b as a function of the distance of the observation wells from the riverbank. According to these results the response time of the observation well G4 to the time variation of the river discharge is the smallest one, despite the fact that its distance from the river is larger than that of the observation wells G8 and G10. The largest response time has the observation well G1. This response time is significantly larger than that of the well G6, which is located approximately at the same distance from the riverbank as the well G1. As it is shown in Fig. 4b, the dependence of the lag time of the wells G6 and G4 strongly deviates from the trend outlined by the observation wells G1, G2, G5, G8 and G10. The importance of the temporal characteristics of the groundwater level for the model calibration is shown in the following section.

4 Preliminary results of simulations

A measure of the deviations between measured and predicted time series of groundwater levels used in groundwater simulations is:

$$Crh_j = 1 - \sum_i^N \left(h_{sim,j,i} - h_{obs,j,i}\right)^2 \Big/ \sum_i^N \left(h_{obs,j,i} - \overline{h}_{obs,j}\right)^2 \tag{1}$$

where $h_{obs,j,i}$ and $h_{sim,j,i}$ are the observed and simulated, respectively, groundwater level on day i at observation well j and $\overline{h}_{obs,j}$ is the mean groundwater level at observation well j for the simulated period. Crh_j expresses the relationship between the variance of the difference between measured and simulated groundwater levels and the variance of the observed groundwater levels.

However, an aspect which is not sufficiently evaluated by the criterion in Eq.(1) is the timing error or phase shift error between observed and predicted time series of groundwater level. The timing error $Cr\Delta t_{j,k}$ can be defined as the time lag (phase shift) between the k-th significant harmonic of the predicted time series of the groundwater level in well j and the corresponding harmonic of the observed groundwater level. The existence, in the predicted time series, of similar harmonics as in the observed time series is a basic condition for the reliability of a transient groundwater model.

For the calibration of the groundwater model of Glafkos aquifer both measures, Crh_j and $Cr\Delta t_{j,k}$, have been used. The model of Glafkos aquifer is based on the code SEAWAT (Langevin et al. 2003), which takes into consideration the density differences between the seawater and the fresh groundwater. The boundary of the modeled area is shown in Fig. 1 and 2. Details about the geometrical, geological and hydraulic parameters of the aquifer used in the model as well as about the boundary conditions, the time dependent hydrologic input variables and the model discretization are given in Ziogas (2011).

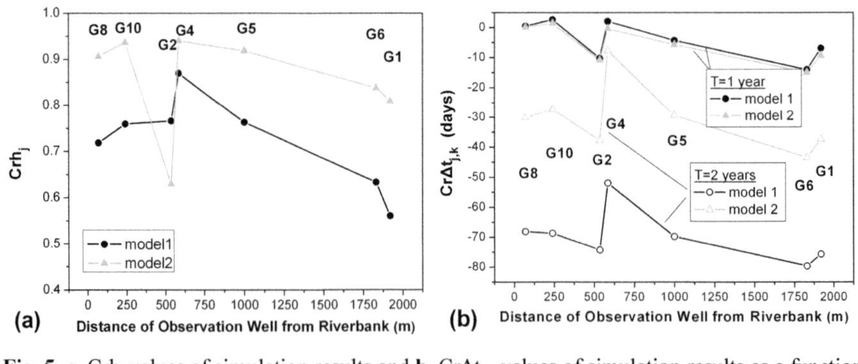

Fig. 5. a. Crh_j values of simulation results and **b.** $Cr\Delta t_{j,k}$ values of simulation results as a function of the distance of the observation wells from the riverbank for simulations performed with different combinations of the model parameters.

Figure 5 shows the values of Crh_j and $Cr\Delta t_{j,k}$ for two models of the Glafkos aquifer, whose main differences are summarised in Table 1. Groundwater recharge in the two models differs considerably concerning both its annual volume as well as its time variation.

As shown in Fig. 5a, the difference between the Crh_j values resulting for the two models is for the most observation wells less than 25%. Increased differences of about 30% and 45% result for the observation wells G6 and G1 respectively. However, as is shown in Fig. 5b the reduction of phase shift for the 2-year harmonic in case of model 2 compared to the phase shift of model 1 is for all the observation wells larger than 45%. This implies that $Cr\Delta t_{j,k}$ is more sensitive against the groundwater recharge mechanisms of the aquifer, i.e. against the volume and the distribution of the groundwater recharge, compared to Crh_j. Consequently, the phase shift between measured and predicted time series represents, in combination with Crh_j, a valuable measure for the evaluation of the groundwater recharge mechanisms.

Table 1. Values of the main hydrologic input variables in models 1 and 2.

Hydrologic year	Groundwater recharge from:		Pumping ($x\ 10^6\ m^3$/year)
	river infiltration ($x\ 10^6\ m^3$/year)	boundary fluxes ($x\ 10^6\ m^3$/year)	
		Model 1*	
2008 – 2009	10.8	8.1	9.3
2009 – 2010	11.9	7.8	9.3
		Model 2**	
2008 – 2009	11.7	8.1	6.8
2009 – 2010	10.9	7.0	9.3

* 45% and 30.5% of the annual infiltration volume is concentrated at the beginning of the hydr. years 2008-2009 and 2009-2010 respectively.

** 49% and 28% of the annual infiltration volume is concentrated at the beginning of the hydr. years 2008-2009 and 2009-2010 respectively.

5 Conclusions

The construction of a network of seven observation wells in Glafkos aquifer provided valuable information about the geological structure of the aquifer. The electrical resistivity logs recorded during the construction of the wells allow the insight into the vertical hydraulic conductivity distribution of the aquifer and the appropriate representation of the geology in the developed groundwater model.

The groundwater level time series obtained from the operation of the monitoring network since July 2008, allow a better understanding of the groundwater dynamics. The harmonic analysis of the time series of the groundwater level and its correlation with the harmonic analysis of the time series of the river discharge shows the importance of the river infiltration for the groundwater recharge.

The phase shift of the predicted time series of the groundwater level compared to the observed time series seems to be an interesting measure for the model calibration in the strongly dynamic Glafkos aquifer. Preliminary results obtained from a SEAWAT groundwater model for the aquifer show that phase shift is more sensitive to the groundwater recharge mechanisms than criteria based on the groundwater level.

Acknowledgments This research was supported in part by the EU-project INTERREG IIIA Greece-Italy 2000-2006 (grand I3101034) and in part by the project UFZ-02/2009 (RA-3205/09) of the Helmholtz-Zentrum fuer Umweltforschung GmbH-UFZ.

References

Bear J (1972) Dynamics of Fluids in Porous Media, Dover Publications, Inc., USA

Driscoll FG (1986) Groundwater and Wells, Jonson Division, Minnesota, USA

Kaleris V, Ziogas A (2011) An attempt to estimate hydraulic conductivity profiles using borehole resistivity logs (in preparation)

Lambrakis NJ, Voudouris KS, Tiniakos LN, Kallergis GA (1997) Impacts of simultaneous action of drought and overpumping on Quaternary aquifers of Glafkos basin (Patras region, western Greece). Environ. Geol.29 (3/4)

Langevin CD, Shoemaker WB, Guo W (2003) MODFLOW-2000, the U.S. Geological Survey Modular Ground-Water Model-Documentation of the SEAWAT-2000 Version with the Variable-Density Flow Process (VDF) and the Integrated MT3DMS Transport Process (IMT).US Geological Survey. Open-File Report 03-426

Mandilaras D (2005) Environmental-hydrogeological study of Glafkos River basin. PhD Thesis. Laboratory of Hydrogeology, Department of Geology, University of Patras, p 435 (in Greek)

Mandilaras D, Lambrakis N, Stamatis G (2008) The role of bromide and iodide ions in the salinization mapping of the aquifer of Glafkos River basin (northwest Achaia, Greece). Hydrol. Process. 22, 611-622

Voudouris K (1995) Hydrogeological conditions in northwest Achaia. PhD Thesis. Department of Geology, University of Patras (in Greek)

Ziogas A (2006) Water budget of Glafkos aquifer and estimation of the capture zone of pumping wells. Master thesis. Hydraulic Engineering Laboratory, Department of Civil Engineering, University of Patras (in Greek)

Ziogas A (2011) Monitoring and management of coastal aquifers with seasonally varying extraction and recharge. PhD Thesis. Hydraulic engineering Laboratory, Department of Civil Engineering, University of Patras (in preparation)

A data-driven model of the dynamic response to rainfall of a shallow porous aquifer of south Basilicata - Italy

A. Doglioni, A. Galeandro, V. Simeone

Engineering Faculty of Taranto, Technical University of Bari, Viale del Turismo 8, 74123, Taranto, Italy. a.doglioni@poliba.it

Abstract The analysis of the dynamic response of a shallow aquifer to rainfall is a hot topic for groundwater resources management. A data-driven methodology namely Evolutionary Polynomial Regression is here undertaken to perform this kind of analysis. The introduced methodology is an evolutionary modeling approach, based on the multiobjective evolution of a population of explicit models. The methodology is here applied to a particular case study involving a shallow porous aquifer located in south Italy, showing some peculiar characters. The outcome of this approach is here discussed focusing on the hydrogeology of the aquifer. Then some hydrological considerations on the results are presented. Finally, the obtained model is consistent with past studies on the same aquifer by other authors and based on different methodologies.

1 Introduction

Water resources management is a paramount problem for dry and semi-arid regions, notoriously affected by periodic dry periods (Siegfried and Kinzelbach 2006). In fact, a precarious balance between rainfall and water resources characterizes these. Custodio (2002) emphasizes that groundwater resources management, depends on the detailed and updated characterization of the aquifer dynamic analysis. Such management should be based on a combination of rainfall analysis and monitoring, aquifer characterization and, in general, system modeling. Evolutionary Polynomial Regression (EPR) is an evolutionary modeling technique (Giustolisi and Savic 2009) successfully applied to multiple problems related to natural systems (Doglioni et al. 2010; Giustolisi et al. 2008). It proved quite effective at modeling the dynamic relationship between groundwater levels and rainfall heights for specific cases, related both to porous and karst aquifers (Doglioni et al. 2011; Giustolisi et al. 2008, Mancarella and Simeone 2008). Although, EPR is a data driven model based on a regressive approach, the returned models are explicit polynomial equations, and this allows for conjecturing on the relationship between the main variables of the investigated phenomena. This allows for highlighting some unclear or unknown relations and to indirectly get information about the physics of the phenomenon at stake. This paper introduces an application of EPR to a very shallow porous aquifer, whereas the dynamic relationship between rainfall and groundwater table is pursued. The aquifer flow across sand strata, while at the topsoil there is a silt layer. Therefore differently from the aforementioned works, in this aquifer groundwater occasionally flows confined under pressure (Polemio et al. 2003), when the piezometric level is higher than the bound between sand and silt layers. Due to the alternate groundwater flow both as a free and sometimes as a confined aquifer, model identification can be challenging. The modeled case-study consists in 25 years data, concerning total monthly rainfall and average monthly groundwater levels measured in a sampling well. EPR-MOGA results are compared with the outcome from a similar investigation attempted on the same aquifer according to a statistical approach (Polemio 1994). Presented results are pursuant with those obtained by past studies. Finally, EPR-MOGA already proved to be particularly fit to those cases where the input to the process and the boundary conditions are not easily accessible, as for the case-study.

2 The Evolutionary Polynomial Regression - EPR

EPR-MOGA is a multiobjective evolutionary modeling technique by Giustolisi and Savic (2009), which allows for the identification of a set of explicit equations starting from a measured dataset. These models show variable performance both in terms of fitness to data and structural complexity of the equations. Therefore, their cross comparison can help the choice of the model which better represent the trade-off between structure of the equation and fitness to data. EPR-MOGA is a two-stage methodology: firstly a structural model identification based on a Genetic Algorithm is made; afterwards, an estimation of the constant values is made, based on the Least Square technique. To this aim, the contribution of the user is particularly valuable, since she/he can assume the main structure of the model. In particular, EPR-MOGA can simultaneously optimize at most three objective functions: minimization of the Sum of Squared Errors, number of monomial terms and of percentage of input, selected among the candidates given by the user. The general expression of the pursued models is:

$$\mathbf{H} = a_0 + \sum_{j=1}^{\mathrm{m}} a_j \cdot (\mathbf{X}_1)^{\mathbf{ES}(j,1)} \cdot \ldots \cdot (\mathbf{X}_{\mathrm{k}})^{\mathbf{ES}(j,\mathrm{k})} \tag{1}$$

where $\mathbf{H}$ is the forecasted value of the water table level, $\mathbf{X}_{\mathrm{k}}$ are the attributes of the model (the past rainfalls and groundwater levels), a_j represents the model parameters, a_0 is an offset or bias value, m is the number of monomial values involved into the equation, k represent the general column selected by the matrix of the inputs, and **ES**(*j*,k) represents the selected exponent.

3 The case study

EPR-MOGA is here applied to a porous aquifer located in south Italy, along the south coast of Basilicata. In particular, the test well is located (Fig. 1) close to an abandoned check-point on the railway, named “Casello 49”, which is on the railway between Taranto and Reggio Calabria. The available timeseries cover 25 years, from January 1951 to December 1975; they are part of the collection of the former National Hydrographic Bureau. The rainfall timeseries consists of daily measures, while the piezometric head (m AMSL) timeseries, referred to the test well, comprise measures made every three days. The annual average rainfall during the investigated time interval is 570.2 mm.

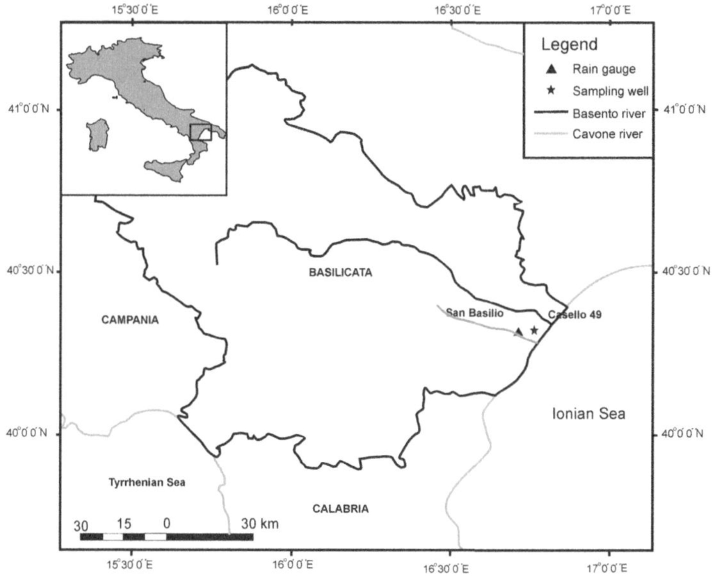

Fig. 1. Location of the investigated site.

The investigated aquifer is located in the flat area between the mouths of the Basento and Cavone Rivers that is part of the large coastal plain of Metaponto (Cilumbriello et al. 2010; Fidelibus et al. 2004; Polemio et al. 2003; Polemio 1994). The aquifer is formed by the Holocenic alluvial sediments, made up by fine grey sands and silts in strata of small thickness (level C) with overlying ochraceous and grey sands and coarse sands (level B), and a final ochraceous silty clay stratum at the topsoil (level A) (Fig. 2). These deposits show in general a medium permeability; locally they may be really slightly permeable. This is due to the irregularly distribution of silty clayey strata (Polemio et al. 2003), which host the aquifer. The presence of clayey levels forces the groundwater to flow at different levels: however, these levels are interconnected. The depth to water table is of about 2 meters. Rainfall directly recharges the backward heteropic terraced sand gravel aquifer, which is highly permeable and supplies the aquifer of the plain. So rainfall indirectly recharges the aquifer through quite quick flow paths.

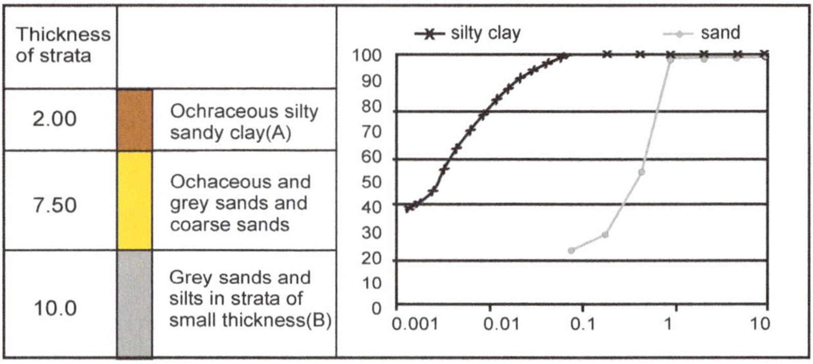

Fig. 2. Strata sequence and grain size distribution of levels A and B.

For the aims of modeling, the total monthly rainfall heights and the monthly average hydraulic heads were computed starting from the available timeseries. This led to a dataset of 300 observations, both for rainfall and hydraulic heads. The use of monthly average values of water table heights was preferred in order to smooth the timeseries, avoiding a potential presence of noise and/or errors not pertaining to the physics of the phenomenon but to unknown extra input. Table 1 provides some details and statistics about raingauge station and sampling well.

Table 1. Data details and main statistics.

	Pluviometric data [mm]	Phreatimetric data [m AMSL]
Station	San Basilio	Casello 49
Gauge height	67 m AMSL	6.80 AMSL
Max	350.8	5.60
Min	0	3.80
Mean	47.5	4.48
Standard deviation	49.5	0.36

Figure 3 represents a time plot of the rainfall and water level timeseries. In order to model the phenomenon, both timeseries were split into two datasets: the training set, about 2/3 of all the data (Ljung 1987), and the test set, the remaining 1/3 of data (Fig. 3). This subdivision is necessary in order to test the generalization capabilities of the returned models. In fact, while EPR-MOGA identifies the models exclusively on the training set, the test set is used after the model identification stage, in order to test models on a “new” set of data. In particular data, from January 1951 to August 1967, are used as training set, while data ranging between September 1967 and December 1975 are used as test set.

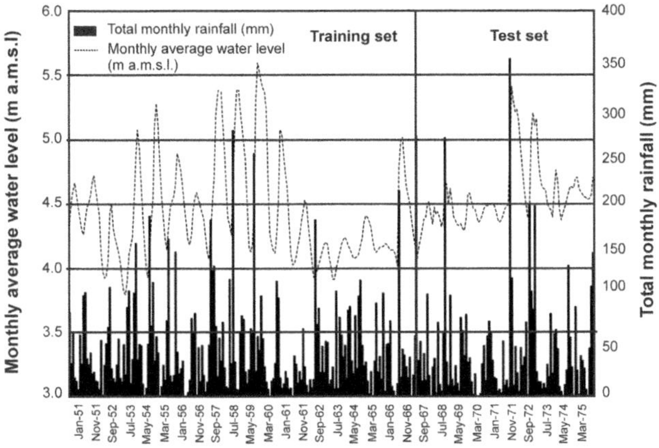

Fig. 3. Modeled data sets and their subdivision.

Looking at groundwater data, it is interesting to observe that water level oscillations are larger when rainfall peaks are registered; however, a lag of about 1-2 months subsists between rainfall and water levels peaks. The oscillations are noteworthy since the water table is quite shallow, being depth to water averagely 2.3 m. The geological nature of the top-soil, mainly constituted by silty-clay deposits, can imply occasional pressurized flow through the aquifer, at high groundwater levels, which can changes the dynamic behavior of the groundwater levels. In fact, when groundwater flows under pressure, it can originate quick increase of piezometric level without a real severe increase of the amount of water in the aquifer. Starting since 1967, groundwater levels show a moderate increasing trend up to the end of the observed data, likely related to the presence of an unknown extra-input, which are currently investigated by ongoing works.

4 Results and discussion

In order to model the aquifer a potential set of candidate input is assumed, as well as the general structure of the models. The candidate input set is constituted by the total monthly rainfall heights up to 8 months before the groundwater level to be predicted and the two measured water levels before the level to predict. The structure of the models is generally assumed as polynomial (equation (1)), where up to

4 monomial terms are allowed, while the set of potential exponents is {0; 0.5; 1; 2}. The result of the modeling search is a Pareto set of 17 models, which correspond to different levels of model complexity and fitness to data. Among these models one was chosen, since it represents a good trade-off between model complexity, a-priori physical knowledge of the investigated aquifer and fitness of model-returned water levels to data both for short-period predictions and for medium-long-period predictions. Moreover, fitness performance was evaluated particularly looking at the results on the test set of data. Here the fitness was accounted using as key indicator the Coefficient of Determination, which is representative of the ratio between the sum of squares errors between the predicted and the measured values and the unbiased estimate of the variance of groundwater levers. CoD is evaluated as:

$$CoD = 1 - \frac{N-1}{N} \frac{\Sigma_N (\hat{H} - H_{\exp})^2}{\sum_N (H_{\exp} - avg(H_{\exp}))^2} \tag{2}$$

where **N** is the number of samples, $\hat{\boldsymbol{H}}$ and $\boldsymbol{H}_{\exp}$ are the EPR returned values and measured, respectively, and avg($\boldsymbol{H}_{\exp}$) represents the average value of measured groundwater levels evaluated for the N samples. On this premise, the chosen model among the 17 constituting the Pareto set follows:

$$H_t = 0.0021289 \cdot P_{t-1} + 0.096254 \cdot H_{t-1}^2 + 2.4205 \tag{2}$$

Figure 4 shows the model-predicted water levels for short-time periods, 1 and 2 months and medium-long periods, 6 and 12 months. Equation (3) provides some information on the behavior of the system, and on the interpretation of its dynamics. Looking at equation (2) it is possible to observe that the structure of the equation is quite simple; whereas the only significant selected input are the total monthly rainfall of the month before the predicted water level and the water level preceding the forecasted one. This imply that the aquifer is responsive to precipitations with one-month lag, while an important role is also related to the persistence of the phenomenon, related to the variable $\boldsymbol{H}_{t-1}$. However, this variable is raised to power 2, implying that the effect of the persistence is somehow amplified. This particular behavior was partly unexpected, since past experience on another porous aquifer of south Italy (Giustolisi et al. 2008; Mancarella and Simeone 2008), returned a more complex model, whereas the persistence term was raised to power 1. Here, the power 2 of the persistence term is related to the high memory effect already emphasized by Polemio (1994) for the same aquifer. This pertains to occurrence of high piezometric levels, which force groundwater to flow in a confined condition. In this condition, piezometric levels become sensitive to aquifer recharge. For this reason, the performance of the selected model tends to decay when the time horizon of predictions increases, as shown by the values of the

CoD, which are reported in the following Table 2. Note that CoD shows negative values for 12-month ahead prediction, which means that the variance of residual exceeds the variance of water levels, thus implying a high error. For long-term prediction horizons, it is necessary to model at the same time free and confined flow behavior, then the model can deteriorate its effectiveness.

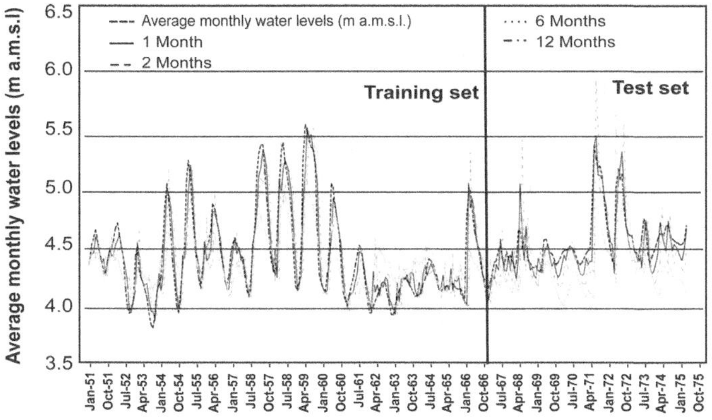

Fig. 4. Model-predicted water levels.

Finally, it is noteworthy that equation (3) involves those variables, $\boldsymbol{P}_{t-1}$ and $\boldsymbol{H}_{t-1}$, which are consistent with the main variables identified by Polemio (1994) for the dynamics of the same aquifer. Indeed, the outcome of this study, based on a regressive technique, agrees with the results from a statistical approach, and from spectral analysis.

Table 2. CoD values for the accounted predictions.

	1 month	2 months	6 months	12 months
CoD training	0.87	0.60	0.20	-0.60
CoD test	0.82	0.41	0.21	-0.27

5 Conclusions

This paper presents an application of EPR-MOGA technique for modeling the dynamic response of a shallow porous aquifer to precipitations. The result shows a good agreement of the model with the expected behavior of the aquifer in short-term prediction. However, for long-term prediction the chosen model did not prove effective. The decay of the performance can be due to the fact that sometimes groundwater flow under pressure and occasionally overlying ground is flooded. These events occur randomly besides being time-confined, i.e. the flow under pressure is relative brief, as well as there is a change in the dynamics of the system. Therefore, the presence of extra phenomena like these biases the procedure, since it is based on the identification of a model from measured data, whereas extra input constitutes marginal information contained in data. On the other hand, EPR-MOGA identified a simple and significant model, which provides some knowledge on the aquifer behavior on the base of the simple variables used to study the dynamic response of the aquifer. This model is consistent with the outcome of past studies made on the same aquifer and based on different approaches (Polemio 1994). It shows that EPR-MOGA can be a really powerful and reliable for the study of the dynamics of aquifers, also in complex scenarios. Finally, it is noteworthy that the identified model can support the management of the aquifer, in particular for the implementation of short-medium term management policies, due to the reasonable good predictions of the water levels returned up to 6-months ahead.

References

Cilumbrello A, Sabato L, Tropeano M, Gallicchio S, Grippa A, Maiorano P, Mateu-Vicens G, Rossi CA, Spilotro G, Calcagnile L, Quarta G (2010) Sedimentology, stratigraphic architecture and preliminary hydrostratigraphy of the Metaponto coastal plain subsurface (Southern Italy). Mem. Descr. Carta Geol. D'It., 67-84

Custodio E (2002) Aquifer overexploitation: what does it mean? Hydrogeol. J., 10(2), 254-277

Doglioni A, Mancarella D, Simeone V, Giustolisi O (2010) Inferring groundwater system dynamics from time series data. Hydrolog. Sci. J., 55(4), 593-608

Doglioni A, Simeone V, Giustolisi O (2011) Karst and porous aquifer dynamic analysis by EPR, in Proceedings of the 4th ASCE-EWRI International Perspective on Water Resources & the Environment (formerly entitled Singapore 2011), Singapore January 4-6, 2011

Fidelibus M, Caporale F, Spilotro G (2004) Studies on different kinds of salinisation in the groundwaters of the Ionian coastal plain of the Basilicata region. In Custodio, Manzano and Araguàs (Eds), 18 SWIM, Cartagena, 2004, Spain

Giustolisi O, Doglioni A, Savic DA, Di Pierro F (2008) An Evolutionary Multi-Objective Strategy for the Effective Management of Groundwater Resources. Water Resour. Res., 44, W01403

Giustolisi O., Savic DA (2009) Advances in data-driven analyses and modelling using EPR-MOGA. J. Hydroinform., 11(3–4): 225–236

Ljung L. (1987) System Identification: Theory for the User. Prentice-Hall Inc., Upper Saddle River, New Jersey, USA

Mancarella D, Simeone V (2008) Modellazione e previsione nei sistemi idrogeologici mediante la tecnica E.P.R. (Evolutionary Polynomial Regression). Giornale di Geologia Applicata, 8, 8-16

Polemio M (1994), Il regime della falda costiera ionica di Metaponto, Proceedings of III Conf. of Geotechnical Engineering Researchers "The role of fluids for geotechnical problems - Il ruolo dei fluidi nei problemi di Ingegneria geotecnica", Mondovì, Italy, Vol. 1, 135-149

Polemio M, Dragone V. Limoni PP, Mitolo D, Santaloia F (2003) Caratterizzazione idrogeologica della piana di Metaponto, qualità e rischi di degrado delle acque sotterranee. Acque Sotterranee, 8,35-49

Siegfried T, Kinzelbach W (2006) A multiobjective discrete stochastic optimisation approach to shared aquifer management: Methodology and application. Water Resour. Res., 42(2), W02402

Evaluating three different model setups in the MIKE 11 NAM model

Ch. Doulgeris[1*], P. Georgiou[2], D. Papadimos[1], D. Papamichail[2]

[1]The Goulandris Natural History Museum, Greek Biotope/Wetland Centre, 14th km Thessaloniki-Mihaniona, Thermi, GR 57 001, Greece.

[2]Faculty of Agriculture, Aristotle University of Thessaloniki, Thessaloniki, GR 54 124, Greece.
* haris@ekby.gr

Abstract In this paper, the NAM model was applied in the Strymonas River catchment to simulate the daily discharge that ends up into the Lake Kerkini. In the NAM model, the catchment is represented by four reservoirs, each one representing different physical elements of the catchment. The model uses precipitation, potential evapotranspiration and temperature as driving forces in the simulation of snow accumulation and melting, interception, actual evapotranspiration, overland flow, interflow, groundwater recharge and baseflow. Optimal calibration of the model parameters was obtained for three different model setups by using meteorological and discharge data. The model was also validated against discharge data. The NAM model was used to estimate the continuous Strymonas River mean daily discharge and the results show that the estimation accuracy is depended by the model setup.

1 Introduction

The use of simulation models should be considered as an important tool in water resources management and related decision-making. Hydrological models are well-developed and a long tradition of application exists. Rainfall – Runoff modeling is the process of transforming rainfall into catchment runoff. Almost all rainfall - runoff models take as input data, at least, precipitation and potential evapotranspiration and calculate as result catchment runoff.

The MIKE 11 is a powerful hydrological modeling system which can be used in water resources management. The system, developed by DHI, was designed to simulate water flow in rivers and open channels. It is composed by several modules namely rainfall-runoff (RR), hydrodynamic (HD), advection-dispersion (AD) etc., which in some cases can be used in combination and in others cases as standalone simulators (DHI 2009). MIKE 11-NAM is a rainfall-runoff model which is part of the MIKE 11 RR module.

MIKE 11-NAM model has been used by many researchers. Refsgaard and Knudsen (1996) validated three different models (MIKE 11-NAM, MIKE SHE and WATBAL) on three catchments in Zimbabwe for decision making on water resources management, and when at least one year's data were available for calibration, all three models performed equally well. Madsen (2000) formulated an automatic calibration strategy for the MIKE 11-NAM model that included the optimization of multiple objectives, measuring different aspects of the hydrograph, and was based on the shuffled complex evolution algorithm. Keskin et al. (2007) applied the MIKE 11-NAM to simulate (in a semi-distributed manner) the runoff from snowmelt to the Yuvacik Basin in Turkey. Hatzispiroglou et al. (2009) simulated single rainfall-runoff events in four catchments of Samos Island using MIKE 11-NAM and model results were in good agreement with the observed data. Recently, Makungo et al. (2010) used the MIKE 11-NAM model and Australian Water Balance Model (AWBM) to simulate runoff hydrographs for the ungauged Nzhelele River.

The objective of this paper is to present an application of a rainfall-runoff model, MIKE 11-NAM, in the Strymonas River catchment to simulate the inflow discharge into the Lake Kerkini, at a daily time step. The model was tested using three different model setups; a) the initial setup using the default model parameters, b) the second setup using the extended groundwater parameters and c) the final model setup using the snowmelt component parameters. For each setup the model was calibrated and validated against discharge observations alongside the Strymonas River. The simulated hydrographs can be used in water resources planning and management of Lake Kerkini, which involves the achievement of multiple and conflicting goals or objectives, such as to meet the water requirements of agriculture and ecosystems of Lake Kerkini and Strymonas River estuary.

2 Study area and data sets

The Strymonas River catchment is located on the Balkan Peninsula and occupies a total area of 16,747 km^2 (Fig. 1). It is a transboundary catchment shared by Bulgaria (50.6%, 8,473 km^2), Greece (35.8%, 5,990 km^2), FYROM (9.8%, 1,641 km^2) and Serbia (3.8%, 643 km^2). The total length of the river is about 390 km and originates from the Southern slopes of the Vitosha Mountain in Bulgaria (2180 m a.m.s.l.). At approximately 290 km, it collects the discharge from several tributaries, the Dragovishtitsa, the Dzerman, the Strumeshnitsa, etc. and leaves Bulgarian territory near Kulata village (85 m a.m.s.l.). After a course of approximately 25 km in Greek territory, it flows into the north-eastern part of Lake Kerkini. It continues downstream from Kerkini, draining the plain of Serres before it discharges into the Strymonikos Gulf. Lake Kerkini is a manmade lake constructed between the period 1933-36 mainly for the protection of Serres plain against floods, caused by the Strymonas River. Soon after, it was used as a reservoir for irrigation water, playing an important role in the welfare of the Serres plain, where irrigated agriculture is probably the most important issue from an economical and social point of view (Gerakis et al. 2007; Halkidis and Papadimos 2007; Doulgeris et al. 2008).

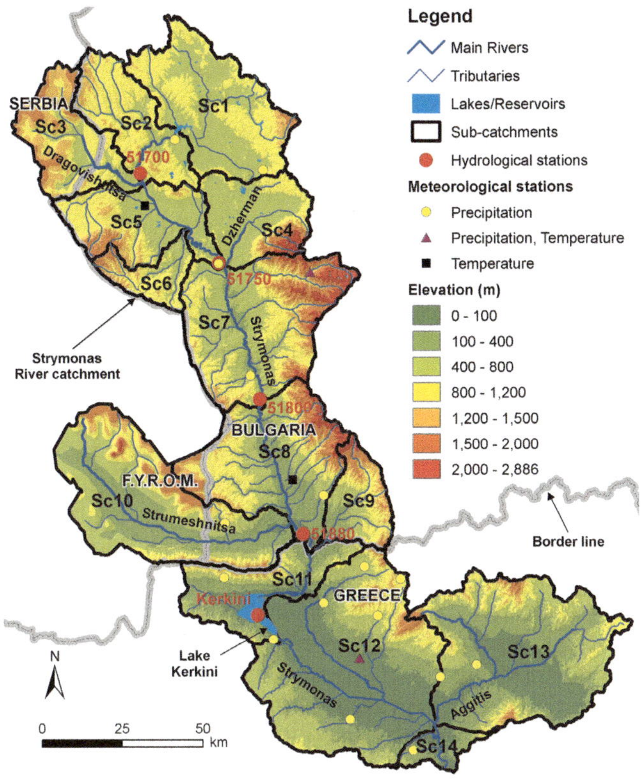

Fig. 1. The transboundary Strymonas River catchment

Precipitation data were acquired from 19 stations: eight in Bulgaria and 11 in Greece. In addition, temperature data were gathered from four stations: three in Bulgaria and one in Greece. The average annual precipitation in the catchment is 686 mm and the mean temperature is 12 °C. Observed discharge measurements were obtained from four monitoring stations (51700, 51750, 51800 and 51880) at a daily time step for the period 2003-2006 along the Strymonas River. The mean annual discharge at the above stations is 8.4, 21.8, 43.5 and 67.4 m^3/s, respectively. The location of the meteorological and hydrological stations as well as the delineation of the sub-catchments are shown in Figure 1 (Doulgeris et al. 2008).

3 The MIKE 11-NAM model

MIKE 11-NAM model represents various components of the rainfall-runoff process by continuously accounting for the water content in four different and mutually interrelated reservoirs. Each reservoir represents different physical elements of the catchment and, depending on the requirements and the data available, the NAM can be prepared in a number of different modes.

The NAM model can be characterized as a deterministic, lumped, conceptual model with moderate input data requirements. By default, it can be prepared with nine parameters representing surface-root zone storage as well as groundwater storage. For a better description of the baseflow, additional (lower) groundwater storage can be defined. In this case, the upper groundwater storage provides the fast responding component of the baseflow, whereas the lower storage usually has a slower response. When using a snow melt component, precipitation is driven to snow storage if the temperature is below the defined base temperature and if the accumulated snow melts at a constant rate defined by a degree-day coefficient (DHI 2009).

Model parameters must be calibrated against a time series of hydrological observations, usually discharge observations versus model results. Model calibration can be done either manually or by using computer-based automatic procedures. In automatic calibration, the calibration objectives are formulated as numerical goodness-of-fit measures that are optimized automatically. In this paper, the calibration scheme includes the overall volume error and the overall root mean square error (Madsen 2000; DHI 2009). These performance statistics are included in a multi-objective optimization problem which is solved by the shuffled complex evolution algorithm (Madsen 2000, Madsen et al. 2002).

A model's performance was evaluated using a variety of standard statistical criteria including the overall water balance error (WB_{er}), the correlation coefficient (R), the Nash-Sutcliffe efficiency (R_{NS}), the index of agreement (d), the modified index of agreement (d_{mod}) and the relative index of agreement (d_{rel}). The above statistical criteria are given by the formulas (Martinec and Rango 1989; Papamichail and Papazafiriou 1992; Krause et al. 2005):

$$WB_{er} = \left|1 - \left(\sum_{i=1}^{N} Q_{s,i} \Big/ \sum_{i=1}^{N} Q_{o,i}\right)\right| \tag{1}$$

$$R = \left(\sum_{i=1}^{N}\left(Q_{o,i} - \overline{Q_o}\right)\left(Q_{s,i} - \overline{Q_s}\right) \Big/ \sqrt{\sum_{i=1}^{N}\left(Q_{o,i} - \overline{Q_o}\right)^2}\sqrt{\sum_{i=1}^{N}\left(Q_{s,i} - \overline{Q_s}\right)^2}\right) \tag{2}$$

$$R_{NS} = 1 - \left(\sum_{i=1}^{N}\left(Q_{o,i} - Q_{s,i}\right)^2 \Big/ \sum_{i=1}^{N}\left(Q_{o,i} - \overline{Q_o}\right)^2\right) \tag{3}$$

$$d = 1 - \left(\sum_{i=1}^{N}\left(Q_{o,i} - Q_{s,i}\right)^2 \Big/ \sum_{i=1}^{N}\left(\left|Q_{s,i} - \overline{Q_o}\right| + \left|Q_{o,i} - \overline{Q_o}\right|\right)^2\right) \tag{4}$$

$$d_{mod} = 1 - \left(\sum_{i=1}^{N} (Q_{o,i} - Q_{s,i}) \Big/ \sum_{i=1}^{N} \left(|Q_{s,i} - \overline{Q_o}| + |Q_{o,i} - \overline{Q_o}| \right) \right) \quad (5)$$

$$d_{rel} = 1 - \left(\sum_{i=1}^{N} \left(\frac{Q_{o,i} - Q_{s,i}}{Q_{o,i}} \right) \Big/ \sum_{i=1}^{N} \left(\frac{|Q_{s,i} - \overline{Q_o}| + |Q_{o,i} - \overline{Q_o}|}{\overline{Q_o}} \right) \right) \quad (6)$$

where $Q_{o,i}$ is the observed discharge at time i, $Q_{s,i}$ is the simulated discharge at time i, $\overline{Q_o}$ is the mean observed discharge, $\overline{Q_s}$ is the mean simulated discharge and N is the total number of time steps.

If the model performs well, WB_{er} values should be close to zero and the values of R, R_{NS}, d, d_{mod} and d_{rel} should be close to unity. The performance of the model is usually checked for the calibration and for the validation periods, separately, and using the calibrated parameters the model can be applied to predict the catchment's runoff at other time periods.

4 Results and discussion

The NAM model was used to estimate the discharge at a daily time step in the Strymonas River. The simulation period was from 1/1/2003 until 31/12/2006, in line with the availability of discharge data. The calibration period was from 1/1/2003 until 31/12/2004 and the validation period from 1/1/2005 until 31/12/2006. The hydrological regime of the simulation period is characterized as normal or wet, including also extreme events of discharge.

Meteorological data (i.e., precipitation, temperature and potential evapotranspiration) and hydrological data (i.e., discharge) were input into the NAM model. Mean areal precipitation in the sub-catchments was computed using the common Thiessen polygon method (Shaw 1988) and daily rainfall data. Mean temperature was computed by the lapse rate approach (Mutreja 1986). In the NAM model, the actual evapotranspiration is calculated from the model and based on potential evapotranspiration and on the model's parameters. Potential evapotranspiration was estimated using the Thornthwaite method (Shaw 1988), since temperature was the only available meteorological data.

Initially, the NAM model was set up using a minimum set of nine parameters (SimS1-default setup) whose values were estimated with an auto-calibration procedure based on available discharge data of the four monitoring stations (51700, 51750, 51800 and 51880). Afterwards, two new parameters were included into the previous setup in order a slower baseflow to be simulated for the catchment (SimS2-extended groundwater setup). The values of the 11 model parameters were estimated using the auto-calibration procedure. Finally, the snow melt component was included in the model (SimS3-final setup) and the calibration of the two new parameters was done manually. For a sound evaluation of the model's results, both graphical and numerical evaluation criteria have been applied. In Fig-

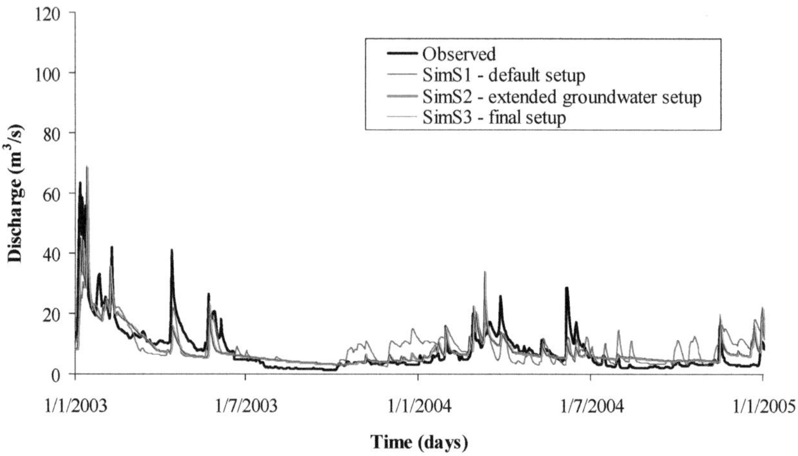

Fig. 2. Discharge in Strymonas River at the station 51700 for the calibration period.

ure 2 is shown for station 51700 the observed and the simulated hydrograph according to the three different model setups for the calibration period.

According to Figure 2, there are more or less systematic deviations of the simulated discharge with the SimS1 setup. Model results improved significantly when the extended groundwater parameters were included in the model (SimS2), compared to the initial model setup (SimS1). More specifically, the baseflow, the runoff peaks and the dates at which they appear are simulated better when SimS2 was used. The improved simulation of baseflow is due to the more detailed description of the groundwater zone. The statistical evaluation criteria are given in Table 1. The WB_{er} was decreased from 12.4% (SimS1) to 1.6% (SimS2) and the other statistical criteria R, R_{NS}, d, d_{mod} and d_{rel} were improved significantly. The final model setup (SimS3) gives almost identical results to SimS2 and the small differences are due to the snow effects in discharge. The comparison of model results for SimS1, SimS2 and SimS3 gives similar to the above conclusions for the stations 51750, 51800 and 51880.

Table 1. Evaluation of NAM model results at station 51700 of Strymonas River.

Statistical criteria	Calibration period			Validation period		
	SimS1	SimS2	SimS3	SimS1	SimS2	SimS3
WB_{er} (%)	12.4	1.6	3.9	11.0	3.1	1.0
R	0.66	0.78	0.78	0.68	0.80	0.81
R_{NS}	0.39	0.59	0.59	0.44	0.63	0.64
d	0.78	0.86	0.86	0.78	0.86	0.87
d_{mod}	0.55	0.67	0.67	0.57	0.67	0.68
d_{rel}	0.36	0.76	0.74	0.58	0.80	0.81

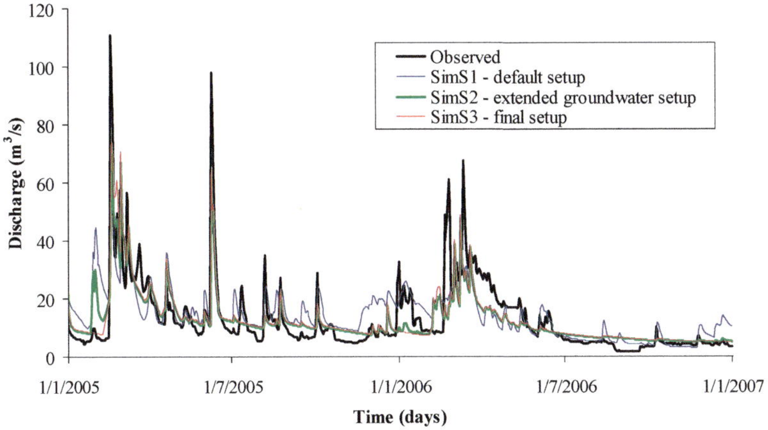

Fig. 3. Discharge in Strymonas River at the station 51700 for the validation period.

To validate the NAM model in Strymonas River, daily discharge was predicted for the period 2005-2006 using the model parameters estimated during the calibration period. In Figure 3 is shown the observed and the simulated hydrograph according to the three different model setups for the validation period. The statistical evaluation criteria are given in Table 1. The model results improved significantly for the SimS2 model setup compared to the SimS1. The WB_{er} was decreased from 11% (SimS1) to 3.1% (SimS2), the R, R_{NS}, d, d_{mod} and d_{rel} were improved significantly and the final calibration setup (SimS3) gives similar results to SimS2.

Using the model parameters of the final model setup (SimS3), the NAM model was also used to compute the discharge (and the yearly accumulated water volume) that inflows into Lake Kerkini (Fig. 4). The mean simulated annual discharge (79.6 m^3/s) and the corresponding water volume (2,510 10^6 m^3) should be managed properly to cover the irrigation demands from Lake Kerkini taking also into account the requirements of the ecosystems of Lake Kerkini and Strymonas River estuary.

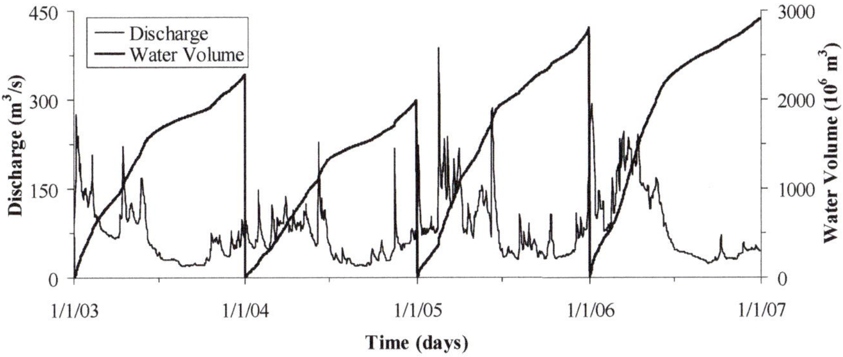

Fig. 4. Simulated discharge and yearly accumulated water volume into Lake Kerkini by using the final model setup (SimS3).

5 Conclusions

The rainfall-runoff model (MIKE 11-NAM) was applied to all sub-catchments of the Strymonas River for a four-year time period at a daily time step. Using both an automatic calibration and a trial-and-error procedure, the model's parameters were calibrated against the observed discharge along the Strymonas River over a period of two consecutive years. The use of groundwater extended parameters (SimS2) as well as snow-melt parameters (SimS3) significantly improved the model's results. Furthermore, the model was also validated for a two years period and satisfactorily predicted the Strymonas River discharge. Therefore, the model is capable to predict the water inflow into the Lake Kerkini, and in combination with its simplicity, it can make it a valuable tool in the design and implementation of water resources management plans of Lake Kerkini.

References

DHI (2009) MIKE 11: A modeling system for rivers and channels. Reference manual. Danish Hydraulic Institute, Denmark

Doulgeris Ch, Halkidis I, Papadimos D (2008) Use of modern technology for the protection and management of water resources in Strymonas/Struma River basin. Technical Report, Greek Biotope/Wetland Centre (EKBY). Thermi, Greece. 82 p

Gerakis PA, Tsiouris S, Tsiaoussi V (2007) Water regime and biota: proposed minimum values of lakes water level and of rivers discharge in Macedonia and Thrace, Greece. The Goulandris Natural History Museum/Greek Biotope-Wetland Centre, Thermi, Greece (in Greek, summary in English)

Halkidis I, Papadimos D (eds) (2007) Technical report of LIFE Environment project: Ecosystem based water resources management to minimise environmental impacts from agriculture using state-of-the-art modeling tools in Strymonas basin. Greek Biotope/Wetland Centre (EKBY), Thermi, Greece (in Greek, summary in English)

Hatzispiroglou J, Moussouliotis A, Zissis T, Anastasiadou-Partheniou E (2009) Application of the NAM rainfall-runoff model in hydrological basins of Samos Island. Proceedings of the common Conference of the 11th Hellenic Hydrotechnical Society and of the 7th Conference of the Hellenic Committee of Water Management. 229 - 236 (in Greek)

Keskin F, Şensoy AA, Şorman A, Şorman ÜA (2007) Application of MIKE 11 model for the simulation of snowmelt runoff in Yuvacik dam basin, Turkey. International Congress on River Basin Management, The role of general directorate of state Hydraulic works (DSI) in development of water resources of Turkey. DSI

Krause P, Boyle DP, Bäse F (2005) Comparison of different efficiency criteria for hydrological model assessment. Adv. Geosci. 5, 89-97

Madsen H (2000) Automatic calibration of a conceptual rainfall-runoff model using multiple objectives. J. Hydrol. 235, 276-288

Madsen H, Wilson G, Ammentrop HC (2002) Comparison of different automated strategies for calibration of rainfall-runoff models. J. Hydrol. 261, 48-59

Makungo R, Odiyo JO, Ndiritu JG, Mwaka N (2010) Rainfall - runoff modeling approach for ungauged catchments: A case study of Nzhelele River sub-quaternary catchment. Phys. Chem. Earth. 35, 596-607

Martinec J, Rango A (1989) Merits of statistical criteria for the performance of hydrological models. Water Resour. Bull. 25(2), 421-432

Mutreja KN (1986) Applied hydrology. Tata McGraw-Hill Publishing Company Limited, New Delhi

Papamichail DM, Papazafiriou ZG (1992) Multiple input - single output linear functional models for river flow routing. J. Hydrol. 133, 365-377

Refsgaard JC, Knudsen J (1996) Operational validation and intercomparison of different types of hydrological models. Water Resour. Res. 32 (7), 2189-2202

Shaw EM (1988) Hydrology in practice. Second edition, Chapman and Hall

Potential solutions in prevention of saltwater intrusion: a modelling approach

A. Khomine, Sz. János, K. Balázs

Department of Mineralogy, Petrology and Geochemistry, University of Szeged, Hungary, khomine@gmail.com

Abstract Damsarkho (Latakia, Syria) coastal aquifer is under severe hydrological stress due to the overexploitation of a shallow aquifer for irrigation. Excessive pumping during the past few decades has led to seawater intrusion into the aquifer. For the purpose of planning and management, SEAWAT, a variable density solute transport computer code, was used to study groundwater volume and quality. The conceptual model is based on field and laboratory data collected between 1960 and 2003. This simulation demonstrates that the use of injection wells or a subsurface barrier would both represent a good method with which to improve water quality and prevent seawater intrusion.

Introduction

In many coastal areas, the growth of human settlements, together with the development of agricultural, industrial and tourist activities, has led to the overexploitation of aquifers. Such overexploitation commonly results in a rise of the freshwater–saltwater interface (seawater intrusion) and thus degradation of the chemical quality of groundwater. The Mediterranean provides many clear examples of the presence of seawater intrusion (Abou Zakhem and Hafez 2003; Arfib and de Marsily 2004; Bonacci and Roje-Bonacci 1997; Chiocchini et al. 1997; El-Bihery and Lachmar 1994; Paniconi et al. 2001b; Petalas and Diamantis 1999), with the Spanish coast not an exception (Calvache and Pulido-Bosch 1994; Giménez and Morell 1997; Iribar and Custodio 1992; Padilla et al. 1997a, b). This phenomenon has also been observed on the Syrian coast north of Latakia (the Damsarkho plain). The model will be used for hydrodynamic simulation and appropriate management of local water resources, with the aquifer system representing a good example of the recent saltwater intrusion of coastal plains.

Potential solutions to saltwater intrusion

A variety of different measures are commonly used to control seawater intrusion and protect groundwater resources, although the target of most is to increase the flow of fresh groundwater and/or reduce the flow of saltwater. As such there are a number of methods with which to prevent saltwater contamination of groundwater, including subsurface barriers and artificial recharge. *Subsurface barriers:* Defined as underground semi-impervious or impervious structures constructed in coastal aquifers, subsurface barriers are used to simultaneously impede the inland infiltration of seawater and increase the groundwater storage capacity of an aquifer. The cost is highly dependent on the depth of cut-off, length of wall and specific material availability. *Artificial recharge:* Todd (1980) defines artificial recharge as the augmentation of the natural movement of surface water into underground formations. A variety of artificial recharge techniques are currently available, including water spreading and recharge wells (Todd 1980). A technique is involving the transfer of surface water into an aquifer. Use of a dual-purpose well is economically preferable to the construction of a specialized recharge well. The general purpose of recharge wells is to overcome the high cost of the water spreading technique in areas where suitable land is scarce and/or expensive. In addition to clogging problems, several other difficulties are associated with the recharge well technique. For instance, a large amount of dissolved air is carried together with recharge water, whilst water quality research has indicated that a variety of bacteria are also found in recharge water. Under certain circumstances, bacteria can grow quickly and eventually reduce the filtering area of the well screen (Civan 2007; van Beek et al. 2009).

Conceptual model of the study area: The Damsarkho coastal plain is located north of the town of Latakia, Syria, and is composed of marine and alluvial sediments which cover an area of about 40 km^2. The study area is characterized by Mediterranean climate, with wet winters and dry summers, while the average annual rainfall varies from 800 to 1000 mm/year (Selkhozpromexport 1979). The Damsarkho plain is characterized by the presence of favourable hydraulic characteristics aquifers, which consist of loose sands, gravelly sands, sandstone, limestone and sandy gravelly clay. With an average thickness of around 25 m, these aquifers are in direct hydraulic contact with seawater.

The main geometric-structural and hydrogeological characteristics of the Damsarkho multi-aquifer system were reconstructed on the basis of geological maps of the region and the borehole lithological sections from 17 wells drilled across the plain. The system is composed of five layers (sandy clay, limestone, sandy clay with gravels, marl and dolomite), although these layers occasionally combine to form a single-layer aquifer.

Mesh discretisation Based on the SEAWAT code (Langevin and Guo 2002), the model developed simulates transient variable density groundwater flow and solute transport for the period from 1966 to 2010, using a database developed by Abed

Rabo (2000). Regularly spaced finite-difference 50 m x 50 m cells on the horizontal plane are used, with the final grid consisting of 120 rows and 120 columns in the horizontal, and 5 regularly spaced layers in the vertical direction (Fig. 1, 2).

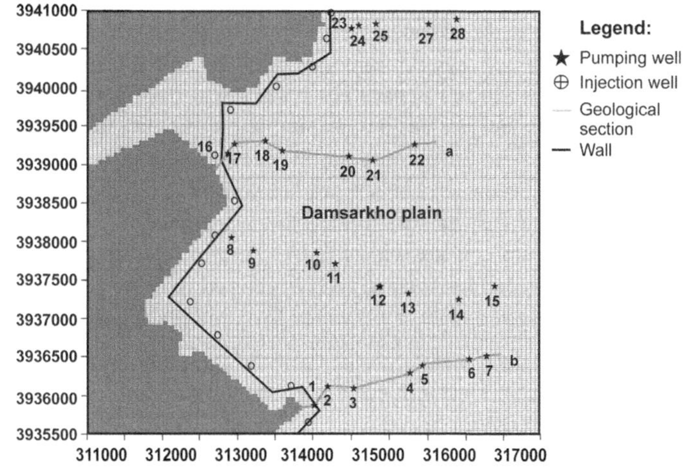

Fig. 1. The grid of modelled Damsarkho plain area showing the locations of geological sections, pumping wells, suggested places of injection wells and wall with the co-ordinates given in metres.

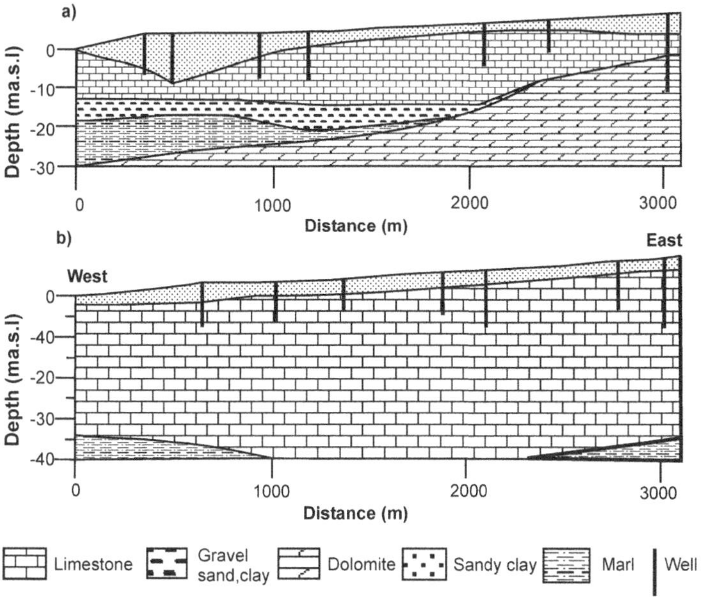

Fig. 2. Geological sections through selected pumping wells. The locations of the sections are shown in Fig. 1. Note that all wells discharge from limestone layer.

Results and discussion

This study simulated the period from 1960 to 2010, using available data (Abed Rabo 2000; Abou Zakhem and Hafez; Syrian Irrigation Ministry, unpublished data) to follow each of the past, current and the future of seawater intrusion. As mentioned previously, although many methods of preventing seawater intrusion are available, two such procedures will be tested separately here. To determine the effectiveness of each method, their expected effects were simulated up to the year 2020, with the results from each simulation then compared with each other. The suggested places of injection wells and subsurface barrier (wall) are shown in Figure 1.

Subsurface barrier After installation of the subsurface barrier in the model (thickness 1 m, hydraulic conductivity 1.5 10^{-9} m/s), it can be observed that the inland movement of seawater intrusion is stopped, and in some places also pushed back seaward. The projected change between 2010 and 2020 is shown in a second layer (Fig. 3).

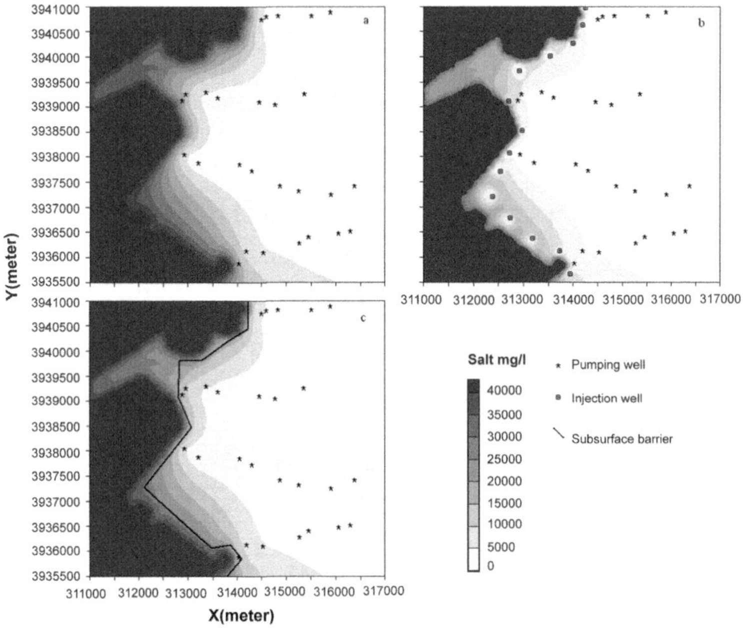

Fig. 3. Simulated change in TDS concentration (mg/L) in the second layer before (a 2010) after installation of injection wells (b 2020) and a subsurface barrier (c 2020).

Injection wells 15 Shallow injection wells were introduced in the model between the sea and production wells, with a screen in the range of -20 to -40 m, a distance of 500 m between them and an injection yield of about 11 L/s. The effectiveness of injection wells depends on several factors; the most important of which being the location and length of screen, the volume of injected water, as well as the hydraulic conductivity of the targeted layer. The projected change from 2010 to 2020 after the introduction of an injection well is shown in a limestone layer (Fig. 4).

Comparison of seawater intrusion prevention techniques

In order to compare the effectiveness of the two solutions to seawater intrusion, salinity changes in second layer were tracked over a period of 10 years at the first pumping well in depth 38m (see Fig. 1 for location of first pumping well), and by the hydraulic balance in the five layers. The results of this comparison are presented in Figure 4 and Table 1. Analysis of salinity changes reveals that in both cases, the salt content is significantly reduced. However, this decrease is greater after introduction of the injection wells than the subsurface barrier. The hydraulic balance enhanced this conclusion, so we can confirm that the injection wells are better solution in prevention of seawater intrusion.

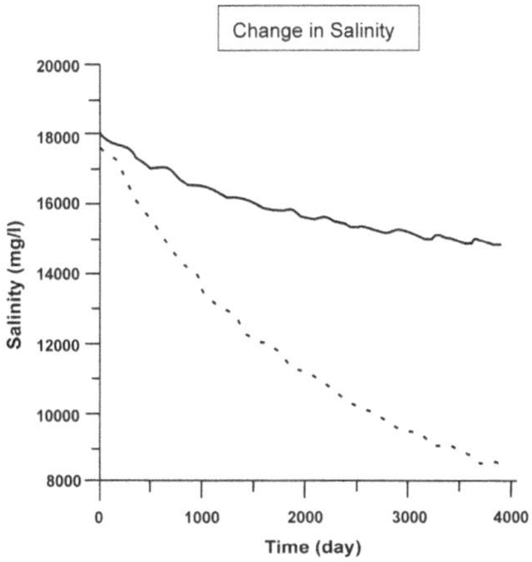

Fig. 4. Comparison of salinity in the bottom of the first well after using the mentioned prevention techniques, Solid line for subsurface barrier and dotted line for injection well solution.

Table 1. Simulated hydraulic balance of the Damsarkho area in August , a, 2010 Before using solutions, b, 2020 after using injection wells, c, 2020 after using subsurface barrier.

scenario tried	Sea boundary (m^3)		Recharge (m^3)	Well (m^3)		East boundary(m^3)		Storage(m^3)	
	In	Out	In	In	Out	In	Out	In	Out
a	7998	49057	21793	-	154132	19034	0	156085	378
b	3317	136555	25184	85929	154141	7130	239	149265	364
c	6510	14489	25172	-	154135	0	43099	180451	251

Conclusion

SEAWAT has proved to be a useful tool with which to simulate the transition zone of the Damsarkho coastal aquifer. The results obtained from the model have been used to identify locations for a potential subsurface barrier and a number of injection wells, as well as to predict the pattern of future intrusion, if the current excessive rate of abstraction continues. From our results, it can be confirmed that both the subsurface barrier and injection systems represent good solutions to the seawater intrusion problem. Installation of a subsurface barrier is suggested as being the most economic, despite its initial high cost, since it would not have to be maintained over time. Although the injection well method is projected as being the most effective practical solution, problems remain regarding the source of the injected water and the potentially high future maintenance costs.

Reference

Abed Rabo R (2000) Study of seawater intrusion into fresh groundwater in the coastal area. Master's Thesis, Damascus University

Abou Zakhem B, Hafez R (2003) Environmental isotopes study of the aquifer system in the coastal area (Syria). AECS-G/ FRSR report no. 285, Syria

Arfib B, de Marsily G (2004) Modeling the salinity of an inland coastal brackish karstic spring with a conduit-matrix model. Water Resources Research 40: W11506. doi: 10.1029/2004WR003147

Badon-Ghijben W (1889) Nota in Verband met de Voorgenomen Put boring Nabij Amsterdam [Notes on the probable results of well drilling near Amsterdam]. Tijdschrift van het Koninklijk Instituut van Ingenieurs, Den Haag, pp 8–22

Bonacci O, Roje-Bonacci T (1997) Sea water intrusion in coastal karst springs: example of the Blaž Spring (Croatia). Hydrol Sci J 42(1):89-100

Calvache ML, Pulido-Bosch A (1994) Modeling the effects of salt-water intrusion dynamics for a coastal karstified block connected to a detrital aquifer. Ground Water 32:767–777

Chiocchini U, Gisotti G, Macioce A, Manna F, Bolasco A, Lucarini C, Patrizi GM (1997) Environmental geology problems in the Tyrrhenian coastal area of Santa Marinella, Province of Rome, central Italy. Environmental Geology 32(1):1-8

Civan F (2007) Reservoir Formation Damage, 2nd Edition. Gulf Professional Publishing, Elsevier

El-Bihery MA, Lachmar TE (1994) Groundwater quality degradation as a result of overpumping in the Delta Wadi El-Arish Area, Sinai Peninsula, Egypt. Environ Geol 24(4):293–305

Giménez E, Morell I (1997) Hydrogeochemical analysis of salinization processes in the coastal aquifer of Oropesa (Castellón, Spain). Environ Geol 29:118–131

Iribar V, Custodio E (1992) Advancement of seawater intrusion in the Llobregat delta aquifer. In: Custodio E, Galofre A (eds) SWIM Study and Modelling of Saltwater Intrusion into Aquifers. CIMNE–UPC, Barcelona, pp 35–50

Kovacs B, Szanyi J (2004) Hidrodinamikai és transzport modellezés I. és II. Miskolci egyetem, Miskoc. Hungary

Langevin CD, Guo W (2002) User's Guide to SEAWAT, A computer program for simulation of threedimensional variable density groundwater flow. U.S Geological Survey, Open-File Report 01-434, Tallahassee, Florida

Padilla F, Secretan Y, Leclerc M (1997a) On open boundaries in the finite element approximation of twodimensional advection-diffusion flows. Int J Numer Meth Engng 40:2493-2516

Padilla F, Benavente J, Cruz-Sanjuliân J (1997b) Numerical simulation of the influence of management alternatives of a projected reservoir on a small alluvial aquifer affected by seawater intrusion (Almunecar, Spain). Environ Geol 33(1):72-80

Paniconi C, Klhaifi I, Giacomelli A, Tarhouni J (2001) A modelling study of seawater intrusion in the Korba coastal plain, Tunisia. Physics and Chemistry of the Earth (B) Hydrogeology, Oceans and Atmosphere 26(4):345–351

Petalas CP, Diamantis JV (1999) Origin and distribution of saline groundwaters in the upper Miocene aquifer system, coastal Rhodope area, northeastern Greece. Hydrogeology Journal 7(3):305-316

Ponikarov (1966) Explanatory notes of the geological map of Syria, scale ½00000, sheets 1-37 XIX, 1-36 XXIV

Selkhozpromexport (1979) Hydrogeological and hydrological surveys and investigation in four areas of Syrian Arab Republic, vol. 1 and 2, Hydrology

Todd DK (1980) Groundwater Hydrology, 2nd Edition. John Wiley and sons, New York

Van Beek CGEM, Breedveld RJM, Juhasz-Holterman M, Oosterhof A, Stuyfzand PJ (2009) Cause and prevention of well bore clogging by particles. Hydrogeology 17:1877-1886

Geophysical research of groundwater degradation at the eastern Nestos River Delta, NE Greece

I. Gkiougkis, T. Tzevelekis, F. Pliakas, I. Diamantis, A. Pechtelidis

Department of Civil Engineering, Democritus University of Thrace, Xanthi, GR 67100, Greece. jgiougis@civil.duth.gr

Abstract This paper deals with the hydrogeological study of the evolution of groundwater salinisation - degradation at the eastern Nestos River Delta, NE Greece, based on geophysical research including 10 geoelectrical soundings. Relevant hydrogeological works included in-situ measurements of groundwater level, pH, Specific Electrical Conductivity (SEC) and temperature, as well as chemical analysis of groundwater samples (July 2009). Important conclusions were drawn based on the analysis of all the available data and the elaboration of piezometric and SEC distribution maps and the designed geoelectrical intersections.

1 Introduction

Seawater intrusion appears within the majority of the coastal aquifers in Greece, which are under hydrogeological investigation, and is responsible for their qualitative degradation. Today, all scientific evidences lead to the conclusion that the intrusion of seawater into coastal aquifers is a phenomenon that regards the majority of the coastal area of northern Greece, including the coastal area of Eastern Macedonia and Thrace (Kallioras et al. 2006).

In saltwater intrusion studies, hydro-geochemistry analysis from monitoring wells and geophysical methods are used. The best geophysical method to assign, particularly in salinity mapping is geo-electrical method (Baharuddin et al. 2009, Loke 2000). Geo-electrical methods are used to detect variations in aquifer electrical resistivity and this in turn has been inferred with changes in groundwater salinity (Al-Bassam and Hussein 2008). There are applications in the literature where both geo-electrical method and groundwater quality studies have been integrated together (Al-Bassam and Hussein 2008; Benkabbour et al. 2004; El-Waheidi et al. 1992; Gnanasundar and Elango 2001; Goldman et al. 1991; Khalil 2006; Samsudin et al. 2008; Sung-Ho et al. 2007).

The construction of several drainage works and the increasing use of irrigation water in the east delta plain of Nestos River, have resulted in seawater intrusion and related deterioration of the groundwater quality (Gkiougkis 2011, Daskalopoulos 2007, Sakkas et al. 1998, Pliakas et al. 2001). This paper deals with the study of geological and hydrogeological characteristics of the coastal aquifer system of the eastern delta area of the Nestos River, based on geoelectrical soundings. The main target was to trace the extent of groundwater salinisation in the study area.

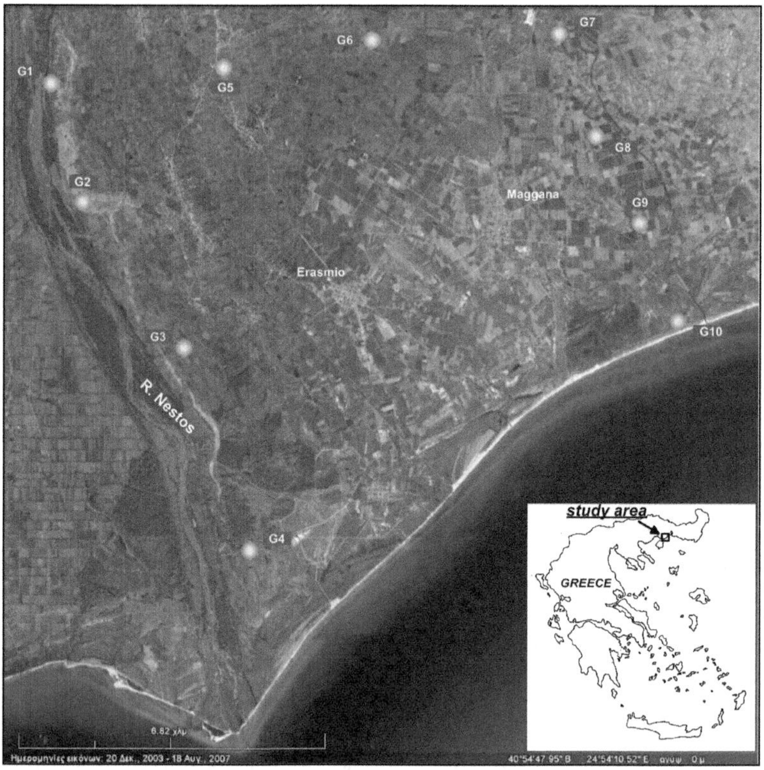

Fig. 1. Geoelectrical sounding measurements sites across the study area borders.

2 Geomorphological and geological framework, hydrogeology

The study area is located south of the city of Xanthi in Xanthi Prefecture, N.E. Greece. It is bounded to the west by the Nestos River and to the south by the Aegean Sea (Fig. 1). The morphology of the region is characterized by a very low relief and shallow pits, to the S and SE, forming small ponds, which are part of the eastern Nestos delta wetlands. The most important surface water body of the area is Nestos River located to the west of the study area. It is also one of the main sources of freshwater recharge for the coastal unconfined aquifer in the study area.

Laspias stream is located across the eastern side of the study area where the degraded industrial and sewage treatment effluents are discharged. Northerly, at a distance of approximately 2 km, a drainage trench is located, which after draining the northerly irrigated land, conveys water into the Laspias stream.

The main cultivation in the area is corn and cotton, which have been irrigated with groundwater from the coastal aquifer. The intensive agricultural activities in combination with the irrational use of fertilizers and insecticides have resulted in severe problems such as soil degradation, exhaustion of the aquifers potential and increase of the salt concentrations in the shallow aquifers particularly in the coastal areas (Diamantis et al. 2009).

The study area is located in a recent sedimentary delta environment of a thickness of some tens of meters created by the Nestos deposits. These alternate sand, clay and silt layering deposits, resulting from a wide range of structural and depositional processes, produce a heterogeneous geological environment. It is worth mentioning the presence of organic clay at some points due to the delta marshes (Gkiougkis 2011; Gkiougkis et al. 2010).

The east delta plain extends to 176.4 km^2, from which the 106.63 km^2 are cultivated (only 60%), while the coastal saline uncultivated lands extend to 45 km^2. Considering the hydrological data of Nestos Delta plain (mean annual rainfall for the period 1965 - 1996: 546.9 mm, infiltration approximately: 15%), the annual infiltration for the 131.4 km^2 area is estimated at 10.8×10^6 m^3. The irrigated lands extend to 89.90 km^2, while the 35 km^2 of them meet irrigation needs from the Nestos River. The rest areas meet irrigation needs by pumping, where considering the type and the extent of the crops as also as well operation data, the annual water consumption is estimated approximately at 27×10^6 m^3 (Gkiougkis 2011). Taking into account the amounts of the two basic parameters of the groundwater hydrologic balance mentioned above (infiltration - water consumption), a lack of about 16×10^6 m^3 water is evaluated (Pliakas et al. 2001, Sakkas et al. 1998). It is estimated that a small part of this lack is replaced by the Nestos riverbed percolation, at rates depending on the run of the buried old streambeds and the distance from the river (Gkiougkis 2011; Gkiougkis et al. 2010).

There are two hydrogeological systems that are formed within the alluvial deposits of the wider study area (Pliakas et al. 2001, Sakkas et al. 1998): (a) The shallow system, consisting of phreatic and mostly of semi-confined aquifers extended down to a depth of about 30 m. Natural recharge to this system comes mostly from the infiltration of rainfall and less from stream bed percolation of the north hilly area. (b) The deeper system, consisting of confined aquifers extended to a depth of at least 190 m. Natural recharge to this system comes to a great extent from river Nestos percolation through buried old stream beds, and from the lateral groundwater inflows coming from the adjoining Vistonis lagoon hydrogeological basin.

Evaluation of the groundwater hydraulic parameters of the study area aquifers after working out the results from the pumping tests of 11 selected wells in the wider study area resulted in values for (Pliakas et al. 2001, Sakkas et al. 1998):

(1) transmissivity (T), ranging from 4.0×10^{-4} to 1.1×10^{-2} m^2/sec, (2) storage coefficient (S), ranging from very low values to 10^{-3}, characterizing the aquifers of the study area as confined mainly westward, and in some sites as semiconfined. Figure 2 shows the piezometric map of the study area in April 2009, before the irrigation period. The major groundwater flow direction is from the northwest towards the south with minor flow from the N.E. and central parts of the study area towards the south (Gkiougkis 2011; Gkiougkis et al. 2010).

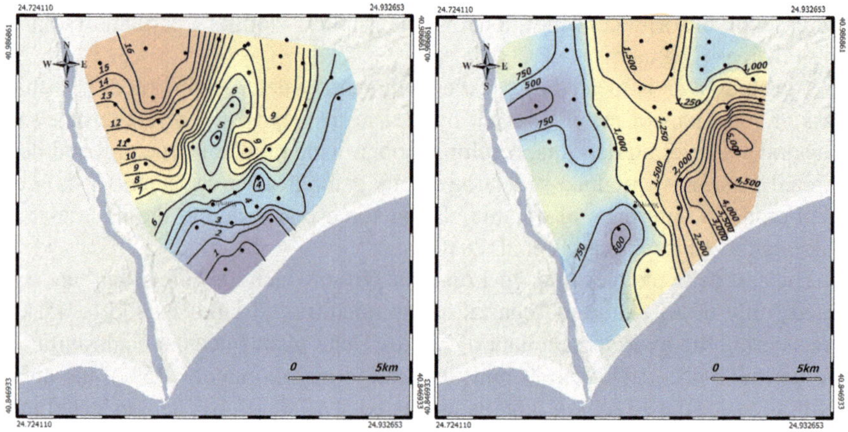

Fig. 2. Left: piezometric map (April 2009), Right: distribution of SEC values μS/cm (July 2009) (Gkiougkis 2011; Gkiougkis et al. 2010).

The Laboratory of Engineering Geology of the Department of Civil Engineering, D.U.Th., Greece, in collaboration with the Hydrogeology Group of the Technical University of Darmstadt, Institute of Applied Geosciences, Germany, conducted a groundwater quality research at the study area on July 2009. Groundwater samples from 30 wells were collected (1/7/2009 and 2/7/2009). Sampling also included in situ measurements of the physicochemical parameters of the groundwater of the study area (pH, SEC-Specific Electrical Conductivity, Temperature). There are quite a few wells within the study area with SEC values above 1,500 μS/cm, reaching values of almost 6000 μS/cm. This could be attributed to seawater intrusion as these wells are in close proximity to the sea. The wells with SEC values lower than 1,500 μS/cm are mainly under constant fresh water recharge from Nestos River (Gkiougkis 2011; Gkiougkis et al. 2010).

3 Geoelectrical research

In the study area 10 geoelectrical soundings were performed, and the main target was to identify the stratigraphy of the area (Fig. 1). The geoelectrical soundings were performed according to the Schlumberger electrodes provision with a maximum current electrodes spread of 800 m (G1, G5) – (AB/2 = 400 m) and 1000 m - (AB/2 = 500 m). The survey points were selected in such a way in order to cover sectors of the study area where only swallow boreholes (<12 m) are present and as a result the stratigraphy is not defined in detail. The survey points are located across Nestos River (3 geoelectrical soundings), across Laspias stream (3 geoelectrical soundings) and across the upper border of the study area, 9.5 km from the nearest seashore of Erasmio (4 geoelectrical soundings).

The outcome of all geoelectrical soundings was corresponding to direct stratigraphy data. The comparison between geoelectrical sounding results and the actual nearest borehole intersections was totally satisfactory. Only in one case (G1) a thick sand-clay layer cannot be detected, mainly due to the positioning of the sounding upon the Nestos river mound. Before the implementation of electrical soundings, every possibility of direct surface or underground geological stratigraphy recognition must be exhausted. The interpretation of a geoelectrical sounding is questionable without any comparison data of surface geology and borehole intersections.

Today the groundwater exploitation is diminished to a minimum. Nowadays most of the local stakeholders, irrigate their cultivations with surface water from nearby Laspias and Megalomacheros streams. Another possible cause of groundwater salinisation, apart from extensive overpumping, might have been the recharge from Laspias stream, whose water electrical conductivity exceeds 3,000 μS/cm, during summer period (Gkiougkis 2011). As for the geoelectrical soundings G9 and G10 the results can be characterized as expectable and logical compared to all recent surveys and researches on the study area. The G9 and G10 point are located near the seashore, where the relevant indicators and quality specifications of the groundwater aquifer system had been worsen due to seawater intrusion. Only the geoelectrical sounding G8 (5 km from the seashore), from 7 m depth up to 85 m depth, came across with saline groundwater conditions, an unexpected result. The above finding has to be investigated further in order to study the possible affection of a geothermal field of the broader area deeper below.

The results of the performance of 10 geophysical (electrical sounding) measurements (Fig. 1, 3) showed 6 distinct main geoelectrical formations categorized according to the values of their specific electrical resistivity, as follows (ASCE 1987, Kallergis 2000): (1) gravel-pebble mixture (50-220 Ohm×m), (2) gravel (25-50 Ohm×m), (3) sand (20-25 Ohm×m), (4) clayey sand (16-20 Ohm×m), (5) sandy clay (11-16 Ohm×m), (6) clay (8-11 Ohm×m), salinization conditions (< 8 Ohm×m), while the dominance of clay and clay-sand materials throughout the whole extent of the study site is very distinct.

The above resistivity values are only valid in the study area, and thus the range of the values is smaller, compared to the values range described in relevant tables, on international bibliography, for the above geological formations.

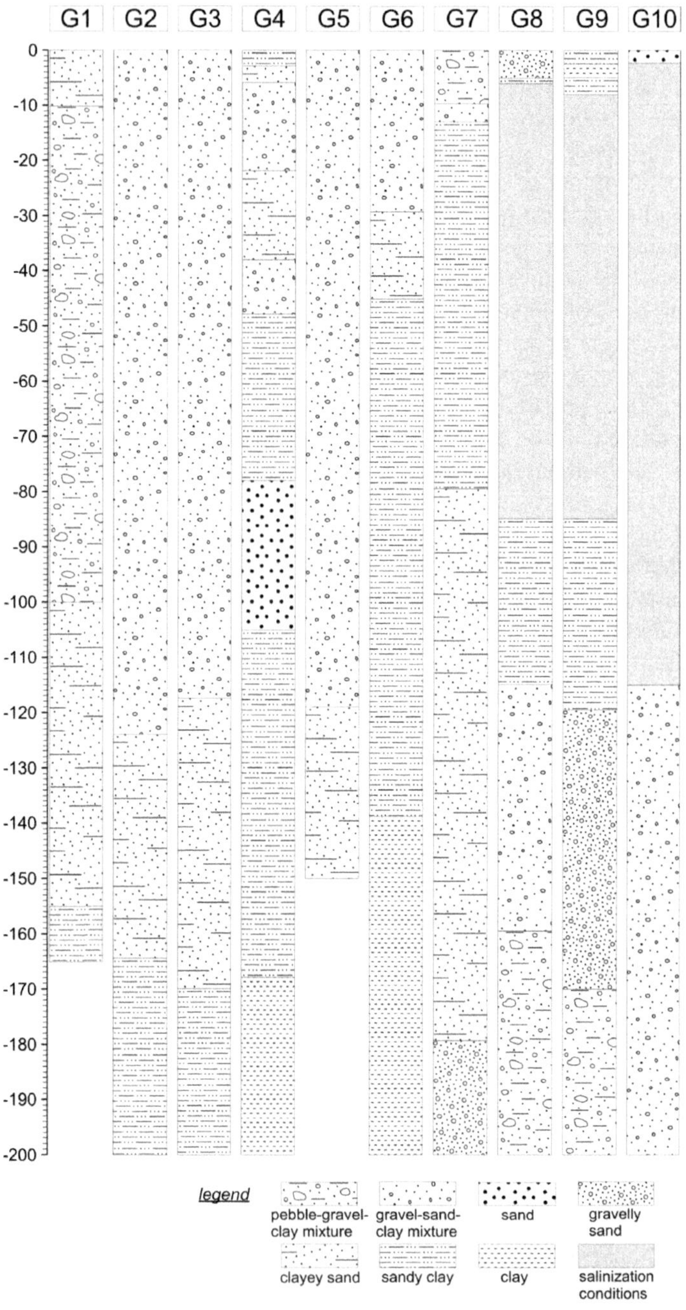

Fig. 3. Geoelectrical intersections (G1-G10).

4 Conclusions

The area of investigation is a typical example of lack of scientific groundwater resources management of coastal aquifers, which has resulted in the aggressive intrusion of seawater wedge at the mainland freshwater aquifer. The continuous over-pumping conditions and the overexploitation of the aquifer resources enhance the qualitative degradation of the aquifer and pose threats for a variety of factors of the area such as environmental, social, economic and agricultural development.

The fluctuations of the specific electrical resistivity values for the same geological formations have a great range. Two totally different geological formations could have similar specific electrical resistivity values. The interpretation of the geoelectrical sounding measurements should be done very carefully. A safe method is to compare the measured values with the actual stratigraphy data for the same area. In the eastern Nestos River Delta, the geoelectrical soundings results were totally satisfactory when compared with the actual geological and hydrogeological data for the specific area.

References

Al-Bassam AM, Hussein MT (2008) Combined geo-electrical and hydro-chemical methods to detect salt-water intrusion: A case study from southwest Saudi Arabia. Management of Environmental Quality, 19[2] 179 - 193

ASCE (1987) Ground Water Management. Third Edition, ASCE Manuals and Reports on Engineering Practice No 40, New York, 263p

Baharuddin MFT, Hashim R, Taib S, (2009) Electrical imaging resistivity study at the coastal area of Sungai Besar, Selangor, Malaysia. J. Applied Sci., 9[16], 2897-2906

Benkabbour B, Toto EA Fakir Y (2004) Using DC resistivity method to characterize the geometry and the salinity of the Plioquaternary consolidated coastal aquifer of the Mamora plain, Morocco. Environ. Geol., 45, 518-526

Daskalopoulos D (2007) Hydrogeological study of an aquifer system at the eastern delta of Nestos River. Master Thesis, Postgraduate Course: Hydraulic Engineering, Department of Civil Engineering, Democritus University of Thrace, Xanthi, (in Greek)

Diamantis V, Gkiougkis I, Kallioras A, Pliakas F, Diamantis I (2009) Use of saline waters for the reclamation of salt-affected soils, Proceedings of Common Conferences, 11th Conference of Greek Hydrotechnical Association and 7th Conference of Greek Committee for Water Resources Management, Volos, 27-30/5/2009 Greece, 313-320 (in Greek)

El-Waheidi MM, Merlanti F, Pavan M (1992) Geoelectrical resistivity survey of the central part of Azraq plain (Jordan) for identifying saltwater/freshwater interface. J. Applied Geophys., 29, 125-133

Gkiougkis I (2011) Investigation of seawater intrusion into coastal aquifers in deltaic environment. The case of Nestos River Delta. PhD Thesis, Department of Civil Engineering, Democritus University of Thrace, Xanthi, (in progress), (in Greek)

Gkiougkis I, Mwila G, Pliakas F, Kallioras A, Diamantis I (2010) Hydrogeological assessment of groundwater degradation at the eastern Nestos River Delta, N.E. Greece. 12th International Conference of the Geological Society of Greece, Patras, Greece, 19-22/5/2010, 4, 1697-1706

Gnanasundar D, Elango L (2001) Groundwater quality assessment of a coastal aquifer using geoelectrical techniques. Journal of Environmental Hydrology, 7 paper 2 1-8

Goldman M, Gilad D, Ronen A, Melloul A (1991) Mapping of seawater intrusion into the coastal aquifer of Israel by the time domain electromagnetic method. Geoexploration, 28[2], 153-174

Kallioras A, Pliakas F, Diamantis I (2006) Conceptual model of a coastal aquifer system in northern Greece and assessment of saline vulnerability due to seawater intrusion conditions. Environmental Geology, Springer, 51[3], 349-361

Khalil MH (2006) Geoelectric resistivity sounding for delineating salt water intrusion in the Abu Zenima Area, West Sinai, Egypt. J. Geophys. Eng., 3, 243-251

Loke MH (2000) Electrical imaging surveys for environmental and engineering studies. A Practical Guide to 2D and 3D Surveys

Pliakas F, Diamantis I, Petalas C (2001) Saline water intrusion and groundwater artificial recharge in east delta of Nestos River. Proceedings of the 7th International Conference on Environmental Science and Technology, University of the Aegean, Dept. of Environmental Studies, and Global Nest, Ermoupolis, Syros, Greece, 3-6/9/2001, 2, 719-726

Sakkas I, Diamantis I, Pliakas F et al. (1998) Groundwater artificial recharge study of Xanthi – Rhodopi aquifers (in Thrace, Greece). Greek Ministry of Agriculture Research Project, Final Report. Sections of Geotechnical Engineering and Hydraulics of the Civil Engineering Department of Democritus University of Thrace, Xanthi, Greece, (in Greek)

Samsudin A, Haryono A, Hamzah U, Rafek A (2008) Salinity mapping of coastal groundwater aquifers using hydrogeochemical and geophysical methods: a case study from north Kelantan, Malaysia. Environmental Geology, 55[8], 1737-1743. doi:10.1007/s00254-007-1124-9

Sung-Ho S, Jin-Yong L, Namsik P (2007) Use of vertical electrical soundings to delineate seawater intrusion in a coastal area of Byunsan, Korea. Environ. Geol., 52, 1207-1219

Piezometric conditions in Pieria basin, Kavala Prefecture, Macedonia, Greece

T. Kaklis[1], G. Soulios[1], G. Dimopoulos[1], I. Diamantis[2]

[1] Laboratory of Engineering Geology & Hydrogeology, Department of Geology, Aristotle University of Thessaloniki, 54124, Thessaloniki, Greece kaklis@geo.auth.gr

[2] Democritus University of Thrace , Geotechnical Section, Dept. of Civil Engineering, School of Engineers, 671 00 Xanthi, Greece

Abstract Pieria basin is located between Paggeon and Symvolon mountains, in Kavala Prefecture. It is characterized by elongated shape with NE-SW direction of development. In the plain area intensive cultivation is applied, which imply to increased water quantities for irrigation. In this paper the piezometry of the study area is analyzed in two periods of measurements according to the lower and the higher water level as well as the correlation between the water level and the irrigation periods. For this reason, systematic water level measurements from March 2005 to November 2007 in regular base were collected from selected boreholes of the basin. Furthermore, data collected in 1996 were also used for comparison

1 Introduction

Pieria basin is located between the Paggeon and Symvolon mountains in Kavala Prefecture (Fig. 1). It covers an area of 190 km^2, with a NE-SW direction. Morphologically is characterized as a semi-mountainous to mountainous basin. Its drainage network is very well developed and the main water stream is Marmaras River which is discharged in the Northern Aegean Sea. The main occupation of the population of the area is the cultivation of the farms locating in the plain area.

Because of the increased rural development of the area many boreholes have been drilled which belong to the local irrigation service of Pieria basin, which is the main irrigation water supplier in the area. It is calculated, from the collected data using the mean value of discharge rate of each borehole, that the total amount of the water used for irrigation, in the year 2004 was 6500000 m^3. The hydrogeological research which took place in the study area includes an inventory of the boreholes, a monitoring network for the observation of the fluctuations of the water level and finally water level measurements from March 2005 to November 2007. There were also collected data of the water level measurements of Pieria basin (TOEB) for the period from January 1996 to November 2004. The main purpose of this paper is to present the fluctuation of the groundwater level during the periods of September 2003 and March 2005 combined with the corresponding abstractions for irrigation in order to be shown the recharge and the discharge rate of the aquifers of the basin. The main reason for the selection of these two periods is because the observations of the water level in September 2003 were the lower ones and respectively in March of 2005 the observed water level at the boreholes had the maximum values in Pieria basin area.

2 Geological setting

Geologically the area is a part of the Rhodope massif. Rhodope massif mainly consists of metamorphic rocks of different ages and tectonic history. This geological unit is a complex with heterogenic lithological synthesis (Brunn 1960; Kockel and Walther 1965; Meyer 1966, 1968; Boneev 1971). The first researcher of the Rhodope massif Kronberg (1966, 1967, 1969, 1970) suggested that is a complex of metamorphic rocks which have possibly polymetamorphosed and folded during the Alpine orogeny period. Various researchers (Papanikolaou and Panagopoulos 1981; Dimadis and Zachos 1986; Kilias and Mountrakis 1990 have distinguished two tectonic units (Sidironero and Paggeon) within the area of the Greek part of the Rhodope massif. The study area is entirely located in Paggeon unit. The geological map of the area is shown in Figure 1.

The geological structure of the area is defined as follows:

a) The lower Gneiss-schist zone (Gn, Gsch, Mr_sch). The presence of gneisses, mica schists, green schists and amphibolites is characterized this geological unit. During the transition to the upper parts we have a rotation of mica schists and marbles. These rotations recommend us the boundary with the upper located carbonated zone of the marbles.
b) Marbles zone (Mr). The marbles in this area are not characterized by intense bedding. In general they are massive and white colored which mainly originated from limestones created in shallow condition. Marbles characterized by banded structure are present in this zone. These banded marbles are located in the lower parts of this carbonated zone in the transition zone from the marbles to the lower located gneiss-schist formation.
c) Igneous rocks - Granitoids (γ): In Pieria basin there is a lack of volcanic rocks as it is in some parts of the Rhodope massif. The granitoids that occurred in the Paggeon Mountain area have been penetrated in the lower tectonic unit of Rhodope massif (Eleftheriadis and Koroneos 2003). In general the presence has a SSW-NNE direction in masses along the areas of Mesolakia, Podohori and Mesorropi. These granitoids emplacement in Paggeon is affiliated with the tectonic stresses that had as a result the formation of the Paggeon anticline (Schenk 1970).
d) Pleistocene formations (Pt.cs): Pleistocene sediments in the area cover the SW part of the basin. In these formations there is a presence of tuffaceous limestones which originated from brackish or fresh water (Sakellariou-Mane 1966).

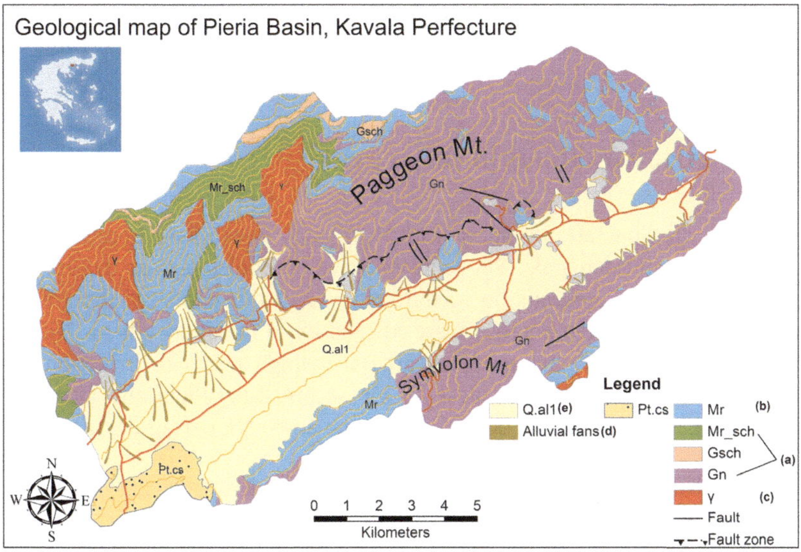

Fig. 1. Geological map of the study area (γ: granioids, Gn: Gneiss, GnSch: Gneiss schists, Mr_Sch: Mica schists, Mr: Marbles, Pt.cs: Pleistocene formations, Q.al1: Holocene deposits) (IGME 1974, Dimopoulos 1978).

These deposits mainly consist of sands, sandstones, clay, conglomerates and tuffaceous limestones. There is a presence of formations which originated from high flow torrential river and by gritty marls, clays, sandstones, sand and conglomerates. Conglomerates supply is exclusively provided by bedrocks (metamorphic rocks) and for these reason the dominant material is roundstones originated by marbles, gneisses and granites (Dimopoulos 1978). The sudden spreading interruption of the Pleistocene sediments to the east of the basin might have an attachment with the faulting in NNW-SSE direction and the displacement of the eastern part.

e) Holocene deposits (Q.al1): These deposits are fluvial terrestrial and they occur within the alluvial fans. The most representative types of these formations are the slide rocks and river deposits. In these Holocene deposits formations we can distinguish three basic types between them (Dimopoulos, 1978):

A lower sequence of loose deposits with roundstones, which is mostly developed at the northern margins of the basin close to the marbles of the Paggeon (with intermediate material the red clay). The second sequence is followed by lateral formations (lateral scree) whose emplacement is due to its transportion with the surface water. The third sequence of formations is similar to slide rocks with larger dimensions of the rocks developed in the mass. Substantially they are fluvio-torrential deposits. These facies are observed at Paggeon foothills with greater frequency than at the S and SE parts of Symvolon Mountain. As a conclusion according to the presence of the slide rocks mostly in the northern part of the area we can accept that in the greater area of Pieria basin we had a regional uplift of the south part of Paggeon Mountain against the northern part of Symvolon Mountain.

3 Piezometric conditions in the area

Pieria basin area has an extremely well developed network of boreholes which cover the irrigation demands. Six representative boreholes, across the area, have been chosen for the presentation of the water level fluctuation of the area from year 1996 to 2007. These boreholes are MP1 close to Podohori village, MPL3 located near Platanotopos, which is an area with high needs in irrigation water, MMS2, MM6, MD6 boreholes close to Messoropi, Moystheni and Domatia respectively and finally MK1 next to Kipia village at the NE part of the plain.

Table 1. Absolute water level (m) above sea level and the water level difference for the periods of September 2003 and March 2005.

Borehole	Water level (m) (asl) September 2003	Water level (m) (asl) March 2005	Difference (m)
MP1	73.4	78.6	5.2
MPL3	96.5	110.4	13.9
MMS2	91.1	95.1	4
MM6	101.3	108.1	6.8
MD6	89.1	94.7	5.6
MK1	117.1	122.5	5.4

The fluctuation of the water level of the aforementioned boreholes from January 1996 to November 2007 is shown in Figure 2. As a remark is that the fluctuation of the water level is mainly during the year, meaning that the lower level is monitored during the irrigation period (summer months to September) and the maximum observed water level is monitored during the winter period till spring until the beginning of the pumping period. Substantially is shown that besides the high pumping rate during the irrigation period the mean water level recurs to the initial level. During the period 2005-2007 the water level of the boreholes MD6 and MM6 especially during the winter increased more than the average mean of the previous years. This is related to the fact that in the Paggeon Mountain area we have particularly high amounts of atmospheric precipitation. According to Figure 2 we can assume that the recharge of the aquifers was extremely quick and intense. Paggeon Mountain is the area that mostly provides feeding in the aquifers. This is happening because of the high amounts of precipitation occurring there and the combination of the existence of the marbles with their positive geometric characteristics that favor infiltration and water recharge.

From the water level measurements, as given in Table 1, and the diagram in Figure 2, is assumed that the average lower levels observed in September 2003 and the respectively higher in March 2005.In Figure 3 is given the spatial distribution of the piezometry (meters above sea level) for the period of September 2003. Is shown that the hydraulic gradient takes values around 3.9% and is higher in the western part of the area close to Platanotopos area. In the eastern part of the study area the hydraulic gradient is reduced and in the area of Mesotopos and Melisokomeio is estimated to be 7‰.

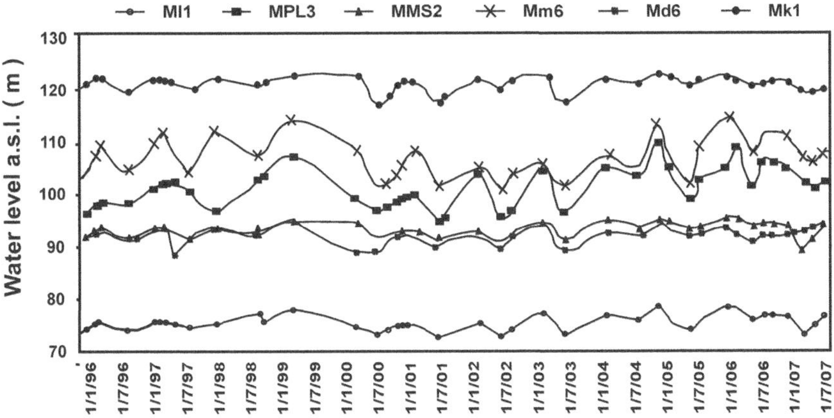

Fig. 2. Water level (m) in boreholes for the period from January of 1996 until November of 2007.

In Figure 4 the spatial distribution of the piezometry in the period that observed the maximum water level measurements (March 2005) is shown. The hydraulic gradient for this period was calculated to be 5.3% in Platanotopos area and approximately 6‰ in Melisokomeio area as measured from the compiled piezometric mps. It is assumed that during the dry period we have a greater value of the hydraulic gradient from Paggeon towards Platanotopos area compared to the wet period in the same area. In Melisokomeio area the hydraulic gradient during the dry period is respectively greater than the wet period. Nevertheless the piezometric curves shape in both periods doesn't differentiate much. This fact indicates the presence of an unified aquifer, as well as an uniformity in the recharge and discharge rate of the aquifers in the whole area. High values of the hydraulic gradient that are observed in Platanotopos area evince the presence of a zone with low permeability and a higher degree of difficulty in recharge of the aquifer in the basin.

The presence of springs with relatively high discharge rates and clay layers indicates the absence of marble occurrence at the basin margins. Based on the data collected for the pumping rates of the boreholes is assumed that these values are higher in Platanotopos area, because the plain and the cultivated area is significantly larger than the Melissokomeio area. The exploitation of the aquifers in this part of the basin (Platanotopos) takes place in a significant higher degree than in the rest of the basin. Groundwater flows are mainly from the northwest towards Aegean Sea.

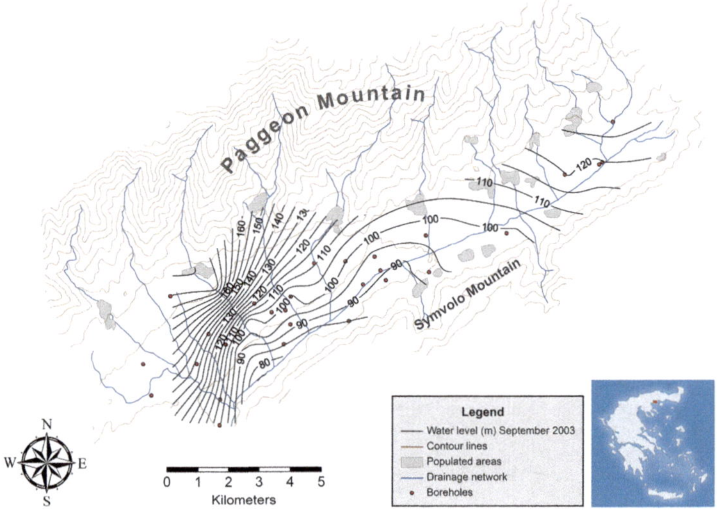

Fig. 3. Piezometric map for the period of September 2003.

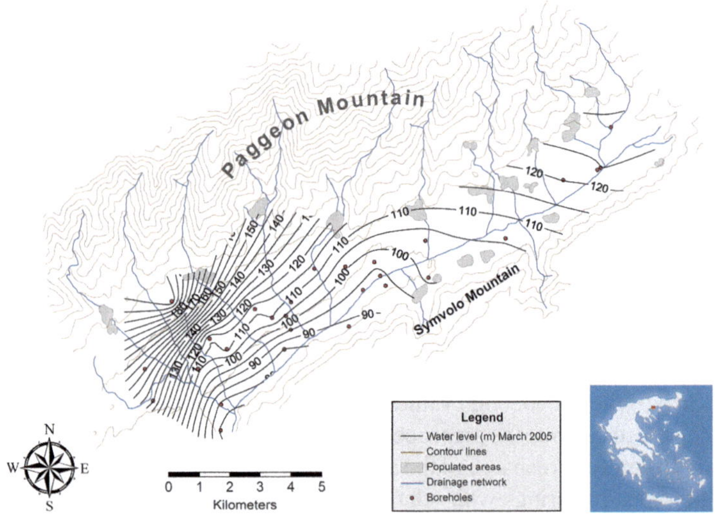

Fig. 4. Piezometric map for the period of March 2005.

Characteristically it is pointed out that the boreholes have been drilled close and parallel to the direction of the river have artesian flow during the whole year, except some periods of time especially in the middle of irrigation period, when the water levels across the whole area decline. The spatial distribution of the water level difference between September 2003 and March 2005 is shown in Figure 5.

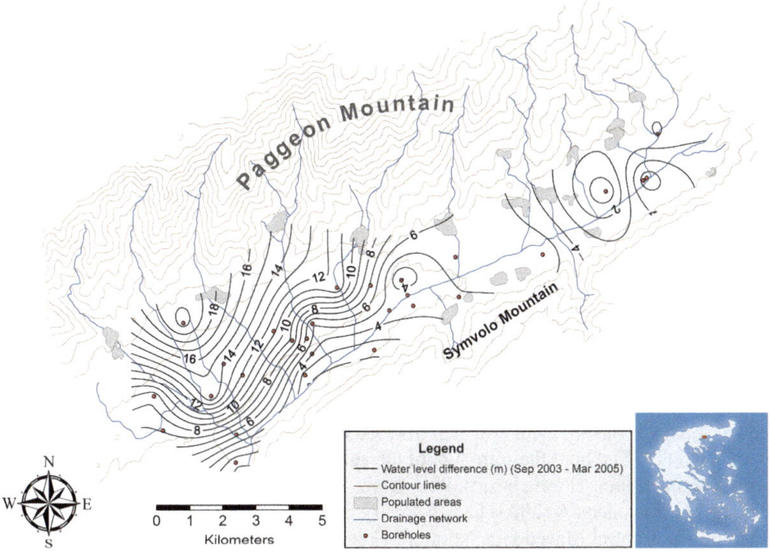

Fig. 5. Spatial distribution showing the difference of the water levels (m) between March 2005 and September 2003.

From these variations it is shown that the high differences of the water level observed mostly in the Platanotopos-Moustheni area. On the other hand these fluctuations are significantly lower in the rest area. The main reason for these greater differences in the first case is the high rate of discharges taking place in this part of the area during the irrigation period. As a consequence of these high pumping rates we observe the high hydraulic gradients mentioned before and the difficulty of groundwater movement for the recharge of the aquifer layers.

4 Conclusions

Pieria basin is characterized by intense exploitation of groundwater, especially during summer, in order to cover the increased demands for irrigation purposes. The main aquifer is developed within the alluvial deposits of Plio-pleistocene age that covering the lowlands of the basin. A decline of Piezometric surface of the aquifers is recorded during the irrigation period due to high rate of pumping. Irrigation period starts from the beginning of April until October. The water level arises during the wet period of the year (October-April) due to recharge mainly from the Paggeon mountain, which receive significantly high amounts of atmospheric precipitation, which because of the hydraulic contact that we have in this area it feeds the aquifers of the plain area (mostly through the marbles system). Hydraulic gradients are higher in the western part of the area due mostly in the over pumping of the boreholes located in this area. The entirely area is developing significant water bearing capacity, despite the large amounts of water that are pumped every year. The recharge rate is direct and fast. It is calculated for the year of 2004, was about 6500000 m^3 only for irrigation needing. The water level recurs in short time by the end of the pumping period. This means that the recharge of the aquifers is much bigger than 6500000 m^3 annually.

References

Boneev E (1971) Problems of the Bulgarian geotectonics. Technical, 204 pp (in Bulgarian)

Brunn JH (1960) Les zones Helleniques internes et leur extension Reflexions sur l'orogenese alpine. Bull. Soc. Geol. France (7)

Dimadis E, Zaxos K (1986) Geologic map of Rhodope massif, scale 1:200000. IGME. Xanthi

Dimopoulos G (1978) About the formationof the aquifers in the basin between Paggeon and Symvolon mountains (Pieria basin), PhD thesis, Thessaloniki, 203pp (in Greek)

Eleftheriadis G, Koroneos A (2003) Geochemistry and Petrogenesis of Post-Collision Pangeon Granitoids in Central Macedonia, Northern Greece. Chemie der erde Geochemistry, 63, 364-389

Institute of Geologican and Mineral Explorations (IGME) (1974) Geological map of Greece scale 1:50000, Sheet Nikisiani by P.Kronberg and P.F.Schenk

Kilias A, Mountrakis D. (1990) Kinematics of the crystalline sequences in the western Rhodope massif. Geol.Rhodopica, 2, 100-116

Kockel FU, Walther HW (1965) Die Strimonlinie als Grenze zwischen Serbo-Makedonischen und Rilla Rhodopen Massiv in Ost-Makedonien. Geol.Jb 83, Hannover

Kronberg P (1966) Petrographie und tektonik im Rhodopen-Kristallin des Tsal-Dag Simvolon und Ost-Paggeon (Griechisch-Makedonien), N.Jb.Geol.Palaont. Mh 7 Stutgart

Kronberg P (1967) Zur Anwendung photogeologischer Methoden in Kristallin Gebieten, N.Jb.Geol.Palaont Abh. 129 Stutgart

Kronberg P. (1969) Gliederung, Petrographie und Tektogenese des Rhodopen-Kristallins im Tsal-Dag, Simvolon und Ost-Paggeon (Grieschish-Makedonien), Geotekt.Forsch. 31, I-III, Stuttgart

Kronberg P. (1970)Geologie der Rilla-Rhodope Masse zwischen Strimon und Nestos, Beih.Geol.Jb., 88, Hannover

Meyer W (1966) Uber das Alter von tektonik, Metamorphose und Plutonismous im Sudteil des Rhodopen – Massivs (Griechenland), Habil, Schrift, Clausthall

Meyer W (1968) Die Faltensachsen im Rhodopen-Kristallin ostlich des Strimon (NE-Griechenland), Geotekt. Forsch, 31

Papanikolaou D, Panagopoulos A (1891) On the structural style of southern Rhodope, Greece, Geol.Balcanica, 11, 13-22

Sakellariou-Mane E (1966) Contribution in the geology of the area of Orfanos Gulf in eastern Macedonia, Scient. Yearbook of the fac. of sciences, Thessaloniki, 10, 1-48

Schenk P.F. (1970) Geologie des westlichen Pangaion in Griechisch-Ostmakedonien, Diss.1966, Beih.geol. Jb H.88, Hannover

Water Balance and temporal changes of the surface water quality in Xynias basin (SW Thessaly)

N. Charizopoulos[1], G. Stamatis[1], A. Psilovikos[2]

[1]Agricultural University of Athens, Lab. of Minearology-Geology, Iera Odos 75, 118 55 Athens, nchariz@gmail.com.

[2]University of Thessaly, Department of Ichthyology and Aquatic Environment, Odos Fitokou, 384 46 Volos.

Abstract The study of the Xynias basin is focused on the investigation of the water balance and surface water quality and their impacts due to anthropogenic and natural factors. The average annual precipitation within the study area amounted to 685 mm of which 139 mm (20.3%) is the total runoff, the 428 mm (62.5%) is estimated to be the actual evapotranspiration and 118 mm (17.2%) is the total infiltration. The chemical analyses showed that water flowing from upper reaches of River Enippeas is classified to the Ca-HCO_3 hydrochemical type and it is associated with the presence of limestone's and flysch formations, while water from the drainage ditch, is classified to the Mg-HCO_3 hydrochemical type and is associated with the development of the ophiolites. The drainage ditch is found to be polluted mainly due to anthropogenic factors.

1 Introduction

Rivers are receivers of their basin waters. They transfer significant amounts of water and dissolved contaminants of natural or anthropogenic origin. During their flow through the hydrographic network, the removable mass subject to constant chemical, biological and physical changes, which can greatly change the physical qualitative. These changes are directly related to geological, morphological and hydrographic factors which prevail in the wider region of the river basin, but also to anthropogenic interference (Tsakiris 1995, Antonopoulos 2001, Voudouris 2008). As part of this investigation runoff was measured and chemical analyses were performed on water samples collected from the basic water axes of Xynias and Skopia basins. The survey aims to highlight the natural and anthropogenic influences that contribute to the qualitative differentiation of the characteristics of surface waters in the basin. The survey is part of a Phd Thesis that is in progress at the Agricultural University of Athens. This paper presents the first results that were obtained from the hydrogeological investigation of the area.

2 Methodology

To investigate the quality of surface waters and to identify the various factors, natural and anthropogenic, affecting them, the following tasks took place: a) hydrolithological recognition and interpretation using geological maps of IGME sheets «Anavra», «Domokos» (Marinos et al. 1957), «Leontarion» (Marinos et al. 1962), «Stylis» (Marinos et al. 1963) και «Lamia» (Marinos et al. 1967), b) surface runoff measurements using a flow-meter (Flow Probe FP101/Global Water). Measurements were taken every 15 days, during the period from 10/2009 to 03/2011, in three stations at the drainage ditch of Xynias basin and two stations at the Skopia basin (Fig. 1, 2) and c) surface water samples were collected and chemical analyses were performed on a monthly basis (4th/2010 to 3th/2011) at the same stations as runoff measurements. Using portable instruments in the field, water temperature (T_w °C), pH, O_2 mg/l and electric conductivity (EC_w μS/cm) were measured. In the Laboratory of Mineralogy and Geology of Agricultural University of Athens, chemical analyses were performed on samples using titration methods, spectrophotometry and flame photometry, and the following parameters were determined: 1) hardness, 2) HCO_3^-, 3) Ca^{2+}, 4) Mg^{2+}, 5) Na^+, 6) K^+, 7) Cl^-, 8) SO_4^{2-}, 9) NO_3^- and 10) NH_4^+.

3 Study area

Geomorphology: The study area is focused on the Xynias basin, which covers an area of 606.69 km^2. It consists of the hydrological basins of Xynias and the upper reaches of the river Enippeas known as the basin of Scopia. The average altitude of the basin is 605 m, with lowest at 300 m at the outlet of the basin of Skopia and highest at 1600 m in the mountains of Anavra to Othrys. The basin of Xynias is the basin of the central drainage ditch, of the former Xynias lake, covering 167.90 km^2 and of 66.19 km perimeter. Morphologically, the area consists of an elongated plain section, quite smooth, enclosing outskirts with relatively small area, except the SW section. To the N dominates the limestone massif of Xerovouni, to the NW the naked height Dolines, to the E the area is quite plain, to south the central ridge of Othrys and finally to the SW and W a series of hills forming a complex but not particularly sharp relief. Previously, much of the lowland part of the area (34 km^2) was covered by the waters of the former Xynias lake. In 1946 the drainage works of the lake had been completed and through the drainage ditch the waters were drained out. The remaining land was released for cultivation. Thus human intervention changed the region from a hydrogeological and hydrological closed basin, to a sub-basin of the Pinios River. The basin of Skopia is the river basin of the upper reaches of the Enippeas River, with total area 438.79 km^2 and 116.06 km perimeter. Enippeas is the longest tributary of Pinios River and has a total length exceeding 280 km. Geomorphologically, this basin consists of a flat central and northern part which is the main part and also the high central peaks of Othrys to south and east and small ground elevations to the north and west. It is separated from the neighboring Xynias basin with an almost imperceptible bulge in “Magoula Mines”.

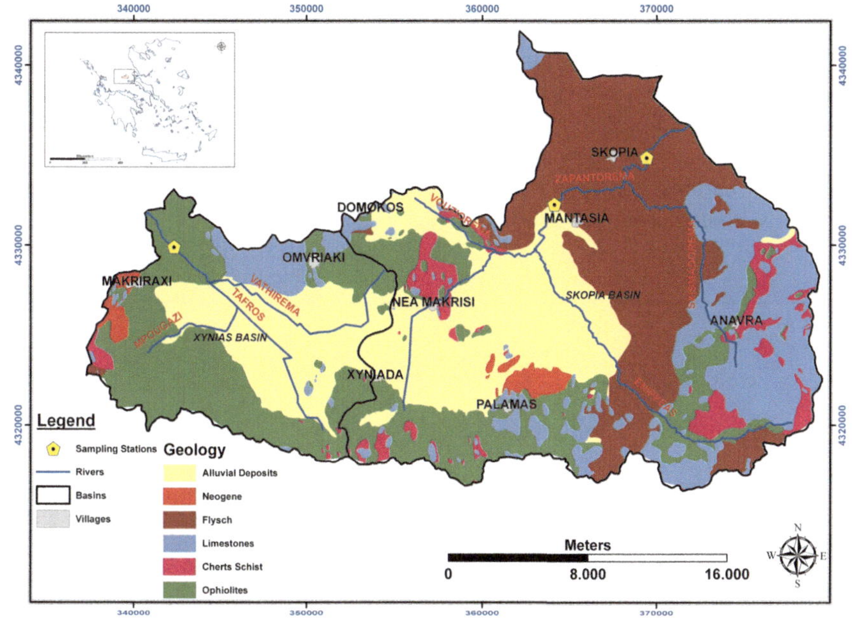

Fig. 1. Geological map of the research area (Marinos et.al. 1957, 1962, 1963, 1967).

Geology: The study area belongs to the geological section of the Eastern Greece and throughout the Sub-Pelagonian unit (Fig. 1). The stratigraphic structures from the youngest to the oldest formations include: a) alluvial and diluvial deposits of the Quaternary consisting of clay, sand, gravel, pebbles and lateral scree, b) marls, clays, sandstones, conglomerates, marly limestones and lacustrine sediments with lignites of the Neogene, c) flysch (Upper Cretaceous) consisting of clay sandstone, clayey limestone and conglomerates, d) platty marly limestones (Upper Cretaceous), e) schist rocks of the Triassic-Jurassic age consisting of a system of cherts schist, marly limestones in thin alternating layers, f) peridotites, serpentinites, diavases, dolerites and tofu basic igneous rocks. The Xynias basin and the entire Othrys mountain have undergone tectonic effects of three different geotectonic units; the Pelagonian, the Parnassos-Giona unit and Pindos. The results of these actions are the faulting and upthrusting which are widely observed. Xynias basin tectonic features are folded structures with synclines and large curvature anticlines, with no scaling observed (Soulios 1975).

4 Hydrological setting

Hydrographic Network: The form of the hydrographic of a basin is in accordance to the lithological composition of the basin that has a direct effect on the permeability of the slopes, the morphological relief, the rainfall and its seasonal distribution. The shape of the drainage network of the basin is mainly dendritic and its total length is 1574.73 km (Fig. 2). The dendritic form justifies the superiority of the surface runoff against infiltration. The hydrogeological formations that appear within Xynias basin are divided into three main groups of rocks, which is a function of their lithological composition and permeability. Into the first group all granular rocks (quaternary deposits) are ordered, in the second group the karstic carbonate formations and in the third group the fissured rocks ophiolites, cherts schist, flysch and post-alpine deposits (Fig. 2).

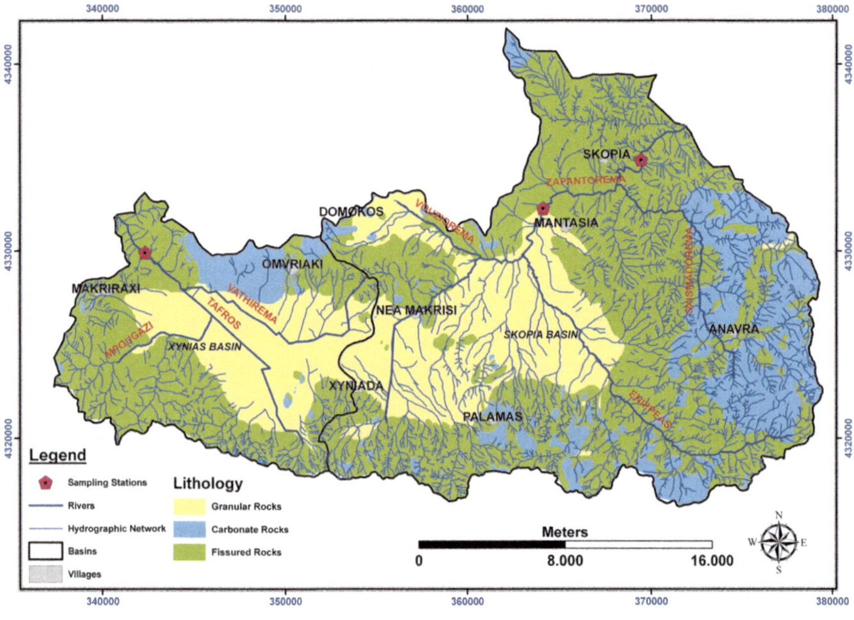

Fig. 2. Drainage network and Lithological composition of the study area.

Hydrological balance: The climatic conditions in the region are characterized by the alternation of wet and cold season, starting in October and a dry and hot one, starting in the middle of May (Karapiperis 1977). For accurate knowledge of the water resources of the study area, is necessary to calculate the water balance given by the governing equation **P=ET+R+I**. For the determination of precipitation (P, mm) of the hydrological balance, monthly rainfall database (years 1995-2010) of 5 stations with good geographic distribution has been set. By using linear regression method their shortcomings have been completed and have held the relation of annual precipitation with the altitude of the stations. The linear relationship Y=0.602*X+290.4 with a correlation coefficient r=0.85 resulting, gives mm rainfall of each altitude in the basin. Applying the rain-degree method calculated the average annual rainfall equivalent to the average altitude (605m) of the basin equal to 685mm of rain. And the distribution is uneven among the elevation zones with higher values in the mountainous area of 651-1253mm rainfall and lowest in the semi-mountainous 471-651mm. Thus the total volume of water precipitation received by the area of the plateau Xynias is $P=415.58*10^6$ m^3. The annual runoff of the basin according to the monthly measurements was calculated to 84.1 m^3 of water, or 139mm (20.29%) of the annual precipitation of the basin area. To estimate the infiltration used infiltration coefficients as those identified in the past (Soulios 1986). Thus, the infiltration coefficients, expressed as a percentage of precipitation, are 20% approximately for the alluvial deposits, 50% for the limestones and 6% for the ophiolites, flysch, etc. Based on these coefficients and the area in km^2 of the hydrolithological formations the total infiltration was estimated at $71.6*10^6 m^3$ or 118mm=17.2% of the annual precipitation. The determination of actual evapotranspiration made using equation of water balance after having calculated the other parameters. Thus ET=P-R-I or ET=685-139-118 or finally ET=428mm=62.5% of the annual precipitation.

5 Quality of surface waters

From the field measurements occurred in surface waters of Xynias drainage ditch and upper reaches of River Enippeas significant differences on the qualitative parameters were found. The values of electrical conductivity ranged in the ditch between 588-940μS/cm while in the Enippeas ranged at lower levels between 422 and 662μS/cm. Similar variation in values occurs in the total dissolved solids. Concentrations in the ditch ranged between 593mg/L and 1094 mg/L, while in the Enippeas ranged between 453 mg/L and 605 mg/L. These differences are mainly related to the lithology of the two sub-basins and especially to human activities. Figure 3 shows the variation of concentrations of main elements of both water currents Xynias and Enippeas. The highest concentrations occur during the wet season, while during the dry season the values are severely limited. The pH values ranged in both water currents at the same levels 7.1 to 8.4 and show the slightly alkaline type of water. Among the gaialkalines in the Xynias basin prevails Mg^{2+}: 52.4-112.6 mg/L, while calcium ranged at lower levels Ca^{2+}: 20.0-97.6 mg/L. The strong presence of Mg^{2+} is obviously associated with the development of ophiolites in the region, which are rich in soluble magnesium minerals (olivines etc). In the Enippeas basin, calcium prevails against the magnesium with values Ca^{2+}: 80.0-98.4 mg/L and Mg^{2+}: 12.4-25.1 mg/L. The source of calcium is associated with the presence of carbonate rocks in the region. Alkalines Na^+ and K^+ present in all, sufficiently low concentrations ranging Na^+: 4.1-24.6 mg/L and K^+: 0.2-3.4

mg/L. Among the anions are dominated by HCO_3^-, with the highest concentrations occur in the Xynias ditch (HCO_3^-: 439.2-591.7 mg/L) and lowest in Enippeas (HCO_3^-: 256.2-347.7 mg/L).

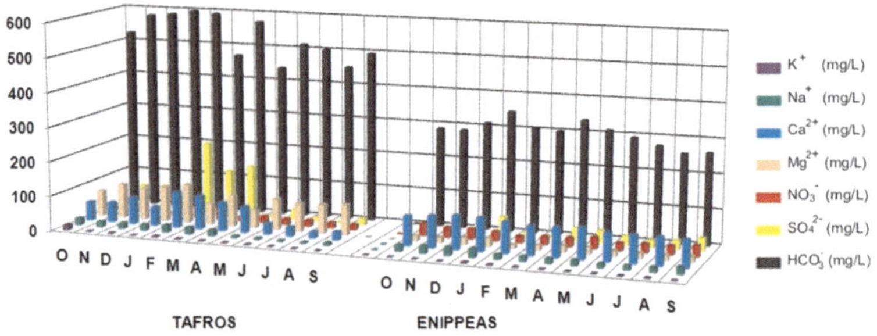

Fig. 3. Variation of concentrations of main elements in the surface waters of Xynias ditch and upper reaches of Enippeas river.

The Cl^- in all, were represented in both water currents with similar concentrations ranging from 7.1mg/L to 28.4 mg/L. The SO_4^{2-} appears in the Xynias ditch in high levels SO_4^{2-}: 3.7-209.7 mg/L, as well as nitrate ions NO_3^-: 13.6-35.6 mg/L, ammonia ions NH_4^+: 0.15-10.58 mg/L and phosphate ions PO_4^{3-}: 0.23-1.72 mg/L. Their presence is obviously related with the strong presence of organic material in alluvial deposits and the various anthropogenic influences, the intensive agriculture, and waste of human settlements and livestock farms. On the contrary, within Enippeas basin, concentrations of these compounds are found at lower levels (SO_4^{2-}: 8.3-56.1 mg/L, NO_3^-: 23.3-36.1 mg/L and NH_4^+: 0.05-0.26 mg/L PO_4^{3-}: 0.08-1.00 mg/L).

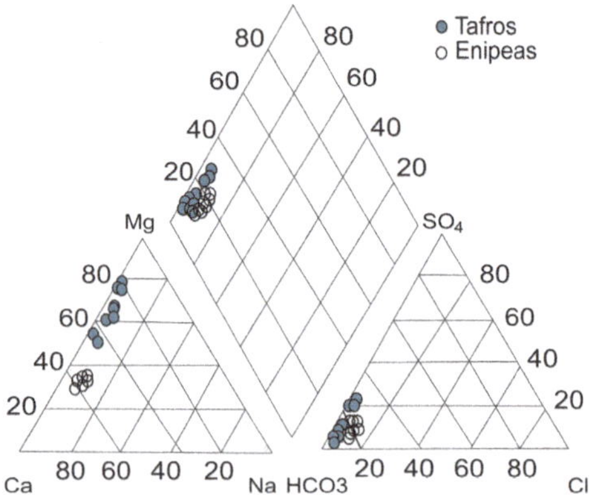

Fig. 4. Classification of surface waters in the Piper diagram.

From the Piper diagram (Fig. 4) was found that the waters of the Xynias ditch and Enippeas have been classified as gaialkalines and their hydrochemical types are Mg-HCO_3 and Ca-HCO_3 respectively. Generally it was found that Enippeas waters have low concentrations of dissolved solids, have a limited burden of anthropogenic factors and have high quality. Unlike the waters of the Xynias ditch, where higher concentrations of dissolved solids were found due to surface pollutants from human activities.

6 Conclusions

The investigation of the hydrological regime, that took place on the Xynias Basin, in combination with the investigation of the qualitative parameters of surface waters, obtains the following results: a) the total precipitation amounted up to 685 mm of which 139 mm (20.3%) is the surface runoff, 428mm (62.5%) is estimated into actual evapotranspiration and 118 mm (17.2%) is the total infiltration, b) Water drainage ditch is classified to the Mg-HCO_3 hydrochemical type associated with the development of ophiolites and reveal quality degradation mainly due to anthropogenic influences, c) River Enippeas waters is classified to Ca-HCO_3 hydrochemical type, associated with the presence of carbonate rocks and flysch formations. The pollution load is detected at very low levels.

References

Antonopoulos V (2001) Quality and contamination of groundwater. Ziti publications, Thessaloniki

Voudouris K (2008) Environmental Hydrogeology, Groundwater and Environment. Jiola publications, Athens

Karapiperis L (1977) The distribution of rainfall to Greek area. Geological Society of Greece Bulletin XI/1:1-28

Marinos G, Anastopoulos J, Maratos N, Melidonis N, Andronopoulos B (1957) Geological map of Greece, Sheets Anavra and Domokos, 1:50 000. Institute for Geology and Subsurface Research, Athens

Marinos G, Anastopoulos J, Maratos N, Melidonis N, Andronopoulos B (1962) Geological map of Greece, sheet, Leontarion, 1:50 000. Institute for Geology and Subsurface Research, Athens

Marinos G, Anastopoulos J, Maratos N, Melidonis N, Andronopoulos B (1963) Geological map of Greece, sheet, Stylis, 1:50 000. Institute for Geology and Subsurface Research, Athens

Marinos G, Anastopoulos J, Maratos N, Melidonis N, Andronopoulos B (1967 Geological map of Greece, sheet Lamia, 1:50 000. Institute for Geology and Subsurface Research, Athens

Soulios G (1975) Hydrogeological study of the Xyniada basin (Fthiotida). Doctoral dissertation Thesis, University of Thessaloniki, p. 99

Soulios G (1986) General Hydrogeology, Volume I. University Studio Press, Thessaloniki

Tsakiris G (1995) Water Resources I. Engineering Hydrology. Symmetry Publications, Athens

Hydraulic connection between the river and the phreatic aquifer and analysis of the piezometric surface in the plain west of Mavrovouni, Laconia, Greece

N. Karalemas

Department of Dynamic, Tectonic and Applied Geology, Faculty of Geology and Geoenvironment, National and Kapodistrian University of Athens, Panepistimioupolis Zografou, 157 84 Athens, Greece

nkaralem@geol.uoa.gr

Abstract This paper concerns the hydraulic relation between the water table hosted in the quaternary formations and the river that crosses the plain west of Mavrovouni. In terms of this study, the groundwater level in the wells was measured in four periods. Also, the river flow was measured at the entry point of the plain and in the area near the river mouth. The water level data were used to create piezometric maps. These maps show that there is an underground drainage line, which could be interpreted as a movement of water from the aquifers to the river, especially at the northwestern part of the plain. However, in the entire area, the river lies at a higher level than the groundwater level; therefore, the only movement that takes places is that from the river to the aquifers. At the region south-eastern of Petrovouni, the piezometric lines for the three out of the four periods of study show a movement of water from the river toward the aquifers. During the fourth cycle, October 2007, the river had run dry, making the feeding of the aquifers impossible. These findings are confirmed by flow measurements of the river, which showed that the river feeds the aquifers as long as it has a flow. Finally, as expected, a variation of the groundwater level measurements between wet and dry seasons was observed and was measured at 1.4 m on average. It should be noted that the measurements took place during a dry hydrological year.

1 Introduction

Northwestern of Gythio lays the largest hydrological basin of the eastern Taygetos. The basin has a general slope toward the Southwest and ends at Laconikos Gulf, western of Mavrovouni (Fig. 1b). The river that crosses the basin is known as "Platy Potami". The main spring that feeds this river is "Agia Marina Spring", which is the largest spring of eastern Taygetos (Karalemas 2006). Downstream of the spring, the river flows for the largest period of the year. Upstream of the area of the village Archontiko, the river crosses alpine formations that are mainly schist. Downstream of that region, the river moves into Aegies plain and enters a narrow valley, at the area of Monachi Sykia. Then, it moves into the plain lying western of Mavrovouni toward South-southeast and flows into Stomio Bay. The area of study of this paper is the last part of the basin.

In terms of this project, piezometric maps were constructed for the study area. The study of piezometric maps generates useful conclusions concerning the groundwater movement, the hydraulic connection between neighbouring water systems, and the hydraulic conductivity at each position of the aquifer (Kallergis 2001, Mimikou and Baltas 2003, Lekkas and Alexopoulos 2005). Furthermore, the river flow at the area where the river enters the plain and at the river mouth was measured. The aim of the aforementioned was to monitor the hydraulic relationship between the river and the water tables hosted in the geological formations that cover the plain. For this reason, the groundwater level in 44 wells, which were initially selected, was measured at four periods. The altitude of the wells was calculated by means of topographic maps of 1:5000 scale of the Hellenic Military Geographical Service.

This work was carried out in terms of the PhD thesis of the author, a thesis that is supported by the Research Project titled “Decision Support System for the Protection and Management of Water Resources of Laconia Prefecture by means of Geographic Information Systems” (PENED 2003). This project is co-financed by 90% by National and Community Funds (25% from the Greek Ministry of Development-General Secretariat of Research and Technology and 75% from E.U.-European Social Fund) and 10% by Development Corporation of Prefecture of Laconia, Regional Development Company of Parnonas and Development Corporation “Parnona - Taygetou”.

2 Geological and hydrogeological setting

The geological map (Fig. 1b) shows that there are four geological formations in the study area, the lowest of which are alpine formations that consist of schist of Phyllite-Quartzite Unit, overlain by minor tectonic napes, composed of marble belonging to the Mani Unit, according to on-site research that was conducted (Latsoudas 1984 considers that these rocks belong to Tripoli Unit). Above these formations post-alpine sediments are found, which are divided to neogene sediments that mainly consist of marls and to quaternary sediments the composition and grain size distribution of which vary.

From the hydrogeological point of view, schistolithic formations have very low permeability, thus are practically impermeable compared to the adjacent formations. Also, the Neogene formations are practically impermeable, because the clay in the marls obstructs the groundwater movement into their mass. On the contrary, the quaternary sediments are loose and consist mainly of sand, gravel and pebbles, and therefore, they are a formation with a remarkable permeability, which varies horizontally and vertically. Consequently, it is expected that the quaternary sediments host water tables, since they overlie impermeable formations. Finally, the marbles are karstic aquifers, which have very high permeability, but extremely limited appearance; as a result they do not play an important hydrogeological role in this area.

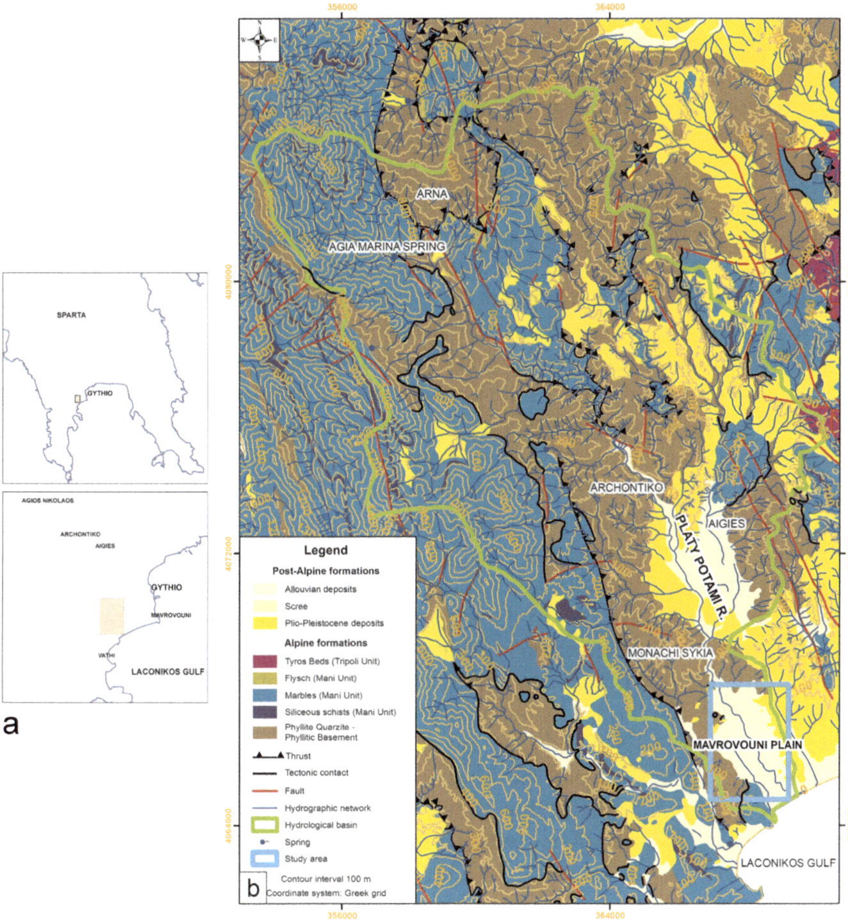

Fig. 1. a. Location map. **b.** Geological map (Psonis and Latsoudas 1983, Latsoudas 1984; Dimadis 1989, modified) of the region area that shows the boundary of the hydrological basin of Platy Potami River.

From all the above, it is concluded that the aforementioned area is ideal for study, since the permeable clastic formations are generally surrounded by impermeable, and therefore are considered as hydraulically isolated (Fig. 3). An exception is the southern part of the plain, where a small area of the quaternary formations is in contact with marble. The marble extend to the sea, but does not actually affect the hydraulic heads of the plain, due to the small area of contact.

3 First measurement period (March 2007)

The first period of measurements took place from 17 to 19 March 2007. For this time of the year higher levels of groundwater were expected. Figure 2a shows the piezometric map of the plain of Mavrovouni with a half-meter interval of subsequent piezometric lines. An underground drainage line, which roughly follows the present riverbed and has general direction NNW - SSE, is observed. Observing the north-northwestern area, the piezometric lines appear dense in both regions separated by this axis, especially in the eastern; this suggests that the permeability of the formations near the river is higher than that of the adjacent formations. In the centre of the plain, southeastern of Petrovouni, the underground drainage line presents a small offset to the West and does not follow the current riverbed. Furthermore, eastern of this axis, there is an underground watershed of the same direction, and even more eastern, an even less distinct drainage line is formed. This setting continues to the Southeast until the centre of the plain, where the underground watershed gradually disappears, the two underground drainage lines are joined and the distance between the piezometric lines becomes greater. However, the absence of wells that are apt for groundwater level measurements in the region eastern of the river does not allow a detailed depiction of the piezometric surface. Even more southern, near the main road leading from Mavrovouni to Areopolis, the piezometric lines tend to straighten (transect virtually the river) and the distance between them becomes smaller. In the southern part of the map, northern of the carbonate formations, an additional couple of an underground drainage line and an underground watershed of short length are found. They are almost parallel to the main underground drainage line. These axes are formed by the piezometric line of 4 m. It must be noted that the piezometric lines of 3.5 m and 4.5 m are not accurate, due to lack of data.

The fact that the piezometric lines roughly transect the contact between the quaternary and the impermeable formations indicates the absence of hydraulic communication between them. In contrast, in the western part of the plain, southern of Petrovouni, the piezometric lines are almost parallel to the border of the plain, possibly due to the limited supply of quaternary formations from a small stream passing nearby. It must be noted that there is not a sufficient number of wells in this region to enable the accurate mapping of the piezometric surface. In the southern part of the map, piezometric lines are expected to form a small angle with the contact between quaternary and carbonate formations. Specifically, according the data, the groundwater level in the quaternary formations is above sea level, while the altitude of the karstic water should be at the sea level. Therefore, the aquifers of quaternary formations are expected to feed the karstic.

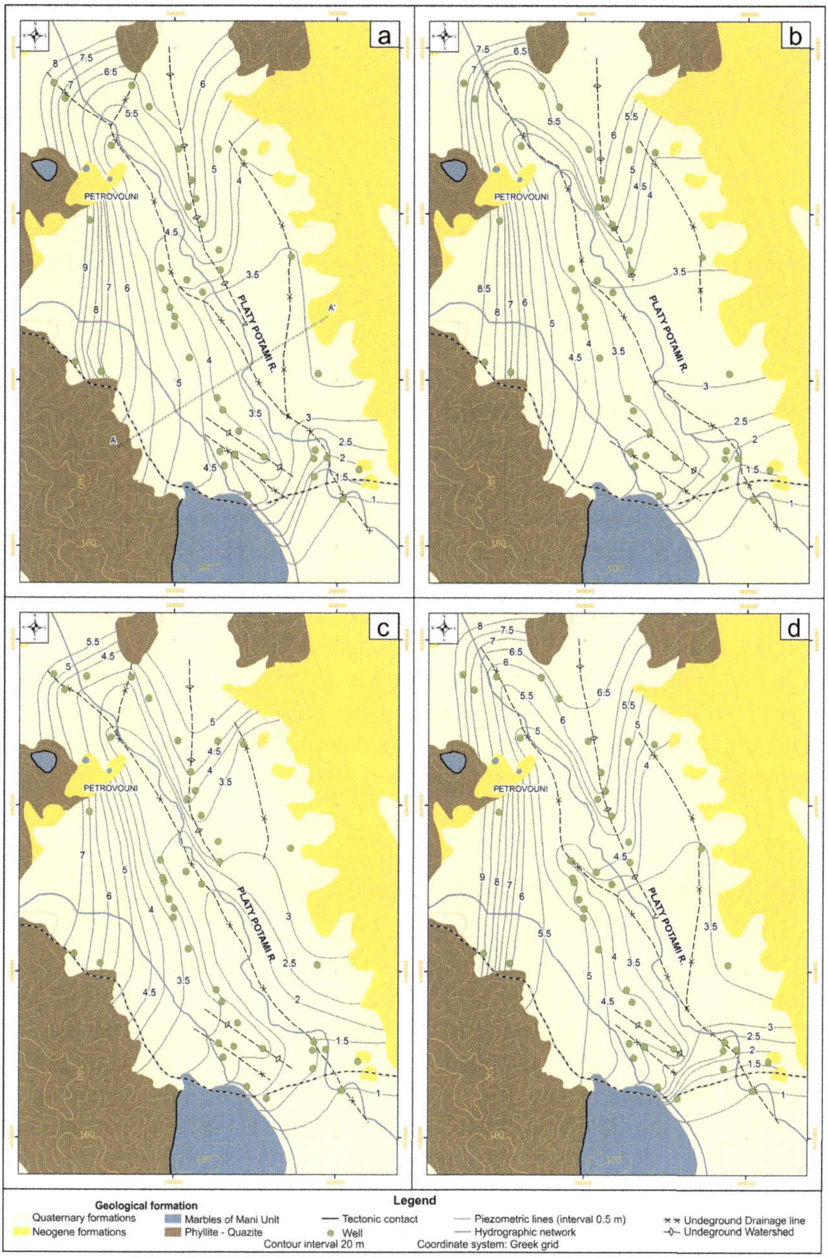

Fig. 2. Piezometric map of **a.** March 2007, **b.** June 2007, **c.** October 2007, **d.** January 2008 of the study area.

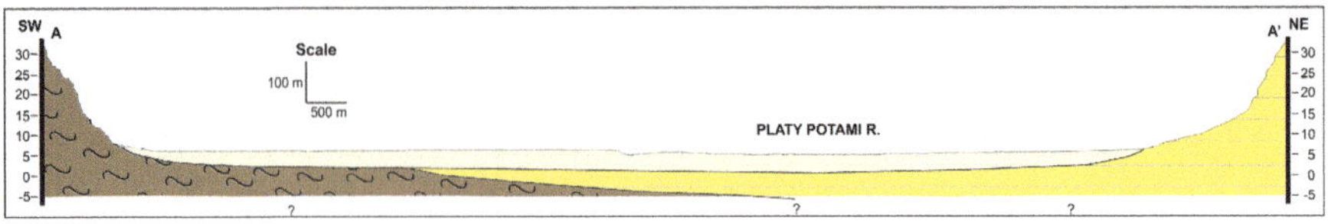

Fig. 3. Geological cross-section.

The flow of Platy Potami River was 665 m^3/h the area entering the plain and 515 m^3/h at the river mouth. Hence, along the plain, the river feeds the water tables that are hosted in the quaternary formations.

4 Second measurement period (June 2007)

The second measurement period took place between 25 and 27 June 2007. The groundwater levels were on average 0.4 m lower than those measured in March.

The piezometric surface in June 2007 is shown at the map (Fig. 2b). The shape of the piezometric surface is similar to that of March. The only difference lies in the southern part of the map, near the carbonate rocks, where the underground drainage line and the parallel underground watershed are less distinct.

The Platy Potami River maintained a very low flow (36 m^3/h) near the entrance to the plain, water that was infiltrating nearby. Also, there was stagnant water downstream of the public road leading from Mavrovouni to Areopolis, because the river mouth was blocked by a large volume of sand dunes.

5 Third measurement period (October 2007)

The third period includes measurements from 10 to 15 October 2007. This season is characterized by lower groundwater levels, because it is the end of the irrigation season and the new hydrological year has not started yet. It is significant that some wells in the northern part of the plain were dry. The piezometric surface in October 2007 is shown in the map (Fig. 2c). In comparison to the aforementioned piezometric maps, the following differences are observed. Firstly, the main underground drainage line is more accurately defined in the southeastern part of the plain. The piezometric surface is on average 0.95 m lower than that of June 2007.

A notable change is the shift to the East of the main underground drainage line at the area southeastern of Petrovouni until it concurs with the riverbed. Finally, the underground drainage line and the parallel underground watershed in the southern part of the map are not very distinct.

The river showed no flow downstream of the village Archontiko. Shortly before the river mouth, stagnant water was observed, similar to that of the previous period, but had a much lower level.

6 Fourth measurement period (January 2008)

The measurements of the fourth period were performed from 14 to 15 January 2008. Even though the hydrological year 2007-2008 was also dry, the groundwater level was higher in January, as expected. The piezometric surface in January 2008 is presented on the map (Fig. 2d). The groundwater level has risen by an average of 1.30 m compared to that of October; as a result, the shape of the piezometric surface in the southeastern part of the plain was restored, as it was in March and June 2007. Also, the shift of the main underground drainage line at the area southeastern of Petrovouni is relocated westwards of the river bed. Finally, the underground drainage line and the parallel underground watershed in the southern plain became more obvious and more similar to the situation that was "captured" last March.

The flow of the river was measured at 927 m^3/h at the area entering the plain and 833 m^3/h at the river mouth. As shown by the measurements, the river was feeding the quaternary aquifers.

7 Discussion - Conclusions

The piezometric maps mentioned in this paper were compared to those of two other studies at the same area (Dimakopoulos 1991, Arvaniti and Gerasimopoulos 2009). The major difference is the absence of the main underground drainage line. It is estimated that this difference is owed to the limited number of wells included in the other studies (6 and 31, respectively).

From the maps of the four successive periods, it appears that the shape of the piezometric surface changes in the central and southern parts of the plain only during the period of October 2007. This variation is possibly due to the groundwater level dropdown, which results in the piezometric surface approximating the permeable-impermeable surface. The variation in the groundwater level of the water table between wet and dry season (March and October 2007) is an average of 1.4 m.

During the studied periods, in the northwestern part of the plain, the curves of the piezometric lines show concave towards downstream; this would normally suggest that the water tables feed the stream. However, in the entire area, the phreatic water table is located at lower elevations than the river bed. Thus, the only possible water movement is that from the river to the water tables. Therefore, the shape of the piezometric lines that forms the underground drainage line, which concurs with the riverbed, is due to the feeding of the formations under riverbed from the adjacent formations. For this reason, this shape is maintained throughout the year. At the area southeastern of Petrovouni, during the wet periods (March-January), the curves of the piezometric lines show concave towards upstream; this proves the feeding of the water tables from the river. These are confirmed by the flow measurements of the river, since feeding from the river to the water tables took place during all the periods that the river had flow.

In the southern part of the plain, near the carbonate rocks, an underground drainage line and a parallel underground watershed appears on the maps created for all periods. These forms are limited in length due to the absence of wells. The fact that these forms exist during the four periods is more likely owed to potential variations in the altitude of the bedrock in the region.

References

Arvaniti E, Gerasimopoulos Ch (2009) Hydrogeological study at the area of Mavrovouni - Gythio, thesis, National and Kapodistrian University of Athens. Athens (in greek)

Dimakopoulos P (1991) Hydrogeological Research of Prefecture of Laconia, Prefecture of Laconia, Athens (in Greek)

Dimadis E (1989) Geological map of Greece, scale 1:50000, Gythion Sheet. Inst. Geol. and Miner. Explor., Athens

Kallergis G (2001) Applied - Environmental Hydrogeology, 2nd Edition. Technical Chamber of Greece, Athens (in Greek)

Karalemas N (2006) Operational Mechanism of springs of Eastern Taygetos. MS thesis, National and Kapodistrian University of Athens, Greece (in Greek)

Mimikou M, Baltas E (2003) Engineering Hydrology, 3rd Edition, NTUA, Athens (in Greek)

Latsoudas C (1984) Geological map of Greece, scale 1:50000, Mavrovounion - Areopolis - Gerolimin Sheet. Inst. Geol. and Miner. Explor., Athens

Lekkas S, Alexopoulos A (2005) Introduction to Hydrogeology, National and Kapodistrian University of Athens, Greece (in Greek)

Psonis C, Latsoudas C (1983) Geological map of Greece, scale 1:50000, Xirokampion Sheet. Inst. Geol. and Miner. Explor., Athens

Groundwater recharge using a Soil Aquifer Treatment (SAT) system in NE Greece

F. Pliakas[1], A. Kallioras[2], I. Diamantis[1], M. Stergiou[3]

[1]Department of Civil Engineering, Democritus University of Thrace, Xanthi, GR 67100, Greece. fpliakas@civil.duth.gr

[2]National Technical University of Athens, School of Mining and Metallurgical Engineering, Athens, Greece.

[3]National Technical University of Athens, School of Applied Mathematical and Physical Sciences, Athens, Greece.

Abstract The design of a proper Soil-Aquifer Treatment (SAT) groundwater recharge system is proposed after studying the geological and hydrogeological regime of the coastal aquifer system at Nea Peramos, NE Greece. The investigation of the qualitative problem of the study area included groundwater level measurements and groundwater sampling and chemical analyses, respectively. The paper also includes the design of necessary maps, such as geological, piezometric and distribution of specific qualitative parameters. The investigation concludes to further research and managerial useful proposals.

1 Introduction

The majority of coastal aquifer systems in Greece are environmentally degraded due to groundwater salinization mainly because of the intrusion of seawater towards the mainland. In Northern Greece, research projects carried out by the Laboratory of Engineering Geology of the Department of Civil Engineering of Democritus University of Thrace have provided evidence that the encroachment of seawater into freshwater coastal aquifers of Eastern Macedonia and Thrace takes place in Rivers Evros, Lissos and Nestos (through both the river course and the aquifer of each delta environment), Lakes Vistonida and Ismarida, and coastal areas of Rhodope, Xanthi and Kavala (Diamantis and Petalas 1989; Kallioras 2008; Kallioras et al 2006; Kallioras et al 2010; Petalas 1997; Petalas et al. 2002; Pliakas et al 2001; Pliakas et al. 2007). The presence of seawater intrusion phenomenon in the above areas is mainly due to overpumping conditions for irrigation purposes.

Laboratory of Engineering Geology of the Department of Civil Engineering, Democritus University of Thrace, elaborates the salinization problem in the study area focusing on designing a Soil-Aquifer Treatment (SAT) system artificially recharging effluent from the Nea Peramos WWTP (former Municipality of Eleftheres which now belongs to Municipality of Paggaion, Kavala, NE Greece). Soil-aquifer treatment refers to processes that occur in the soil (vadose zone) and aquifer that act to remove or reduce chemical and biological constituents of concern.

Soil-aquifer treatment can be an important step in the treatment train for reuse of wastewater. SAT systems for treatment and storage of waters of impaired quality must be tailored to local hydrogeology, quality of input water and climate (National Research Council 1994).

2 Description of the study area - geological and hydrogeological characteristics

The area of investigation is located in Prefecture of Kavala, Municipality of Paggaion and covers the hydrogeologic basin of Nea Peramos (Fig. 1). The study area covers approximately 20 km^2 and it contains small settlements with small populations which increase during the summer months due to tourist visitors. The geomorphology of the study area is characterised as flat, with elevation of less than 20 m throughout the entire study site, while the most important surface water body of the area is the stream of Nea Peramos, which flows through the area and discharges into the sea from SW to NE. The main agricultural activity is the farming of vineyards, while there is a number of fruit-bearing trees and olive trees in the area, all playing a vital role in the economy of the local communities. The climate of the area is typical Mediterranean with dry and hot summers. The main geological unit of the surface area is dominated by quaternary deposits, while coastal sands and alluvial deposits appear in the form of strips at the eastern and western boundaries, respectively (Fig. 1). The aquifer system of the study site includes an upper layer of sedimentary deposists of a varying depth up to 20 m (composed of terrestrial deposits and alluvial fans) and coastal sands at the region adjacent to the seashore (Pliakas et al. 2007). The underlying metamorphic geological formation is mainly composed of granodiorite which is strongly foliated. The phenomenon of seawater intrusion involves the unconfined aquifer layer.

The area of investigation includes a number of groundwater wells, which are distinguished into three distinct categories according to their depth and hence the exploitation of certain aquifer layers (Geoservice 2000; Pliakas et al. 2007): (i) shallow wells which pump from the unconfined aquifer only, (ii) deep wells which pump exclusively from the underlying granodiorite aquifer and (iii) deep wells which have open screens to both aquifer layers and exploit both geological formations.

The quantitative regime of the unconfined aquifer was investigated through measurements of groundwater levels, after the design of an appropriate monitoring network of 20 groundwater wells. Piezometric maps (Fig. 2) prior and after the irrigation period show (Pliakas et al. 2007): (i) a constant recharge axis from NW to SE direction, fact which is attributed to the discharge of the groundwater potential of the mountainous zone, (ii) that the unconfined aquifer of investigation discharges into the sea along three main axes with direction W-E, NW-SE and N-S (seawater intrusion is indicated along the above axes, as it is proved at the chemical analyses section of the paper in question).

The research also included the qualitative investigation of the unconfined aquifer, part which was achieved by the design of an appropriate qualitative monitoring network of 15 wells and sampling of groundwater (15/07/2006). The research also included in situ measurements of physicochemical parameters of the groundwater of the area such as electrical conductivity, pH and temperature (Pliakas et al. 2007). The chemical analyses took place at the Laboratory of Engineering Geology of Civil Engineering Department of Democritus University of Thrace.

Fig. 1. Geographical location and geological map of the study area (IGME 1983, modified).

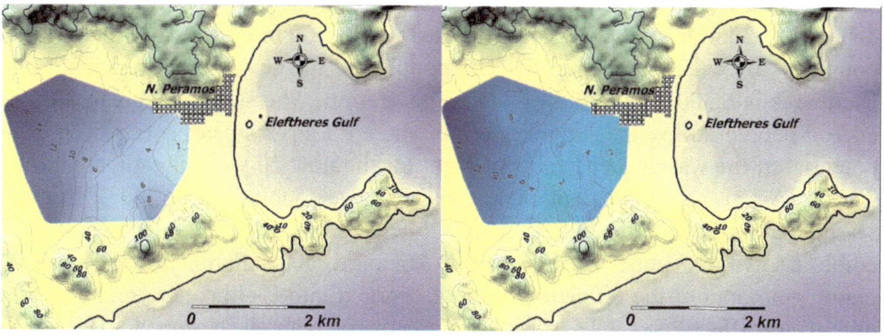

Fig. 2. Piezometric maps of the study area (left: May 2006, right: October 2006) (Pliakas et al. 2007).

According to water standards from Kallergis (2000), the groundwater samples of the aquifer of investigation appear to have good quality, with the exception of some samples from wells located within the coastal zone. The latter ones have Revelle values between 1 and 2, fact which characterizes the groundwater samples as being slightly polluted by seawater intrusion conditions. The above ascertainment can be supported by the fact that these samples also have high chloride values and values of electrical conductivity greater than 1500 μS/cm. It should be noted that the above groundwater wells are located along the NE-SW main discharge axis of the unconfined aquifer (main salinization axis) (Fig. 3).

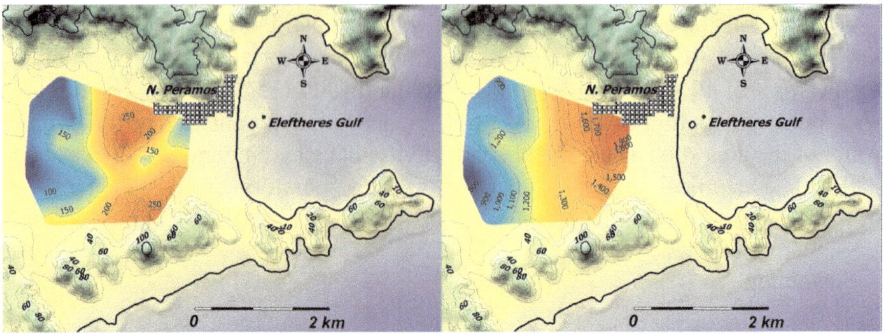

Fig. 3. Left: Distribution of chloride ions (mg/L) of the groundwater samples of the study area (sampling date 15/07/2006) Right: Distribution of electrical conductivity (μS/cm) of the groundwater samples of the study area (sampling date 15/07/2006) (Pliakas et al. 2007).

3 Indicative proposal of groundwater artificial recharge using a SAT system

The Wastewater Treatment Plant of former Municipality of Eleftheres is constructed at a local topographical peak (with an altitude of 10–20 m) located at a distance of 2 km from the coast and approximately 400 m upstream of the western boundary of N. Peramos (Fig. 4). The population of the former Municipality of Eleftheres is 6,300 people during the winter period, whereas it is increased up to 24,700 people during the summer period where a touristic peak is observed. The potential of the wastewater treatment plant for the next 20 years is estimated to be 30,000 IA (eq. pers.) (Markantonakos and Papavasilopoulos a, b, 2003). It is assessed that part of the treated effluent will be used for irrigation purposes of non-edible cultivation (under relative quality standards for suitability), while the remaining amount of the effluent will be discharged into Eleftheres torrent after the use of specific filters. The wastewater treatment plant is designed in order to achieve a high degree of organic pollutants, ammonia, nitrates and phosphorus removal; so that the potential or risk for contamination or pollution to the receiving water body (Eleftheres torrent) will be minimized. Additionally, the application of UV distillation and purification results in effective removal of bacterials and pathogens within the water. In general terms, the design and operation of the wastewater treatment plant of Eleftheres is such, so that the risk of negative environmental impacts in the surrounding area will be minimized.

The selection of appropriate location for a proposed SAT plant is one of the most important stages during the assessment of such water engineering works.

Such a selection is based on different criteria according to the geological and hydrogeological regime of the proposed site whereas consideration of the relative agricultural areas (public or private) should also take place. With reference to the latter, it has to be noted that the financial cost assessment should also consider possible costs of expropriations of private properties, which in many cases might cause significant raise of the total budget of the project. One of the most important and objective requirements for the design and implementation of a SAT system is the presence of a thick unsaturated zone which underlies the SAT installation in order to ensure all necessary aerobic processes and the removal of certain bacteria. For this reason, the so-called California Guidelines (Hultquist et al. 1991) suggest that the minimum vertical distance between the bottom of the recharge basin and the water table should be at least 3 m. However it has been recorded that artificial recharge in the Netherlands with the application of bank infiltration systems, has been designed and constructed under controlled conditions below the water table.

The measurements of the water table of the coastal phreatic aquifer of N. Peramos and the interpretation of its fluctuation in different periods of the hydrological year (11/1/2006, 21/3/2006, 5/5/2006, 16/10/2006) (Pliakas et al. 2007) revealed that the water table in each of the groundwater wells E035, E113 και E03 (Fig. 4) meets the aforementioned criteria with respect to the minimum vertical distance between the bottom of the recharge basin and the water table.

According to previous geological and hydrogeological investigations for the study area (Geoservice 2000; Pliakas et al. 2007) it is believed that the alluvial and quaternary deposits which compose the unsaturated zone in groundwater wells E035 και E113, NW of the wastewater treatment plant, satisfy the demands (in terms of soil type and hydrogeological conditions) for an effective and proper infiltration of the treated effluent and hence the recharge of the underground phreatic aquifer. Groundwater well E03 is adjacent to the foot hill where the outcrops of the granodioritic formation appear and therefore is not recommended for the installation of a SAT system at that specific location. The combination of all the aforementioned conclusions lead to the decision that the installation of the proposed SAT system at the coastal area of Eleftheres basin should be located at a distance of 1 km NW from the wastewater treatment plant (Fig. 4) (Stergiou 2009). The proposed pilot SAT system should include the following (Fig. 4):

- The assessment and dimensioning of the effluent tank (dimensions of the tank bottom, depth and tank volume) in order to effectively receive the treated effluent from the wastewater treatment plant.
- The excavation and formation of 2 systems of shallow recharge basins A and B (Fig. 4); each system containing 3 recharge basins of individual dimensions: surface area 15×30 m, depth 0.50 – 0.80 m, in order to be filled with water not more than 20 – 30 cm in height so that the clogging of the basin bottom (and hence the effectiveness of the SAT system itself) will be minimized.
- Coating of the bottom of the recharge basins with the use of gravels or medium grained sand (0.10 – 0.20 cm)

- Installation of proper water pipes for the transfer of water from the effluent tank to the recharge basins as well as proper water pipes which will be connecting the adjacent recharge basins of the SAT system.
- The installation of a monitoring network of groundwater wells and piezometers in order to inspect on a frequent basis the qualitative and quantitative effectiveness of the SAT system.

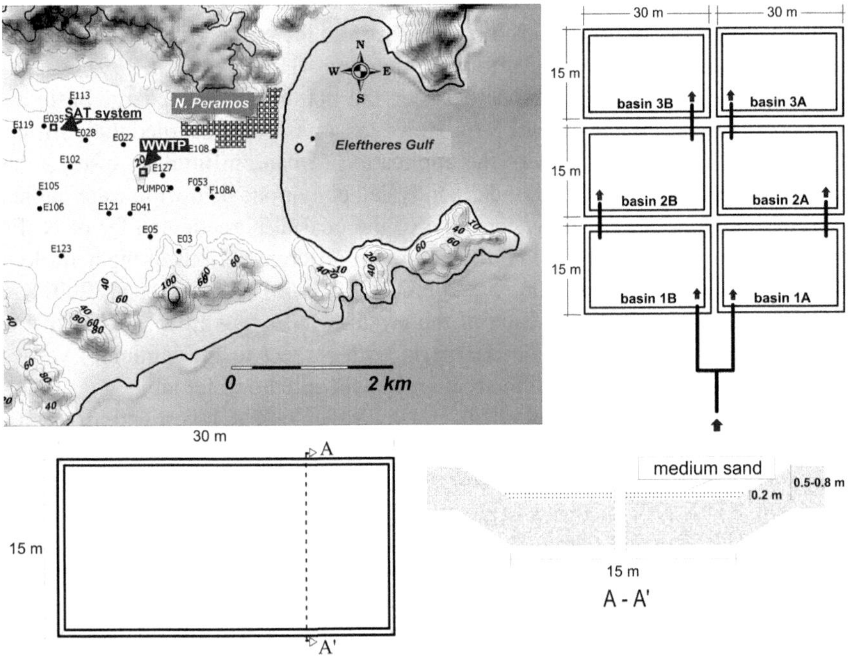

Fig. 4. Quantitative and qualitative groundwater monitoring wells network of the study area. Indicative proposed location of recharge basins for the application of SAT system. Positioning and dimensioning of 2 systems of recharge basins (with 3 basins each). Plan and profile of a representative recharge basin (Pliakas et al. 2007, Stergiou 2009).

4 Conclusions – Managerial proposals

This paper studies the design of a proper Soil-Aquifer Treatment (SAT) groundwater recharge system proposed after studying the geological and hydrogeological regime of the coastal aquifer system at Nea Peramos, NE Greece. The piezometric surface of the study aquifer layer shows groundwater level values which are greater than the mean sea level, due to the recharge conditions of the investigated aquifer, fact which eliminates the possibility of active seawater intrusion. The quality of groundwater at the western part is good, while the groundwater of the eastern coastal part appear slightly saline as revealed by their Revelle values.

Laboratory of Engineering Geology of the Department of Civil Engineering, University of Thrace, elaborates the salinization problem in the study area focusing on designing Soil-Aquifer Treatment (SAT) systems artificially recharging effluent from the Nea Peramos WWTP (Municipality of Paggaion, Kavala, NE Greece). In the framework of this research and based on relevant experience of the above laboratory regarding the management of aquifers recharge and according to the relevant recommendations, specifications, propositions and guidelines of scientific organizations, committees and agencies in Greece and worldwide (ASCE 1987, 2001, CMD 2009, EU 2006, US EPA 1992, US NRC 1994), the following research works and managerial proposals were concluded: (1) study of the hydrogeological regime of the aquifer system and feasibility study of applying effluent recharge through SAT systems using recharge basins, (2) definition and testing of all the necessary qualitative characteristics of the effluent based on documentation and experience of other researches and applications worldwide (SS, TDS, SAR, TOC, BOD, pathogenic microorganisms, organic compounds, DBPs, heavy metals, nutrients, nitrogen, phosphate), (3) selection of suitable off-channel or in-channel sites for the recharge basins, (4) testing an experimental pilot recharge application and evaluation of the recharge effectiveness, (5) design of a network of wells monitoring the quality of the recovered water, (6) determination of the proper testing parameters, (7) confrontation of potential clogging problems, (8) determining the recovered water uses, (9) environmental impact assessment for the broader region, (10) establishment of safety countermeasures in the designed recharge facilities.

References

ASCE (1987) Ground Water Management. 3rd Edition, ASCE Manuals and Reports on Engineering Practice No 40, New York

ASCE (2001) Standard Guidelines for Artificial Recharge of Ground Water. Environmental and Water Resources Institute, EWRI/ASCE 34-01, Virginia, USA

CMD No 39626/2208/2009 Determination of measures for the protection of groundwater against pollution and deterioration, in compliance with the provisions of Directive 2006/118/EC on the protection of groundwater against pollution and deterioration, Greece

Diamantis J, Petalas C (1989) Seawater intrusion into coastal aquifers of Thrace and its impact on the environment. Toxicological and Environmental Chemistry, V. 20-21, 291-305

EU (2006) Directive 2006/118/EC of the European Parliament and of the Council of 12 December 2006 on the protection of groundwater against pollution and deterioration

Geoservice (2000) Cystallization technologies for prevention of saline water intrusion, CRYSTECHSALIN (EVK1-2000-631): Geological-hydrogeological and geochemical survey of the Greek test site in Eleftheres basin, Greece. Report 1, Research Directorates, 5th Framework Program, Energy, Environment and Sustainable Development, European Commission, 64p

Hultquist RH, Sakaji RH, Asano T (1991) Proposed California regulations for ground water recharge with reclaimed municipal wastewater. Proceedings of the 1991 Specialty Conference, Environmental Engineering, American Society for Civil Engineers. Reno, Nev. July 1991, 759-764

Institute of Geology and Mineral Exploration (IGME) (1983) Geological Map of Greece, Scale 1:500,000, 2nd edition, Athens

Kallergis G (2000) Applied Environmental Hydrogeology, 2nd Edition, T.C.G., Vol. 2 (in Greek)

Kallioras A (2008) Groundwater resources management of aquifers subjected to seawater intrusion regime. The case of western coastal plain of the Prefecture of Rhodope. Doctoral dissertation submitted to the Department of Civil Engineering, Democritus University of Thrace, Xanthi, Greece, 2008, 333p (in Greek)

Kallioras A, Pliakas F, Diamantis I (2006) Conceptual model of a coastal aquifer system in northern Greece and assessment of saline vulnerability due to seawater intrusion conditions, J. Environ Geol. 51[3], 349-361

Kallioras A, Pliakas F, Diamantis I (2010) Simulation of groundwater flow in a sedimentary aquifer system, subjected to overexploitation. Water, Air, & Soil Pollution, Springer, 211[1-4], 177-201

Markantonakos P, Papavasilopoulos E (2003a) Central Wastewater Treatment Plant of Eleftheres Municipality - Preliminary Study. Region of Eastern Macedonia and Thrace, Prefecture of Kavala, Municipality of Eleftheres, Athens (in Greek)

Markantonakos P, Papavasilopoulos E (2003b) Sewerage System and Central Wastewater Treatment Plant of Eleftheres Municipality – Environmental Impact Study. Region of Eastern Macedonia and Thrace, Prefecture of Kavala, Municipality of Eleftheres, Athens (in Greek)

National Research Council (1994) Ground water recharge using waters of impaired quality. Washington, D. C.: National Academy Press

Petalas C (1997) Analysis of aquifer systems in the heterogeneous coastal part of prefecture of Rhodope (in Greek). PhD Dissertation. Department of Civil Engineering, Democritus University of Thrace, Xanthi, Greece, 288p (in Greek)

Petalas C, Pliakas F, Diamantis I (2002) Seawater intrusion problem in coastal aquifers in East Macedonia and Thrace and methods of prevention. Technika Chronika, Sci. J. of T.C.G., I, 22[1-2], 31-43 (in Greek)

Pliakas F, Diamantis I, Petalas C (2001) Saline water intrusion and groundwater artificial recharge in east delta of Nestos River. Proceedings of the 7th International Conference on Environmental Science and Technology, University of the Aegean, Global Nest, Syros, Greece, 3-6/9/2001, Vol. 2, 719-726

Pliakas F, Kallioras A, Diamantis I, Giougis I (2007) Seawater intrusion in a coastal phreatic aquifer of Kavala Prefecture, Northern Greece. Proceedings of the 10th International Conference on the Environmental Science and Technology (CEST2007), University of the Aegean, Dept. of Environmental Studies, and Global Nest, 5-7/9/2007, Cos Island, Greece, Vol. B, 626-633

Stergiou (2009) Groundwater recharge by using waters of impaired quality – Relative application at N. Peramos, Kavala, Greece. Undergraduate Thesis submitted to the Department of Civil Engineering, Democritus University of Thrace, Xanthi, Greece, 124p (in Greek)

U.S. Environmental Protection Agency EPA (1992) Guidelines for Water Reuse. Technology Transfer Manual EPA/625/R-92/004. U.S. EPA. Washington, D.C. 247p

U.S. National Research Council (1994) Ground Water Recharge Using Waters of Impaired Quality. National Academy Press, Washington, D.C., 283p

Enhancing Protection of Dar es Salaam Quaternary Aquifer: Groundwater Recharge Assessment

Y. Mtoni[1,2], I.C. Mjemah[3], M. Van Camp[1], K. Walraevens[1]

[1] Laboratory for Applied Geology and Hydrogeology, Ghent University, Krijgslaan 281 S8, 9000, Ghent, Belgium.
[2] National Environment Management Council (NEMC), P.O. Box 63154, Dar es Salaam, Tanzania
[3] Sokoine University of Agriculture (SUA), P.O. Box 3038, Morogoro, Tanzania

Abstract Water balance for Dar es Salaam Quaternary coastal aquifer (DQCA) was calculated according to the method of Thornthwaite and Mather. Monthly potential evapotranspiration estimates were calculated from 39 years (1971-2009) of routine meteorological data using Penman-Monteith, Thornthwaite, Hargreaves and Hamon methods. Results were compared in order to show possible differences that could be attributed to the methods. Determination of groundwater recharge rates gave a mean value of 184 mm $year^{-1}$ which is equivalent to 71.39×10^6 m^3 $year^{-1}$ indicating that 16.5% of the long term average annual precipitation of 1114 mm ends up as groundwater recharge. Groundwater abstraction from DQCA was estimated to be 69.3×10^6 m^3 $year^{-1}$. These results are alarming in two aspects: high abstraction rate and the increasing trend of borehole drilling. Groundwater abstraction which is nearly equal to the amount of the groundwater recharge is far greater than the least conservative sustainable yield calculated at 70% (equivalent to 49.97×10^6 m^3 $year^{-1}$).

Introduction

Dar es Salaam is the fastest growing city in Tanzania with a population of about 4.0 million and with high concentration of business, industries and institutions. The high population growth and deterioration of surface water supply has resulted in an incredibly increased dependency on groundwater from Dar es Salaam Quaternary coastal aquifer (DQCA). The threats to this aquifer system among other things include seawater intrusion due to overexploitation of the aquifer and nitrate contamination caused by the use of pit latrines and urban agriculture (Mjemah 2007; Mtoni and Walraevens 2010; Mtoni et al. 2010). The objective of this study was estimation of a sustainable yield for the DQCA through determination of the amount of groundwater recharge, based on the estimation of water budget for the aquifer.

Geological and hydrogeological setting

The study area (Fig. 1) is located in Dar es Salaam City which is part of the coastal area of Tanzania. The latter consists of Precambrian basement rocks and overlying sedimentary formations: Karoo, Jurassic, Cretaceous, Tertiary and Quaternary. The study area is mainly located in the Quaternary deposit (Fig. 2). As indicated in Figure 2, the study area and its surroundings comprise three major parts distinguished by the geological formations: the central coastal plain with Quaternary fluviatile deltaic sediments, the Mio-Pliocene deltaic clay-bound sands and gravels in the northwest and southeast and the Lower Miocene fluviatile sandstones of Pugu Hills in the west of the study area.

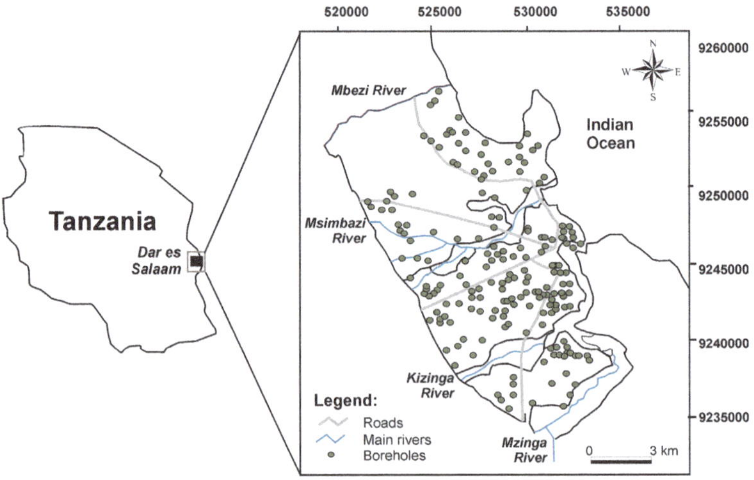

Fig. 1. Location of the study area: a) over 4000 boreholes are estimated to have been drilled only in the inner part of the city; b) Boreholes visited with yield > 10 m^3/h.

The hydrogeological regime of Dar es Salaam City is characterized mainly by two aquifers both of Quaternary age. The upper unconfined sand aquifer is superimposed on the lower semi-confined sand aquifer, having thickness ranging from 1-10 m and 5-50 m respectively; both being separated by a clay aquitard with a thickness ranging from 1-30 m (Mjemah 2007).

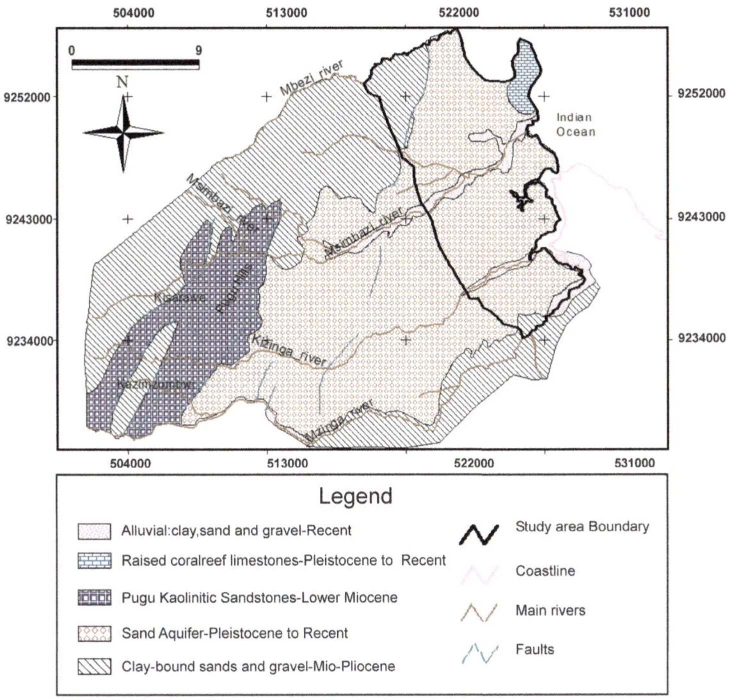

Fig. 2. Geological map of the broader area (map modified from Mjemah et al. 2009).

Climatic conditions

The average amount of rainfall in Dar es Salaam City is approximately 1114 mm per year (average calculated based on data from 1971-2009) with considerable variation. The major precipitation (78.6%) occurs in two periods: long and heavy rains occurring in March to May with peak rainfall in the month of April, while short and light rains occur in October to December (Fig. 3). The dry season begins in June until September, and is characterized by limited rainfall. The maximum mean monthly temperature is 32.4°C and the minimum mean monthly temperature is 18.3°C. The lowest mean monthly temperature is recorded between June and August. Radiation has maximum mean monthly value of 20.76 MJm^{-2} while the minimum is approximately 16 MJm^{-2}; observed to decrease during the rainy season. Radiation shows a similar trend with potential evapotranspiration (PET) while a weak similarity is observed for wind speed. The mean monthly wind speed varies from 4.44 ms^{-1} to 6.79 ms^{-1}. The high mean monthly PET values are experienced during the early months of the year (i.e. January and February).

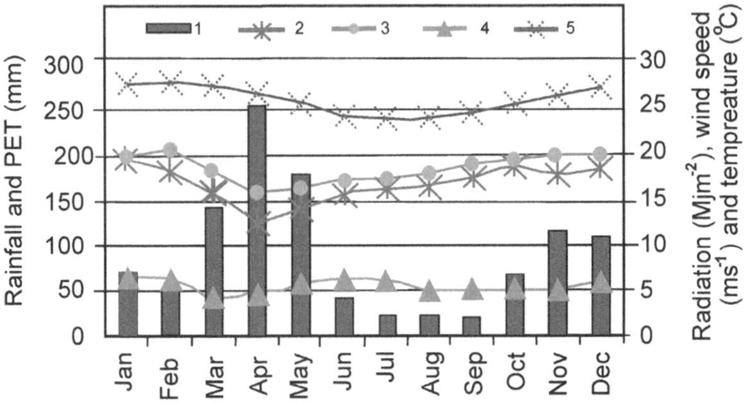

Fig. 3. Variation of average monthly of: 1) rainfall (mm), 2) potential evapotranspiration (PET) (mm), 3) radiation (MJm^{-2}), 4) wind speed (ms^{-1}), and 5) temperature (°C) for the period from 1971 to 2009.

Soil moisture balance and ground water recharge

The study area is characterized by vegetation cover (grass) of shallow roots (0.5 m) and a lithology of fine sand (vadose zone), which, according to Thornthwaite and Mather (1957), this assumes a water holding capacity of 10% over the entire root zone. This gives a water capacity of the root zone of 50 mm. This refers to the plant available water (PAW) calculated as the difference between field capacity (FC) and permanent wilting point (PWP). Excess soil moisture to the FC is drained to the groundwater reservoir in the form of groundwater recharge. The different terms of the soil moisture budget were computed using the concept of water balance of the unsaturated zone according to Thornthwaite and Mather (1955; 1957). It consists of monitoring of the accumulated potential water loss (APWL) and the amount of water stored in the soil (S_B). The latter is calculated as $S_B=PAW*e^{-APWL/PAW}+WPWP$ (with WPWP: the water content of the root zone at PWP = PWP * root depth). Calculations to determine S_B and APWL were performed for each month using monthly mean precipitation (P) and PET. Soil water content at PWP for sand is assumed to be 6% by volume (Raes et al. 2010). This gives 30 mm available water content of the root zone at PWP. Soil moisture was calculated by balancing precipitation and evapotranspiration using FC and PWP as parameters. The runoff factor for this study was taken as 8% of the precipitation as suggested by previous studies (Mjemah 2007; Mjemah et al. 2011). The hydrological year has been considered from the beginning of October (the beginning of the rainy season) to the end of September of the next year (i.e. the end of dry season). The initial PAW in October 1971 is assumed to be 0 mm.

The average annual recharge computed on a monthly basis using the evapotranspiration calculated respectively by Penman-Monteith (PM_PET), Hargreaves (HS_PET), Thornthwaite (TH_PET) and Hamon (HN_PET), gave the values of 166 mm/year, 174 mm/year, 198 mm/year and 199 mm/year respectively. The four methods gave a mean of 184 mm/year. Figure 4 shows the comparison between the average values of monthly potential evapotranspiration calculated using the four methods; Hargreaves and Penman methods show a comparable trend, as opposed to Thornthwaite and Hamon methods. Mostly, Thornthwaite and Hamon show lower values, except in April where all methods show more or less the same value.

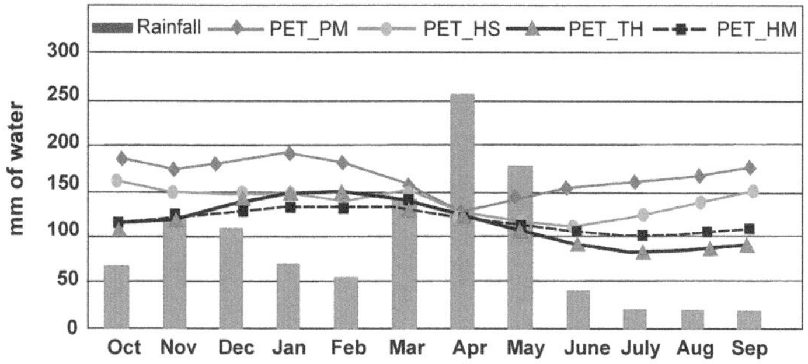

Fig. 4. Comparison of average monthly potential evapotranspiration (PET) computed using different methods for the period 1971/1972-2008/2009.

Figure 5 shows the variation of natural groundwater recharge rate calculated for each year using average PET estimated by different methods. The highest groundwater recharge rate is 504 mm/year, which occurred in 1994/95, while in 1979/80, 1984/85, 1999/2000 and 2002/2003, very little or no groundwater recharge occurred.

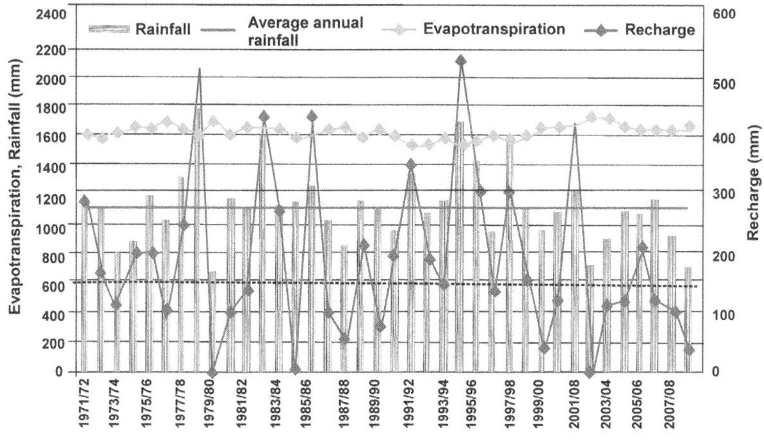

Fig. 5. Variation of annual potential evapotranspiration (average of different methods), rainfall and groundwater recharge.

Water balance and sustainable yield

Determination of groundwater recharge rates gave a mean value 184 mm/year being equivalent to 71.39 x 10^6 m^3 /year, an amount which represents 16.5% of the long term average annual precipitation of 1114 mm ending up as groundwater recharge. Sustainable yield is considered as a fractional of natural recharge and was calculated according to the concept proposed by Ponce (2007): average percentages estimated at approximately 40%, the least conservative approximately 70%, and reasonably conservative approximately 10%. These estimates are presented in Table 1.

Table 1. Estimates of sustainable yield according to the concept of Ponce (2007).

Estimated recharge	40% of recharge	10% of recharge	70% of recharge
71.39 x 10^6 m^3 /year	28.56 x 10^6 m^3 /year	7.14 x 10^6 m^3 /year	49.97 x 10^6 m^3 /year

Estimation of water budget

Groundwater provides over 50 percent of the present water supply for Dar es Salaam City. Since 1997, industries (e.g. beverage, pharmaceutical, chemical, etc), big hotels, institutes, colleges and urban farming are increasingly depending on groundwater. Groundwater survey in the study area was conducted in different periods between 2009 and 2010. The survey which took place in the most part of the "inner city" (Fig. 1) was used as a guide to estimate the amount of groundwater use in the Dar es Salaam area that extends from the Kisarawe/Pugu Hills west of Dar-es-Salaam City to the Indian Ocean in the east, and from the Mzinga River in the south to the Mbezi River in the north (Fig. 2). Five hundrend boreholes were visited in the study area. Dar es Salaam is situated largely on a sand aquifer and most boreholes are drilled between 20-60 m. The aquifer shows high yield of more than 6 m^3/h in median value. Most of the visited boreholes used for water supply have yields ranging between 10-15 m^3/h and few of them have yields between 15-30 m^3/h.

Mjemah et al. (2011) estimated boreholes yields from Dar es Salaam Quaternary aquifer at 8.56 x 10^6 m^3/year using data collected in 2005. This estimation was based on investigation of 1300 registered boreholes. However, many unregistered boreholes exist and most of them were drilled illegally. Mato and Mjwahuzi (2010) report the existence of over 3500 registered boreholes in the city. Out of 500 boreholes randomly visited during this study, about two third of them are not found in the database and about 30 percent are not operating. Based on the record of 3500 boreholes registered by the year 2010, it is likely that there will be about

7000 unregistered boreholes. This makes a total of about 10,500 boreholes drilled in Dar es Salaam. Using the experience of the field survey, we can assume that only about 70 percent of these boreholes are actively operating and the rest are abandoned due to either high salinity in water, drying up or low yield of boreholes and/or pump problems.

According to Mjemah et al. (2011) boreholes/wells used for domestic purposes are assumed to be pumping for 1 h/day, at every day of the year, while wells used for public water supply are assumed to pump for 6 h/day, during 365 days/year. The industrial wells are assumed to pump for 6 h/day, during 260 days/year. The yearly totals per well were then summed whereby 1300 borehole/wells yields were estimated to be 8.56 x 10^6 m^3/year. Based on the same approach, for 10,500 boreholes/wells one will expect a yield of about 69.3 x 10^6 m^3 /year. This abstraction which is nearly equal to the amount of the groundwater recharge is far greater than the estimated sustainable yield of the aquifer.

Conclusion and Recommendations

Water balance for Dar es Salaam Quaternary aquifer was calculated on the basis of long term average monthly precipitation, evapotranspiration and combined soil and vegetation characteristics, according to the method proposed by Thornthwaite and Mather. The water balance of the catchment suggests for the average sustainable yield (calculated at 40% of natural groundwater recharge) a value of 28.56 x 10^6 m^3 $year^{-1}$ with the current use approximately 69.3 x 10^6 m^3 $year^{-1}$.

Groundwater levels fluctuate seasonally. During the dry period, the water table falls due to pumping, evapotranspiration and discharge to streams. The major source of renewable groundwater in the aquifer is rainfall. During years of drought or below average rainfall, the reduced rate of recharge is opposed to the increased pumping, resulting in decreased discharge to streams.

In conclusion, so far this study has provided an increased understanding of a number of important factors including the need for further research. Conclusive factors were that: i) the estimated water balance shows a danger of aquifer overexploitation: as a result of immense industrialization and high population growth, groundwater is heavily relied on in Dar es Salaam City to serve as alternative to surface water which is seriously deteriorating; ii) increase of borehole drilling that goes unchecked affects the protection of the groundwater resource; iii) minimizing the adverse effects of scarcity requires optimal as well as sustainable patterns of groundwater management; iv) a proper management of the aquifer system is required; and v) a systematic groundwater model should be developed to control seawater intrusion and to provide more insight in understanding the hydrological behaviour of the system in relation to the current imposed stress on the recharge flux.

References

Mato RAM, Mjwahuzi (2010) Groundwater Governance Case Study: Tanzania, Groundwater use, characterization and vulnerability. http://xa.yimg.com/kq/groups/22477246/889666431/name/Aquifer+characteristics.pdf, Accessed on November 29, 2010

Mjemah IC (2007) Hydrogeological and Hydrogeochemical Investigation of a Coastal aquifer in Dar es Salaam, Tanzania. PhD thesis, University of Gent, Belgium, 222p

Mjemah IC, Van Camp M, Walraevens K (2009) Groundwater exploitation and hydraulic parameter estimation for a Quaternary aquifer in Dar es Salaam Tanzania. Journal of African Earth Sciences 55 (2009) 134-146

Mjemah, IC, Van Camp, M, Martens K, Walraevens K (2011) Groundwater exploitation and recharge rate estimation of a quaternary sand aquifer in Dar-es-Salaam area, Tanzania. Springer, Environmental Earth Sciences Journal. 63 (3), pp. 559-569 (DOI: 10.1007/s12665-010-0723-z)

Mng'ong'o OS (2004) A Browning Process, The case of Dar es Salaam City, Tanzania, Doctoral Dissertation, Department of Infrastructure, KTH, Stockholm, Sweden. (ISBN 91-7323-086-3)

Mtoni Y, Walraevens K (2010) Saltwater Intrusion in the Quaternary Aquifer of the Dar es Salaam Region, Tanzania. Proceedings of 21st Saltwater Intrusion Meeting (SWIM) held at Azores University, Portugal, 21-26 June 2010. Pp. 158-161. (ISBN 978-972-97711-5-6)

Mtoni Y, Mjemah IC, Walraevens K (2010) Towards Effective Pollution Control of Dar es Salaam Quaternary Sand Aquifer: A Challenge to Achieve Sustainable Development. In: Proceedings of the IAHR International Groundwater Symposium 2010, Valencia, Spain, September 22-24, 2010

Ponce MV (2007) Sustainable Yield of Groundwater http://ponce.sdsu.edu/groundwater_sustainable_yield.html. Accessed on March 13, 2011

Raes D, Steduto P, Siao TC, Fereres E (2010) Reference Manual, Chapter 2–AquaCrop, Version 3.1 January 2010. http://www.fao.org/nr/water/docs/aquacrop3_1/AquaCropV31Chapter2.pdf
Accessed on February 23, 2011

Thornthwaite CW, Mather JR (1955) The water balance. Laboratory of Climatology, No. 8, Centerton New Jersey

Thornthwaite CW, Mather JR (1957) Instructions and tables for computing potential evapotranspiration and the water balance. Int J Climatol 10(3):183-311

Analysis of surface and ground water exchange in two different watersheds

M. Bogdani-Ndini

Institute of Energy, Water and Environment, University of Polytechnics, Tirana, Albania

e-mail: ndinimiriam@yahoo.com

Abstract Water regime of Albanian rivers is a Mediterranean typical one. During the wet period, it flows 85-90% of the annual flow and the dry period represents only 10-15% of the annual amount. In this paper the water regime of the Semani and Vjosa River is analyzed. These are two rivers with total different hydrogeological characteristics. Vjosa river watershed is mainly composed of massif calcareous rocks that are streaky and karstified. A totally different view is in the other river, in Seman where the impermeable rocks dominate. Even the distribution of the precipitation is quite different. In the Vjosa River the amount of the precipitation varies from 1500 mm to 2500 mm per year and in Semani River only 1100 mm per year. The flow in the wet period is mainly a result of the precipitation and the minimum discharge occurs during the dry period representing the base flow that is the contribution from the groundwater. In this point of view the flow in the dry period is also an indicator of the underground water resource. During the dry period the watershed gives what it received and what has cumulated during the wet period. This is more evident in the case of a karstic watershed. The recession curves were analysed for all the hydrometric stations in both river basins and the parameters of these curves are evaluated. These results are analyzed and compared between the two different watersheds reflecting the differences on water exchange of surface and ground water. Finally an assessment of groundwater resources in both hydrogeological basins worked out.

1 Principal characteristics of the basins

The Vjosa River is the biggest river in South Albania and the second in the country. Its catchment's area is about 6710 km^2 and is 272 km long. About 1/3 of its watershed is situated in Greek territory (Fig. 1).The mean altitude of the watershed, from its spring to the mouth of the river, is to 855 m above the sea level. The slope of the Vjosa watershed is almost uniformly over all its longitude (27-29 %).

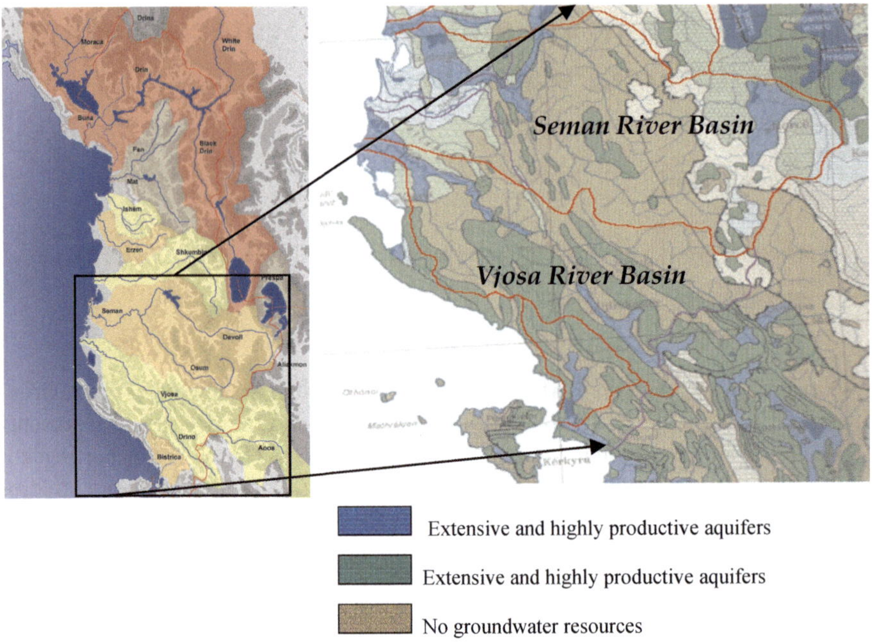

Fig. 1. Schematic and geological map of the study area (after Eftimi R.)

The Semani River in the south part is confined with Vjosa River. Its catchment area is 5658 km^2 that is about 1/5 of the Albanian territory. The two main branches are Devolli and Osumi Rivers, with catchment areas, respectively 3130 km^2 and 2142 km^2. The relief is mountainous with a total slope of 1900 m. The mean altitude of Semani watershed is 863.4 m a.s.l and of Devolli watershed 1082 m a.s.l.

From the hydrogeological point of view the Vjosa River watershed is composed mainly of stratified limestone, where karst is a powerful phenomenon. The limestone rocks are intensively karstified and many surface and underground karst forms facilitate the surface water infiltration.

In general Vjosa watershed is characterized by a high percentage of rocks with high permeability. The main part of this formation is located on the left side of the watershed; meanwhile the right part is almost composed by non-permeable rocks. In this side, Vjosa is bored with Semani River. This is an indicator of the high percentage of the infiltration of water in the ground and of the possibilities of a high drainage of the water from this catchments area.

The Semani River traverses almost all the tectonic zones characterized by a complicated geological structure. Devolli springs out from limestone rocks but on its way down the rock formation of the watershed are flyches characterized by a low permeability. The Osumit River too, flows in a territory composed of carbonic rocks and flysch. So this watershed, from the hydro-geological conditions, is considered the most poor considering all the others watershed in Albania.

Even in terms of precipitation regime, these two watersheds are quite different. The Vjosa basin, it is characterized by abundant precipitation. The main type of precipitation is the rainfall. The mean layer of precipitation in the Vjosa basin is 1670 mm. In the upper part of the basin the amount of precipitation is 1387 mm, and during the summer this zone has the lower amount of precipitation; only 106 mm. In the Drino basin the mean multi-annual value is 1920 mm.

The mountain chain on both sides of the Drino River receives a large amount of precipitation, over 2500 mm. So, these are the main parts of the watershed "feeding" the river and having a considerable influence on the water regime and its distribution within the territory of the watershed (Dakova 1998).

In terms of the seasonal distribution the precipitation regime is a typical Mediterranean with scarce amount of precipitation during the summer, that take 7 to 10 % of the annual weight and abundant precipitation during the winter, 45-50% of the total annual precipitation. This precipitation falling during the wet season is accumulated (stored) in aquifers and drains into the river during the dry season.

A total different view is in the Semani watershed. In Devolli watershed there is the less amount of precipitation, only 762 mm per year. In Osumi this amount is higher with values up to 1020 mm - 1070 mm per year. This amount of precipitation comes down during the fall-winter season. During the summer this amount represents only 12%-14% of the total annual of precipitation.

2 Water storage of the basin

Water potential of the underground basin is determined by its hydro-geological properties and by the amount of infiltrated precipitation. In the carbonate rocks environment karst phenomena has an important role. Its presence explains the main characteristics of groundwater (Fröhlich et al. 1994).

The region of the Vjosa watershed in general is characterized by good water storage. This is due to the presence of rocks with good characteristics on water storage, and due to the high level of tectonics in it (Fig. 1). This has conditioned the high infiltration into the karstic limestones. But on the other hand, the complex geological-structural composition, the stressed mountainous character, the well developed hydrographic network, influence for the groundwater in the region to be not uniformly distributed. The heterogeneity of the rocks, composing this region in general, makes possible the presence of different kind of groundwater as porous, karstic or fissured and all their combinations (Dakoli and Xhemalaj 1996). The opposite view is for Semani watershed where the rocks composing this watershed do not have storage capacities as flysch formation the groundwater is scarce.

The main features of a karstic region are either the absence or the scarce surface hydrographic network and the good underground network, a high module of the underground flow and in many cases its underground hydrographic watershed is not the same with the surface one. This is also due to the drainage of groundwater in adjacent watersheds. A good example of it is the karst massif of Mali Gjere.

3 Recession of groundwater resources

Recession represents the process of diminishing discharge in a water flow as a result of the drainage of the water storage layers of its basin during the non-influencing regime in the absence of precipitation. Precisely, the way in which this discharge diminishes is the way that represents the most important characteristic of a particular water basin (Dakova 1998).

The rate of underground nutrition of a river is determined by the total amount of precipitation, by their annual distribution and evapotraspiration (Kirchner 2009). So the amount of precipitation occurring during the wet period called **decisive precipitation** is important for the recharging of karstic water.

The precipitation occurring during the fall mainly fills the waterways and the capillaries of the unsaturated zone, so they have a small contribution to the karstic waters but they prepare the conditions for the infiltration during the successive months. Even during the months of June-October the precipitation do not influence so much on the recharging of karstic waters. During this period the water in the river flow is a result of groundwater stored during the preceding periods from precipitation (Vogel and Kroll 1992). During the dry season when the infiltration is close to zero, the water regime is called **non-influenced regime**.

The recession of underground resources can be expressed in an analytic form, making more practical the evaluation of different issues. Whatever this analytic expression is its expresses in a simple way the physical process of the recession of the storage layers. The most common analytic form of this process recommended (Tallaksen, 1995) for the evaluation is the exponential one, as it follows:

$$Q = Q_0 * e^{-\alpha\tau} \qquad (1)$$

Where Q_0- initial discharge, α- slope of the retention curve and ι- time

The recession coefficient α- represents the slope of the curve and is a characteristic for a certain spring. From this coefficient one can have a preliminary result of the dimensions of the water recession layer. Consequently the water volume stored in the rock can be evaluated. So the smaller the recession coefficient is, the higher the water volume stored on the underground layers is (Srinivas et al. 1999). In the same time from this coefficient the porosity and the permeability of the rock formation can be evaluated. The higher this coefficient is (the faster the water is drained from the underground formation), the smaller the effective porosity is (or the storage coefficient) and the higher the permeability is. As a conclusion, the retention coefficient is a very important characteristic of the watershed.

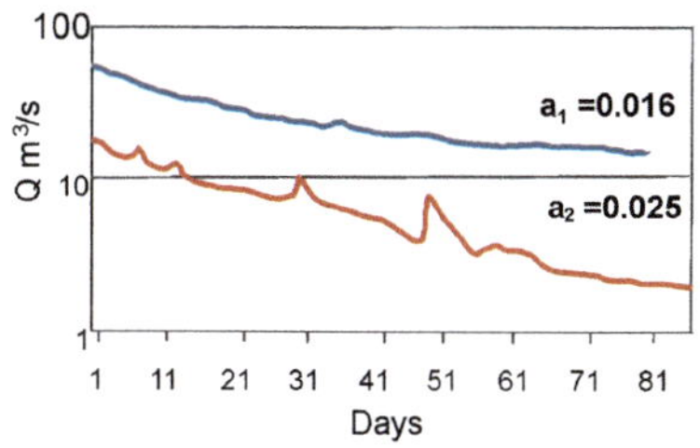

Fig. 2. The hydrograph of retention from the formation with poor water storage capacity (Semani river) (red line) and karstic formation (Vjosa river) (blue line).

The shape of the recession hydrograph, (Fig. 2) especially of a flow recharged by karstic rock formations, represent in reality the structure, function and hydrodynamic parameters of its water storage system (Eftimi 1979).

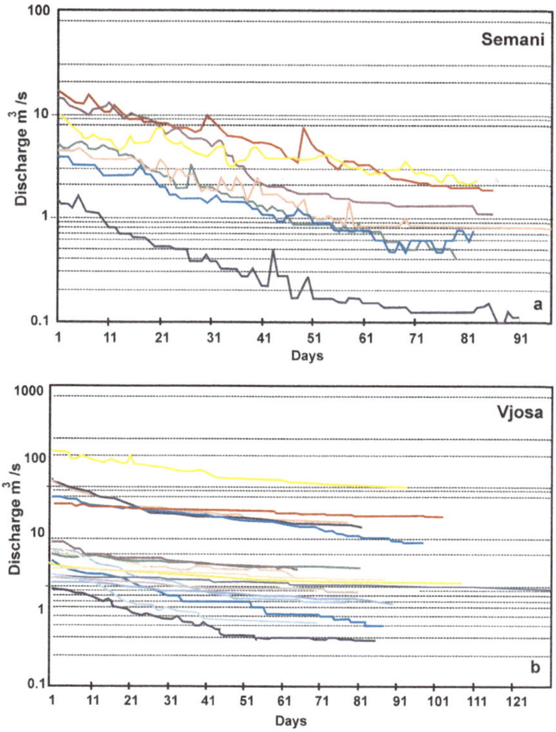

Fig. 3. a. The characteristic recession curves for Semani and **b**. Vjosa basin.

For both Rivers, there were selected those years where the recession process has normally occurred, that means those years when during the dry period no precipitation happened or was not significant and does not disturb the process of gradual diminishing of water discharge. It was intended to select those years when the precipitation during the wet period was of high values and the least possible during the dry period. These conditions secure an optimal "recharge" of the underground reservoirs during the winter-wet season, and a not "disturbed" emptiness during the summer- dry season (Bogdani 1997).

This was the logic to determine the characteristic curves of recession for a particular gauging station (Fig. 3).

4 The assessment of groundwater storage

The underground water resources of a water basin can be approximately estimated, using the recession analysis as above (1). The calculation is considered as "approximate" because, first it is difficult to definitively evaluate the impact of summer precipitation on these water resources and second, the end of recession curve never corresponds with the end of water resources. In other words, even in the smoother part of the curve, at the end of the recession period, there is always a certain water resource on the water basin. What can be exactly estimated by these recession curves is the water volume stored in the basin during the wet season and drains out of it through the end of the dry season (Eftimi and Kolaneci 1986).

So the integral of the recession curve gives the quantity of water stored in the underground layers from "0" time selected as origin. This integral can be applied in the limit zero-infinite. In this case the recession curve is supposed to continue gradually to the infinite, without being influenced by precipitation.

The volume coming out from the calculations represents the water volume stored in the underground formations during the wet season that drains during the non influenced water regime plus that part of that water which has not had the opportunity to drains because of the precipitation that disturbed this regime. In this case the underground reservoirs start to fill up without being totally empty. These are the ***potential resources*** of a water basin.

In the case where the integral will be calculated for the limits 0-t, where the regime is interrupted by the precipitation, the water volume represent only that part of underground water resources draining into the river valley during the regime without precipitation (Gustard *et al.* 1989) making up the effective resources. In both cases the water volume is calculated from the following equations:

$$V = \int_{t1}^{t2} Q_t dt = \int_{t1}^{t2} Q_0 * e^{-\alpha\tau} dt = \frac{Q_0}{\alpha} * e^{-\alpha t} \Big|_{t1}^{t2} \quad (2)$$

$$V_{pt} = \int_{t=0}^{\infty} Q_t dt = \int_{t=0}^{\infty} Q_0 * e^{-\alpha\tau} dt = \frac{Q_0}{\alpha} \quad (3)$$

Table 1. The potential and effective water volume for Vjosa and Semani watersheds.

Gauging Station	Year	Surface (km^2)	V(potential) (m^3)	V(effective) (m^3)
VJOSA				
Vjosa-Çarshove	1978	2175	$3.4788*10^8$	$1.2896*10^8$
Vjosa-Permet	1978	2820	$4.3622*10^8$	$1.9792*10^8$
Vjosa-Dorez	1960	5424	$11.4654*10^8$	$5.9697*10^8$
Perroi Çarshove	1981	90.8	17 869 272	6 536 592
Perroi Langaric	1981	297	41 726 066	19 719 504
Kardiq-Palokast	1980	182	50 958 135	23 662 800
Drino-Hormove	1986	1280	$2.2732*10^8$	$1.6311*10^8$
Bença – Bençe	1960	133	87 872 301	22 330 080
Sh.-Kuç	1980	64.8	42 337 241	22 289 472
Sh.-Lepenice	1978	332	55 234 996	34 299 072
Sh.-Vodice	1960	548	$1.052*10^8$	92 024 640
Smokthina	1978	32.8	42 885 818	15 655 680
Kalas – Tatzat	1986	29.1	$1.0411*10^8$	21 146 832
Kalas-Blerimas	1957	228	18240708	16 912 800
Bistrica – Krane	1960	108	$6.5212*10^8$	$1.9961*10^8$
Pavla – Bogas	1960	337	$1.2534*10^8$	47 907 072
SEMANI				
D.-Miras	1960	934	4 802 548	4 178 606
D.-Sheqeras	1960	786	$2.4593*10^8$	15 245 280
D.-Gjinikas	1978	828	48 440 502	32 415 120
D.-Kokel	1960	1884	87 114 050	78 494 400
O.-Qafzez	1974	1065	11 963 077	10 345 183
O.-Corovode	1960	1106	41 041 123	33 860 866
O.-U.Vajgur.	1960	1122	50 625 000	44 879 336
Prr.-Çorovodes	1965	1198	16 314 353	15 275 575

The water resources are calculated for the pre-selected typical years, that means those years where the dry season has /or has not such low precipitation that does not influence the recession curve.

In the case where the curve is composed by two micro regimes, the calculation is done for both of them. It is calculated the volume for the first phase from t=0 up to t_1 time and then for the second phase considering the beginning of this one at t_1

time up to the t_2 time. So there is a Q_0 of the second phase not in the originate axe of ordinate but on the axe where t_1 begin up to the t_2 time that represents the end of the recession curve. This is the way to calculate the water resources in the water basin that do not drains from integrating from the time when the curve is interrupted to the ∞ assuming as the slope the α coefficient of the second phase.

Using the above approach, the potential water volume (the total amount of water in aquifer) and effective water volume (the amount of water in aquifer that in specific condition flow as surface water), are calculated and these values are showed on the Table 1.

References

Bogdani M, (1997) Analyze of low flow in the rivers of south Albania. Thesis for Doctorate. Tirana, Albania (In Albanian)

Dakoli H, Xhemalaj Xh (1996) Hydro geological balances for some karstic massifs. Conference of Water Management-Tirana, Albania (In Albanian)

Dakova, S. (1998) Some Considerations of Base Flow. Low Flows Expert Meeting, The University of Belgrade, 10-12 June 1998, Belgrade, Yugoslavia

Eftimi R, (1979) A short view of underground water in Albania. Studies Resume (In Albanian)

Eftimi R, Kolaneci M (1986) Water regime of Selita spring. Hyd-Met Studies Nr.11 Tirana, Albania (In Albanian)

Fröhlich, K., Fröhlich, W., Wittenberg, H. (1994) Determination of Groundwater Recharge by Baseflow Separation: Regional Analysis in Northeast China. In FRIEND: Flow Regimes from International Experimental and Network Data, ed. by P. Seuna, A. Gustard, N.W. Arnell and G.A. Cole, IAHS Publication No. 221, 69-75

Gustard, A., Roald, L.A., Demuth, S., Lumadjeng, H.S., Gross, R. (1989) Flow Regimes from Experimental and Network Data (FRIEND), Volume I, Hydrological Studies, Institute of Hydrology, Wallingford

Kolaneci, M., (1983) Flow prediction in Vjosa River. Thesis for Doctorate. Tirana, Albania A.G.S (1985) Hydrogeological Map of Albania

Kirchner, J. W2009 Catchments as simple dynamical systems: Catchment characterization, rainfall-runoff modeling, and doing hydrology backward, Water Resources Research, 45, W02 429

Ndini M, Kolaneci M, (2004) Potential Groundwater Resources of Vjosa Watershed AJNTS, N_0 16.2004

Srinivas A, Venkateswara B Rao& V.V.S. Gurunadha (1999) Recharge process and aquifer model of a small watershed. Hydrological Sciences Journal Vol. 44 Nr. 5

Tallaksen, L.M., (1995) A Review of Baseflow Recession Analysis. J. Hydrol., 165, 349-370

Vogel, R.M. and Kroll, C. N., (1992) Regional geohydrologic-geomorphic relationships for the estimation of low-flow statistics, Water Resources Research, 28, 2451-2458, doi:10.1029/92WR0100

Evaluation of multivariate statistical methods for the identification of groundwater facies, in a multilayered coastal aquifer

E. Galazoulas, C. Petalas, V. Tsihrintzis

Laboratory of Ecological Engineering and Technology, Department of Environmental Engineering, Democritus University of Thrace, Xanthi, GR 67 100, Greece.
egalazou@env.duth.gr

Abstract The multilayered coastal aquifer system of the southern Rhodope province in North Greece was selected to examine the applicability of multivariate statistical methods in the investigation of groundwater variability and to evaluate their performance in the presence of overlapping processes. The aquifer system which forms the main source of water supply to numerous productive wells in the area, drilled at various depth horizons, faces serious degradation of groundwater quality. A widespread sampling survey was conducted at the end of the wet season 2010 and 16 physicochemical parameters were determined for every sample collected. Multivariate statistical techniques (cluster and factor analysis) were applied twice, first including all the sampling points, and second after the removal of deep boreholes samples. The results showed that the above methods were able to separate a mixture of processes in the aquifer system (saltwater intrusion, agricultural runoff, mineral dissolution, ion exchange, connate water influence), and spatially distribute them in the study area. They also highlighted the need for knowledge of technical information regarding the boreholes and of sound hydrogeological data in the area, since the inclusion in the analysis of deep connate groundwater masses caused masking effects.

1 Introduction

Identification of groundwater regimes during a hydrochemical survey is often a difficult assignment because of the large number of samples created and the complexity of the hydrochemical procedures met in groundwater environments (Appelo and Postma 2005). Especially in the case of coastal aquifers, discrimination between different salinization sources is troublesome due to the conjunction of various processes (Morell et al. 1996). Multivariate statistical techniques can fill this gap, as they present two very important attributes. First, they can significantly reduce the number of variables used without loosing significant information, and second, they can produce comprehensive classification of either variables or cases

(Davis 2002). There are various studies in the literature using these methods in groundwater data (e.g., Melloul 1995; Abu-Jaber et al. 1997; Jayakumar and Siraz 1997; Voudouris et al. 1997; Kumar et al. 2009). In all these studies, the authors were able to productively identify the hydrological procedures exhibited in a variety of groundwater environments and to differentiate among them.

In cluster analysis, variables or cases are grouped together according to a variety of distance measures and amalgamation rules, while in factor analysis the eigenvectors of the correlation data matrix are calculated and the ones representing significant proportions of data variability are selected following certain criteria. The eigenvectors represent the principal factors of the original dataset which are all uncorrelated and orthogonal to each other (Davis 2002).

The multilayered coastal aquifer located at the southern Rhodope Province, Greece, which is facing saltwater intrusion and anthropogenic contamination problems over the last 30 years, was selected as a case study area. The aim of this study is to apply certain multivariate statistical techniques (cluster and factor analysis) in order to identify and spatially distribute the various hydrochemical procedures, to enhance observations from previous studies (Petalas 1997; Petalas and Diamantis 1999), and to evaluate their performance in the case of overlapping processes.

2 Study area description and hydrogeological setting

The study area is situated in the southwest coastal zone of Rhodope Province, in Northeastern Greece (Fig. 1). The climate of the area is semi-arid, with an average rainfall of 554 mm/year (2003-2009). The wet season (October – April) is followed by the dry period between May and September, which is the main irrigation period. Over-utilization of the regional groundwater resources has led to a remarkable piezometric drawdown and intensive saltwater intrusion (Petalas 1997). The hydrogeology of the area has been investigated thoroughly in the past (Petalas 1997; Petalas and Diamantis 1999). Examination of geological, hydrogeological and hydrochemical data showed the presence of a complex multilayered aquifer structure deposited during the upper Miocene. Two main aquifers can be identified in the area. The shallower semiconfined aquifer which is formed by intercalations of loose sandstones, siltstones and clays and the deeper confined aquifer which is formed within deposits of alluvial origin underneath the semiconfined layer. The above aquifers are separated from each other by a thick clay unit. At the base of the system, a thick green clay layer forms a baseline isolating the productive aquifers from the underlying deposits where connate water masses are present. The depth of this barrier unit is approximately 100 meters below mean sea level.

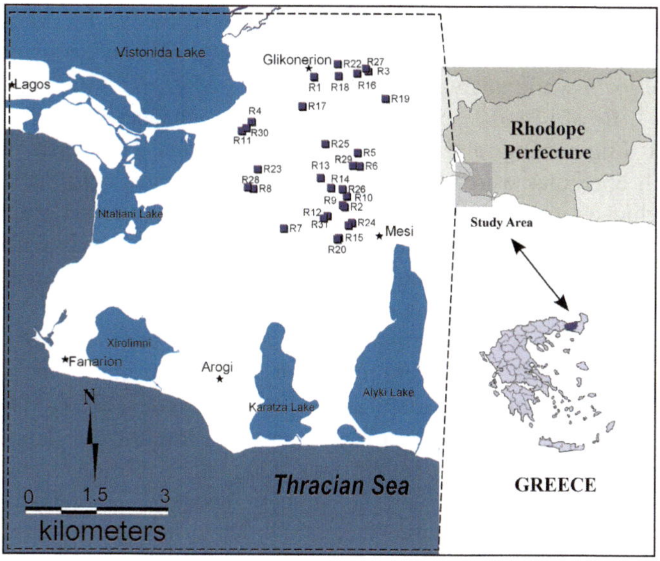

Fig. 1. Location of the study area and sampling points.

Pumping wells in the study area can be divided into three groups. First, wells that are screened to only one of the productive aquifers identified before; second, wells open to both the shallower and the deeper parts of the aquifer system; and third, wells that are screened below the green clay baseline. This final group of wells is quite developed during the last years due to the need for better quality water in highly contaminated zones. Groundwater recharge at the top aquifer is through infiltration of surface runoff, direct percolation of rainfall, groundwater inflow along the western and northeastern boundaries, and infiltration of irrigation water. The bottom aquifer is recharged mainly by inflow from contiguous aquifers.

3 Sampling procedures, chemical analysis and statistical treatment

For the purpose of this study, an excessive sampling survey was conducted during the beginning of the irrigation season (May 2010). A network of productive boreholes (Fig. 1) was selected in order to cover equally and representatively the area of interest, and special care was taken so that samples were collected from boreholes corresponding to different screen depths. In certain occasions, wells laying very close to each other were selected (e.g. R15 and R24, or R2 and R9), because they depicted dissimilar physicochemical properties, or because technical data about them were unknown. Sampling, preservation and transport of the samples was conducted following the standard procedures provided by APHA (1995). Physicochemical parameters (Electrical Conductivity, pH and Temperature) were measured *in situ* while analyses of Na^+, K^+, Ca^{2+}, Mg^{2+}, Cl^-, SO_4^{2-},HCO_3^-, NO_3^-, NO_2^-, NH_4^+, Br^-, Li^+ and F^- concentrations were conducted in the Laboratory of Ecological Engineering and Technology, Democritus University of Thrace, Greece. All boreholes were pumped for at least 30 minutes before a sample was taken to achieve representative samples of groundwater attributes. In total 31 samples were collected, producing the (31 x 16) whole dataset matrix with five samples corresponding to deep boreholes. Thus the number of samples analyzed in the reduced dataset was 26.

For the application of cluster analysis on both data sets, the simple Euclidean distance and the Ward's clustering method were selected due to their good performance and their applicability in previous studies (Hussein 2004; Mahlknecht et al. 2004). Prior to the application of factor analysis the Mayer – Kaiser – Olkin (KMO) measure of sampling adequacy was examined. The results showed that the dataset is suitable for factor analysis in both datasets. The varimax rotation scheme was applied on the initial factors extracted, in order to achieve higher loadings on them and increase the total proportion of variance explained. After the rotation, significant factors were selected according to certain criteria (eigenvalues and scree-plot inspection) (Davis 2002).

4 Results and discussion

Cluster analysis identified three main clusters on both datasets based on the visual interpretation of the graphs.

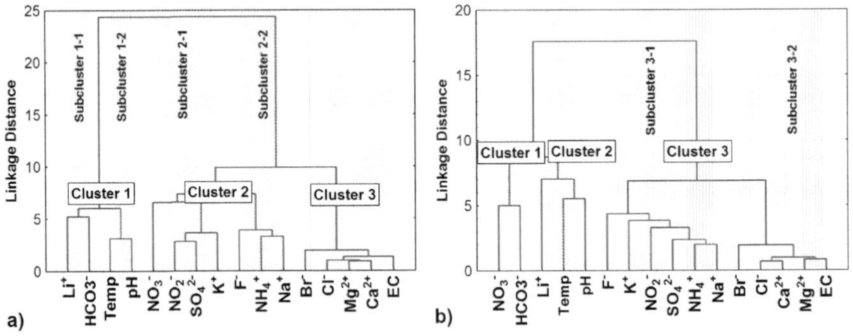

Fig. 2. Cluster analysis **a.** on the whole dataset. **b.** on the reduced dataset.

Cluster analysis results on the whole dataset are shown in Fig. 2a. The first cluster includes the variables Li^+, HCO_3^-, Temp and pH. Since Li^+ presence in the area is related to thermal waters (Petalas 1997) and is grouped along with the Temp variable, this cluster is indicative of the influence of deep thermal connate water masses. The additional presence of HCO_3^- and pH in the same cluster is explained by dissolution of the calcite mineral, a process which consumes hydrogen ions and increases bicarbonate concentration, according to the equation (Hem 1985):

$$CaCO_3 + H^+ = Ca^{2+} + HCO_3^-$$

All the samples from deep boreholes in the area appear marginally oversaturated in calcite, due to the long residence time of groundwater at these depths. The second cluster encloses two subgroups of variables. The first (subcluster 2.1) contains the ions NO_3^-, NO_2^-, SO_4^{2-}, K^+ and is related to anthropogenic contamination through the application of agricultural fertilizers. The variables enclosed in the subcluster 2.2 (Na^+, NH_4^+, F^-) form a group of elements which generally exhibit a strong relationship with calcium. The first two ions participate in cation exchange phenomena with Ca^{2+} between aquifer matrix and groundwater in coastal aquifers (Gimenez 1994). On the other hand F^- has no other known origin except the dissolution of fluorite (CaF_2) mineral whose solubility is affected by the concentration of Ca^{2+} in groundwater (Boyle 1992). Saturation indices of fluorite mineral in the area indicate balanced dissolution of the mineral, at the groundwater environment. In general cluster 2 is associated with a mixture of processes taking place in the groundwater system. Finally, cluster 3 includes the variables Br^-, Cl^- , Mg^{2+}, Ca^{2+} and EC. The above grouping is associated with the intrusion of saltwater in the aquifer, which significantly increases groundwater electrical conductivity and has higher concentrations in chloride and bromide ions compared to freshwater. This process is coupled by the concomitant inverse cation exchange phenomena (absorption of Na^+ and release of Ca^+) between the saline water and the clays in the aquifer matrix (Appelo and Postma 2005).

The results of cluster analysis application on the reduced dataset are shown in Fig. 2b. Cluster 1 contains nitrate and bicarbonate anions which indicate the presence of recharging waters, since bicarbonate is the major freshwater anion in the area, while nitrate is the most stable form of dissolved nitrogen produced by agricultural fertilizers. This process could not be identified in the previous analysis because of the high concentration of deep borehole samples in the bicarbonate ion and the lack of significant nitrate concentrations at these samples. Although deep samples are removed from the analysis, variables associated with them are grouped in cluster 2. This grouping shows the regional uplift of deep connate water masses to the above confined aquifer. The cluster 3 of the whole dataset is completely the same with the subcluster 3-2 of the reduced dataset. The variables in both clusters are associated with saltwater intrusion and inverse cation exchange phenomena. Agricultural contamination, cation exchange and fluorite mineral dissolution processes are now depicted in subcluster 3-1 as a single group.

Factor analysis results in the whole dataset indicate the presence of three significant factors and thus three controlling processes in the groundwater environment (Fig. 3a, b), which explain 82.94% of the total data variability. The first one, which explains 55.8% of the data variance, depicts high loadings on the variables

EC, K^+, Na^+, Ca^{2+}, Mg^{2+}, Cl^-, SO_4^{2-}, $NO_2^-$and Br^- and is associated with the salinization processes in the aquifers, including both saltwater intrusion and anthropogenic contamination. Factor 2 corresponds to 20.6% of the total variance and is negatively loaded with the variables pH, Temp and Li^+ and positively loaded with the nitrate anion.

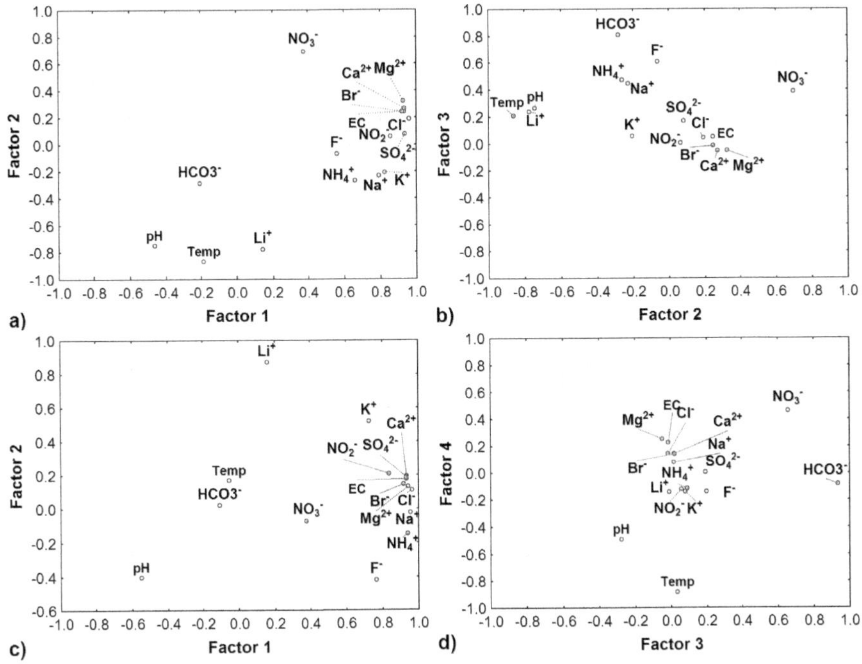

Fig. 3. a, b. Rotated factors loadings of the whole dataset, **c, d**. Rotated factor loadings on the reduced dataset.

This opposition reflects the contradicting influence of deep water masses and recharging waters represented by the nitrate ion. The bicarbonate ion, which is a major constituent of both processes, fails to load on this factor and is individually loaded in the third factor, which explains 6.6% of the variability in the dataset. Scores for this factor showed high values for both shallow and deep wells. Thus, the inclusion of deep borehole samples in the analysis fails to associate bicarbonate ion with a certain process. Variables F^- and NH_4^+ are not strongly correlated with any of the extracted factors but they are moderately loaded in the first and the third of them. This means that their variability is possibly affected by multiple processes of the aquifer system.

Fig. 3 (c, d) shows the factor loadings achieved for every variable, after the removal of the deep borehole samples. A four factors model was selected for the reduced dataset following the eigenvalue criterion, explaining 87% of the total data variability. Factor 1 corresponds to 61% of the variance and depicts high loadings on the variables EC, Na^+, K^+, Ca^{2+}, Mg^{2+}, Cl^- , SO_4^{2-}, NO_2^-, NH_4^+, Br^- and F^-. This factor is again associated with a variety of processes taking place in the aquifer system resulting in high salinization. Factor 2 is highly loaded by the lithium ion, moderately high with the potassium ion, and explains 10% of the data variability. This factor reveals the upconing of deep water masses to higher horizons. This time high factor scores were representative of shallow wells. The last two factors explained approximately 9% of the total variance each and show high loadings for HCO_3^- in factor 3 and for the Temp variable for the fourth factor. Factor 3 showing high loadings for the bicarbonate ion and moderately high for the nitrate ion is similar to previously identified groups and corresponds to recharging waters and shallow boreholes. Finally, the fourth factor showing negative loadings for Temp and pH while the NO_3 loading is moderately positive high depicts the opposing processes between the deeper and shallower horizons in the area. All the communalities achieved were close to 1 except for the values of the variables Li^+ and HCO_3^- which were slightly lower. The nitrate ion does not depict high loadings in any of the extracted factors but moderately high in three of them. This is an indication of the participation of the ion in multiple processes.

To identify the areas mostly affected by every one of the above factors, the areal distribution of factor scores were examined. The results showed discernible patterns in the processes illustrated before, especially for the second and third principal components which are more distinctive. The second factor shows high scores for wells in the lower confined aquifer highlighting zones of possible deep water upconing. The sampling stations depicting high factor scores for the third factor represent very shallow wells easily reached by recharging waters.

5 Conclusions

The use of multivariate statistical techniques for the identification of the governing processes affecting the groundwater environment is proved fruitful in the study area. Cluster and factor analysis separated known procedures (saltwater intrusion, agricultural impact, natural recharge) and highlighted new ones hitherto unknown (dissolution of fluorite). Masking effects due to the superposition of overlapping processes appear on both methods, when deep water samples are included in the analysis because bicarbonate ion is the major constituent of both connate water masses and recharging freshwater. In the case of cluster analysis, bicarbonate ion is grouped along with deep sample indicative variables and the recharging process is not directly visualized. Factor analysis, including the deep samples, loads a single factor with the bicarbonate ion but the factor scores are significantly high for both shallow and deep wells. Clusters and factors associated with thermal connate water masses are included in the analysis even after the removal of deep groundwater samples. This indicates the regional upconing of deep groundwater to the above confined aquifer.

References

APHA, AWWA, WPCF (1995) Standard Methods for the Examination of Water and Wastewater, 19th edn. APHA: Washington, DC

Abu-Jaber NS, Aloosy ASE, Ali AJ (1997) Determination of aquifer susceptibility to pollution using statistical analysis. Environ. Geol. 31, 94-106

Appelo CAJ, Postma D (2005). Geochemistry, Groundwater and Pollution, second ed. Balkema, Rotterdam

Boyle DR (1992) Effects of base exchange softening on fluoride uptake in groundwaters of the Moncton Sub-basin, New Brunkswick, Canada. In: Kharaka, Y.K., Matest A.S. (eds.) Water-rock interaction, pp. 771-774. Proc. 7th Int. Symp Water-rock interaction. A.A. Balkema, Rotterdam (1992)

Davis JC (2002) Statistics and Data Analysis in Geology, John Willey and Sons, New York

Gimenez E (1994) Caracterizacion hidrogeoquımica de los procesos de salinizacion del acuıfero detrıtico costero de la Plana de Castellon. Tesis Doctoral, Universidad de Granada, 1994. p. 390 [in Spanish]

Hem JD (1985) Study and interpretation of the chemical characteristics of natural water (3d Ed).U.S Geological Survey Water-Supply Paper 2254

Hussein MT (2004) Hydrochemical evaluation of groundwater in the Blue Nile Basin, eastern Sudan, using conventional and multivariate techniques, Hydrogeol. J. 12, 144-158

Jayakumar R, Siraz L (1997) Factor analysis in hydrogeochemistry of coastal aquifers – a preliminary study, Environ. Geol. 31, 174-177

Kumar M, Kumari K, Singh U, Ramanathan AL (2009) Hydrogeochemical processes in the groundwater environment of Muktsar, Punjab: conventional graphical and multivariate statistical approach, Environ. Geol. 57, 873-884

Mahlknecht, J., B. Steinich, and I. Navarro de León (2004) Groundwater chemistry and mass transfers in the Independence aquifer, central Mexico, by using multivariate statistics and mass-balance models. Environ. Geol. 45, 781-795

Melloul AJ (1995) Use of principal components analysis for studying deep aquifers with scarce data-application to the Nubian Sandstone Aquifer, Egypt And Israel, Hydrogeol. J. 3, 19-39

Morell I, Gimenez E, Esteller MV (1996) Application of principal components analysis to the study of salinization on the Castellon Plain Spain. Sci. Total Environ. 177, 161–171

Petalas CP (1997) Analysis of Aquifer systems in the Heterogeneous Coastal Part of Prefecture of Rhodope (in Greek). PhD Dissertation. Department of Civil Engineering, Democritus University of Thrace, Xanthi, Greece, 288 pp

Petalas CP, Diamantis IB (1999) Origin and distribution of saline groundwaters in the upper Miocene aquifer system, coastal Rhodope area, Northeastern Greece. Hydrogeol. J. 7, 305-316

Voudouris K, Lambrakis N, Papatheothorou G, Daskalaki P (1997) An application of factor analysis for the study of the hydrogeological conditions in Plio-Pleistocene aquifers of NW Achaia (NW Peloponnesus, Greece), Math. Geol. 29, 43-59

Delimitation of the salinity zone of groundwater in the front between the municipalities of Moschato and Glyfada of the prefecture of Attica

Ch. Mpitzileki[1], I. Koumantakis[2], E. Vasileiou[2], K. Markantonis[2]

[1] MSc in Water Resources Science and Technology, Rural and Surveying Engineer in the Department of Technical Works in the Prefecture of Attica, Syggrou Av. 80-88, P.C. 11741, Athens, Greece, xmpitzileki@gmail.com

[2] N.T.U.A., School of Mining Engineering and Metallurgy, Section of Geological Sciences, Laboratory of Engineering Geology and Hydrogeology, Iroon Polytexneiou 9, P.C. 15780, Zografou, Athens, Greece, elvas@metal.ntua.gr, markantonis@metal.ntua.gr

Abstract In the coastal zone between the municipalities of Moschato and Glyfada in the south – western area of the prefecture of Attica the aquifers appear saline as it was identified in the past. The main reasons for the intrusion of the sea, is the low water table elevation because of the recharge decrease and the groundwater overexploitation. Samplings and measurements regarding the groundwater table in the studied area were taken, in "low - water" period (September - October 2009). The degradation of the quality of the groundwater and the intrusion of sea water were established. According to chemical analysis, high concentrations of ions of chloride (Cl^-) and sodium (Na^+) were measured in the regions of Moschato and Kallithea. Comparisons to the previous twelve year period in 1997, the hydrogeological conditions are improved, as concerning water table level and the salinity of water. In the areas where the recharge of the aquifer is satisfactory, high concentrations of ions of calcium (Ca^{2+}) and high concentrations of acid carbonic ions (HCO_3^-) are observed in the municipality of Alimos. Regarding irrigation, the groundwater of the study area is characterized mostly as high risk of salinity and low risk of alkalinity (C_3S_1 by Wilcox). The sea intrusion and the need for groundwater protection in the area, demand rational management of water resources. It is proposed that pumping and the application of artificial recharge must be decreased.

1 Introduction

The study area is about 50 km^2 and is located at the southwest section of Attica in the coastal zone consisting of municipalities Moschato and Glyfada being also the south part of the hydrological basin of Athens.

Previous investigation for the hydrogeological status of Athens (Koumantakis 1997, Koumantakis et al. 1997) proved the intrusion of sea water to the groundwater.

In this study a research takes place for delimiting the salinity zone, considering, geological-hydrogeological conditions of the area, water table measurements and the chemical analysis of water samples in the period September-October 2009.

2 Geology - Hydrogeology

The geological formations of the area are consisted mainly of quaternary and neogene period deposits with locally appearances of schist, ophiolithe, limestone (Fig. 1).

Especially for the quaternary deposits appeared in loose condition, riverside origin with presence of side fallen -very thick- fragments at places, consisting of gravels, sand, clay, silt in several mixing ratio. All the alluvial-quaternary deposits are medium permeability formations in general. Because of the urban areas the recharge zone reduced at the west part of Ymittos Mountain.

The aquifers that are developed in these formations are shallow - unconfined or deeper confined at the lower part of the area. They appeared with low recharge because the urban areas reduced surface water flow from Ymittos Mountain and the aquifer is poor of capacity.

Neogene deposits especially of marly limestone and sandstone are appeared at the coastal regions Alimos, Elliniko, Glyfada and slightly at municipalities of Palaio Faliro, Kalithea, and Moschato. At the region of Alimos, sands and silts from the period of Late Miocene are present. Formations of the Neogene period are impermeable to semi permeable at places, caused by intercalations of sandstone-conglomerate-limestone, with low interest of exploitation but a significant role as cut off shutter to prevent the seawater intrusion. As a result, from the alternations of impermeable and permeable rocks in bedding or in lens or in irregular shape, is the presence of confined aquifers. All these accordingly to cumulative thickness of aquifers, permeability and the conditions of recharge observed between very poor to medium.

3 Piezometric Status

Piezometric measurements performed at 23 boreholes and 2 pits (Mpitzileki 2010). Points selected according the layout of the coastal zone Moschato-Glyfada.

The evaluation of piezometric maps based on the considering that the aquifer system is uniform although there is some alternation of several aquifers. As a part of the previous theory there is a hydraulic communication within wedges of the several aquifers so this consideration does not corrupt the whole situation. According to the measurements of water table, a map compiled for the period between September and October 2009 (Fig. 1b).

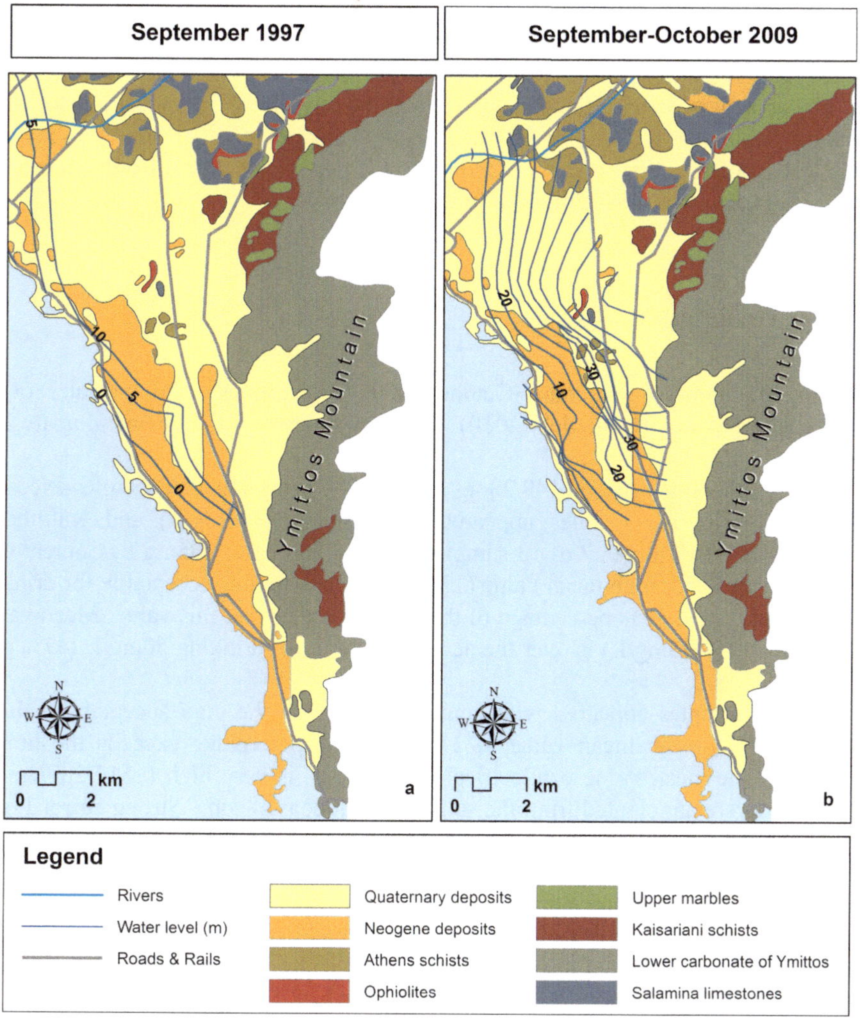

Fig. 1. a. Piezometric map of September 1997. **b.** Piezometric map of September-October 2009.

The conclusion of the piezometric map is the low water level at the coastal zone of Elliniko with the possible intrusion of seawater in one section where the hydraulic load became zero. The intrusion of subsaline water, documented with chemical analysis of water samples, controlled mainly by geological structure of the area (neogene marls along coastal zone).

As long as we are leaving from the coastal zone water level raised and when we are reaching to the edge of south part Ymittos Mountain raised up to 40m. It is necessary to make a special reference about the water level progress between the last twelve years (September 1997, September-October 2009) (Fig. 1a, b).

From the comparison of the maps the piezometry for the last twelve years is appeared slightly well possibly because of the decrease of pumping. Especially for the period of September 1997 (Fig. 1a) two areas with zero water level (dry) located, the bigger one settled at the regions of Moschato-Kallithea as a result of over-pumping, and the second smaller at regions Elliniko-Glyfada, while the rest of the area piezometric status appeared with positive load. In opposition for the period September - October 2009 the zero water level (dry) area Moschato-Kallithea not appeared and the condition is better, while a small reduction appeared at the second area Elliniko-Glyfada.

4 Hydrochemistry

During the period of September-October 2009, 47 samplings of groundwater took place at study area (Mpitzileki 2010). Statistics process of the chemical analysis results at Table 1.

The mean value of Cl^- is 149.28mg/L. The 32% percentages of samples exceed 200mg/L, with max values appeared to Elliniko (848 mg/L) and Kallithea (805mg/L). At Kallithea 7 of 13 samples with Cl^- 200 up to 805mg/L. Content of Na^+ appeared with high mean value (175.02mg/L), with max acceptable for drinking 175mg/L and 40% percentage of the samples exceeding this value. Mean value of Mg^{2+} (52.16mg/L) is over the acceptable limit for drinking 50mg/L (47% of samples exceed).

Eleven samples appeared with content of K^+ over the max acceptable value 12mg/L. NO_3^- with mean value 48.37mg/L, 40% percentage exceeds the limit 50mg/L. The mean value of electrical conductivity is very high (1511.5μS/cm). All these were anticipated from the salinisation at coastal zone. Strong correlation exists between values EC-Na^+ (y=0.1708x-83.181, R2=0.8199), EC-Cl-(y=0.2574x-239.85, R2=0.7456) (Fig. 2), Na^+-Cl^- (y=1.4724x-108.42, R2=0.868) as a result of groundwater subsaline.

At the region of Moschato appeared high values of Ca^{2+} (108.69mg/L), Na^+ (168.60mg/L) and Cl^- (114,60mg/L). Compared with measurements took place at 1997, there is a reduction of Cl^-, SO_4^{2-}, NO_3^- values. The groundwater has been significantly polluted by human factor.

At municipality of Kallithea appeared high values of Na^+ (221.40mg/L) and Cl^- (236.46mg/L). Compared with measurements of 1997 an increase of SO_4^{--}, Cl^- and the electrical conductivity can be observed, opposite to the values of NO_3^- that are reduced.

Groundwater has charge qualitative analysis caused of subsaline and human pollution.

At municipalities of Alimos and Palaio Faliro high values of Na^+ (228.88mg/L), Mg^{2+} (66.04mg/L), Cl^- (187.42mg/L) measured. Compared with readings of 1997 we have a reduction of SO_4^{--}, Cl^-, NO_3^- and Na^+, in opposition the electrical conductivity (1765.67μS/cm) increased.

Table 1. Statistic analysis of the chemical analysis results.

Parameter	Maximum value	Minimum value	Mean value (m)	Standard deviation (s)
Ca^{++}	146.18	18.42	84.61	30.26
Na^+	600.00	6.70	175.02	129.13
K^+	38.80	0.70	7.51	7.42
Mg^{++}	133.89	1.46	52.16	26.48
Cl^-	848.00	3.00	149.28	204.07
SO_4^{--}	325.00	30.00	128.49	68.06
HCO_3^-	733.22	145.18	410.40	111.08
NO_3^-	198.00	2.40	48.38	35.43
pH	9.69	7.22	7.70	0.43
E.C.	3700.00	291.00	1511.50	684.45

At municipalities of Glyfada and Elliniko high values of Na^+ (103.42mg/L) related with measurements of 1997 while all the other parameters reduced. High values of Cl^- and electrical conductivity are measured to Elliniko, Alimos, Palaio Faliro, Kallithea, Moschato in opposite at Glyfada lower values. Less charge areas observed, far from the sea, near Vouliagmenis avenue and the main part of Glyfada.

According to Wilcox diagram (Fig. 3) more samples belong to category C_3S_1 (4 at Moschato, 7 at Kallithea, all at Elliniko, 1 at Alimos and 4 at Glyfada), water with high salinity and low alkalinity. In category C_3S_2 belonging all the samples of Palaio Faliro, 4 Kallithea, 3 at Alimos and 1 at Moschato, water with high salinity and medium alkalinity. Two samples from Kallithea and one from Elliniko classified at category C_4S_3, with very high salinity and improper for irrigation use.

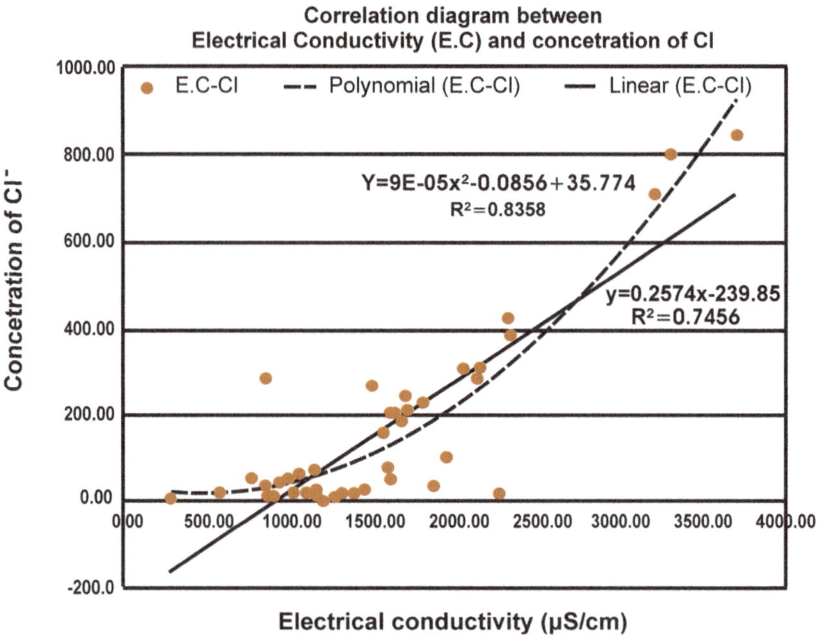

Fig. 2. Correlation diagram between Electrical Conductivity and concentration of Cl^-.

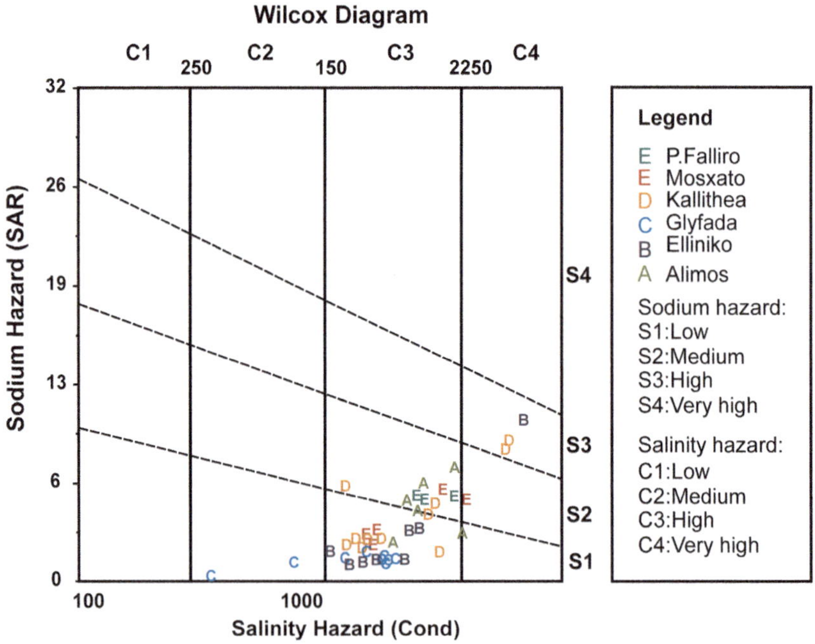

Fig. 3. Wilcox diagram.

5 Conclusions- Recommendations

According to the measurements of 1997 (Koumantakis et al 1997; Koumantakis, 1997) at the coastal zone Moschato-Glyfada an extended salinity area is developed. Recent measurements at the region from Moschato up to Glyfada showed a small decrease of the salinity area. The improved situation of water level at September 2009 is possibly caused by pumping reduction. Significant reason to the controlled intrusion of seawater is the geological structure at the coastal zone composed of neogene marls. Regions with high altitude and far from the sea are not facing problems with salinity.

Salinity approved with the quality deterioration of groundwater which certified by samplings in the area at September 2009 (Mpitzileki 2010). Zones limited by municipalities of Moschato-Elliniko appeared with high contents Na^+, Cl^- and high electrical conductivity. High contents at places of nitrate, sulphate ions certify pollution from human factor. Groundwater of Moschato, Kallithea, Elliniko, Glyfada can be useful for watering under conditions of salinity control. Groundwater of Palaio Faliro, Alimos demands satisfactory washout and drainage while there are individual instances where the water is unsuitable for watering.

On the grounds of development and protection for underground water bearing systematic monitoring needed. Possible application of technically enrichment could help the additional decrease of the salinity zone as this has been done with the same way at similar circumstances in our country (Korinthos, Argolida).

Additionally could be a development of rain water with storage in tanks for use in dry periods or with drainage underground for technically enrichment. Finally for covering the water needs the processed liquids waste recommended to reuse as previously tested for suitability according to a recent legislation.

References

Koumantakis I. Stavropoulos Ks. Dimitrakopoulos D. (1997) "Hydrogeological status and operating condition of ground water in the basin of Attica". Minutes of the 4th Hydrogeological Conference of the Greek Hydrogeological Committee, Thessaloniki 1997

Koumantakis I. (1997) Research of hydrogeological conditions and of the operating procedures of the ground water in the basin of Attica, Main Issue, (A′ and B′ Phase), research program in the years 1996 and 1997 performed by the research team of the Section of Geological Sciences in cooperation with the Organization of Planning and Environmental Protection Of Athens

Mpitzileki Ch. (2010) Delimitation of the salinity zone of groundwater in the front between the municipalities of Moschato and Glyfada of the Prefecture of Attica, Postgraduate Thesis for the Postgraduate Coarse "Water Resources Science and Technology" NTUA

Hydrogeological conditions of the upper part of Gallikos river basin

C. Mattas, G. Soulios

School of Geology, Aristotle University of Thessaloniki, Thessaloniki, P.C. 54124, Greece, cmattas@geo.auth.gr, gsoulios@geo.auth.gr

Abstract The upper part of Gallikos river watershed basin covers an area of 868.1 km^2. The largest part of the basin administratively belongs to the Prefecture of Kilkis. A small part, along the southern border, belongs to the Prefecture of Thessaloniki. The present paper examines the hydrogeological and hydrological conditions of the area. It also describes the geological, lithostratigraphical and tectonic conditions which comprise a determining factor for the quantitative characteristics of surface and groundwater. The recharge and operation mechanism of the aquifers is investigated, as well as their interaction and the role of the surface water. River discharge, water table level, precipitation and water consumption data are presented and the hydrologic balance of the area is calculated.

1 Introduction

The upper part of Gallikos river basin covers an area of 868.1 km^2 (Fig. 1). It is located in the eastern part of Kilkis Prefecture following a NNE-SSW orientation. It has a pear-like shape and its mean altitude has been calculated at 357.7 m. The hydrographic network is very well developed with the main branch being approximately 45.5 km long. It is a seventh class branch according to the Strahler classification. Its flow direction is NNE-SSW and it is characterised as torrential. The population within the limits of the study area is 39,353 inhabitants (according to the 2001 census), a significant percentage of which is employed in the agricultural sector. The cultivated areas are approximately 1,556 acres (source: Land Reclamation Survey of Kilkis).

The water demands in the area are mainly covered by the groundwater exploitation. The distribution of the population, and therefore the water demands, is not uniform. The majority of the population, cultivated lands and industrial activities are concentrated in the central and southern part of the basin (south of Kilkis city).

2 Geological and tectonic setting

The upper part of Gallikos river basin (Fig. 1) belongs to the Circum Rhodope geotectonic zone and Serbo-Macedonian massif. The main pre-alpine formations mainly consist of paleozoic gneisses, schists, amphibolites and quartzites, which form the crystalline basement of the area (Mercier 1966). At the central-southern part of the basin, carbonate rocks (limestones) of Triassic and Jurassic age outcrop (Meladiotis 1984). The quaternary deposits consist of lacustrine, terrestrial, fluviatile and fluviatile-torrent sediments, with sands, clays and conglomerates.

The tectonic setting of the area is very complicated (Mountrakis 1985) and follows the general tectonic conditions of the Circum Rhodope and Serbo-Macedonian. In general, an inversion of the layers at the western boundary of the Serbomacedonian and Circum Rhodope contact is observed, which results in the tectonic placement of older Serbomacedonian formations (Vertiskos series) over the younger Perirhodopic sediments. Due to the activity of the Circum Rhodope reverse faults, many irregular tectonic contacts are created.

3 Hydrogeological setting

According to their lithological composition and tectonic impact, the geological formations have different hydrogeological behavior. They are characterized either as permeable or as impermeable. The quaternary deposits and the limestones are considered as permeable. The gneisses, schists, amphibolites and quartzites of the study area basement are considered as impermeable.

The main aquifer is developed at the plain part of the basin within the quaternary deposits (Mattas 2009). They cover approximately 40% of the area extent (Fig. 1) with a mean thickness of 33 m. The mean value of the hydraulic conductivity (k) is estimated at $2.4*10^{-4}$ m/sec. The exploitable discharge of the boreholes according to the data of their lithological sections and pumping tests, range from 4 m^3/h to 300 m^3/h. The mean value is calculated at 89.6 m^3/h.

The highest values of boreholes discharge are recorded in the south plain part of the basin, where the quaternary deposits are mainly developed, i.e. they have their biggest thickness and their largest surface development with no presence of other formations.

The majority of the boreholes have been drilled along the river bed. Their water level is near the ground surface. The discharge rate is high for small drilling depths (15-20 m). The river through the infiltration is the main recharge source of the aquifers of the quaternary deposits, in combination with the direct infiltration from the precipitation.

The limestones outcrops cover a very small area, just the 2.5% of the basin, but their hydrogeological importance is remarkable. Their mean thickness is 80 m and the mean hydraulic conductivity (k) is calculated at $2.4*10^{-3}$ m/sec. The discharge rate at the boreholes drilled in the limestones ranges from 40 m^3/h to 90 m^3/h, with

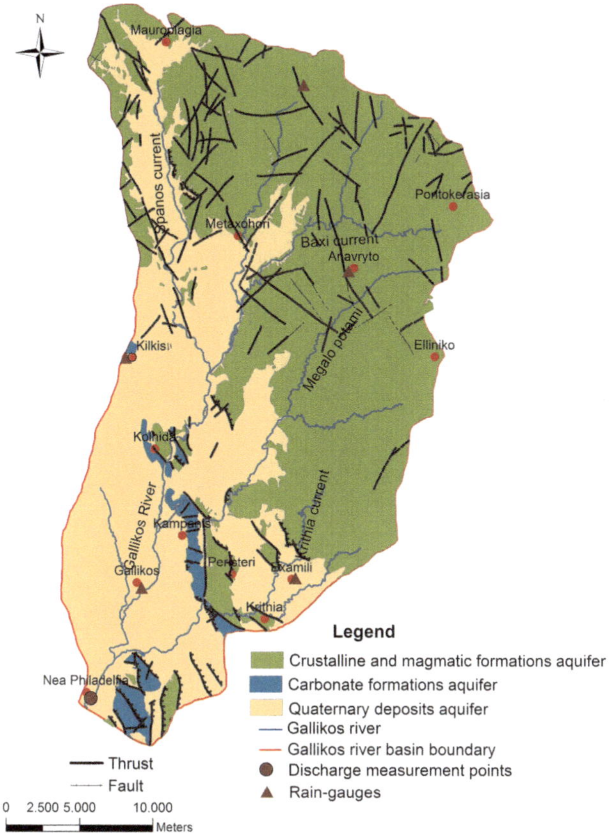

Fig. 1. Simplified hydrolithological map of Gallikos river basin.

a mean value of 60 m^3/h. The depth of the boreholes drilled in these formations is usually more than 150 m, penetrating also the surface sediments or the crystalline bedrock.

The available data from the drilling exploration and water level measurements showed that even in the originally impermeable formations, low potential aquifers developed, due to the intensive tectonic activity that creates a secondary porosity. This has as a result the formation of preferential groundwater flow paths.

The impermeable geologic formations where gneisses prevail cover the biggest part of the basin (57.5%). The mean hydraulic conductivity is calculated at $4.8*10^{-6}$ m/sec. The available data acquired from the lithological sections of the boreholes show that the thickness is at least 230 m. The mean discharge rate is 19.5 m^3/h. In most cases, the depth of the boreholes exceeds 110 m. Several of the boreholes that were drilled in these formations did not encounter any aquifers. Hence, the mean values of the hydrogeological parameters and the data from the wells are not representative for the entire area covered by these formations, but only for the part

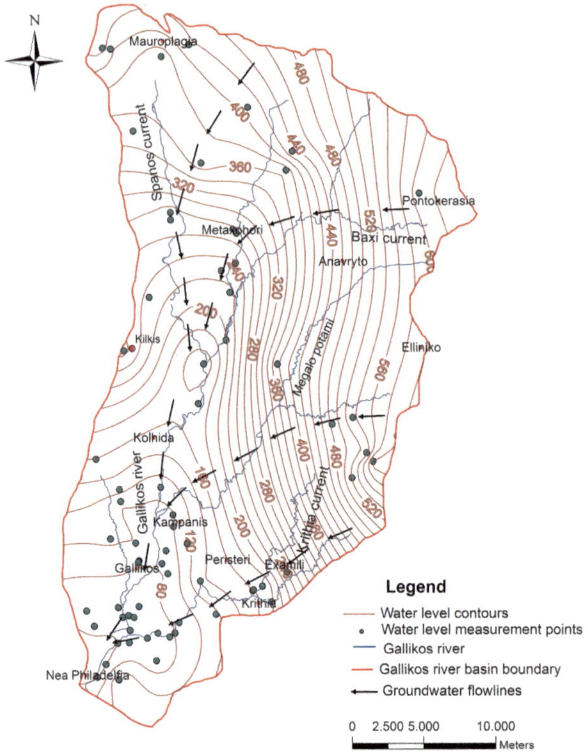

Fig. 2. Water table map during April 2005.

with the most favorable conditions. It should be mentioned that the yields decrease with time and water-level drawdown, since the pumped water comes from the fissures and joints. The effective porosity in such formations is very low (Matthess-Ubell 1983). Regarding the recharge of this aquifer, the role of the river is essential, thereby most of the boreholes were drilled along the river bed.

The presence of the groundwater in the basin is controlled by the tectonic setting, the topography and the river. The relatively small thickness of the quaternary deposits, their intersections with metamorphic and carbonate rocks and the layer inversion at the western boundary of Serbomacedonian and Circum Rhodope zones, in combination with the lepioid structure that creates many irregular tectonic contacts, indicate hydraulic communication between the aquifers in the study area.

Therefore, the water level contours were drawn for the entire basin (Fig. 2). The examination of the water table level shows that there is a clear axis demonstrating the groundwater flow direction. This axis follows the main tributaries of the hydrographic network and is drained towards the south, at the exit of the basin (Nea Philadelfia area). The hydraulic gradient is high (3.7%) at the eastern part of the basin, but it has a typical value (0.75%) at the western part.

It seems that there is lateral recharge from the northern and eastern part of the basin, but it not considered of great importance, despite the long recharge front. This can be attributed to the small hydraulic conductivity of the formations.

4 Hydrological setting, hydrological balance

The significant surface occurrence of the impermeable metamorphic and magmatic formations has favoured the development of an intense and dense hydrographic network, which is recharged directly from the precipitation and not from springs. These conditions, in combination with the small river bed depth, results frequently in flooding phenomena, especially when extreme weather phenomena occur. On the contrary, in dry years recession phenomena in the river are observed. The function of Gallikos river could be characterized as a torrential. Superficial runoff is observed at the exit of the basin (Nea Philadelfia area), even when the river branches have recessed before they reach it. This is attributed to the subterranean flow that spouts in Nea Philadelfia and participates again, to the surface cycle through hydrographic network.

The Laboratory of Engineering Geology and Hydrogeology/School of Geology/ AUTH, installed a water level meter of continuous recording in the area of Nea Philadelfia, in order to measure the water depth of the main Gallikos river branch at the exit of the basin.

The location was selected in order to have a non turbulent river flow and as much as possible unchanging river bed shape and dimensions.

During the recording period, the river discharge was periodically measured. The correlation between water depth-discharge (Fig. 3) resulted in a logarithmic equation that relates the two parameters. The hydrograph of Gallikos river (Fig. 4) was constructed for the hydrologic year 2005-2006 (1/10/05-30/9/06).

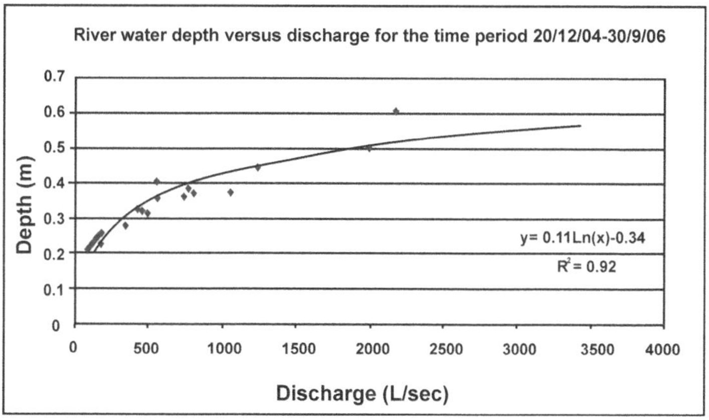

Fig. 3. Depth of river water versus discharge of the main branch of Gallikos river for the time period 20/12/04-30/9/06 at Nea Philadelfia area.

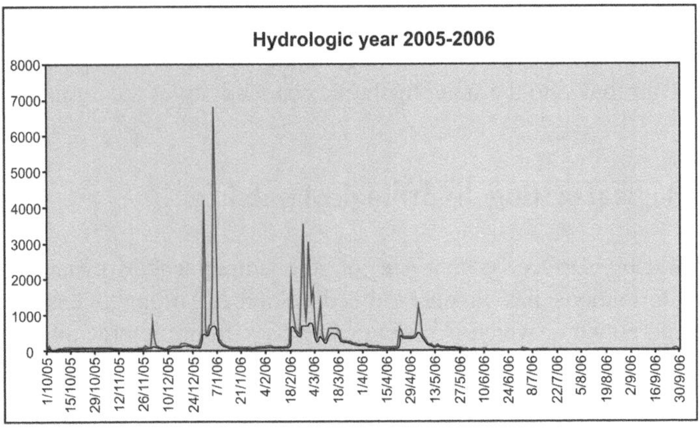

Fig. 4. Hydrograph for the hydrologic year 2005-2006. Discharge lt/sec.

The total discharge for the above time period, was calculated to be Q= $56.1*10^6$ m^3. According to he hydrograph analysis, the superficial runoff is R= $45.3*10^6$ m^3 (80.8%) and subterranean flow is I_R= $10.8*10^6$ m^3 (19.2%).

A network of cumulative monthly rain-gauges was operating in the area, under the supervision of the Kilkis Prefecture (Fig. 1). This network was later complemented by the Laboratory of Engineering Geology and Hydrogeology/ AUTH.

The elevation of the rain-gauges (H) was correlated with the precipitation (P) height according to the measurements of the period 1/10/05-30/9/06. The equation that connects (H) and (P) is:

$$P(mm)=0.36H(m)+370.3$$

Since the mean basin elevation is 357.7 m, the annual volume of the precipitation is $433.8*10^6$ m^3.

According to the population data (2001 census) and assuming a mean water consumption of 250 L/day/inhabitant, the annual water demands for domestic use are calculated at $3.6*10^6$ m^3. The annual irrigation demands are estimated at $3.4*10^6 m^3$, taking into account the type and extent of cultivation, the irrigation network losses and the estimated over-consumption by the farmers. The irrigation returns are estimated to be about 15% or $0.5*10^6$ m^3.

Finally, the water amount that is consumed for animal husbandry and industrial activities or for any other water use was estimated to be $1.0*10^6$ m^3 per year.

The lateral (northern and eastern) recharge that apparently takes place, according to the water table level, is estimated at $15.8*10^6$ m^3, using the Darcy's formula. The same quantity is estimated for the groundwater losses at the southwestern part of the basin.

The water demands are covered almost entirely by the exploitation of the groundwater. The groundwater reserves do not change significantly over long-term periods (more than a year). This means that the consumed water, which is $8.5*10^6$ m^3, as well as the irrigation returns, must be added to the $10.8*10^6$ m^3 that is the amount of the water that infiltrated to the aquifers and came up to the surface.

According to the above data, the resulting hydrological balance is presented in Table 1.

Table 1. Hydrological balance of Gallikos river basin for the hydrologic year 2005-2006.

	P	E	I	R
(mm)	499.7	425.4	22.2	52.1
(10^6 m^3)	433.8	369.2	19.3	45.3
(%)	100	85.1	4.5	10.4

The hydrological balance of the area establishes the significance of the impermeable crystalline rocks presence, which favors the superficial runoff instead of infiltration.

According to the Turc formula (Turc 1951) the evapotranspiration is 440.3 mm. The measured data give a value of 425.4 mm. The deviation between calculated and measured values is negligible (3.5%).

5 Conclusions

The main aquifers of the upper part of Gallikos river basin are developed within the quaternary deposits and carbonate rocks. The river constitutes an important factor of the aquifers recharge. The water demands in the area are covered by the exploitation of groundwater aquifers. The boreholes were mainly drilled along the main river bed, following a NNE-SSW axis, which coincides with the groundwater flow direction. The significant surface development of the primarily impermeable crystalline basement formations favors the superficial runoff than infiltration. According to the hydrological balance, the evapotranspiration is 85.1%, the infiltration is 4.5% and the superficial runoff is 10.4%

Acknowledgments The authors would like to thank the General Secretariat of Research and Technology and Kilkis Prefecture for their financial support in the frame of the PENED 2001 Programme.

References

Mattas C (2009) Hydrogeological research Gallikos river basin. PhD, School of Geology, A.U.TH

Matthes G, Ubell K (1983) Algemeine Hydrogeologie. Grundwasserhaushalt, Gebr. Borntr. Berlin-Stuttgart, 438 p

Meladiotis I (1984) Geological research of the eastern part of Thessaloniki-Giannitsa plain and especially the area among Axios and Gallikos rivers where exploitable aquifers are developed. PhD, School of Civil Engineering, A.U.TH

Mercier J (1966) Stude geologique des zones internes des Hellenides en Macedoine Centrale (Grece), Ann. Geol. Des Pays Hell. (20):1-596

Mountrakis D (1985) Geology of Greece. University Studio Press, Thessaloniki

Strahler A (1952) Hypsometric (area-altitude) analysis of erosional topography, Geol. Soc. Amer. Bull. (63):1117-1142

Turc L (1951) Nouvelle formule pour le calcul du bilan de l' eau en function des valeurs moyennes annuelles des precipitations et de la temperature. C.R. Ac. Sc. vol. 233

A methodological approach for the selection of groundwater monitoring points: application in typical Greek basins

A. Panagopoulos, Y. Vrouhakis, S. Stathaki

National Agricultural research Foundation – Land Recalamation Institute, 57400, Sindos, Greece. panagopoulosa@gmail.com

Abstract This paper aims at proposing a methodology for the compilation of groundwater monitoring networks. The methodology is based on the implementation of a simple algorithm that assists the objective selection of the most appropriate sites out of a large inventory, which is originally drawn. The proposed algorithm was developed in the framework of a research project related to the study of possible residues of 1,3 Dichloropropene in groundwater, in four representative basins of Greece. It considers a number of criteria in order to ensure that the most appropriate and representative sites are selected for the scope of this particular project. This algorithm aims at the establishment of specific criteria the adoption of which may minimize the subjectivity incorporated in the selection of monitoring points, especially when different field operators perform this exercise at geographically distant basins. Subject to the necessary amendments in order for the criteria considered to meet the specific targets of a given project, it is suggested that the proposed algorithm may be applicable at all cases of groundwater monitoring networks establishment.

1 Introduction

Groundwater resources management relies amongst others on accurate, reliable and reproducible data. Long discussions have been made and scientific efforts are still under way towards the direction of addressing issues such as groundwater quality and quantity in a continent wide perspective, employing commonly acceptable methodologies and approaches. The need for developing Europe-wide common background reference data is becoming more pronounced (Quevauviller 2008), and several attempts have made towards this direction (Wendland at al. 2008). Data integrity, comparability and accuracy is an essential issue even on national level, and to a large extent it depends on the principles adopted and applied in setting-up groundwater monitoring networks.

Compilation of monitoring networks is a time-consuming, laborious and very responsible task that requires knowledge and expertise. Several textbooks, protocols and research works exist on the principles that one should consider in setting up a monitoring network in order to ensure accuracy, reliability and representativeness of collected data (Tulipano et al. 2005; Tebutt 1992; Babiy 1984; Panagopoulos 1995; Panagopoulos et al. 2008). Clear definition of the scopes that a monitoring network should serve is of paramount importance to its successful design. Sound scientific justification of the sites selection ensures its reliability, representativeness, acceptance and approval from the scientific community and the decision makers. When networks need to be designed in different geographical locations to monitor the same parameters, extra caution is required to ensure that all networks are compiled under exactly the same philosophy and framework of principles. Normally, the selection of monitoring points is done manually based on specific principles and the expertise of the scientific team that undertakes this task. Although this is a typical and commonly accepted procedure, it does not safeguard that selection is not subjective.

In this exercise, a monitoring network was compiled in each of four basins spread around Greece, aiming at assessing the fate and transport of 1,3-Dichloropropene (1,3-D) in groundwater. 1,3-Dichloropropene is a plant protection product the use of which is widely spread throughout Europe and Greece as well (Terry et al. 2008). In Greece, this study focused on the following representative basins where the product is applied and that are characterized by a multitude of climatic, hydrogeological, land use and water use conditions: Chryssoupoli basin in northern Greece, Trifilia basin in Peloponnese, Tymbaki basin in southern central Crete and Ierapetra basin in southeastern Greece.

Selection criteria for setting-up monitoring networks are dictated by standard requirements that such networks should meet, but also have to fulfil special specifications drawn from each individual study. In this case the following criteria were considered amongst others: proximity to the 1,3-D application zone, reasonable coverage of the study basin, representative coverage of major hydrodynamic mechanisms, focus on the same aquifer system, existence of reasonable amount of information on well construction, use of monitoring points for production of potable water (as a first priority). It has to be stressed that in setting up the monitoring networks, practical issues, such as ease of accessibility and operability of the monitoring points throughout the year were taken into account.

2 Materials and methods

For each study basin the existing well inventories are retrieved. These inventories were compiled in the framework of older consultancy works, development projects, research studies, or simply working maps for practical purposes. A uniform digital data base was then compiled on the basis of the collected material and each point of the inventory per basin was checked in order to make the necessary amendments and corrections. Upon completion of this stage of the work a "base" well inventory was created for each study basin that comprised of: 419 points for Chrysoupoli basin, 67 points for Trifilia basin, 128 points for Tymbaki basin and 35 points for Ierapetra basin. Analysis of this database returned a pre-selection inventory consisting of 16 points per basin, for which detailed information were collected and compiled. Out of this inventory, up to 5 points are selected to form the monitoring network of this study. Selection of the monitoring points is based on 14 criteria analysed using a simple algorithm, in order to minimize the subjectivity and maximize objectivity in the selection procedure.

In this algorithm each criterion is assigned a specific weighting factor and a class, in order to compute a rank for each candidate monitoring point. Hence, subjectivity is reduced and constrained to the selection of the evaluation criteria, the declaration of the required weighting factors and the corresponding classes. The degree of subjectivity is thus constant for all four monitoring networks that were designed in Greece in the framework of this study. Arbitrary scales are assumed for weighting factors (scale 1-10) and corresponding classes (scale 1-5) and assigned values are integers. Class categories for each criterion are assigned in a way to reflect typical situations in Greece. In every study basin the compiled algorithm is applied on every candidate monitoring point and the top 5 rank points are finally selected to form the monitoring network. The exact form of the algorithm is as follows:

$$R = \frac{\sum(WF \times C)}{\sum(WF)}$$

Where R: is the computed rank, WF: the weighting factor and C: the class.

The considered criteria are illustrated in Table 1 and the corresponding classes are tabulated per criterion in Table 2.

Table 1. Considered criteria and assigned weighting factors in the selection algorithm.

No	Criterion	Weighting factor
1	Distance from application	9
2	Existence of borehole lithological section	10
3	Depth to water level	8
4	Water level measurement date	7
5	Well depth	5
6	Depth of screen intervals	8
7	Work type	2
8	Accessibility-sampling permit	5
9	Current use of work	10
10	Operation status	7
11	Aquifer(s) tapped	9
12	Property status	4
13	Relative position to 1,3-D application zone	7
14	Well-head protection status	10

Table 2. Established classes per identified criterion.

Classes / Parameters	5	4	3	2	1
Distance from application	0-20m	21-100m	101-200m	>200m	Unknown
Existence of borehole lithological section	Reliable	Reliable-near	Descriptive	Descriptive-near	Non-existent
Depth to water level	<6m	6-20m	21-50m	>50m	Unknown
Water level measurement date	2005	2000-2004	1995-1999	Before 1995	Unknown
Well depth	<21m	21-50m	51-100m	>101m	Unknown
Depth of screen intervals	Only <25m	Discrete <25m>	Continuous	Only>25m	Unknown
Work type	Housed b'hole	Open-air b'hole	Housed LD well	Open-air LD well	B'hole/LD well housed in greenhouse
Accessibility-sampling permit	A-SP-ET-YR*	A-SP-YR	A-SP-ET	A-SP	A or SP or ET or YR
Current use of work	Potable	Combined	Irrigation	Livestock	Industrial
Operation status	In full operation	Completed-no power	Completed-faulty equipment	Completed-not equipped	Under construction-servicing
Aquifer(s) tapped	Unconfined	Multiple	Single confined	Multiple confined	Unknown
Property status	Municipal	Local Irrigation Organisation	Private	Armed forces	Unknown
Relative position to 1,3-D application zone	Core of main zone	Margins of main zone	Downstream of wide zone	Offstream of wide zone	Upstream of wide zone
Well-head protection status	High	Average	Acceptable	Low	Insufficient

*A: Accessible, SP: Sampling permit granted, ET: Existing sampling tap, YR: Year round operation.

3 Results

Application of the algorithm yielded the five top ranked monitoring points per studied basin. Figure 1 illustrates the distribution of the selected top ranked monitoring points in one of the studied basins, following application of the algorithm presented in the previous chapter.

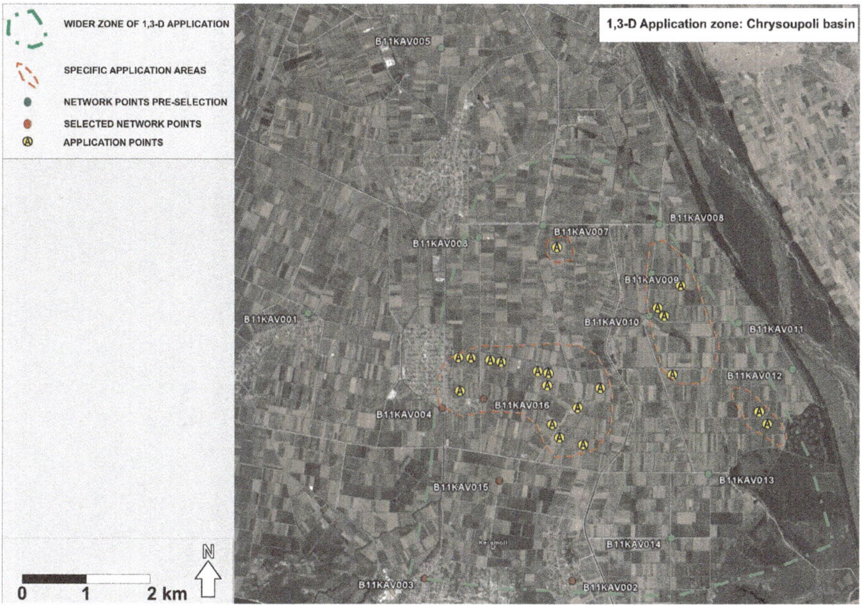

Fig. 1. Selected monitoring points at Chryssoupoli basin.

Table 3 illustrates the encoded analysis of each of the considered 14 criteria for the pre-selected monitoring points in Chryssoupoli basin and table 4 contains the resultant ranks for each of the considered pre-selected monitoring points for the same basin.

Table 3. Tabulated selection criteria for Chryssoupoli basin.

	Distance from applications	Lithological section	Depth to water level	WL measurement date	Well depth	Screens depth	Work type	Accessibility-sampling permit
B11KAV001	2500m	Reliable	2.60m	1987	177m	Only>25m	Housed borehole	A - SP - ET - YR
B11KAV002	*1700m*	*Reliable*	*<0.5*	*1982*	*144m*	*Only>25m*	*Housed borehole*	*A - SP - ET - YR*
B11KAV003	*3000m*	*Reliable*	*<0.75*	*1983*	*140m*	*Only>25m*	*Housed borehole*	*A - SP - ET - YR*
B11KAV004	*50m*	*Reliable-near*	*U/N*	*U/N*	*U/N*	*U/N*	*Housed borehole*	*A - SP - ET - YR*
B11KAV005	3600m	Reliable	8.80m	1986	142.25m	Only>25m	Housed borehole	A - SP -YR
B11KAV006	1200m	Reliable	6.5m	1986	123m	Only>25m	Housed borehole	A - SP - ET - YR
B11KAV007	400m	Reliable	7.20m	1986	136m	Only>25m	Housed borehole	A - SP - ET - YR
B11KAV008	1000m	No	U/N	U/N	U/N	U/N	Housed borehole	A - SP - ET - YR
B11KAV009	500m	No	U/N	U/N	U/N	U/N	Housed borehole	A - SP - ET - YR
B11KAV010	400m	Reliable	3.5m	1986	123m	Only>25m	Housed borehole	A - SP -YR
B11KAV011	1000m	No	U/N	U/N	U/N	U/N	Housed borehole	A - SP - ET - YR
B11KAV012	800m	Reliable	2.60m	1986	133.45m	Only>25m	Housed borehole	A - SP - ET - YR
B11KAV013	1200m	Reliable	2.40m	1990	172.48m	Only>25m	Housed borehole	A - SP - ET - YR
B11KAV014	2000m	Reliable	U/N	U/N	133.50m	Only>25m	Housed borehole	A - SP - ET - YR
B11KAV015	*1200m*	*Reliable*	*1m*	*1991*	*142m*	*Continuous*	*Housed borehole*	*A - SP - ET - YR*
B11KAV016	*50m*	*Reliable*	*4.10m*	*1990*	*152m*	*Only>25m*	*Housed borehole*	*A - SP - ET - YR*

Table 3. *Continued.*

	Current use	Operation status	Aquifer tapped	Property status	Relative position	Well- head protection
B11KAV001	Potable	In full operation	Multiple confined	Municipal	Offstream of wide zone	Average
B11KAV002	*Potable*	*In full operation*	*Multiple confined*	*Municipal*	*Downstream of wide zone*	*High*
B11KAV003	*Potable*	*In full operation*	*Multiple confined*	*Municipal*	*Downstream of wide zone*	*High*
B11KAV004	*Potable*	*In full operation*	*Multiple confined*	*Municipal*	*Margins of main zone*	*High*
B11KAV005	Irrigation	In full operation	Multiple	Local Irrigation Organization	Upstream of wide zone	High
B11KAV006	Irrigation	In full operation	Unconfined	Local Irrigation Organization	Upstream of wide zone	High
B11KAV007	Irrigation	In full operation	Multiple confined	Local Irrigation Organization	Upstream of wide zone	High
B11KAV008	Irrigation	In full operation	U/N	Local Irrigation Organization	Margins of main zone	High
B11KAV009	Irrigation	In full operation	U/N	Local Irrigation Organization	Margins of main zone	High
B11KAV010	Irrigation	In full operation	Multiple	Local Irrigation Organization	Offstream of wide zone	High
B11KAV011	Irrigation	In full operation	U/N	Local Irrigation Organization	Downstream of wide zone	High
B11KAV012	Irrigation	In full operation	Multiple confined	Local Irrigation Organization	Downstream of wide zone	High
B11KAV013	Irrigation	In full operation	Multiple confined	Local Irrigation Organization	Downstream of wide zone	High
B11KAV014	Irrigation	In full operation	Multiple	Local Irrigation Organization	Downstream of wide zone	Average
B11KAV015	*Irrigation*	*In full operation*	*Multiple*	*Local Irrigation Organization*	*Downstream of wide zone*	*High*
B11KAV016	*Irrigation*	*In full operation*	*Multiple confined*	*Local Irrigation Organization*	*Margins of main zone*	*High*

Table 4. Resultant ranks at Chryssoupoli basin.

Well code	Rank	Well code	Rank	Well code	Rank
B11KAV001	3.63	B11KAV007	3.33	B11KAV012	3.58
B11KAV002	3.82	B11KAV008	2.69	B11KAV013	3.58
B11KAV003	3.82	B11KAV009	2.69	B11KAV014	3.28
B11KAV004	3.71	B11KAV010	3.62	B11KAV015	3.83
B11KAV005	3.46	B11KAV011	2.60	B11KAV016	3.84
B11KAV006	3.59				

Similar results were obtained for each of the other three studied basins. As deduced from the study of Table 4, the resultant ranks vary considerably between the analysed candidate points, however it is noticeable that a considerable number of monitoring points do present similar rank values.

4 Discussion-Conclusions

A simple algorithm was synthesised to facilitate the need for selection of the most appropriate monitoring points. The proposed algorithm was applied in four diverse basins of Greece targeting at selecting the most appropriate groundwater monitoring points that would best facilitate the requirements of an environmental fate and transport study for 1,3-D that is widely used in specific basins of the country. A number of criteria were considered on the basis of the specific requirements and targets of the monitoring networks and analysed on multi-criteria analysis adopting common standards. This way it was ensured that common objective assessment of each and every monitoring point was performed, regardless the field operator that provided the initial data for characterisation and the distance between the candidate points.

Even though the proposed approach introduces a high level of objectivity, it still incorporates subjective elements that lay on the appointment of the selection criteria, the assignment of weighting factors in each of the criteria and of course the classification categories identified for each individual criterion. Still, it is suggested that this approach provides in a systematic way an assessment of the "candidate" monitoring points on the basis of a common framework. This means that even if the assessment is subjective, the degree of subjectivity remains common for the entire process and therefore the "error" in this context is minimised. Towards minimising the subjectivity of the proposed approach, a Delphi analysis of the weighting factors and classes of each considered parameter could prove to be essential.

It has to be stressed that as in every methodology, its correct and accurate implementation requires that adequate and reliable data are available and also that deep knowledge of the hydrogeological system for which the monitoring network will be compiled, exists. Furthermore, it is of paramount importance that the correct requirements and targets sought by the monitoring network are identified well in advance the monitoring points' selection procedure. It is proposed that the presented algorithm may be used under the appropriate alterations and adaptations in setting up monitoring networks that would meet a wide spectrum of targets.

Acknowledgments This study was compiled in the framework of a research project funded by Dow AgroSciences, under the Responsible Care® program.

References

Babiy L (1984) Reliability of information in hydrochemical investigations. In: Eriksson E (eds.) Proc. of Conf. Hydrochemical balances of freshwater systems pp.179-185. Uppsala

Panagopoulos A, Stathaki S, Vrouhakis Y (2008) Monitoring residues of 1,3-dichloropropene and its metabolites in groundwaters in Greece: methodological approach and lessons learned. In: Panagopoulos A, Vizantinopoulos S, Vrouhakis Y (eds.) Proc. Conference water quality in the framework of the European and the national legislation pp.50-60. Crete

Panagopoulos A. (1995) A methodology for groundwater resources management of a typical alluvial aquifer system in Greece, Ph.D.Dis., The University of Birmingham, U.K

Quevauviller Ph (2008) From the 1996 groundwater action programme to the 2006 groundwater directive-What have we done, what have we learnt, what is the way ahead? J. of Environmental Monitoring 10, 408-421

Tebbut T.H (1992) Principles of water quality control. Oxford

Terry A., Carter A, Humphrey R, Capri E, Grua B, Panagopoulos A, Pulido-Bosch A, Kennedy S (2008) A monitoring programme for 1,3-dichloropropene and metabolites in groundwater in five EU countries. J. Pest Management Science 64:9, 923-932

Tulipano L, Fidelibus M.D, Panagopoulos A (2005) Groundwater management of coastal karstic aquifers, Final report. Luxembourg

Wendland F, Blum A, Costiers M, Gorova R, Griffioen J, Grima J, Hinsby K, Kunkel R, Marandi A, Melo T, Panagopoulos A, Pauwels H, Ruisi M, Traversa P, Vermooten J, Walraevens K (2008) European aquifer typology: a practical framework for an overview of major groundwater composition at European scale. J. Environmental Geology 55:1, 77-85

Stochastic Modeling of Plume Evolution and Monitoring into Heterogeneous Aquifers

K. Papapetridis, E.K. Paleologos

Dept. of Environmental Engineering, Technical University of Crete, Greece.

Abstract A Monte Carlo stochastic model was developed to simulate contaminant transport from an instantaneous source inside a landfill into heterogeneous, two-dimensional aquifers. Probabilities of detection P_d were calculated for different arrangements of monitoring wells. The frequency of sampling is critical in heterogeneous aquifers with bi-annual or annual sampling decreasing average percentage P_d by almost 20%, and 50%, respectively, relative to that by monthly sampling. It appears that at a minimum sampling should take place twice a year, with the once- in-a-month sampling appearing the optimum choice considering the effort involved and the improvements in detection. In heterogeneous aquifers a large number of monitoring wells sampled infrequently does not perform any better in terms of detection than a lower number of wells sampled regularly.

1 Introduction

Successful detection of aquifer contamination via monitoring wells is a complicated problem with many factors, such as the heterogeneity of the geologic environment, the quantity and nature of the contaminants, the number and location of the monitoring wells, and the frequency of sampling, all contributing to the uncertainty of early detection. The current article addresses these issues by investigating the case of instantaneous leakage from a landfill facility into a heterogeneous aquifer, focusing at the importance of sampling frequency.

The major factor that influences the likelihood of early detection of an aquifer's contamination by a landfill leak is its dispersion into heterogeneous geologic formations. The stochastic Monte Carlo framework has been used to address the spread and the evolution of a plume in a 2-D aquifer in order to determine the maximum detection probability of various monitoring wells linear arrangements.

Additional uncertainty arises from lack of knowledge about the leak itself. The location of the source, the quantities and chemical composition of the contaminants, and the time when a leak originated are questions with significant uncertainties involved. Simulation studies (Meyer et al. 1994, Storck et al. 1997, Bierkens 2005, Yenigul et al. 2005, Papapetridis and Paleologos 2010, 2011) usually assume conservative contaminants, with continuous or instantaneous leakages, and with the source's location randomly selected within a landfill's area.

The frequency of a sampling program that is implemented at a monitoring well system is another component that defines the likelihood or not of detecting contaminant concentrations above regulatory threshold values. The EU Directive 1999/31/EU on "the landfill of waste" states that sampling for monitoring purposes should be conducted *"...At a frequency to be determined by the competent authority and in any event at least once a year..."* The dependence of the probability of detection on the sampling frequency can be explained if one considers the sub-region of the plume characterized by concentrations, which are above the threshold regulatory limits. This sub-region continually changes in space and time, and hence for instantaneous leaks infrequent sampling may result in obtaining samples at the wells that are used as regulatory check points when this critical part of the plume has already travelled elsewhere. This is in contrast to continuous leaks of constant flow rates where monitoring has been shown to be insensitive to the frequency of sampling (Yenigul et al. 2006).

2 Model Description

Our study involved the stochastic simulation of groundwater flow and contaminant transport in 2-D heterogeneous aquifers (Meyer et al. 1994, Yenigul et al. 2005, Papapetridis and Paleologos 2011). Monitoring wells were assumed to fully penetrate the aquifer resulting in vertically-averaged concentration measurements. Sources of uncertainty of the physical problem involved the heterogeneity of the two-dimensional geologic field, the dispersion of the contaminant, and the initiation point, size, and duration of a leak. The contaminant was assumed to be conservative, and fully water soluble.

The heterogeneity of the geologic environment was addressed through the hydraulic conductivity, which was simulated as a log-normal, stationary, second order, isotropic stochastic process (Freeze 1975, Gelhar 1986, Elfeki 1996) using the Spectral Turning Bands Method (STUBA) (Mantoglou and Wilson 1982). Contaminant transport into the subsurface heterogeneous environment was simulated using the particle tracking method based on the 'random walk' approach (Uffink 1987, Tompson et al. 1987).

Uncertainty concerning the potential location within a landfill, where a leak may have originated, is related to the lack of information on possible points of failure of a landfill's bottom liner. We have assumed that any single point of the landfill is an equal-probable source of leakage, which takes place once as a single failure event taking place at zero time (when the simulation begins), resulting in an instantaneous ejection of contaminants into an aquifer. The contribution of the unsaturated zone in the contaminant's movement and dispersion was neglected.

Four linear configurations of monitoring-well installations were examined consisting of 3, 6, 12 and 20 wells, equally spaced from each other. Wells that were located at the ends of each arrangement were placed half the distance from the landfill's top and bottom edges so that the efficiency of the monitoring system would be maximized (Yenigul et al. 2005). Monitoring of aquifer was simulated for a 30 year period. The flow field was considered 1000 m long and 400 m wide. The variance of the log hydraulic conductivity varied among the values of 0.0 (homogeneous aquifer), 1.0 (heterogeneous aquifer) and 2.0 (strongly heterogeneous aquifer), while the mean of the log hydraulic conductivity was taken to be equal to 2.3. The correlation length, λ, was considered constant and equal to 20m for both directions x and y (isotropic medium).

A rectangular landfill facility was located between x-coordinates of 10 m and 60 m, and y-coordinates of 140 m and 260 m. The total pollutant mass was equal to 1000 gr, simulated by 8000 discrete particles. The initial concentration of the point source, $C_o = M_o/(nV_o)$, was 4000 mgr/L, where M_o the initial mass, $n = 0.25$ the effective porosity, and $V_o = 1$ m^3. The threshold concentration C_{TH}, which was detectable from the monitoring wells was set at 0.35% of the initial concentration, corresponding to 112 particles. This corresponded to a level of critical contamination from nitrate, cyclohexanon, or diethyleneglycol, which would require remedial procedures (Yenigul et al. 2005). Three different transverse dispersion coefficients, a_T, were examined, equal to 0.01 m, 0.05 m and 0.5 m. The longitudinal dispersion coefficient, a_L, was calculated by the relation $a_L = 10a_T$ (Gelhar 1986, Spitz and Moreno 1996). A detailed description of the simulation model and input parameters can be found at the work of Papapetridis and Paleologos (2011).

3 Results and Discussion

The number of wells that are used in a monitoring arrangement has a great influence on the likelihood of detecting or not potential contamination events from a landfill facility. Having performed 3000 simulations for each case, it was evident from our numerical experiments that in all cases of hydrogeologic heterogeneity and dispersion the more wells utilized for detection purposes the greater the detection probability. This was the major factor that affected the probability of detecting contaminants. Even when the hydrogeologic environment was favorable, such as the case of a homogeneous aquifer, the only way to increase the performance of a linear monitoring-well arrangement was by increasing the number of wells. It was observed that the use of 20 monitoring wells provided extremely high detection probabilities, which in some cases, at least at the numerical level, reached full detection. On the other hand, the minimum requirement of 3 monitoring wells stipulated in the 1999/31/EU directive "on the landfill of waste," did not return detection probabilities over 19%. The implication of this very low performance is that approximately four out of five cases of instant leakage from a landfill would remain undetected if such a well arrangement is to be used.

Variability of the geologic field is a factor that greatly affects detection probability. During field experiments using tracers it was observed that plumes have complex structures and in some cases they were split into separate parts which drifted off in different directions (McLaughlin et al. 1993). As a result of a plume's irregular movement narrow bands of no detectable concentrations were created, rendering monitoring very hard. Numerical simulations have verified that as hydrogeological heterogeneity increases detection probability decreases (Yenigul et al. 2005, Papapetridis and Paleologos 2010, 2011).

In addition to a field's heterogeneity, dispersion greatly affects the form that a contaminant's plume takes (Meyer et al. 1994, Yenigul et al. 2005). The longitudinal dispersion causes elongation of the plume in the direction of groundwater flow, while the transverse dispersion causes it to widen. This means that the farther a plume travels, the more it spreads and dilutes into the ground. The dispersion of pollutants as they travel into an aquifer results in opposing situations with regards to monitoring. As the plume evolves the contaminated area increases, making it more likely for a plume to be detected by a monitoring system. On the other hand though, as the plume evolves the concentration drops, making it more difficult to obtain high concentration samples, and hence to detect at a distance from the source. Detailed discussion on the effects of heterogeneity and dispersion on the detection probability of a linear arrangement of monitoring wells can be found at the work of Yenigul et al. (2005), and Papapetridis and Paleologos (2010, 2011).

Contamination is detected by a monitoring set, when a plume's concentration at any well is sampled above threshold limit. However, considering the irregular form a plume can take and the simultaneous dilution that takes place it is clear that a detection 'window' is created as contamination propagates. Consequently, in cases of instantaneous contamination of ground water, the monitoring wells sampling frequency affects the overall efficiency of detection probability.

The dependence of the detection probability on the sampling frequency is investigated here for the case where the contaminant's mass is released instantaneously into an aquifer. Numerical results refer to sampling frequencies of once a day, once a month, once every two-months, quarterly, once every four-months, bi-annually, and once a year. In each of the following Figures (Fig. 1-3) the detection probability P_d is plotted against the sampling frequency, for three different heterogeneity levels (variance of Y=lnK) and transverse dispersion coefficient a_T.

In homogeneous geologic environments of low dispersivity (a_T =0.01m) it was observed (Fig. 1) that for every monitoring arrangement the detection probability does not change but slightly up to the level of bi-annual sampling. In the case of 20 wells P_d remains at the 98-99% level for all frequencies, decreasing to 95% only when annual sampling is performed. If a medium dispersive environment is assumed (a_T=0.05m), then for 20 wells P_d decreases from 99% to 94% at bi-annual sampling, and to 85% at annual sampling. The same trend holds true for 3, 6, and

12 wells systems, where the average percentage decrease between monthly and bi-annual sampling was 6%, and between monthly and annual sampling 15%. However, in the case of high dispersion (a_T=0.50m) the effect of sampling frequency on P_d is pronounced. When 20 wells are used P_d drops from 37%, when monthly sampling is performed, to 31%, when biannual sampling is followed, to eventually 4%, if samples are taken annually. The average percentage decrease in detection, for all monitoring arrangements, was 20% if sampling decreased from monthly to biannual, and 91% if sampling decreased from monthly to annual. It is obvious that in a highly dispersive field sampling at the once-a-year level nearly cancels all monitoring efficiency.

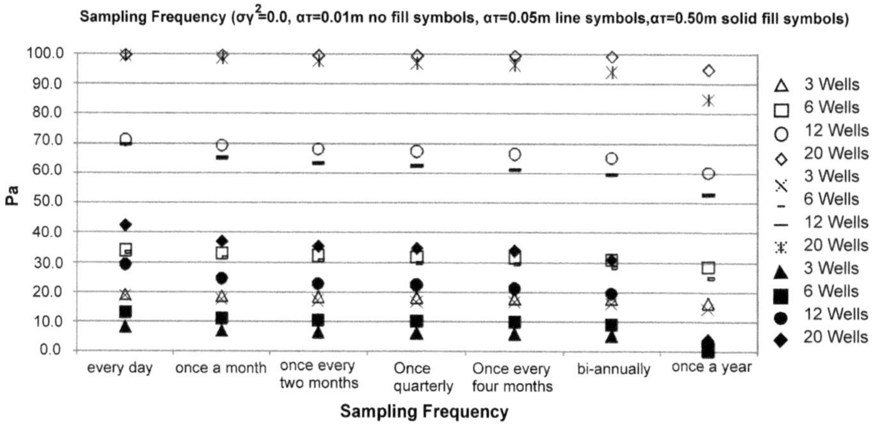

Fig. 1. Detection probability P_d vs sampling frequency for the homogeneous case of hydraulic conductivity σ_Y^2 =0.0, and a_T=0.01m, a_T=0.05m, a_T=0.50m.

In cases of medium heterogeneity (σ_Y^2 =1.0) and of medium dispersion (a_T=0.05m), Figure 2, the average percentage decrease in P_d was 7% if sampling was done bi-annually rather than monthly, and dropped to 22%, if it was done annually. In the case of a_T=0.50m it was observed than the detection attained by bi-annual sampling dropped by 31% relative to that accomplished by monthly sampling, and by 70% if annual sampling was performed.

In Figure 3 a strongly heterogeneous field with variance of lnK equal to 2.0 was examined against the sampling frequency. In this case it was observed that up to medium values of transverse dispersion coefficient (a_T=0.05m) P_d decreased for all monitoring systems with a decline in sampling frequency. The average percentage decrease in detection probability, relative to that attained by monthly sampling, was 10% and 27% if biannual or annual sampling was performed, respectively. In the case of a_T=0.50m (high dispersion) it was observed that for all monitoring systems a decline from monthly to biannual sampling reduced P_d by 37%, while a decline from monthly to annual sampling reduced P_d by 66%. Especially, in case of annual sampling it is observed that the reduction at P_d is 5% and 25% less than the corresponding situation for medium heterogeneous and homogeneous fields of the same dispersion, respectively. This means that field heterogeneity counteracts dispersion, providing this way larger P_d for the same monitoring well arrangements.

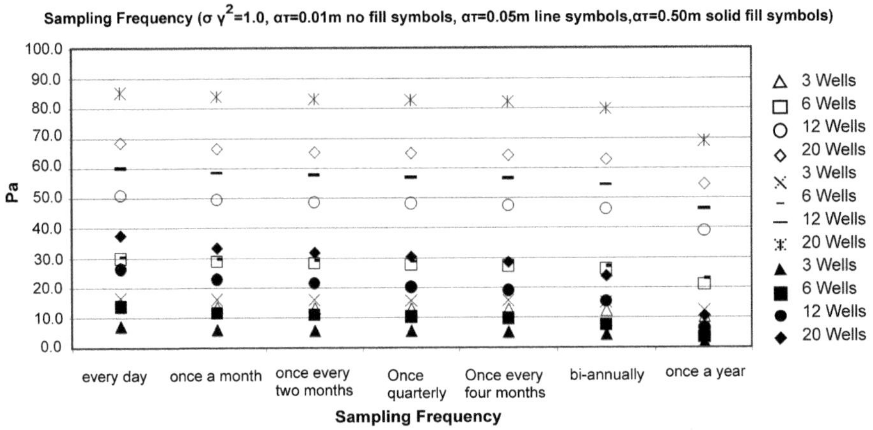

Fig. 2. Detection probability P_d vs sampling frequency for medium heterogeneous case of hydraulic conductivity $\sigma_Y^2 = 1.0$ and a_T=0.01m, a_T=0.05m, a_T=0.50m.

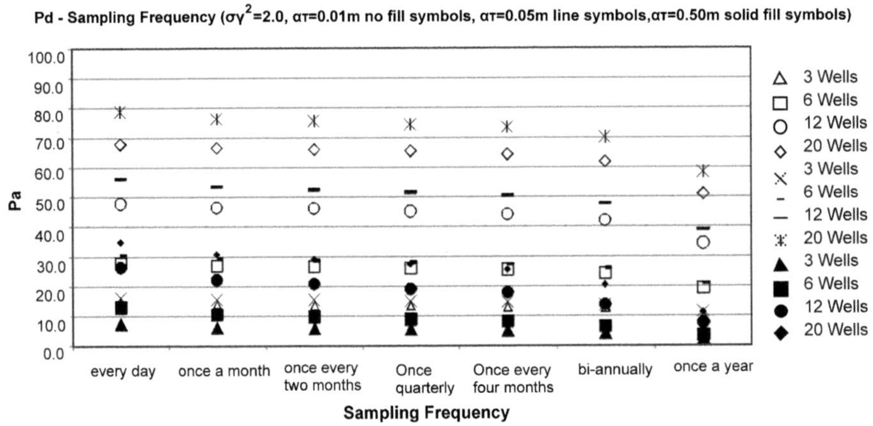

Fig. 3. Detection probability P_d vs sampling frequency for strong heterogeneous case of hydraulic conductivity $\sigma_Y^2 = 2.0$ and a_T=0.01m, a_T=0.05m, a_T=0.50m.

If sampling once every two months is considered then for low dispersive fields of even strong heterogeneity the average percentage decrease in detection, relative to monthly sampling, was less than 2%, while for high dispersions the decrease was less than 7%. In all cases considered once-a-day sampling provided the best results, although, in most cases, it would be a capital and human resources intensive practice to provide for daily monitoring considering the time period that a landfill needs to be monitored and that, even if some leachate is to escape, not all contamination cases are so severe as to justify such a practice.

The effect of sampling frequency on the detection probability is more pronounced when the aquifer is heterogeneous and dispersion is increased (Fig. 2, 3). As a general rule it appears that under all conditions at least bi-annual sampling should occur at a monitoring system. If one wants a higher detection probability then sampling at a frequency of once a month or even once every two months appears to be the optimum choice for most well arrangements, considering both the effort involved if one were to proceed with a much more intense sampling and the improvements on detection attained at this level. It is interesting to note that in heterogeneous aquifers a large number of monitoring wells (such as the case of 12 or 20 wells considered here) if sampled infrequently (for example, once a year) does not perform much better in terms of detection than arrangements having a lower number of wells, but which are sampled more regularly.

4 Conclusions

A Monte Carlo stochastic model was developed to simulate contaminant transport into heterogeneous, two-dimensional aquifers. Pollution originated from a random, instantaneous point source within a landfill facility. Different arrangements and distances of monitoring wells from a landfill were considered, and the corresponding detection probabilities P_d, at different sampling frequencies, were calculated. It was shown that in all cases of hydrogeologic heterogeneity and dispersion the more wells utilized for detection purposes the greater the P_d.

Frequency of sampling is critical in heterogeneous aquifers of high dispersion. Bi-annual sampling decreased the detection probability relative to monthly sampling by about 20% and 35%, for all well arrangements, at transverse dispersion coefficient values of 0.05m and 0.50m, respectively. Once every two months sampling decreased P_d by about 2% and 7% relative to monthly sampling for the same dispersion cases. As a general rule it appears that under all conditions at least bi-annual sampling should occur at a monitoring system. If one wants a higher detection probability then sampling at a frequency of once a month or every other month appears to be the optimum choices for most well arrangements, considering both the effort involved, if one were to proceed with a much more intense sampling, and the improvements on detection attained at this level. It is interesting to note that in heterogeneous aquifers a large number of monitoring wells if sampled infrequently does not perform much better in terms of detection than arrangements having a lower number of wells, but which are sampled more regularly.

Acknowledgments This work was supported by a research grant by the Technical University of Crete.

References

Bierkens MFP (2005) Designing a monitoring network for detecting groundwater pollution with stochastic simulation and a cost model. Stoch. Environ. Res. Risk Assess 20, 335-351

Elfeki AMM (1996) Stochastic characterization of geological heterogeneity and its impact on groundwater contaminant transport. Balkema Publisher, Rotterdam

Freeze RA (1975) A stochastic conceptual analysis of one-dimensional groundwater flow in non uniform homogeneous media. Water Resour. Res. 11, 5, doi:10.1029/WR011i005p00725

Gelhar LW (1986) Stochastic subsurface hydrology from theory to applications. Water Resour. Res. 22, 9S, doi:10.1029/WR022i09Sp0135S

Mantoglou A, and Wilson JL (1982) The Turning Bands method for simulation of random fields using line generation by a spectral method. Water Resour. Res. 18, 5, doi:10.1029/WR018i005p01379.

McLaughlin D, Reid L B, Shu-Guang L, Hyman J (1993) A Stochastic Method for Characterizing Ground-Water Contamination. Ground Water 31, 2, 237-249

Meyer PD, Valocchi AJ, Eheart JW (1994) Monitoring network design to provide initial detection of groundwater contamination. 30, 9, doi:10.1029/94WR00872.

Papapetridis K, and Paleologos E (2010) Stochastic modeling of subsurface pollution in 2-D heterogeneous aquifers and detection probability for linear configuration of monitoring wells. 2nd International Conference Hazardous and Industrial Waste Management, Crete 2010

Papapetridis K, and Paleologos E (2011) Contaminant detection probability in heterogeneous aquifers and corrected risk analysis for remedial response delay. Water Resour. Res. (in review)

Storck P, Eheart JW, Valocchi AJ (1997) A method for the optimal location of monitoring wells for detection of groundwater contamination in three-dimensional heterogeneous aquifers. Water Resour. Res. 33, 9, doi:10.1029/97WR01704

Spitz K, and Moreno J (1996) A practical guide to groundwater and solute transport modeling. Wiley-Interscience Publication, New York

Tompson AFB, Vomvoris G, Gelhar LW (1987) Numerical simulation of solute transport in randomly heterogeneous porous media: motivation, model development, and application. Rep. UCID-21281, Lawrence Livermore Natl. Lab., Livermore, California

Uffink, GJM (1990) Analysis of dispersion by the random walk method. Ph.D. Thesis, Delft University of Technology, The Netherlands

Yenigül NB, Elfeki AMM, Gehrels JC, Akker C, Hensebergn AT, Dekking FM (2005) Reliability assessment of groundwater monitoring networks at landfill sites. Journal of Hydrology 308, 1-17, doi:10.1016/j.jhydrol. 2004.10.017

Yenigül NB, Elfeki AMM, Van den Akker C, Dekking F M (2006) A decision analysis approach for optimal groundwater monitoring system design under uncertainty. Hydrol. Earth Syst. Sci. Discuss. 3, 27-68

Hydrogeological conditions of the lower reaches of Aliakmonas and Loudias rivers aquifer system, Region of Central Macedonia, Northern Greece

N. Veranis[1], A. Chrysafi[2], K. Makrovasili[2]

[1]Institute of Geology and Mining Exploration, Regional Unit of Central Macedonia, 1 Frangon st, Thessaloniki, GR 54626, nveranis@thes.igme.gr
[2]Aristotle University of Thessaloniki, Geology Department

Abstract The aquifer system of the lower reaches of Aliakmona and Loudias rivers covers an area of 1638 km^2 and is mainly hosted within the Quaternary and Neogene sediments. It comprises of intercalations of confined to semi-confined aquifers consisting of sand, gravel, conglomerates, marl-limestones, pyroclastic formations, sandstones which are separated by layers of clays or marls. Based on data of piezometric records we concluded that over the last 20 years the water-balance reached to a deficit of about $28*10^6$ m^3/year. That resulted to a decline of the piezometric surface, which is more intense in the eastern, northern and western parts of the aquifer system, where irrigation water comes from groundwater pumping, than in central and south parts, where the irrigation waters are mainly originated from the Aliakmonas river.

Keywords: aquifers, water balance, piezometric surface decline

1 Introduction

This paper is based on field data which came up after the study of the aquifer systems of the Region of Central Macedonia, which was carried out by the Institute of Geology and Mineral Exploration of Greece, within the third European Community Support Framework (2003-2008). The area of the aquifer system belongs to both the Aqueous Region of Western Macedonia and the Aquifer Region of Central Macedonia but according to the hydrogeological criteria is a single aquifer system (Veranis 2010). Previous studies include only a few parts of the aquifer system (Rapti 1995). In this investigation, for the first time, the hydrological characteristics of the area as an aquifer system are mentioned one of which focusing on the piezometric study and the water balance of the aquifer system.

The aquifer system takes up an area of 1638 km^2 and its larger part (948 km^2) is lowland with altitude from 0 to 20 m. The hill zone (20-200 m) surrounds the lowland and takes up an area of 654 km^2, while a small mountainous zone of 36 km^2 in the northern part of it varies, having altitudes from 200 to 550 m. The semi mountainous and the hill part is transected by a hydrographic net of dendritic form. In the lowland man's intervention, which is attributed to the construction, since 1930, of the drainage ditch of Loudias river, of the ditch 66 and of the net of the irrigation canals as well as the ditches for the rain waters.

2 Geological structure

The cartographic units of the aquifer system consist of the Quaternary and Neogene sediments. The Quaternary sediments are divided into Holocene and Pleistocene aged sediments. The Holocene sediments consist of fine-grained sands, clays, loamy clays, conglomerate and sands at the beds of the torrents as well as materials of torrent - like benching with small thickness. The cones of screes slopes consist of disconnected, non benching gravels and pebbles as well as deposits of the dried lake of Giannitsa which consist of clays and clay-sands. Last but not least, recent cones of trench deposits are traced. The Pleistocene sediments are located in the low land and hill area of the western part as well as the northern plan at the feet of Paiko mountain. Cones of screes of red clays, limestone breccias, boulders, cones of trench deposits disconnected or mildly consolidated colluvial and river-trench deposits. The thickness of the Pleistocene sediments is estimated about 100 m.

Of upper to mid Pliocene- lower Pleistocene age tuffs, toffites and mud flows are occur in the hill side between Anydro and Mandalo villages as well as at the west-north part of Skydra town. The Miocene sediments are located in the eastern and northern-eastern part of the aquifer system and consist of: (a) Sandstone-marly series which is an interchange of fine-grained sands or clays with the intervention of marly lenses, marly sandstones, marly limestones and more rarely of conglomerates. (b) Marls and red clays. (c) Sands, gravels and marls. These sediments are covered by limestones, marly limestones, marls and sandstone-conglomerates.

The thickness of the Quaternary and Neogene sediments rages from a few decades of meters at the margins of the sedimentary basin to approximately 1500 m in the central and eastern part. From the study of the lithological sections of the boreholes constructed for the hydrocarbons exploration, it resulted that the sedimentation started at the Eocene and the whole thickness of the sediments is more than 3400 m. The hydrogeological research is limited to the depth of 650 m, because in even deeper levels the temperature of the water is more than 33-35^{0}C and they are characterized as thermal waters.

3 Hydrolithological and hydrogeological characteristics

The borders of the lower reaches of Aliakmona and Loudias rivers aquifer system at the northern and western part are determined by the contact between the loose sediments and the metamorphic rocks of the basin or the volcanic rocks of Aridea area. At the eastern part it is separated from the granular aquifer system of Axios river through the surface and subsurface water divide. At the southern part the border with the granular system of Kolindros is located in the contact between the Neogene sediments of Kolindros aquifer system and the Quaternary sediments of the low reaches of Aliakmona- Loudias system respectively.

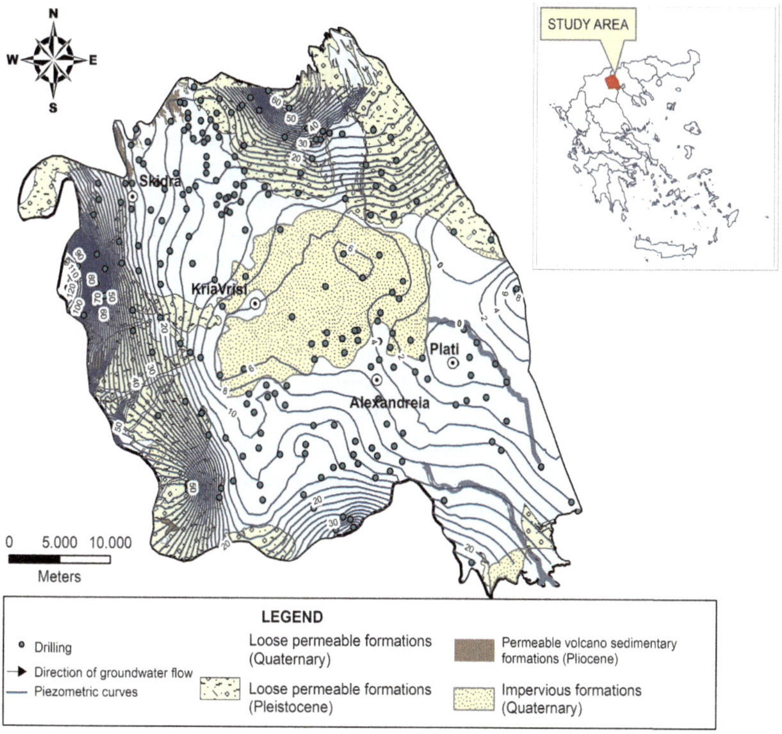

Fig. 1. Hydrogeological map of the aquifer system lower reaches of Aliakmona and Loudia rivers.

The aquifer system consists of aquifer intercalations, consisting of sands, gravels, conglomerates, volcanoclastic rocks, marly limestones and sandstones, which are separated from impermeable to semi-impermeable beds as clays or marls. The aquifer system is confined to semi-confined. The unconfined aquifers are located along the rivers beds and in many places they are exhausted. According to their hydrolithological behavior the different units of the system are classified into four categories (Fig. 1):

- *Micro-macropore permeable sedimentary formations covering an area of 1330 km^2.* Sedimentary deposits of Quaternary age and river trench deposits of Pleistocene age. Their hydrogeological interest is considered important. Due to the various particle composition and the different participation of the clay material, the sediments show differences in their hydrolithogical behavior laterally. The hydraulic conductivity (K) ranges from 10^{-4} to 10^{-6} m/s, the yield of the bore-holes range from 30 to 350 m^3/h, the value of the transmissivity (T) ranges from $1*10^{-4}$ to $2.9*10^{-2}$ m^2/s and the specific capacity (Sc) ranges from 0.5 to 80 m^3/h/m.
- *Macropore permeable formation covering an area of* 50 *km^2*. Permeable formation consisting of volcanoclastic materials with conglomerate intercalations that have a remarkable aquifer capability. The thickness of the formation ranges from 25 to 270 m and overlies an impermeable clay formation (Rapti 1995). The yield of the boreholes range from 55 to 240 m^3/h, the specific capacity (Sc) ranges from 1.96 to 36.4 m^3/h/m, the transmissivity (T) ranges from $1.17*10^{-2}$ to $2.4*10^{-4}$ m^2/s and the hydraulic conductivity varies from $1.1*10^{-3}$ to $2.7*10^{-6}$ m/s. The storage coefficient (S) ranges from $1.5*10^{-3}$ to $1.36*10^{-4}$. Beside their appearances in the N-NW part, the volcanoclastic materials have an important contribution as intercalations in the clastic sediments at the northern, western and central part of the aquifer system and consist of a positive factor for the development of important aquifer ability.
- *Impermeable and low permeable sedimentary formations of small area (8 km^2).* They are consolidated conglomerates with a great percentage of clay and compact layers of red clays which are located at the northern-western part of the aquifer system and along the bed of Aliakmona river.
- *Impermeable formation of large area (*250 km^2). They are located at the area of the drained lake of Giannitsa, they have a small thickness and consist of clays. Under these impermeable formations important aquifers are found within sands and conglomerates separated by clays.

The depth of the bore-holes ranges from 30 to 400 m with a geometrical mean of 150 m (n=580). The thickness of the unsaturated zone ranges from 2 to 145 m with a geometric mean of 21 m and consists, at a great scale (60%) from semi-permeable materials as the alternation of clay with sand, gravels, conglomerates or mixtures of them. At the percentage of 22% they consist of impermeable materials (clays, marls) and the rest 18% consists of permeable materials (sand, conglomerates).

The total thickness of the aquifer beds ranges from 4 to 145 m, with a geometric mean value of 32 m. The percentage participation of the aquifer system at the total length of the transected beds in the bore holes ranges from 4 to 60% with geometric mean of 24%. For a total thickness of sediments of about 650 m, studied up to date for hydrogeological purposes, the thickness of the aquifer system is 150 m.

4 Piezometric study

The depth of the piezometric surface ranges from +0.2 (artesian) to 125 m with a geometrical mean of 4.10 m. The larger rates are traced at the hill side which is the periphery of the aquifer system and the smaller ones at the low land respectively. There is a difference between the hydraulic load (Dφ=1.5-4.5 m) among the different depths of the aquifers. The piezometric surface at the deep aquifers is at a smaller depth than at the shallow ones respectively.

A seasonal variation of the piezometric surface is mentioned. At the western part of the system, where the Pleistocene sediments with the lower factor of hydraulic conductivity occur, the ranging (Ds) is from 2.0 to 10.9 m, while at the low land central and southern part, where the Quaternary sediments are prevailed and it consists the discharge area of the system, the ranging is from 0.2 to 2.0 m respectively and the mean value is Ds=2.32 m. In Figure 2 is presented the seasonal variations of some representative boreholes of the aquifer system. A decline of the piezometric level is obvious in the western and north-eastern part of the system (Fig. 2).

The larger hydraulic slopes are mentioned at the eastern, northern and western part (Fig. 1). In the low land the piezometric curves are less dense which means that a better transmissivity and hydraulic conductivity are observed, than at the hill side, where the Pleistocene sediments prevail, the hydraulic slope and the seasonal ranging of the piezometric surface have higher rates. Generally, a radial flow to the central part of the drained lake of Giannitsa, which is the discharge area of the system, is noticed. The greater hydraulic slopes are noticed at the N-NW part of the system (20-25%$_0$) but generally they range from 7.0 to 1.5%$_0$ with the smaller rates at the Platy area, reaching 0.8 to 0.5%$_0$.

At the area of the Ditch 66 the piezometric curves are parallel which means that there is no hydraulic connection between the ditch and the aquifer system. In the area upstream of the Ditch 66 it seems to exist a small discharge of the aquifer system to the water trenches of East Vermio mountain. At the SW part, Aliakmonas river, at its entering the plain of Imathia at the area of Ammos and Varvares villages feeds the aquifer system. Along the axis of Vergina, Kouloura and Stavros villages, which is the lowest topographical point, there seems to be a discharge of the aquifer system towards the area of the drained lake of Giannitsa. To the southern part, Aliakmonas river transects under an angle the piezometric curves which means that it is recharged from the southern part of the system and it charges it towards the NE part (Kresic 2007). At the southwestern part, from the watertrenches of the hill side there a small provision of the system seems to exist. At the area, south of the Koryfi and Prasinada villages there occurs a discharge of the subsurface waters towards Aliakmonas river. At the central and southern part of the system the direction of the groundwater flow is stable SSW → NNE to the bed of Loudia river. At the areas of Platy, Arachos, Mikro Monastiraki villages along the national road of Chalkidona- Pella and even southern from it, negative absolute altitudes of the piezometric surface are noticed (Fig. 1).

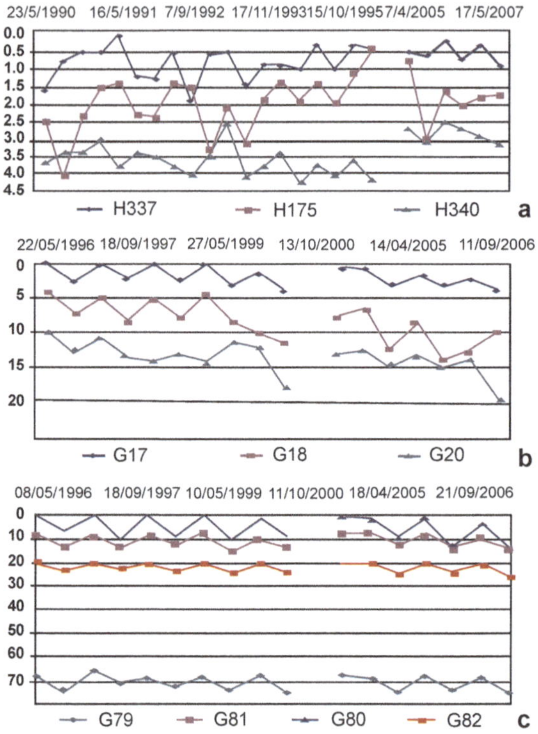

Fig. 2. Seasonal variations of the piezometric surface of the lower reaches of Aliakmona-Loudia rivers aquifer system. A=central part, B=east part and C= west part of the aquifer system.

In the past, the small flow speed of Loudias river contributed to the creation of the appropriate circumstances for the entrance of the sea to land during the tide, but this procedure was blocked by two dams which were placed within its bed (Panoras and Chatsigiannakis 1992). After the water supply of Aliakmonas river ($516*10^6$ m^3/yr) at the draining canals, which takes place every year at the beginning of the irrigation period during May, the piezometric surface height is increased. Long duration data (1990-2007) from water level measurements at 75 irrigation boreholes present a decline of 25cm/y of the mean regional piezometric level, a fact which is attributed to over pumping. On the contrary, at the central and southern part of the system and especially at the area between the rivers of Aliakmonas and Loudias, where the irrigation takes place from surface waters, even the upward of the piezometric level is noticed at some boreholes (Veranis 2010).

5 Water balance

Inflows: From the interpretation of the meteorological stations data of the broader area, it resulted that the mean yearly precipitations at the area of the aquifer system during the period 1980-2007 is 549 mm. The mean value of the real evapotranspiration is estimated to 412 mm/y. The sum of infiltration and surface runoff is 137 mm/y (Veranis 2010). The mean value of the infiltration calculations is based on the distribution of the rocks at the digital geological map and the bibliographical data (Soulios 2004) and is 67 mm/yr (11%). The surface infiltration is estimated by the equation of the water balance and is 70 mm/yr. Besides the immediate infiltration, the recharge of the aquifer system takes place due to the percolation of the surface waters of the drainage ditches, through the irrigation canals and the subsurface side percolation from the karst aquifers at the Paiko and Vermion mountains. An importance recharge takes place due to the irrigation water returns, which is about 15-25% of the consumption (ENM & Special Consultants 2008).

According to data from the lithological sections at 300 bore-holes, the hydrogeological mapping and the bibliography data (Soulios 2004), it is estimated that the mean value of the active porosity (n_e) is 7% (3 -18%). The renewable reserves (V_1) as well as the water quantities (V_2) pumped from the permanent reserves are estimated from the equation: $V=E*Ds*n_e$, where E=the area of the aquifer system, Ds=mean value of the piezometric level fluctuations per year. $V=1638km^2*2.32m*0.07=266*10^6$ m^3/yr. From these quantities the: $V_2=1.638km^2*0.25m*0.07=28*10^6$ m^3/yr are pumped from the permanent reserves and the: $V1=1.638km^2*2.07m*0.07=237*10^6$ m^3/yr from the yearly renewable reserves respectively.

Outflows: The outflows are divided into the quantities of water pumped and the natural discharges. The consumption of the ground water pumping is for the drinking water ($25.2*10^6$ m^3/yr), the industry, crafts ($8.8*10^6$ m^3/yr) and livestock ($2.7*10^6$ m^3/yr). The irrigated areas during 2007-08 at the aquifer system were about 900000 acres, from which the 500000 acres are irrigated by surface waters and the rest 400000 acres from pumping of ground waters. As far as the area is concerned the pumping of the ground waters for the irrigation takes place in the eastern, northern and western part of the aquifer system, while in the central and larger than the southern part are irrigated from surface or source waters which come from areas outside the aquifer system. Taking into account the distribution of the several cultivations in the area of study, the bibliographical data for the necessary quantities of water in relation to the kind of cultivation and the air temperature, with the method of Penman- Monteith, it is estimated that the mean quantity of waters needed is 550 to 580 m^3/acr, (ENM and Special Consultants 2008), so the consumption for the irrigation water is about $228*10^6$ m^3/yr. The result of over-pumping is the negative water balance, which is limited at the areas where the irrigations take place due to ground water pumping and it is estimated that they are about $28*10^6$ m^3/yr. Loudias river is the receiver of the natural discharges of the aquifer system, of the excess of the irrigation water of the canals and of the returns of the irrigation water to the aquifer and presents flow during the summer period only with a supply of 15-20 m^3/s.

6 Conclusions

The granular aquifer system of lower reaches of Aliakmona and Loudias rivers consist a multi-aquifer system confined to semi-confined, hosted within Quaternary and Neogene sediments.

At its larger part, it is characterized by an important aquifer capability with bore-holes yield of 40-300 m^3/h. The geometric mean value of the specific capacity of the bore-holes is 3.7 m^3/h/m (0.17-87.5), the transmissivity (T) is $1.2*10^{-3}$ m^2/s ($5.67*10^{-5}$ to $1.9*10^{-2}$) and the hydraulic conductance (K) is $4.1*10^{-5}$ m/s ($3.3*10^{-6}$ to $1.33*10^{-3}$) and the mean yearly fluctuation of the piezometric surface is 2.32 m.

The hydrological balance of the aquifer system is negative and is about $28*10^6$ m^3/yr. Due to the negative water balance of the aquifer system, the mean value of the piezometric surface decline is 0.25cm/y and the higher values are located in areas where the irrigation water derives from the groundwater.

References

ENM Experts and Special Consultants (2008) System Development Tools and Management of Water Resources Water of West Macedonia, Central Macedonia, Eastern Macedonia and Thrace Regions. Draft Water Resources Management of Water Districts (in Greek) (SAE061/3-20002 SE06130000)

Kresic, N (2007) Hydrogeology and Groundwater modeling, 2nd ed. CRC Press. 807 pp

Panoras, A. - Chatzigiannakis, S. (1992) Estimation of supply and water quality of the tidal river Loudias (in Greek). Hydrotechnica Vol. 2, Issue 1, p. 25-38

Rapti, D (1995) Hydrogeological investigations in Skydra area (in Greek) (Central Macedonia). Ph.D. Thesis, Geology Department, Aristotle Univ., 266 pp

Soulios, G (2004) General Hydrogeology (in Greek). STUDIO PRESS, Thessaloniki

Veranis, N (2010) Hydrogeological Study of the aquifer systems of Central Macedonia Region, Volumes 1 and 2A. CSF Project (2003-2009) "Inventory and Assessment of Hydrogeological Character of the aquifer systems of the country" (in Greek). RUCM-IGME, Thessaloniki

Estimation of Hydrological Balance of "Rafina's Megalo Rema" basin (Eastern Attica) and diachronic change of the surface water quality characteristics

P. Champidi, G. Stamatis, K. Parpodis, D. Kyriazis

Agricultural University of Athens, Laboratory of Mineralogy-Geology, Iera Odos 75, GR-118 55 Athens, pxampidi@yahoo.gr

Abstract This article deals with the geological and hydrogeological characteristics of "Megalo Rema" basin which flows into the Gulf of Rafina in Eastern Attica, the hydrological balance is estimated and the surface water quality is analyzed. The estimation of the hydrological balance brings up proportion of surface runoff R:121.4mm=28.2%, real evapotranspiration ET_{real}:243.9mm=56.5% which responds to Mediterranean values and groundwater flow I:66.1mm=15%. The monthly observation of the surface water quality indicates high concentrations during the dry period and in reverse the wet period. The waters are alkaline; TDS values range from 1100 to 1485 mg/L and belong to the hydrochemical types Ca-Mg-Na-Cl-HCO_3 and Ca-Mg-Na-Cl-HCO_3-SO_4. On the whole, the elements with geogenic origin are detected at the outfall of the river, while the trace elements deriving from human activities are detected in higher concentrations uphill of the river, by the pollution sources. The concentrations of Mn, Cr, Ni, Pb, Cd and Zn exceed the maximum limits of the European directive for drinking water. Their presence is attributed to geogenic factors, as well as to soiling by fertilizers and pesticides' application and by the area's urbanization.

1 Introduction

The "Megalo Rema" basin is located in the middle and north part of Mesogeia valley and the South part of Penteli's foot. After constructing the International Airport of Athens "Eleytherios Venizelos" the motorway "Attiki Odos" and the "Suburban railway", a demographic blowup has turned up (El. Stat 2001), which was sequent with the water need increase as well as the wastes quantities which end up to the environment, since all the towns of the area in issue are lacking of drainage system. Hydrochemical studies carried out in the past (Kounis 1979; Alexakis and Kelepertsis 1998; Kounis 1998; Stamatis et al. 2006; Champidi 2011) indicated the water deterioration due to human activities and bedrock dissolution. The significant reduction of the precipitation of the last decade (Baltas et al. 2010) has negative results in the replenishment of surface and ground water with fresh water, leading gradually to desertification (Kosmas and Danalatos 1994). The aim of the research is the estimation of the hydrological balance of the basin and the emergence of the physical and anthropogenic impacts which contribute to the surface waters' quality modification. The research was carried out within the framework of PhD Thesis, worked out in the Agricultural University of Athens.

2 Study area

Geomorphology: The study area covers 160 km^2 in the Southeastern part of Attica and is located, within Eastern region of South Euboea Gulf, within Northern and Western region of the masses of Penteli and Hymettus mountains correspondingly and within Southern section of the hill-sequence Spata- Artemis (Fig. 1). The relief is intense in the Northern bounds, where Penteli's slopes develop and smoother in the Southern part, which is covered by a big terrain of Mesogeia valley. The river "Megalo Rema" runs along the area, producing two major branches, one deriving from the Northeastern slopes of Hymettus, at Glyka Nera, and a second which is stored by Penteli's slopes. The two rivers join at Pallini, ending up in Rafina Gulf.

Geology: The study area is a part of the Attic-Cycladic massif and its geological structure is dominated by two systems: a) the crystalline basement, extended from the Paleozoic to the Upper Cretaceous and b) the clastic Neogene and Quaternary sedimentary nape (Fig. 1). The compact crystalline bedrock is consisted of schists and bnate rocks, specifically calcite and dolomite marbles. The Neogene and Quaternary formations are composed of marly limestones, marls, clays, sandstones, conglomerates and other incoherent materials. In general, the geological formations of Attica are classified into three series: a) the autochthonous series of schists and marbles composing the Triassic – Jurassic bedrock and the core of the Attica mountains, b) the allochthonous series of Cretaceous schists and sandstones and c) the Tertiary and Quaternary deposits (Lepsius 1983). The autochthonous series appear in the west margins of the basin and is composed of marbles (Upper and Lower Jurassic) and Kesarianis' schists (Jurassic) of about 800 m total thickness, while the formations of the allochthonous series are developed mainly in the north and locally in the south and west part of the Rafina's basin and are comprised of Upper Cretaceous limestones and schists, total thickness of about 250 m, with ophiolites and marbles intercalations. Thin ore layers of mixed sulfide minerals are identified in the base of the Cretaceous carbonate rocks. The major part of the basin is covered by Neogene sediments, marly and travertine limestones, marls, clays, marly sandstones, conglomerates, and other incoherent materials, total thickness of about 50 – 150 m. The terrestrial and fluvial Quaternary deposits (sands, gravels, clays and pebbles), cover up parts of the drainage and coastal network (Katsiavrias et al. 2007).

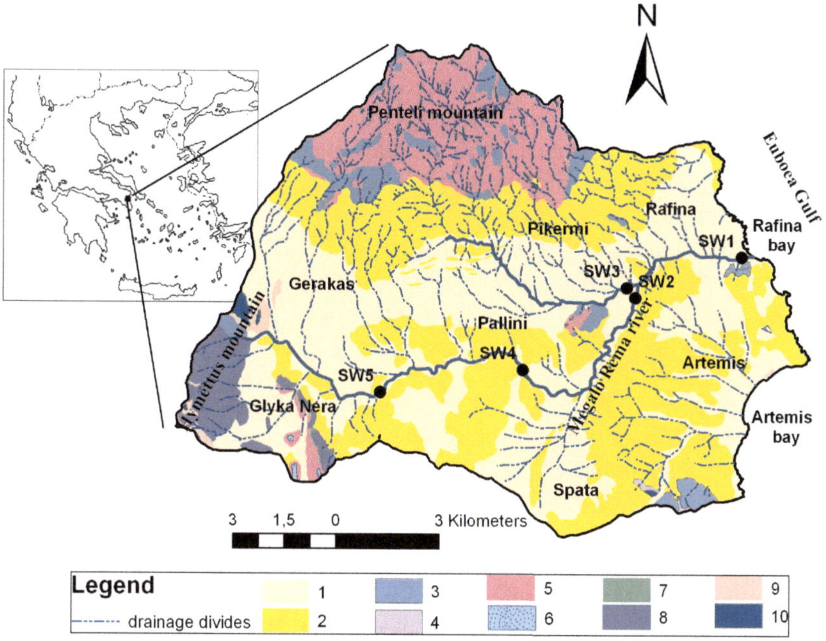

Fig. 1. A simplified geological map of the study area (modified by Katsiavrias et al 1977); 1: Quaternary; 2: Neogene; 3: Limestones (Upper Cretaceous); 4: Schist and Phyllites (Lower Cretaceous), 5: Schists (Lower Cretaceous); 6: Marbles (Lower Cretaceous); 7: Ophiolites (Lower Cretaceous); 8: Upper Marble (Jurassic); 9: Kesarianis Schists (Jurassic); 10: Lower Marble (Jurassic), Sw1 – Sw5: Sampling points of surface water.

The whole area is controlled by an E-NE and W-NW directed fault system, as well as by small appearances of NW-SE normal faults, that are responsible for the Mesogeia's basin displacement. The west part of the basin is exclusively controlled by the Hymettus' marbles, while the central part took advantage of the huge pre-alpine syncline folding, of NE-SW's direction. The following basin's evolution has been influenced by the postalpine brittle tectonics (Jacobshagen 1986; Katsikatsos 1992).

Hydrogeology: Into the loose Quaternary deposits and the Neogene units, it is developed an upper phreatic aquifer of relatively low capacity due to clay material presence which affects negatively their hydraulic behavior. The localized schists compose the impermeable bedrock. The aquifer's enhancement is retained by the instant infiltration of rainfall and by the lateral flow of the consolidated formations. The upper aquifer is utilized by a significant number of wells and boreholes, whose discharge rates don't exceed the 5 m^3/h for the wells, while for the boreholes they range between 15 and 35 m^3/h (Kounis 1979; Stamatis et al. 2000; Champidi 2011). The schists are impermeable, since their fractures are blocked by their own erosion products. The limited capability of leaching water which characterizes these formations results in the greater precipitation proportion going in favor of runoff. The carbonate formations, limestones and marbles, which are noticed at the basin margins, as well as the lenticular marble intercalations which are observed within the schists, comprise the signal area aquifer which is exploited by a big number of boreholes, with flow rates exceeding $50m^3/h$. In the coastal zone, the aquifers are in direct contact with the sea, which leads to their quality deterioration due to groundwater mixture with the seawater. In addition, the intensive exploitation of the aquifers for irrigation use has resulted in the seawater intrusion in the inland zone.

3 Hydrological Balance

The region is characterized by the typical Mediterranean climate of Attica. The mean annual temperature is 18.71°C, whereas the absolute maximum reaches 39.3°C and the absolute minimum -3.35°C. In Figure 2, are presented the parameter of the hydrological balance, as they came up by the estimation. The mean rainfall is 431.42mm with dry periods, usually extending from May to September (Hellenic National Meteorological Service, 2009). The annual hydrogeological balance for the period October 2005 - September 2006, was created based on weekly measures of the water flow in "Megalo Rema" outfall and by estimating the evapotranspiration based on the empirical type of Burdon and Papakis (1963) since a significant part of the geological formations belongs to carbonate rocks:

$$ET = \frac{P(\text{October} - \text{March})}{2} + P\,(\text{April} - \text{Sept})$$

where P is the monthly precipitation.

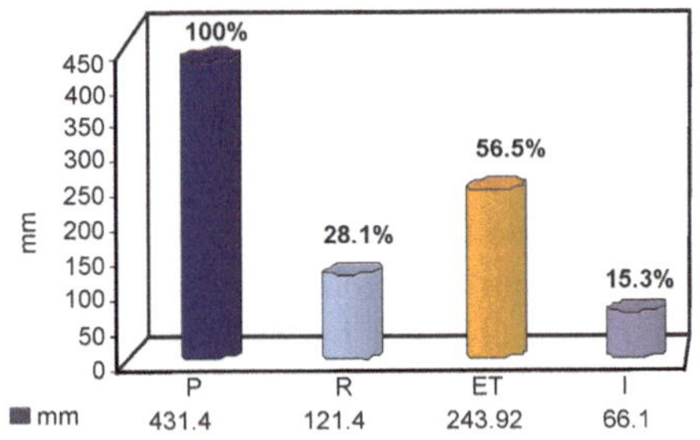

Fig. 2. The hydrological balance's parameters distribution.

According to the results, the surface runoff (R) accounts for a relatively high proportion, 28% of the total precipitation, the evapotranspiration accounts for 57% of the total precipitation which responds to Mediterranean values, and the groundwater flow (I) accounts for 15% of the total precipitation (Champidi 2011).

The relatively high noted surface runoff is attributed mainly to waste inflow from the urban areas and from the several industries working in the area.

4 Hydrochemistry

The surface water samples were collected once a month during the hydrological year 2005-2006 from 5 different points (SW1- SW5) as seen in figure 1. The temperature, the electric conductivity and pH were measured "in situ" by portable apparatus (WTW LF 330/SET and WTWT PH 330i/SET). The chemical analysis of water samples were carried out in the Laboratory of Mineralogy and Geology of the Agricultural University of Athens using titration methods, spectrophotometry (Hach model DR-3000), flame photometry (INTECH/420) and atomic adsorption (AAS GBS 908) methods. The following parameters were determined: Hardness, Ca^{2+}, Mg^{2+}, Na^{+}, K^{+}, HCO_3^{-}, Cl^{-}, $SO_4^{=}$, NO_3^{-}, NH_4^{+}, PO_4^{3-}, SiO_2 and the heavy metals Sr, Fe, Mn, Cu, Cr, Ni, Pb, Cd and Zn. In Figure 3, there are presented in diagrams the average concentrations of the major elements, for each sample point. The waters are slightly alkaline, TDS values range from 1100 to 1485 mg/L and they are part of the hydrochemical types Ca-Mg-Na-Cl-HCO_3 and Ca-Mg-Na-Cl-HCO_3-SO_4. The surface waters of the study area are subjected to quality deterioration, attributed to rock dissolution as well as to human activities. The physichochemical parameters and elements are, for the majority of them, higher than the upper limit of E.U. and W.H.O. guidelines for drinking water (EUC 98/83; WHO 2006). The concentrations of the major cations and anions are characterized by high values. The highest values are determined at the outfall of the major stream (SW1), indicating their geogenic origin and the summative accumulation downhill. The highest concentration of NH_4^{+} and PO_4^{3-} are also spotted at the outfall and are probably attributed to urban wastes and dripping of non septic sinks towards the major stream (Sikora et al. 1976; Kallergis 2000). The nitrates NO_3^{-} are determined with higher values uphill of the river and are attributed to nitrogen fertilizers' application, that's why they were detected to the transit points of the river, deriving from agricultural areas (SW2, 4 and 5). In the branch which is supplied by the metamorphosed units of Penteli (SW3), the concentrations were significantly lower.

The chemical synthesis of this branch, which is supplied by the metamorphosed units and the weathered mantle of Penteli is significantly different from the other samples due to different lithology and the limited land use. The majority of the trace elements excluded Fe and Cu, present high concentrations, overcoming the upper limits of the directive EUC-98/83.

The heavy metals' concentrations are detected lower at the outfall and higher uphill of "Megalo Rema" stream. Their intense presence has to do probably with the soil deterioration, occurred by the mixed sulphure rocks' metallifery (Alexakis and Kelepertsis 1998; Serelis et al. 2010) and by the application of fertilizers, pesticides and soil improving compounds (Kabata-Pendias and Pendias 1992; Ross 1994; Alloway 1995; Siegel 2002; Kabata-Pendias and Mukjerjee 2007). Corresponding the monthly elements' concentrations with the water flow and summing up the monthly values, there were calculated the total carriage polluting loads of the river towards Rafina Gulf.

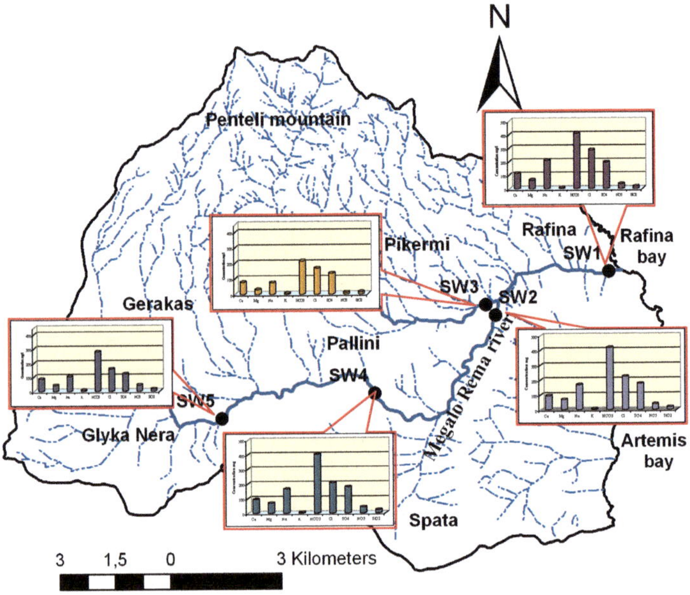

Fig. 3. Quality characteristics of the surface waters.

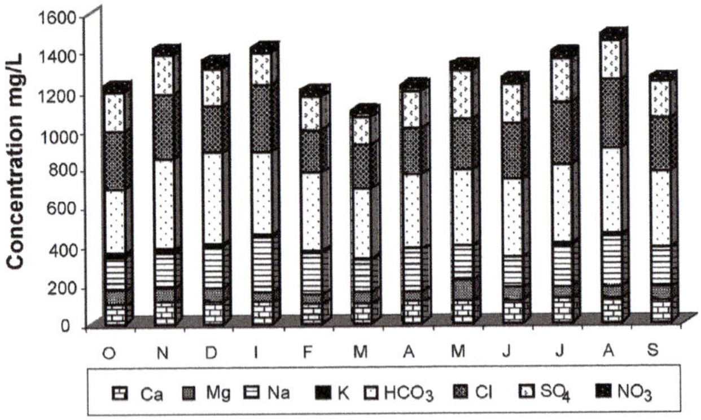

Fig. 4. Concentration values of the major elements' range at the outfall of "Megalo Rema" (point SW1).

It is noted that every year, within "Megalo Rema" river, there are transmitted significant loads of nutrient elements NO_3^-: 284.6 tons, NH_4^+: 7.91 tons and PO_4^{3-}:1.7 tons, which contribute to eutrophication occurrence in Rafina Gulf.

5 Conclusions

From the hydrological and hydrochemical research, carried out in the basin of "Megalo Rema" in Eastern Attica, the following conclusions can be drawn: i) The area is characterized by the typical Mediterranean climate with high temperatures, low precipitation and long-term drought periods ii) Out of the total of 431.4mm annual precipitation, the roof 56.5% is slipped by evapotranspiration, 28.2% outpours as surface runoff and the remaining 15% goes for the groundwater flow, iii) The relatively high surface runoff is attributed to waste inflow from the urban areas, the agricultural applications and mainly from the several industries working in the area, iv) The surface water quality is characterized by seasonal fluctuations and intense peripheral impacts. The anthropogenic factors are illustrated with the vivid presence of the ions NO_3^-, NH_4^+ and PO_4^{3-}, Cl^-, SO_4^{2-} Na^+ and K^+ and the heavy metals, v) From the monthly observation of the quality characteristics it is noted that the higher concentrations of elements with geogenic origin are found at the outfall while the polluting elements are found with higher values near the pollution source, vi) The waters are alkaline, TDS vary between 1100-1485 mg/l and have unspecified hydrochemical type Ca-Mg-Na-Cl-HCO_3 and Ca-Mg-Na-Cl-HCO_3-SO_4, vii) High concentrations of heavy metals are observed and some of them (Mn, Cr, Ni, Pb, Cd and Zn) exceed the upper limits of the European directive for drinking water, viii) Within "Megalo Rema" there are annually transmitted significant loads of nutrient elements NO_3^-: 284.6 tons, NH_4^+:7.91 tons and PO_4^{3-} :1.7 tons, which contribute to eutrophication occurrence in Rafina Gulf.

References

Alexakis, D and Kelepertsis A (1998) The relationship between the chemical composition-quality of groundwater and the geological environment in the East Attiki area, Greece, Mineral Wealth 109: 9-20

Alloway BJ (1995) Heavy metals in soils 2nd edition, Blackie Academic & Professional, London, p.368

Baltas EA, Dervos NA and Mimikou MA (2010) Impact of changing rainfall conditions on surface and groundwater resources in an experimental watershed in Greece. Global Nest, 12:2, 119-125

Burdon D and Papakis N (1963) Handbook of karst Hydrogeology with special Reference on Carbonate Aquifers of the Mediterranean Region. I.G.S.R., p. 276, Athens

Champidi P (2011) Natural and anthropogenic impacts on water and soil quality of East Attica, PhD Thesis, Agricultural University of Athens, Department of Science, in press. (In Greek)

E.U. Council (1998) Council Directive 98/83 about water quality intended for human consumption, in Official Paper of the European Communities: EC, Brussels, L330, 32-54

http://www.statistics.gr (Webpage of Hellenic Statistical Authority)

Jacobshagen, V (1986) Geologie von Griechenland.- pp 363, Berlin-Stuttgart, (Borntraeger)

Kabata-Pendias, A. and B. Mukherjee.2007. Trace Elements from Soil to Human, Springer-Verlag, Berlin Heidelberg. p.550

Kabata-Pendias, A and Pendias H (1992) Trace Elements in Soil and Plant, 2nd ed. CRC Press, Boca Raton, Ann Arbor, London. p.365

Kallergis, GA (2000) Applied Hydrogeology, vol. B- Environmental Hydrogeology (2nd ed.), Technical Chamber of Greece, Athens. p. 331 (In Greek)

Katsiavrias N, Papazeti E, Zorapas N, Stefouli M (2007) Geological map of Mesogeia, Attica (Issue Plaka-Koropi) 1:50,000. Institute of Geology and Mineral Exploration, Athens

Katsikatsos, G (1992) Geology of Greece - 451 pp, Athens

Kosmas CS and Danalatos NG (1994) Climate change, desertification and the Mediterranean region. NATO ASI Series, Soil Responses to Climate Change, M.D.A. Rounsevell and P.J. Loveland (eds., Springer-Verlag, Berlin Heidelberg, Vol. I 23:25-38

Kounis, G (1979) On the hydrogeological conditions and the capability of watering the new Athens airport in Spata, G 1666, Y 867, Athens. (In Greek)

Kounis, GD (1998) A study for the hydrogeology of Attica. Institute of Geology and Mineral Exploration, Athens. (In Greek)

Lepsius, R. (1893) Geologie von Attica. Ein Beitrag zur Lehre vom Metamorphismus der Gesteine. 1956, Berlin 1893 and translated by G.Bougioukas (1906). Maraslis Library, Athens, p. 592

Ross SM (1994) Sources and forms of potentially toxic metals in soil-plant systems. In: Ross S.M. (eds) Toxic Metals in soil-plant systems. Wiley, Chichester, pp. 4-25

Serelis KG, Kafkala IG, Parpodis K and Lazaris S (2010) Anthropogenic and geogenic contamination due to heavy metals in the vast area of Vari, Attica. Bulletin of the Geol. Soc. of Greece, XLIII: 5:2390-2397

Siegel, FR (2002) Environmental Geochemistry of Potentially Toxic Metals. Springer-Verlag, Berlin Heidelberg, p. 218

Sikora LG, Bent MG, Corey RB and Keeney DR (1976) Septic nitrogen and phosphorus removal test system. Groundwater.14:5, 309-314

Stamatis G, Gatsis I, Psomiadis E and Parcharidis I (2000). Satellite data and GIS to the verification of coastal water pollution in Vavrona Gulf (Attica, GR). Proceedings of Iternational Conference "Protection and restoration of the environment", Thassos, Juli 2000, A:535-542

Stamatis G, Lambrakis N, Alexakis D and Zagana E (2006) Groundwater quality in Mesogeia basin in eastern Attica (Greece), Hydrogeological Processes, 20, 2803-2818

WHO (2006): Guidelines for drinking water quality.-vol. 1, 3d eds, World Health Organization, Geneva. p. 515

Karst Hydrogeology

Dynamic Characteristics of Soil Moisture in Aeration Zones under Different Land Uses in Peak Forest Plain Region

F. Lan[1,2], W. Lao[2], K. Wu[2]

[1] Institute of Wetland Ecology, School of Life Science, Nanjing University, Nanjing, 210093, China

[2] Institute of Karst Geology, Chinese Academy of Geological Sciences, Guilin, 541004, Guangxi, China

Abstract In order to research soil moisture of a limestone peak forest plain in Laibin, Guangxi Province, the investigation had chosen three lands, a maize field covered by mulching film, a common maize field and a set-aside land, from March, 2006 to April, 2007, measuring soil moisture content and potential at the depths ranging from 0-450 cm in profiles by neutron probe and tensionmeter. The results showed that the content of soil moisture in aeration zone varied intensely at depth of 10 cm, and attenuate at 100 cm and 450 cm. On the other hand, soil moisture content and dynamic variation presented seasonal features, Between Mar. and April, 2006, the mean soil moisture content was 36.7 %, which was the biggest value in that year. In this period, it was general higher and increased progressively. From May to August, the mean was 34.1%, and both maximum and minimum appeared during this time. From September to November, the mean was 32.6%, mainly showed a trend of decreasing gradually. From December 2006 to February the next year, the mean was 31.4%, which was the lowest value in that year. The ZEP of soil moisture in aeration zone existed 338 days during one year. Depth of ZEP, from November 2006 to February 2007, was 170cm, and less than 100cm in the other months. In all observation time, transpiration of soil moisture was 865.66mm, accounting for 88.5%. And infiltration was 145. 47mm, taking up 11.5%.The field capacity was 30.3%.

1 Introduction

The moisture of aeration zone was the belt associating surface water with groundwater. Soil moisture was considerable part of terrestrial water body. Within global freshwater resource, soil moisture content of aeration zone was 1.65×10^{13} m^3, more than 5×10^{11} m^3 flux of rivers (Korzom 1978), and had the intensive correlation with human production and livelihood. Meanwhile, it was source maintaining terrestrial plants alive and fundamental condition which was essential in agriculture. Laibin, Guangxi located in limestone peak forest plain. Owing to the extraordinary surface and dual-structure underground, surface water availability was reduced and overland flow frequently flown into underground down swallow holes and bergschrund (Yuan Daoxian and Cai Guihong 1988). Therefore moisture had significance in cultivating crop which can plant in arid in Laibin. It had also remarkable sense for directing agriculture productions, researching moisture dynamic of aeration zone in limestone area. Correct evaluation on water resources had the same importance. The study area is located in the middle of Guizhong basin, Guangxi, 2 km far away from Laibin city, at elevations of 80 to 130m. There were two soil types, Mllrendzina and Braunrendzinea. The pH values ranged from 6.5 to 7.5. The total annual solar radiation was up to 1750.2 h. Average temperature was 20.65℃ over the year, Jan. (10.9℃) lowest and Jul. (28.6℃) highest. Mean precipitation was 1347.8 mm, most 2060.6 mm in 1994 and least 820.3 mm in 1989. There was more rainfall from May-August, occupying 62.3% of rainfall throughout one year. Accumulate temperature days is 297, required daily mean temperature >10℃. The climate characteristic, synchronization between heat and humidity, provided well conditions for crop, mainly containing rice, maize and sugarcane.

2 Methods

Fix a system respectively used to monitor moisture aeration zone in a maize field covered by mulching film, a common maize field and a set-aside land. Each system consisted of 4 parts; 1) a neutron probe, 5 m depth; 2) tensionmeter zone (24), 5 m depth of profiles;3) water lifting zone (14), 5 m depth; 4) non-weighting lysimetre (6.67 m^2 size, 2.5 m depth) and observation room which were underground. The points, observing soil moisture content and potential, were sited down profiles, 5 m depth. The arrangement rules as follows; 1/10 cm within 1m, 1/20 cm ranged from 1-3 m, 1/50 cm >3m. Frequency was 1/day in the period of rainfall and growing, 1/2 days in arid and non-growing. Deviation of potential was within 1 mm, and error between two records on moisture content by neutron probe inclined 5 units at the same time and point.

3 Results and discussions

3.1 Variation of space and time on moisture and potential

Soil moisture was the fundamental requirement to crop growth particularly in arid area. It measured the ratio between water contained and soil weight. Soil potential described energy moving least moisture from some level to standard reference one, which represented unit amount. And it indicated how much water was available to pant (Jing Enchun 1994). Investigate and record them via the instruments within 0-450 cm, in the three lands beginning March 2006, ending April, 2007. Using software SPSS to statistic and analysis the data (Table 1 and Fig. 1).

Table 1. Variation of moisture content in different depth.

Depth (cm)	Type of soil	Maximal content (%)	Minimal content (%)	Amplitude varied (%)
0-10	Gray-black, Cultivate soil	34.6	13.9	20.7
10-140	Brown-yellow, gravelly clayey silt	39.1	21.6	17.5
140-320	Brown-red, gravelly clayey silt	49.3	36.1	13.2
320-450	Brown-red clay	50.4	45.9	4.5

Soil physiochemical properties varied with depth of profiles from Table 1. It had no well diversion down 10-320 cm, conversely, maximal and minimal moisture had risen successively 9.8%, 14.5%.which showed that depth was main domination in the middle of profiles. From cultivate soil to gravelly clayey silt and to clay, maximal and minimal soil water respectively increases 15.8% and 32%. So, for the whole aeration zone, soil property is more important than depth of soil profile, the former influence more than the latter.

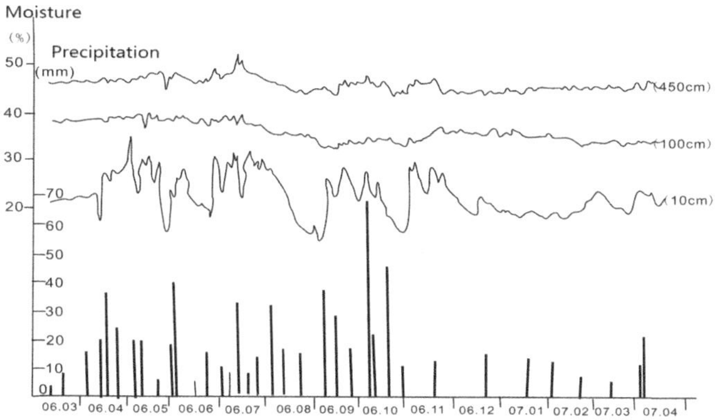

Fig. 1. Variation of moisture of soil profile in one year (from Lao 2011).

Diversification of moisture content was fierce at 10 cm, and undistinguished at 100 and 450 cm. The weaker intensity was, the deeper profile was. Otherwise, its variation was affected by precipitation. Because of abundant rainfall between April and October, 2006, moisture content varied under different depths, more marked at 10 cm than at 100 and 450 cm. The less amplitude of potential modified, the deeper profiles were. Ranged from 0-1 m, diversification was strong, mean -475~ -70cm H_2O. Mean was -250~ -160cm H_2O and -360~ -340 cmH2O, which belonged to transition (1-3 m) and relative stable zone (3-4 m) respectively.

Moisture content and regime were outstanding seasonal in varied depths against time. Precipitation and times increased from March to April, 2006, but less amount and briefer interval. Average content was 36.7%, max within one year. Max amplitude varied reached 17.5%. For more rainfall, larger humidity and lower temperature in the period, the infiltration caused by rain was more than depletion, especially the soil above aeration zone. Therefore, the content was larger generally throughout zone in the time, and laid out the tendency rising gradually.

In the autumn (September - November), times and amount decreased rapidly. However, evaporation remained intensive, and mean moisture was 32.6%. Moisture had the trend low gradually, dominant. Its regime was impacted mainly by individual rainfall, and extent modified turned smaller, 15.4% max.

In winter (December - September next year), due to more wind and less rainfall, it was arid. The moisture maintained evaporation, and amount was smaller and less gradually in every level throughout zone. Mean moisture was only up to 31.4%, and the moisture variation did least in the period, 5.8% max. In the meantime the shortage did heaviest and mean was least.

3.2 Feather of moisture ZFP in aeration zone under variation land-use types

The soil water moved following Darcy's Low on unsaturated flow. The flux could count through Partial Differential Equations under the conditions, moisture was incompressible and no supplement, and its density should be constant (li Baoqing et al. 1990,Li Qian et al.).

$$Q(z_2)-Q(z_1)=\int_{z_1}^{z_2}\theta(z,t_2)dz-\int_{z_1}^{z_2}\theta(z,t_1)dz \qquad \text{equation (1)}$$

In the left of equation, $Q(z_2)$, $Q(z_1)$ parented the total volume in per cross-sectional area, when soil water flowed separately through Z_2 and Z_1 from t_1 to t_2 . calculus equaled moisture changes between Z_2 and Z_1 on right of the equation, θ mean moisture, $\theta(z,t_1)$, $\theta(z,t_2)$ presented respectively water cut curve of soil profiles; $\theta(z, t)$ could measure by neutron probe.

The moisture flux for all positions, q, could be defined through unsaturated permeability coefficient and water potential gradient in correlate site.

$$q=-k(\theta)\partial\varphi/\partial z \qquad \text{equation (2)}$$

In above equation, $k(\theta)$ was unsaturated permeability coefficient, φ was water potential. If the water potential gradient equaled zero, $\partial\varphi/\partial z = 0$, the flux was zero, site of ZF, making up Zero Flux Plane, ZFP. Usually, the water potential distribution was continuous function, correlation with depth. So, the location of points, which required $\partial\varphi/\partial z = 0$ and sited in the curve of water potential distribution ($\varphi(z,t)$), was that of ZFP. According to it, acquire the amount of crop water requirement.

Summing up, the formation and development of ZFP had the following characteristics: (1) Since alternation frequent between humid and arid, layer permeability strong, large limestone developing, the interval was short when ZFP reappearing and missing, stability poor and depth shallow. (2) During the period of ZFP formation and down growth, precipitation was more. For example, in the maize grown time, total was 510.4 mm. When the amount overtook 176.3 mm, it should destroy ZFP. And the ratio it happened went up to 34.5% of all. The volume was 334.1 mm, rainfall percolated into soil above ZFP. It held 65.5%, what could improve the soil moisture storage and have superiority to maize growth. But, it cannot infiltrate down into underground to form water resource of limestone.

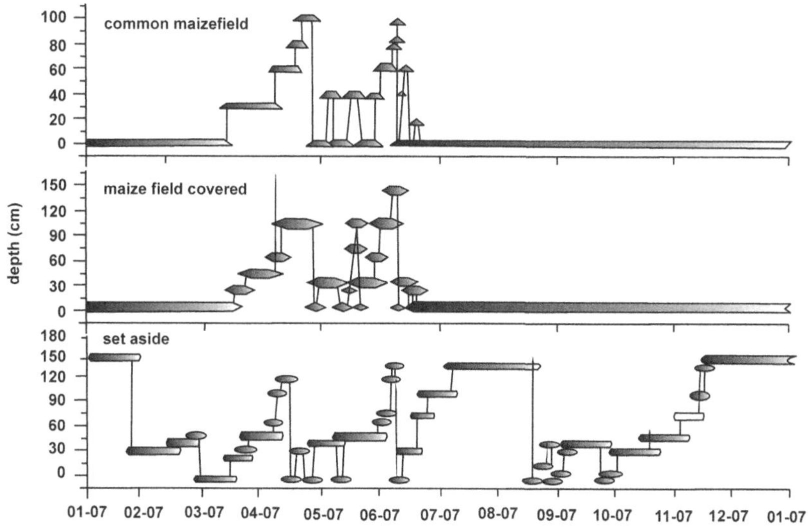

Fig. 2. The depth of ZFP under different land uses (from Lao 2011).

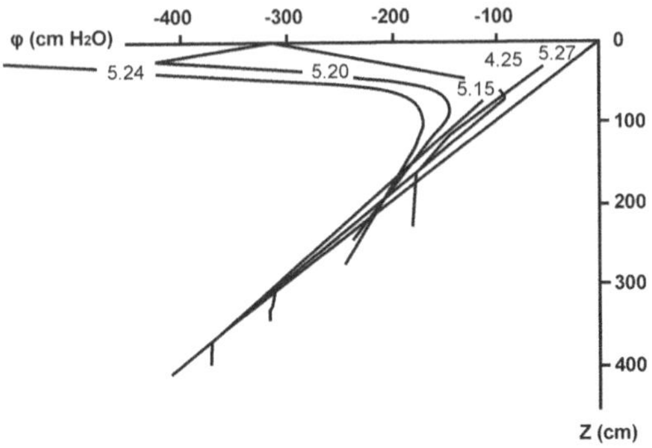

Fig. 3. Motion of Zero Flux Plane of soil (from Lao 2011).

3.3 Feature of soil moisture in evaporation and seepage and calculation on soil field capacity

Observe the evaporation and seepage of soil moisture, though the analogue non-weighting lysimetre, settled in the experiment site (Fig. 3). Use the fundamental theory; keep the water level stable in the overhead of threshold (1.9 m) what was the limit of phreatic evaporation in the site. It equaled the lowest depth of ZFP evaporation summed the altitude of capillary tube sustaining up, 1.7+0.2. When the soil moisture moved down but cannot reach the threshold, the water would be evaporated; on the contrary, it would recharge the groundwater in aeration zone.

The statistical results displayed through consistently observing; the number was 229, in which soil moisture could sustain evaporative, and it accounted 70% of all. The largest evaporation emerged in a range from July, 2006 to January, 2007.
The number of day rainfall was few, intensity weak, interval of consistent rain short, period of continuous drought prolonged, evaporation frequency high. The number of day was 48, when soil moisture percolating down to recharge underground water of aeration. Its rate was 14.7%, and it happened in June, October 2006 and March 2007. The numbers of days were 16, 11 and 6 successively. Mean infiltration monthly was 13.10 mm, 90 mm most, 0 least.
Field capability equaled the most soil moisture content when gravitation all was discharged. It was also the soil storage within 24 h after some intensity rainfall, and varied due to the differences of soil types and humus content. It would stable relatively to the same type of soil, and was a remarkable index of soil retention. Meanwhile, it was important to accounting soil resource available. When the soil moisture content was below 70%-75% of field capacity, the field should be irrigated to maintain the normal growth. Hence, it, studying fiddle capacity, had the realistic significance to guide developing of local corp.

Table 2. Field moisture capacity in different depths of soil profile.

Type	Time of rainfall	Precipitation (mm)	Depth of profile (cm)									
			10	20	30	40	50	60	70	80	90	100
			Field capacity (%)									
Zone covered	26th, May	51.1	24.2	28.3	25.9	23.5	24.7	26.5	24.8	24.4	24.9	24.1
	7th and 8th, July	47.4	23.0	32.3	29.0	28.5	29.3	30.0	29.2	27.8	27.8	27.7
	14th and 15th, July	38.4	23.0	26.8	25.9	25.0	26.2	27.1	26.1	26.1	24.4	25.4
Set-aside	26th, May	51.1	29.4	32.5	31.9	31.8	33.4	36.1	36.6	37.4	37.7	36.8
	7th and 8th, July	47.4	28.3	31.7	33.0	33.5	34.3	36.5	37.8	37.9	36.7	37.4
	14th and 15th, July	38.4	28.8	30.5	30.4	31.6	33.2	35.3	36.1	36.7	36.7	36.5
Zone common	26th, May	51.1	27.5	29.6	30.3	30.0	28.7	29.4	30.4	33.4	34.3	35.2
	7th and 8th, July	47.4	25.0	29.5	30.8	29.7	28.8	29.2	30.5	31.5	34.1	34.1
	14th and 15th, July	38.4	27.0	28.0	28.1	28.6	28.4	28.7	30.6	32.5	33.5	34.0

4 Conclusions

There were some characteristics of soil moisture in aeration zone in limestone peak forest plain. As follows:

- From up to down, water contents and water potential of the aeration zone vary sharpest at 10-cm depth. Soil moisture content and dynamic variation presented seasonal features, Between Mar. and Apr., 2006, the mean soil moisture content was 36.7 %, which was the biggest value in that year. So, in order to promoting the harvest of upland crops, it is necessarily to irrigate in this period.
- The development of ZEP of soil moisture depth was varied with different land-use types. 20-100cm in the field covered, 30-140cm in common field and 10-170cm in set-aside land. On the other hand, there was a close correlation between development of ZFP of soil moisture depth and rainfall. The less rainfall was, the deeper ZFP of soil moisture depth developed.
- The soil water evaporated from the second half of the year to January of the next year. During this period, high temperature and little rainfall lead to a total 542.22mm evaporation, which occupied 62.6% of soil evaporation throughout our observation period. At the same time, the characteristic of field capacity varied in different land-use types. The mean field capacity was 26.4% from 0-100cm in field covered, 30.4% in common field and 34.2% in set-aside land. Along the profile, it kept well stable from 0-100cm in the field covered. This feature is beneficial to crop growth.

Reference

Korzom V I (1978) World Water Balance and Water Resouces of the Earth[M]. Paris, UNESCO

P.R. China. Environment science of limestone. Chongqing (1988)

P.R. China. Experimental research on soil moisture flux. Beijing (1994)

Baoqing Li, Changming Liu, Keding Yang (1990)Application study on ZFP. Geography Research., 9:39-50

Qian Li, Jun-jie Leng, Peiling Gao (2006) Study on Field Evapotranspiration by ZFP. Journal of Arid Land Resources and Environment. 20(2):176-179

Situation and Comprehensive Treatment Strategy of Drought in Karst Mountain Areas of Southwest China

X. Qin, Z. Jiang

Institute of Karst Geology, CAGS, Guilin, Guangxi, P.R. China. 541004)

Abstract There were a long term drought in the karst areas of southwest China from 2009 Autumn to 2010 Spring, which reached severe levels of drought once a hundred years, The drought distributed in southwest of China, with a total area of 100,0000 km^2. 4.3486 million hectares farmland was affected by drought. The affected population hit 51.049 million ones, led to serious problems of drinking water for 16.09 million people and 11.055 million livestock. Besides in the karst area, water often leak into underground rivers. The main four factors contributing to the severe drought are the climate getting drier, rocky desertification, much hydraulic engineering and heavy water pollution. To mitigate the drought and water shortage in southwest China, China government collected huge fund, technique people and equipment in order to drill water-wells for water supply, which play an important role for the drought-resistance. Moreover, we formed many successful examples of exploitation of karst water. However, we should have a complete treatment strategy of drought in future. Systematic regional hydrogeologic survey in scale 1:50000, construction of water conservation and water sources, scientific management of water resources in karst basin and emergency response system of the drought-resistant are the key aspects.

1 Introduction

In recent years, with the climate changes and the increasingly ecological environmental disruption, as a kind of natural disaster, the drought becomes more and more serious, which not only has produced a grave impact on agricultural production but also has led to serious problems of drinking water for people and livestock (Zhu et al. 2006). In China, the drought mainly occurs in North China Plain, Northwest China and Southwest China, but there is a tendency to get drier in Southwest China. Ma Guozhu et al. made an analysis of drying trend over China from 1951 to 2006 (Ma and Ren 2007), by using the monthly precipitation and monthly mean air temperature data set from 160 observational stations over China in 1951-2006.The results showed that: in Southwest region, the precipitation and surface wetness index presented a linear decreasing trend, and the precipitation and surface wetness index greater than the linear trend of the North China plain and Northwest, -0.042/10yrs and -30.4 mm/10yrs respectively. In southwest China occured a long term drought from 2009 Autumn to 2010 Spring, which reached severe levels of drought once a hundred years. Total drought area is about 100,0000 km^2.The severe drought was mainly occurred at east of Yunnan, southwest of Guizhou and northwest of Guangxi, as shown in Figure 1. Geomorphologically, the heavy droughts occur mainly in peak cluster depression, karst plateau, karst tectonic basin and karst valley terrains, and the area with rock desertification (Fig. 2).

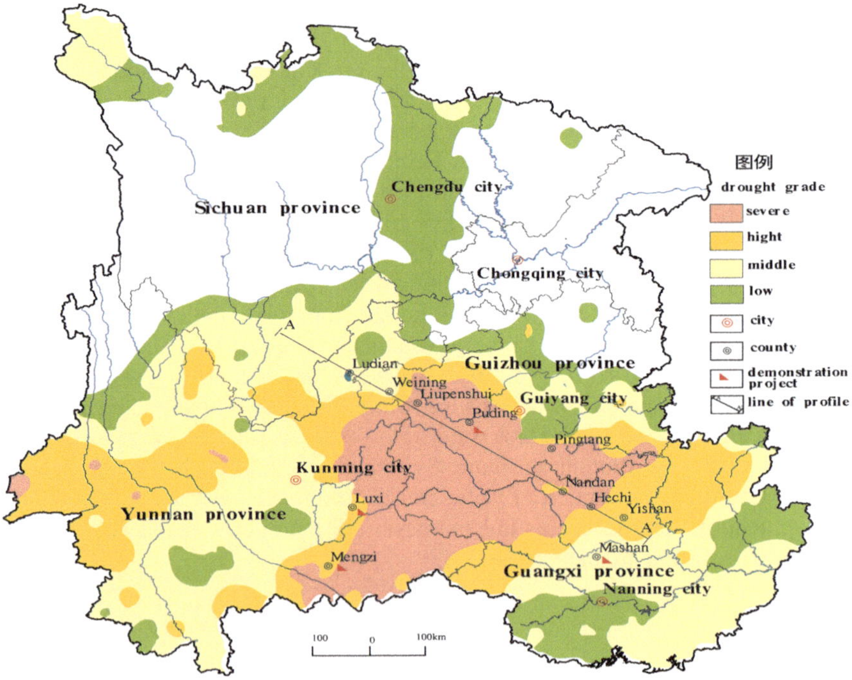

Fig. 1. Distribution of the drought area in Southwest China (from Qin 2011).

The causes of frequent drought in Southwest China have been some studied. In addition to the climate anomalies, there are other three important causes: One here is contiguous karst areas, with 434,300 km^2, and double spatial structure formed after prolonged intensive karstification. Although the average annual rainfall is 1000-1800mm, surface water leaks into the ground soon after the rain, making surface dry. This type of drought is called “Karst Drought”. Second is uneven distribution of rain, more than 70% rainfall occurs in the spring and summer seasons, seldom rainfall in the autumn and winter seasons. Third is economic ackwardness in the karst areas and the local people suffered from shortage of funds and equipments for mining of groundwater.

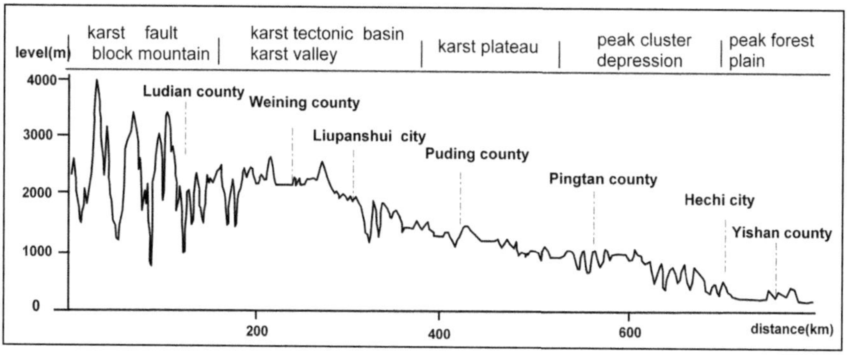

Fig. 2. Geomorphologic profile of the drought area (from Qin 2011).

However, why the drought becomes more and more serious in the karst area of Southwest China? How to deal with the severe drought? No good answer so far. In this paper, the authors introduce the distribution, characteristic and situation of the 2009 -2010 drought in the Southwest China, show some examples of drought treatment, and try to get causes of the drought and the effective countermeasures.

2 Cause analysis of drought in Southwest China

2.1 Atmospheric circulation anomalies lead to climate drought

The rainfall in the Southwest China mainly comes from water vapor transmission of Indian Ocean and Bangladesh Gulf. However, activity of India - Burma geo-tectocline was very weak and disadvantage of water vapor transmission in 2009. In addition, since autumn of 2009, The Tibet plateau created a strong cold high pressure gas and blocked the warm-moist air flow coming from the Indian Ocean and Bangladesh Gulf. On the other hand, the northern cold air could not reach Yunnan-guizhou plateau, cold air and warm air are hard to exchange and form precipitation.

2.2 Large-scale water conservancy project in upstream lead to the situation of water resources in downstream severely

With the industrial and agricultural development and population growth, in order to meet the water needs, more and more water accumulation structures were built in upstream, which would lead to water shortages in downstream.

2.3 Desertification leads to the lowering of water regulation and storage capabilities of ecological environments

Karst rocky desertification lead to forest coverage reduced. In Guangxi, Guizhou and Yunnan province, where karst rocky desertification is serious, the average of forest coverage is only 21.0% in karst area, which is 50.6% of the non-karst areas. For serious degradation of vegetation, with double space structure on karst surface and underground, and funnel, cracks and underground drainage network developed dramatically in karst area. Surface runoff quickly import underground river system and flow away, so its infiltration coefficient is higher, which can be as high as 0.5-0.6 in bare peak and depression terrain. Moreover, with surface soil depleted, the ability of surface water conservation is severely reduced, and water holding capacity is poor (Jiang and Yuan 2003).

2.4 Water pollution worsening the shortage of drinking water in karst area.

The development and utilization of water resources is also impact by serious pollution of surface water and groundwater in some areas. "Three wastes" produced by industrial and mining enterprises and solid waste pollutants stacked on the ground, which will infiltrate into the underground river under the leaching process through carbonate's surface dissolved ditches, erosion fissures, sinkholes, collapse pits, cracks etc. In some places, sewage even directly discharged into the underground river.

3 Some examples for exploitation of karst water

Although surface water in karst areas of Southwest China is generally deficient, its groundwater resources are abundant. According to the data conducted by the China Geology Survey, there is a total $1850 \times 10^8 m^3/yr$, most of them keep in underground streams. Moreover, there are many epikarst springs and fissuring water. So to exploit groundwater resources reasonably in the light of karst geological and geomorphologic characteristics and conditions are effective drought- resistant measures (Shi et al. 2005).

3.1 Drilling underground water for water supply

To mitigate the drought and water shortage in southwest China in spring of 2010, China government collected huge fund, technique people and equipment from other provinces and cities to come there In order to construct drilled and dug water wells. For example, the Ministry of Land Resources organized more than 10,000 technique people from 17 provinces and cities come there drilling underground water for water supply. They totally drilled 2703 bore holes, formed 2348 wells, get water 360,000m^3/d and resolve the drinking water problems of more than five million people.

3.2 Exploit epikarst water resources in Nongla

Nongla village of Mashan County, Guangxi, is a high peak-cluster depression where groundwater buried deep more than 300m. Exploitation of groundwater is difficult and expensive for the farmers who live scattered. Building water tanks in the exit of epikarst spring to store water, and use diversion canal for agricultural irrigation or water pipe for life. In solving the problem of water, local people made agricultural structure adjustment here. Besides planting corn in the depression, rehabilitation of karst hills is undertaken by severely limiting human activities on the summits of karst cones, so permitting forest regrowth; by mid-slope contour terrace construction that permits farming yet limits downslope wash of soil; and by protective cover around the foot of hills to permit sustainable high agricultural production. 30 years late made the peak-cluster depression become a good forest park, with some luxury buildings.

3.3 Wulicong Doline Reservoir

Wulicong Doline Reservoir is in Mengzi County, southern Yunnan, It is on a karst tectonic basin which is the biggest dry karst land in southern Yunnan underlain by Triassic and Devonian carbonate rocks and Cambrian clasolite rocks. With the curtain grouting in karstic belt to form a reservoir in Wulichong doline, storage of 79.49 million m^3 of water commenced 1995 (Wang 2006). In combination with a long aqueduct (seven artificially draw water tunnels, with total 9.5 kilometers long), which delivers water to it from Nanxi River, Since 1997 began water delivery to Mengzi city, 8161 x 10^{4}m^3 in water supply every year (Fig. 3).

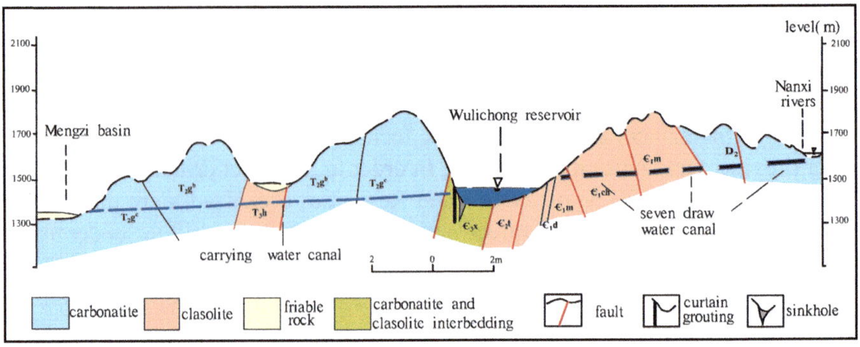

Fig. 3. Profile map of the Wulicong Doline Reservoir (from Qin 2011).

3.4 Maguan joint reservoir with surface water and ground water

Maguan town ship is in the middle Guizhou province. It is on a karst plateau underlain by Triassic carbonate rocks, stopping underground streams in cave exit to form a joint reservoir with surface water in Chongtou depression, and ground water, with a long aqueduct, which delivers water to it from Yangpizhai depression. Achieved the storage of $10^3 \times 10^4$ m^3 of water in low year (reliability p=95%) (Wang 2006). In addition, the project prevented inundation for Maguan basin. This was shown in Figure 4.

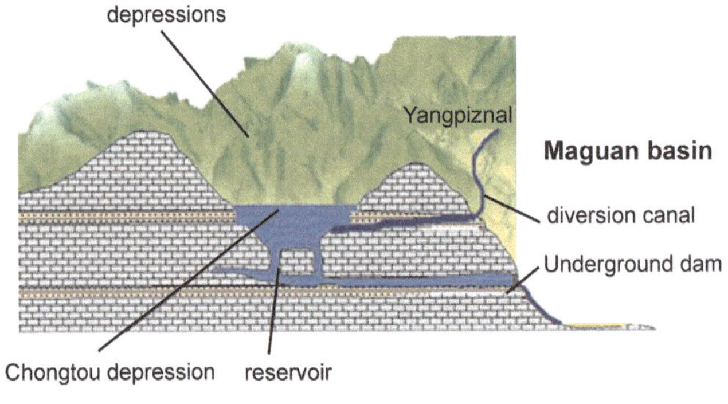

Fig. 4. Profile of the Maguan joint Reservoir (from Qin 2011).

4 Conclusions and discussion

The karst of South China is a fragile environment controlled by topographical and geological conditions, as well as human activities. As a consequence of complicated geological structure, stratigraphic features, and a deeply dissected drainage system, it is divided into many small karst basin units with different groundwater circulation system. To mitigate the drought and water shortage in karst regions, simply increase investment or drilling water engineering are not enough. We should find out the effective way. At first, we should carry out hydro-geologic survey to understand the karst water's types, resources and circulative characteristic, particularly the burial conditions, store rules and way of recharge, runoff and discharge in different types of karst groundwater, based on comprehensive research of the background of the karst geology and karstification rule. Then, we could put forward corresponding targeted programs of development and utilization and specific engineering measures. Finally, according to these specific programs we can increase the investment into the project correspondingly; otherwise it would be hard to succeed. Here are some tips:

1) To carry out 1:50000 hydrogeologic survey systematically.
2) Strengthen the construction of water conservation and water sources.
3) Implementation of scientific management of water resources in karst, development and protection
4) To build the emergency response system of the drought- resistant.

References

Z. Zhu, X. Zhao, Ch. Wang, L. Hou (2006). The rules of drought and the development of water-saving agriculture in southwest China. Ecology and Environment. 15(4): 876-880

Z. Ma, X. Ren (2007).Drying trend over China from 1951 to 2006. Advances in climate change research. China geological survey. 3(4):195-201

Z. Jiang, D. Yuan (2003). Strategy and measures to stop rock desertification in karst areas of southwest China. Proceedings with rocky desertification and karstic groundwater studies of China.13-19

Y. Shi, L. Wang, W. Zhu (2005).The model of water resource utilization in karst area in southwest china .Science and technology review. 23(2):52-55

Y. Wang (2006). Underground stream resources development and utilization condition and the typical engineering research at Yunnan [A] . China geological survey. Karst groundwater resource development and utilization at Southwest China [C]. Beijing: geological publishing house, 11-34

M. Wang (2006). Development and utilization of karst groundwater resources at Guizhou province [A]. China geological survey. karst groundwater resource development and utilization at Southwest China [C]. Beijing: geological publishing house, 2006:35-61

Study on epikarst water system and water resources in Longhe Region

W. Lao, F. Lan

Institute of Karst Geology, CAGS, Guilin, Guangxi 541004, China

Abstract The paper starts with analysis on regional karst water system and elaborates the structure of the epikarst water system and water cycle process in Longhe region. According to hydro-dynamic conditions, several confluence type or scattered flow type of epikarst water subsystems are subdivided. Based on statistic analysis on long-time observation data of the epikarst springs, it is concluded that the quantity of natural epikarst water resources in the area is $31.9\times10^4 m^3$/yr about 9.5% of rainfall can be drawn. Among the quantity of natural epikarst water resources, the exploitable water resource from epikarst springs is $5.9\times10^4 m^3$/yr and the effective amount is 7 732.4m^3/yr.

1 Introduction

The development of the epikarst in south of China have the characteristics of a shallow karst water cycle and underground pipes coupling (Jiang and Yuan 1999, Jiang 1998), Structure and flow dynamic features of the epikarst are constrained by the hydrogeologic structure of the epikarst land cover conditions rainfall and so on (Jiang 1998, Lao et al. 2002, Qin and Jiang 2005), which made differences on features of the epikarst water in different region. The research on features of the epikarst water provides an effective and convenient avenue for solve the problem of drought and water shortage in karst peak-cluster area (Zhihua et al. 2003, Zou et al. 2005, Wang 2007). The research on features of the epikarst water is by a regulation coefficient to evaluation the number of resources in different Spatial scale, this research is good for strengthening empirical and analysis of the epikarst water system. The study area is located in Pingguo County, Guangxi Province, covers an area of 2.66km^2, located in the subtropical monsoon climate zone, the mean annual temperature of 20.2 - 22.6 °C, frost-free period more than 345 days, the mean annual rainfall 1500mm, rainfall abundant but in the year rainfall is unevenly distributed in different seasons, season from May to September each year is the wet season, April, October and November are normal, December to next April is the dry season, with characteristics of rain of the hot season. The area is a typical peak and depression topography, formed by the large number of cones which have various heights and bottom are connected, and the closed depressions of various shapes of polygons during them. Depressions bottom have a height of

150-400m, and the height of the Peak is more than 500m. However, surface water system is not developed. There are about 530 people in the area; income mainly relies on planting, cultivation and export of labor services living in poverty.

2 Karst water system

According to the relationship among supply, discharge, savings and drainage of the karst water, study area could divided into two Karst water systems, one called Longhuai, another is Longhe, both of them have independent aquifer system and have flow system corresponds with them, Longhuai system covers an area of 2.08km^2, Longhe system covers an area of 0.58km^2. The study area have the characteristics of dual structure, some lots have the characteristics of triply structure. Based on the main features on structure, each system could divide into two subsystems, epikarst water system and phreatic karst water system.

Karst water supply source in study area is nature precipitation, there are two main forms, one is throught surface grikes in the form of surface diffused supply water for epikarst water, this form is widespread in the area, another is precipitation and a part of epikarst groundwater through cloups, ponors, regards, shafts, deep and large grikes in the form of concentrated inflow supply water for pheatic zone karst water. The discharge of karst water is mainly nature discharge, in the form of concentrate discharge and diffuse discharge.

3 Epikarst water subsystem

3.1 Distribution and development of epikarst

Subtropical climate in study area provide favorable conditions for the development of epikarst. A long term field investigation found that most epikarsts are developing on the shallow surface of hard pure carbonate rocks, the most typical is the epikarst on Paleozoic limestone which located in south of China. The exposed strata in study area are Permian and Carboniferous. Climate is hot and humid, so epikarst development is widely distributed. Field investigation found that relatively flat lots in study area.

3.2 Epikarst water subsystem structure

Epikarst water subsystem in study area has the main features like these:

1) Source of supply, discharge, storage and drainage of the system are lapie, water-eroded groove, dissolution pore, solutional cavity, karren and so on.
2) Epikarst water subsystem is mainly divided into convergent discharge pattern and diffuse underflow pattern.
3) Epikarsts that distribute in most mountain slope areas around depression have unitary structure. Coverage area of epikarst aquifer subsystem that located in the bottom of depression of Longhuai, Longlie, Longheshang, Longhe, Jieluo present characteristics of dual structure (epikarst and the Layer of soil or vegetation) or triply structure (epikarst and the Layer of soil and vegetation) (Fig. 1, 2).

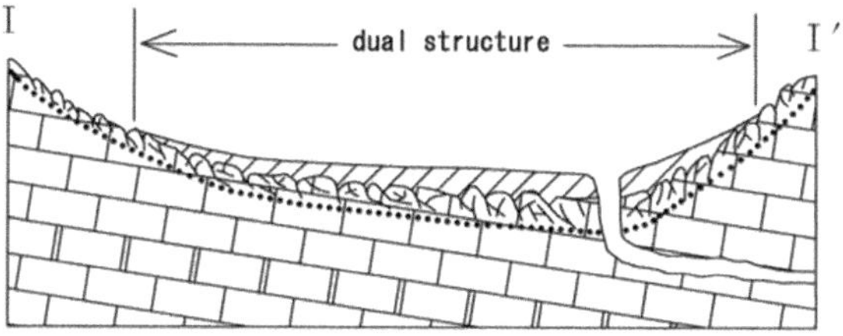

Fig. 1. Section showing the structure of the epikarst water subsystem of Longlie depression (from Wenke Lao 2011).

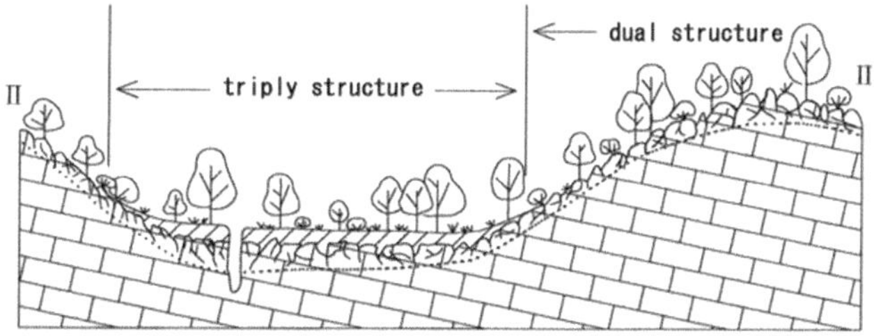

Fig. 2. Section showing the structure of the epikarst water subsystem of Longhe depression (from Wenke Lao 2011).

1) Bottom of epikarst water subsystem have no distinct impermeable layer, because of relatively complete soluble rock zone constitute local impermeable layer which is the bottom boundary of each epikarst water second subsystem.
2) Between epikarst water subsystem and phreatic karst water subsystem, usually developing a certain thickness of the weak developed karst zone, which presenting double porosity.

4 Characters of epikarst water resources

4.1 Dynamic characteristics of epikarst springs flow

According to data collation and statistical process on flow rate of Longhe spring, Buyang1 spring, Buyang2 spring in study area from May 1, 2003 to April 30, 2005 (Fig.3), the flow epikarst springs present the following features:

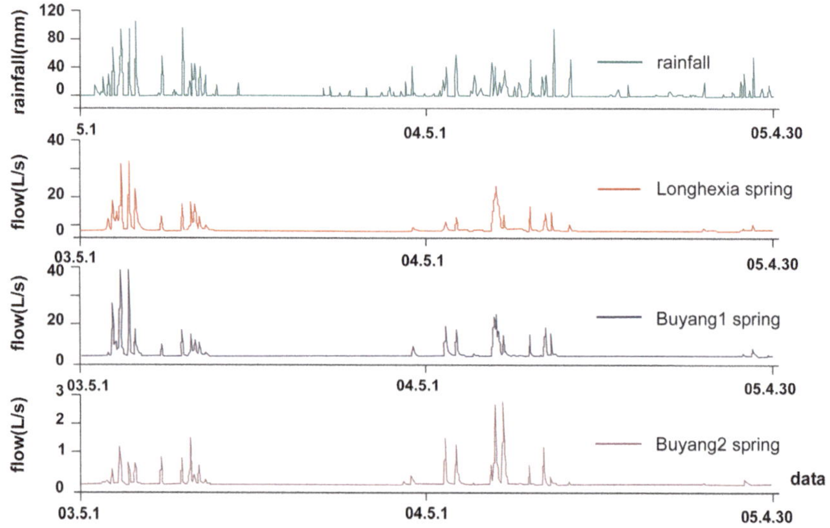

Fig. 3. The relationship curves of epikarst springs' flux and rainfall in Longhe.

(1) The flow rate of epikarst springs are sensitive to precipitation in Longhe area, lag time is short, changing the basic synchronization. According to field observations, the springs' max flow rate usually emerge 1h-2h late after max of precipitation. On the hand, dynamic curve showed a steep peak, it means epikarst have a weak regulation ability on storage water resources.

(2) Usually, the spring flow rate is less in this area. Average flow rates of Longhe spring, Buyang1, Buyang2 are 0.75L/s, 1.02L/s, 0.06L/s. Most of them are dying up from November to February of the next year. According to investigation, we found that Buyang1 spring have a better vegetation and soil coverage, so it has a strong regulation on storage capacity to water resources.

4.2 Distribution features of epikarst water resources

Distribution features of epikarst water resources in epikarst areas are shown in Fig.4, from this fig, we know that:

1) Distribution of epikarst water resources is very uneven during a whole year. The high flow period of Longhe spring, Buyang1 spring, Buyang2 spring is from May to September, respectively account for 94.71%, 94.91%, and 95.03% in a year. While only 1.57%, 1.00% and 0.72% from October to April of the next year.
2) For space, the availability epikarst groundwater resources mainly distribute in convergent epikarst areas.

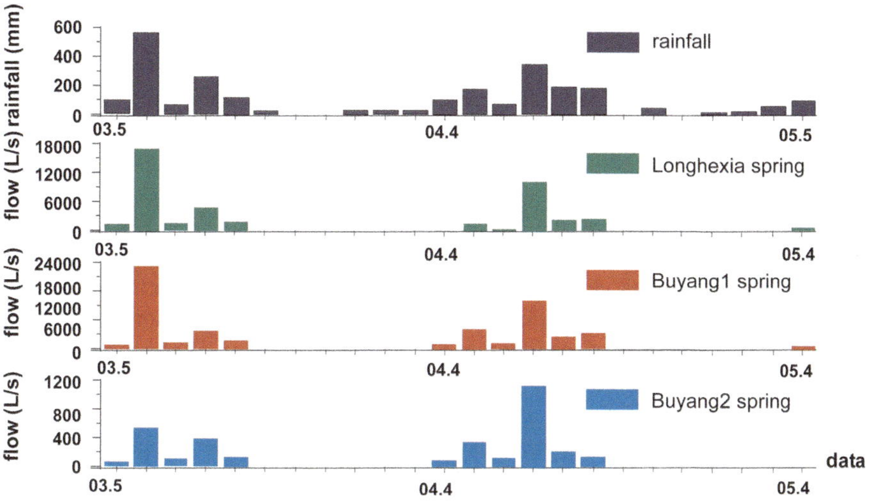

Fig. 4. The monthly distribution of epikarst zone karst water resources in Longhe.

5 Evaluation of the epikarst groundwater resources

5.1 Total natural resources

There many methods to calculation natural resources of epikarst groundwater such as coefficient method, runoff modulus method, discharge method and a new method calling storage coefficient method, etc. According to hydro geological conditions in our study region and observational data available, the authors use infiltration coefficient method to calculate the amount of natural groundwater resources of epikarst. The calculation of infiltration coefficient is:

$$\lambda = Q_1/(P \cdot S) \qquad \text{equation (1)}$$

Where, λ: is the infiltration coefficient of epikarst; Q_1: the amount of total excretion during the period (m^3), P: the calculated total rainfall during the period (m) and S: the recharge area of epikarst springs (m^2)

$$Q_2=\lambda \cdot P \cdot S \qquad \text{equation (2)}$$

Where, Q_2: is the natural groundwater resources of epikarst zone (m^3/yr); λ: the infiltration coefficient of epikarst zone; P: the mean annual rainfall (m/yr) and S: the computing unit (block paragraph) area (m^2).
Through calculating calculation natural resources of epikarst groundwater in each block paragraph by equation (2), the results showed that the amount of natural groundwater resources of Longhe karst water system is $31.90\times10^4 m^3/yr$, Longhuai system is $21.21\times10^4 m^3/yr$, Longhe system is $10.69\times10^4 m^3/yr$.

5.2 Total effective use of resources

Exposed surface epikarst springs are located in relatively high peak and slopes areas, and the springs draining all its water down to the surface, the water flow by itself and can be utilized easily. In theory, the amount of natural excretion of epikarst could be the amount of effective use, in fact, due to precipitation in different seasons during the uneven distribution and the size, number of storage facilities and other issues, often there are vast amounts of water in the wet period has been abandoned and lack of recoverable water resources in water period and dry season. It is more reasonable and practical to evaluate the effective amount of surface karst springs by sum of its gross drain and water quantity that was stored by storage facilities during the rainy seasons. The formula is:

$$Q_3 = 86.4\sum_{i=1}^{m} q_i + nQ_4 \qquad \text{equation (3)}$$

Where, Q_3: is the amount of effective use of epikarst water resources (m^3/a); q_i: the *i* day's daily average flux of surface karst springs during the supply(or demand) period (L/s), i= 1,2,3......m; Q_4: the volume of water storage facilities (m^3); m: gross days during the water supply (or demand) period and n: the use of storage facilities during the rainy season the number of recycling savings

The calculated results show that the total amount of natural epikarst ground water resources in Longhe region is $31.90\times10^4 m^3/yr$, $21.21\times10^4 m^3/yr$ for Longhuai system and $10.69\times10^4 m^3/yr$ for Longhe system.

6 Conclusions

Epikarst water system in Longhe, Guohua, Guangxi have mountain Peaks in southern China exposed surface with the typical characteristics of karst. In the low coverage of vegetation and soil, the region epikarst small storage capacity of water resources, the flow of epikarst springs are generally small and highly volatile, flow rate is sensitive to precipitation, the amount of water is small and uneven distributed in space and time. This is bad for development and utilization of epikarst water resources. The current economic and technological condition made the effective utilization is only 5%. So, improve the vegetation coverage, enhance soil and water conservation, to improve water resources storage capacity of epikarst, increase the flow of epikarst spring is the fundamental way to improve the effective utilization of epikarst groundwater resources.

References

Z. Jiang, D. Yuan (1999) Dynamics Features of the Epikarst Zone and Their Significance in Environments and Resources. Acta Geoscientica Sinica, 20(3):302-308

Z. Jiang (1998) Features of Epikarst Zone and Formation Mechanism in South China. Tropical Geography, 18, 4:322-326

W. Lao, Z. Jiang, J. SHI, B. Liang (2002) Hydrogeologic Structure and Feature of the Epi-karst in Luota. Carsological Sinica. 21(1):30-35

Z. Jiang, R. Wang, J. Pei, S. He (2001) Epikarst zone in south china and its regulation function to karst water. Carsological Sinica, 20(2):106-110

X. Qin, Z. Jiang (2005) A Review on Recent Advances and Perspective in Epikarst Water Study. Carsological Sinica, 24(3):250-254

P.R. China. A Preliminary Evaluation of the Capacity to Storage Groundwater of Epikarst Zone in Southest Karst Mountain of China (2003) Study on Karst Groundwater and rock desertification in China. Nanning, Guangxi Province, 148-154

S. Zou, W. Zhang, X. Liang, W. Luo, B. Liang (2005) Quantitative Calculation of Regulating Coefficient for Epikarsk Zone-Case Study of Zhaojiawan, Luota, West of Hunan. Hydrogeology & Engineering Geology (4):37-42

W. Wang (2007) Evaluation Method for Epikarst Water Resources Based on Analytic Hierarchy Process- A Case Study of the Daxiaojing Drainage Area. Geology of Guizhou, 24(1):17-21.26

Hydrogeochemical Characterization of carbonate aquifers of Lepini Mountains

G.Sappa, L. Tulipano

Dipartimento di Ingegneria Civile, Edile ed Ambientale, Sapienza-Università di Roma, Via Eudossiana, 18, 00186, Rome, Italy; giuseppe.sappa@uniroma1.it

Abstract Groundwaters of carbonate aquifers located in the southern part of the Latium Region have been characterized by hydrochemical tracers for outlining the hydrogeological conceptual model of these aquifers. The preliminary results allow a hydrogeochemical characterization of ground waters, which helps defining main processes of geochemical evolution of the area. This study is based on hydrochemical characterization and their interpretation of 12 springs emerging from the Lepini Mts., monitored from 2003 to 2004, in the Southern part of Latium Region (Central Italy).

1 Introduction

Groundwater of Lepini Mts. is the most important water resource in the south part of Latium Region. In the last few decades many studies were carried out on these aquifers, but they never included hydrogeochemical properties of groundwater: however, the hydrogeochemical characterization of ground waters is nowadays one of the most powerful tools to outline hydrogeological conceptual models. Therefore, with the aim of contributing to the knowledge of the groundwater hydrodynamic features of Lepini carbonate aquifers a hydrochemical survey was carried out over a one year period on the springs of the area.

2 Geological and hydrogeological setting

The Lepini Mts. are the northern part of the Volsci chain, oriented parallel to the Apennines, leaning against the Latium coastline, including also Ausoni Mts. and Aurunci Mts. The Lepini calcareous massif comprises an important karst aquifer that can be regarded as an isolated hydrogeological unit, nevertheless there is still uncertainty about its SE margins (Fig. 1).

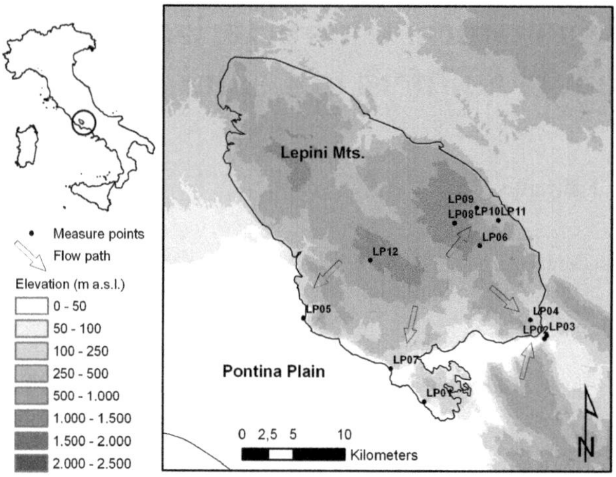

Fig. 1. Location of study area, springs and direction of main flow paths on topographic map.

The top of mounts and hills (up to 500-800 m a.s.l) are calcareous and the valley consists of igneous rocks, often with younger alluvial fan and colluvial deposits. The Pontina Plain, situated in the southwestern part of the Latium Region, is a coastal plain developed along an extensional marine boundary. The Plain is situated between the Lepini-Ausoni mountains of the Central Appenines and the Tyrrhenian Sea. Although it is generally between 0 and 40 m asl, it has several areas where the land is below sea level (Serva and Brunamonte 2007). The aquifer inside the Lepini massif may be classified as “unconfined with an undefined bottom surface”. The saturated zone of the aquifer reaches 1,000 m of thickness in the inner areas of the relief. However, near the margins of the ridge, it may be easily investigated through boreholes which are more close to the surface (Boni et al. 1986). The most important tectonic unconformity of the entire hydrogeological unit is the so-called “Carpineto line”, which is relevant for the study of groundwater circulation. The recharge areas of the aquifer generally coincide with outcropping karstic limestone where recharge takes place mainly through fissures.

3 Methodology

The main springs sampling surveys were carried out in Lepini Mountains from 2003 to 2004 (Table 1). Chemical analyses were performed on the collected water samples for geochemical modeling. Water temperature, electrical conductivity and pH values were determined in the field. A Metropes C2-100 column was used to determine cations, while a Metropes A Supp 4-250 column was used for anions. The Chemical analyses were carried out at the Geochemical Laboratory of Sapienza, “University of Rome”.

Table 1. Main characteristics and chemical composition of sampled waters.

Sample Codes	Elevation (m a.s.l.)	Date (dd/mm/yy)	Ca^{2+} (mg/L)	Mg^{2+} (mg/L)	Na^{+} (mg/L)	K^{+} (mg/L)	NH_4^{+} (mg/L)	Cl^{-} (mg/L)	SO_4^{2-} (mg/L)	HCO_3^{-} (mg/L)	F^{-} (mg/L)	NO_2^{-} (mg/L)	NO_3^{-} (mg/L)	Li^{+} (mg/L)	Sr^{2+} (mg/L)	Br^{-} (mg/L)	PO_4^{3-} (mg/L)	TDS (mg/L)	EC (µs/cm)	T (°C)	pH
LP01	5.00	28/02/03	75.80	31.40	85.90	6.10	<0.01	131.88	35.48	337	0.19	<0.01	4.07	<0.01	0.79	0.30	<0.01	707.82.	860.00	12.00	7.77
LP02	42.00	28/02/03	66.14	15.30	9.00	1.00	<0.01	10.52	5.47	246	0.04	<0.01	4.19	<0.01	0.50	<0.01	<0.01	357.66	405.00	12.00	7.75
LP03	64.00	28/02/03	81.90	4.17	12.70	1.40	<0.01	17.71	8.23	248	0.04	<0.01	1.45	<0.01	0.30	<0.01	<0.01	375.6	473.00	15.00	7.65
LP04	185.00	28/02/03	68.27	11.65	8.60	0.79	<0.01	11.23	4.51	253	0.04	<0.01	4.10	<0.01	0.20	<0.01	<0.01	362.19	403.00	14.00	7.88
LP05	12.00	28/02/03	111.00	44.70	221.00	15.80	<0.01	338.40	85.44	448	0.37	<0.01	<0.01	0.01	1.81	0.73	<0.01	1264.7	1,540.00	14.00	7.65
LP06	840.00	28/02/03	52.31	6.59	4.19	0.31	<0.01	4.81	3.02	192	0.08	<0.01	1.97	<0.01	<0.20	<0.01	<0.01	265.28	281.00	14.00	8.12
LP07	5.00	28/02/03	86.90	31.90	88.00	5.58	<0.01	117.48	40.20	379	0.10	<0.01	1.79	<0.01	0.63	0.15	<0.01	750.95	844.00	15.00	7.92
LP08	1,110.00	12/05/04	49.70	4.10	3.30	0.30	<0.01	4.70	3.10	171	0.00	<0.01	1.20	<0.01	<0.01	<0.01	<0.01	237.4	296.00	11.00	7.60
LP09	740.00	12/05/04	59.30	1.40	2.90	0.10	<0.01	3.90	3.10	183	<0.01	<0.01	1.40	<0.01	<0.01	<0.01	<0.01	255.1	316.00	11.00	7.63
LP10	360.00	12/05/04	34.20	6.50	4.90	2.30	<0.01	8.60	2.60	128	0.10	<0.01	0.60	<0.01	<0.01	<0.01	<0.01	187.8	252.00	12.00	7.25
LP11	360.00	12/05/04	15.40	4.30	4.00	1.50	<0.01	7.50	1.70	67	0.10	<0.01	0.30	<0.01	<0.01	<0.01	<0.01	101.8	138.00	13.00	6.91
LP12	1,065.00	13/05/04	70.20	2.70	4.80	0.20	<0.01	7.50	4.10	226	<0.01	<0.01	1.40	<0.01	<0.01	<0.01	<0.01	316.9	396.00	10.00	8.10

The PHREEQCI software, version 2.10.0.0 (Parkhurst and Appelo 1999), was employed to compute aqueous speciation and fluid-mineral equilibria. The saturation indexes (SI) are approximated due to analytical and activity concentration uncertainties; they are assumed to be ± 0.5 accurate. For the present study, geochemical evolution of groundwater has been evaluated by means of Mg^{++}, Ca^{++}, K^{+}, Na^{+}, Cl^{-}, SO_4^{--}, and HCO_3^{-}, as they were the best indicators of groundwater flowpaths.

4 Results and Discussions

Chebotarev diagram of the samples from Lepini Mts. is shown in Figure 2. LP05 sample belongs to sulphate-chloride-alkaline water type, while samples of LP01, LP02, LP03, LP04, LP06, LP07, LP08, LP09, LP10, LP11 and LP12 belong to bicarbonate-alkaline earth water type. However, LP01 and LP07 samples show a tendency to sulphate-chloride-alkaline composition. Schoeller diagram (Fig. 3) highlights different groups of groundwater.

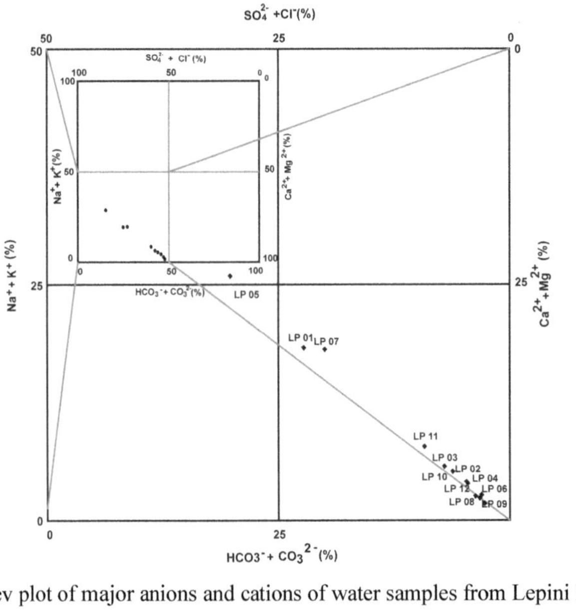

Fig. 2. Chebotarev plot of major anions and cations of water samples from Lepini Mts.

LP02, LP04, LP06, LP08, LP09, LP10, LP11, LP12 samples, located in the central-eastern part of the Lepini Mts., could be classified differently as they indicate low-Mg^{++}-HCO_3^{-}-Ca^{++} values having the average TDS content of 270 mg/l. The average elevation of these springs is 480 m a.s.l; the LP09 and LP12 springs, included in the group, locate at an average elevation 900 m a.s.l. On the contrary,

LP01, LP03, LP05, and LP07 samples show a mean TDS content of 730 mg/l, higher than that of the previous group. These samples belong to the calcium-bicarbonate water type, except the LP05 sample that belongs to the sulphate-chloride-alkaline water type. However, they show relevant concentrations of Cl^-, $SO4^{--}$, Na^+ and K^+. Probably, spring waters belonging to the first water type interacted with calcareous and calcareous-dolomitic rocks, while the chemical features of the springs belonging to the second water type could be connected to those of thermal waters rich in H_2S and CO_2 (Acqua Amara and Acqua Solfa group of springs) occurring in the area. These waters may be related to the outcrops of Roccamonfina volcanic system, which is quite close to the second group of springs.

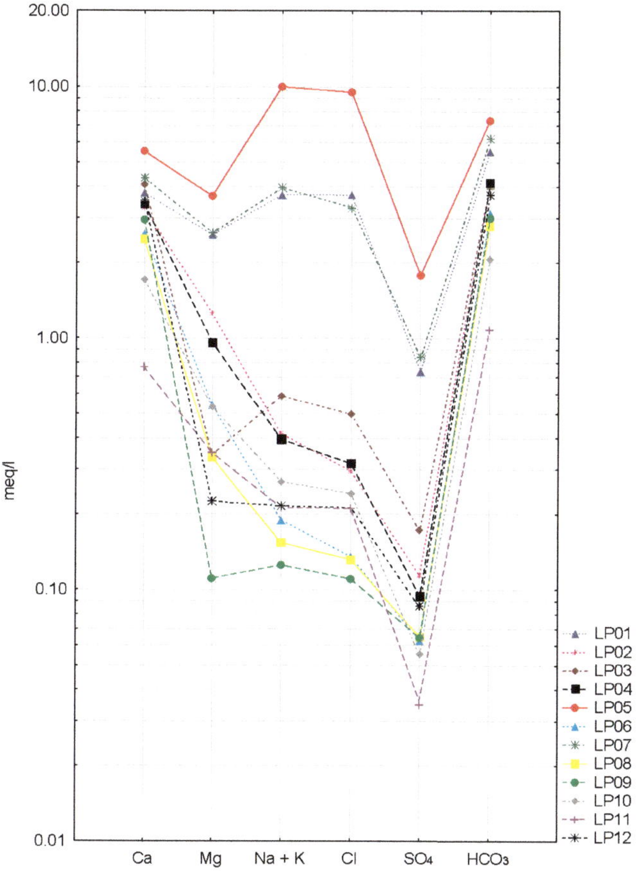

Fig. 3. Schoeller plot of major anions and cations of water samples from Lepini Mts.

Chemical data of the Lepini spring waters show that geochemical evolution occurs mainly by interaction with carbonate rocks and the results of geochemical modeling (Fig. 4) show that the samples LP08, LP09, LP10, LP11 and LP12 are poorly mineralized. These conditions only arise when the groundwater flowpath is short, as observed in thin perched aquifers. The geochemical modeling indicates that LP01, LP02, LP04, LP05, LP06, LP07, and LP12 waters are saturated in calcite and dolomite, which implies strong mineralization and long flowpaths. Water samples of LP01, LP05 and LP07 springs, located at the border with the Pontina Plain, have high saturation indexes in dolomite and high Ca^{++} and Mg^{++} concentrations: chemical data indicate that ground waters follow the longest groundwater paths and have the highest residence times. Waters samples of LP03, LP08 and LP09 springs are saturated in calcite and undersaturated in dolomite. LP10 and LP11 spring waters are undersaturated in both calcite and dolomite, having the lowest Ca^{++} concentrations, which indicate poor mineralization and short flowpaths.

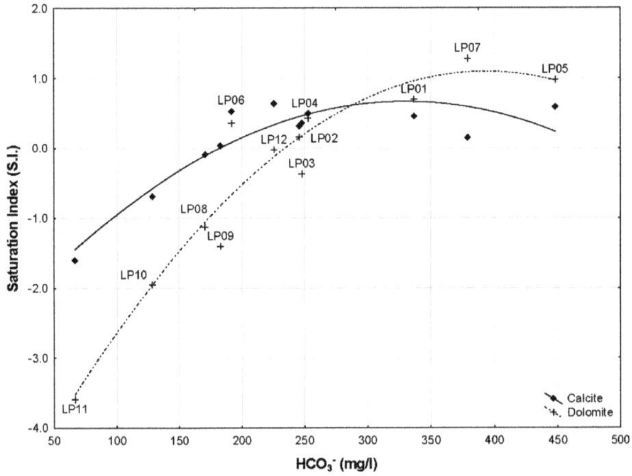

Fig. 4. Saturation Index (S.I.) of calcite and dolomite versus HCO_3^- concentration of Lepini Mts. spring waters.

The distribution of Cl^-, in spring waters of Lepini Mts. (Fig. 5) shows that Cl^- concentrations increase near the border of Pontina Plain. This fact could be related to the effects of seawater intrusion on Pontina Plain aquifer (Fidelibus and Tulipano 1986). Saltwater movement inland can be the main reason for salinity increase; however, other sources may be responsible (Custodio 2002). Moreover, higher Cl- concentrations were observed from different stations in this region from the previous works. These results may imply the possibility that the homogeneity of the isotopic results along the Tyrrhenian coast may, at least partially, be affected by a contribution of seawater spray carried by westerly winds (Longinelli and Selmo 2003).

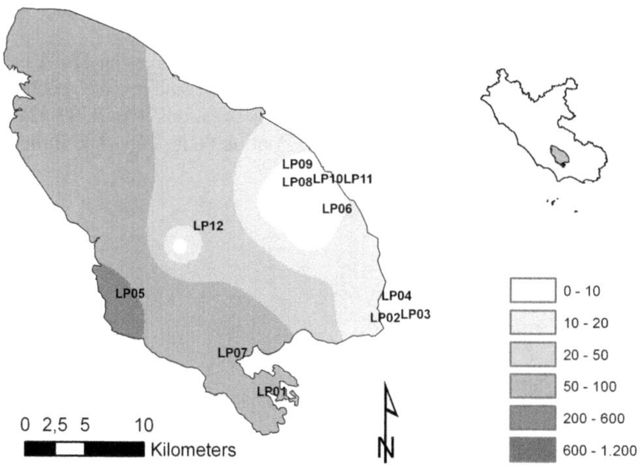

Fig. 5. Isopleths of Cl^- concentrations in mg/l.

5 Conclusions

Geochemical investigation techniques were applied to different groundwater samples, coming from the main springs of Lepini Mts., to understand the hydrogeological behavior of the carbonate aquifers of Southern Latium, Italy. On the basis of the hydrochemical characterization of the area, 3 groups of springs were distinguished in Lepini Mountains. Springs located in the central-eastern area are recharged by perched aquifers; groundwater from these springs is poorly mineralized suggesting short flowpaths. On the contrary, springs located in the southern area are saturated in calcite and dolomite, which implies strong mineralization and long flowpaths. The sampled waters from springs, located at the border with the Pontina Plain, are rich in Cl^-, $SO4^{--}$, Na^+ and K^+, implying a relation with the outcrops of Roccamonfina volcanic system, which is close to the springs.

References

Boni C, Bono P, Capelli G (1986) Schema Idrogeologico dell'Italia Centrale. Memorie Società Geologica Italiana, 36, 991-1012, Roma

Custodio, E. (2002) Coastal aquifers as important natural hydrogeological structures, in: Groundwater and Human Development, M. Bocanegra and M. Massone, eds., Barcelone, pp. 1905-1918

Fidelibus MD, Tulipano L (1986) Mixing phenomena owing to sea water intrusion for the interpretation of chemical and isotopic data of discharge waters in the Apulian coastal carbonate aquifer (southern Italy), Proc. 9th Salt Water Intrusion Meeting, Delft, pp.591-600

Longinelli, A., Selmo, E. (2003) Isotopic composition of precipitation in Italy: a first overall map. J. Hydrol. 270, pp. 75-88

Parkhurst DL, Appelo CAJ (1999) User's guide to PHREEQC (version 2) – A computer program for speciation, batch-reaction, one-dimensional transport, and inverse geochemical calculations. US Geological Survey-Water Resources Investigations Report. 99-4259, pp. 312

Serva L, Brunamonte F (2007) Subsidence in the Pontina Plain, Italy, The Bulletin of Engineering Geology and the Environment, 66, 125-134

Salt ground waters in the Salento karstic coastal aquifer (Apulia, Southern Italy)

M.D. Fidelibus[1], G. Calò[1], R. Tinelli[1], L. Tulipano[2]

1 Water Engineering and Chemistry Dept (DIAC), Polytechnic of Bari, Via Orabona 4, Bari, IT 70125, Italy, d.fidelibus@poliba.it; g.calò@poliba.it; r.tinelli@poliba.it;

2 Civil, Construction and Environmental Engineering Dept (DICEA), Sapienza University, Via Eudossiana 18, 00184, Rome, Italy, luigi.tulipano@uniroma1.it

Abstract The chemical and isotopic features of salt ground waters in the Salento coastal karstic aquifer (Apulia, Southern Italy) reveal that they are very different from modern seawater. Chemical data allow identifying the role of water-rock interaction (Ca^{2+}-Mg^{2+} exchange due to dolomitization, Na^{+}-Ca^{2+} or Mg^{2+} base-exchange, SO_4^{2-} reduction, solution of evaporite salts) in the modification of the chemical characteristics of original seawater. $^{87}Sr/^{86}Sr$ isotope ratios and Sr indicate that salt waters are very ancient and, given the structural features of this aquifer and the paleogeography of the region, they probably date back to the Flandrian transgression. New findings in a thermal area overlooking the Adriatic Sea suggest new scenarios for the role of Ca-Cl_2 brines, until now unknown.

1 Introduction

In coastal aquifers, groundwater salinization can originate from salt sources different from modern seawater (Kapelj et al. 2003; Richter and Kreitler 1993). Among potential salt sources, slow-moving saline fluids, some even thousands of years old, are frequently recognized (Darling et al. 1997; Han et al. 2011). In the coastal aquifers of the Mediterranean area, saline fluids can derive from seawater intruded into aquifers during the sea level rise occurred about 10-15 ka ago, ensuing the last glacial period (Cotecchia et al. 1974; Yechieli et al. 2009); after intrusion, seawater underwent geochemical diagenesis due to water-rock interaction.

Saline fluids show chemical and isotopic features different from those of modern seawater, as do the wider range of saline fluids and brines found in sedimentary basins (Hanor 1983; Land and Prezbindowsky 1985). Generally, in relation to increasing residence time, saline fluids and brines show significant alteration of major and minor (Sr, Li, Br) ion concentrations (Hanor 1983). Moreover, they differ from seawater for stable (D, ^{18}O, ^{13}C, ^{11}B) and radioactive (^{14}C, ^{36}Cl) isotope concentrations, and $^{87}Sr/^{86}Sr$ isotope ratio. Particularly saline fluids in carbonate aquifers, due to dolomitization and sulphate bacterial reduction, show decreasing

Mg/Ca and SO_4/Cl ratios as residence time increases (Ng and Jones 1995; Oetting et al. 1996). The characteristics of saline fluids can be inferred through the extrapolation of saline end-members from the chemical composition of brackish waters (Barbecot et al. 2000). Fidelibus and Tulipano (1996) derived the characteristics of a saline fluid with high Sr and Li concentrations and a very low Mg/Ca ratio from those of the brackish coastal spring of Apulia's karstic coastal aquifers (Southern Italy). In these aquifers, samples of saline fluids were taken at the observation-wells of a net set up in the 1960s for the control of seawater intrusion (Cotecchia et al. 1974; Fidelibus and Tulipano 1991, 1996; Tulipano and Fidelibus 1984). Their Mg/Ca ratios and Sr concentrations respectively range from 6 (modern seawater) to 2 and from 6.8 to 18 mg/L; Li is enriched up to 300 μg/L, with sulphates significantly varying around the seawater value and a very low ^{14}C (pcm), giving apparent ages of 15 to 17 Ka. Later, Barbieri et al. (1999) inferred the existence of very old saline fluids within Apulia's karstic aquifers through their observation of $^{87}Sr/^{86}Sr$ ratios and Sr concentrations.

No further information was until recently available on this area due to lack of perforations that reached salt waters. New findings only came in over the last few years from sampling of salt waters performed in municipal and private wells, drilled in the Salento karstic coastal aquifer (Apulia, Southern Italy). They allow further views regarding the origin and digenesis of saline fluids in the Apulian Region.

2 Geological and hydrogeological framework

The Salento Peninsula (Fig. 1) is one of the three coastal karstic aquifers in the Apulian Region. To the NE, the structural setting features a series of reliefs and tectonic depressions, with faults having mainly NNW-SSE direction. To the SW, the fault system determines rhomboidal basins and ridges: major and minor folds respectively show NNW-SSE to NW-SE, and ENE to NE trending axes; faults (of normal sub-vertical, strike-slip and oblique-slip types) follow directions from NNW-SSE to NW-SE. In some areas, the Mesozoic carbonate basement, about 6,000 meters thick, is covered by Eocene to Quaternary transgressive sediments of variable thickness and sequence, depending on the structural features of the same basement. Throughout the Tertiary and the Quaternary, at regional scale, Salento was subject to the combined effects of isostatic and tectonic uplift, and of sea level cyclical fluctuations. The most depressed areas, varying from time to time, acquired the characteristics of subsident sedimentary basins, interested by various and diverse depositional cycles. This gave rise to many sedimentary successions, whose litho-stratigraphic and sequential characteristics vary depending on the bathymetry and degree of evolution of each basin. The Cretaceous formation constitutes the regional coastal karstic aquifer. Karst drains, mostly sub-horizontal, highly karstified, and connected by fissures, give a homogeneous and high degree of permeability. Freshwater floats on salt water of marine origin; discharge occurs through brackish focused/diffuse coastal springs. The recharge is around 880 Mm^3/y. Hydraulic heads reach 4 m AMSL in the NW and SE areas, with a maximum thickness of the freshwater lens of about 120 m; the mean hydraulic gradient is 0.02 ‰. Groundwater exploitation, starting from the 1950s, has been causing progressive groundwater salinization.

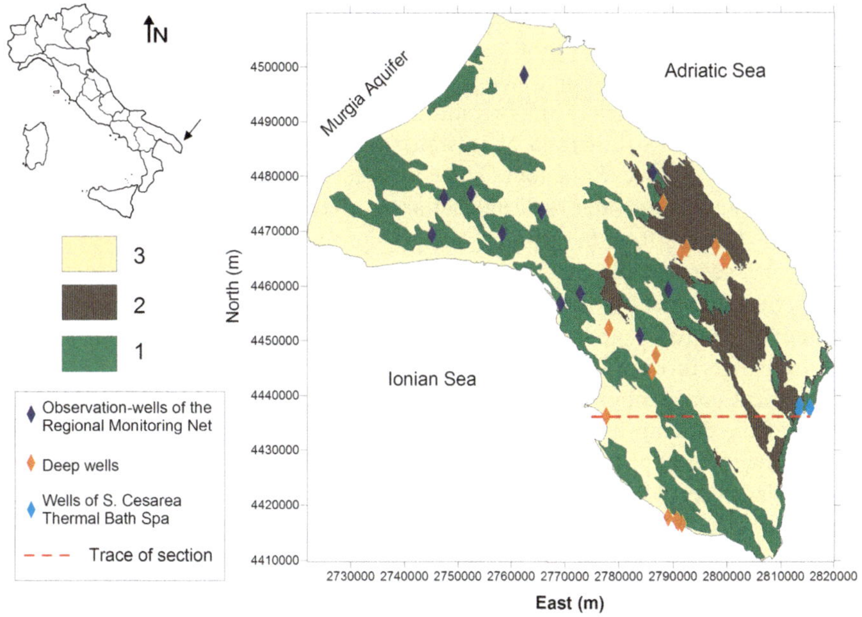

Fig. 1. Geological scheme of the Salento Peninsula (Apulia, Southern Italy) and location of wells (Italian Local Reference System, Gauss Boaga, Rome40). 1: Limestones and dolomitic limestones (Cretaceous); 2: Calcarenites and marly calcarenites (Miocene); 3: Calcarenites, sands and clays (Pliocene-Quaternary). Trace of geological section.

3 Methodology and results

This study is based on chemical and isotope data related to saline/salt ground waters and modern seawater sampled in the Salento aquifer over the last 40 years. Historical data derive from 13 observation-wells (OW) used for the control of seawater intrusion (Fig. 1). New data come from 26 deep wells (DW) aimed at draining storm water and supplying salt water to touristic settlements and fish farms; other new data come from 4 wells drilled to serve the Thermal Health Spa of S. Cesarea (SCW). DWs and SCWs reach salt ground waters underneath the transition zone; before their equipment, such wells were available to carry out temperature and conductivity logs, and take samples of saline/salt fluids.

Chemical analyses of major ions, and sometime of Li and Sr, are available for all samples. Analyses of ^{14}C, δD and $\delta^{18}O$ are available for a total of 12 historical and new samples. The following considerations are largely based on the evaluation of major and minor ions, and $^{87}Sr/^{86}Sr$ ratios related to 15 water samples from DWs, SCWs, and Adriatic and Ionian Seas.

The main application of the $^{87}Sr/^{86}Sr$ ratio methodology in the context of seawater intrusion relates to the identification of saline end-members of mixing (Jørgensen and Banoeng-Yakubo 2001).

Such method is founded on the fact that strontium enters the lattice of aragonite, calcite, fluorite, gypsum, anhydrite and barite: if minerals form in equilibrium with seawater, their $^{87}Sr/^{86}Sr$ ratio is the same as that of coeval seawater. Variations in the $^{87}Sr/^{86}Sr$ ratio of seawater are known to have occurred through the whole Phanerozoic. Sr isotopes show no detectable fractionation by any natural process involving water-rock interaction or mixing and ground waters acquire their $^{87}Sr/^{86}Sr$ ratio from the aquifer rock: the longer the residence time of ground waters, the nearest their chemical equilibrium with the rocks. In coeval ground waters, Sr enrichment is determined by rock mineralogy, while the $^{87}Sr/^{86}Sr$ ratio depends on the rock age: the $^{87}Sr/^{86}Sr$ ratio of salt waters may be influenced by base-exchange, re-crystallization and dolomitization of carbonate rocks.

Chemical compositions of saline/salt ground waters sampled in the DWs and SCWs of the Salento aquifer show differences from modern seawater similar to those shown by salt ground waters from OWs (Fidelibus and Tulipano 1996). TDS (Total Dissolved Solids) ranges from 26 to 44 g/L, reaching 48 g/L in SCWs; average seawater TDS is 40.2 g/L. The correlation between Mg and Ca (Fig. 2A) shows that for all samples except one, even when TDS is lower than seawater TDS, Ca is enriched up to 64 meq/L (SCW) compared to the average seawater value (17.2 meq/L). Mg is mainly depleted (up to a minimum of 70 meq/L); however, some DWs intercept salt waters having Mg concentrations up to 210 meq/L. These variations result in Mg/Ca equivalent ratios different from 7 (average seawater). Other significant variations refer to sulphates (Fig. 2B), which show both enrichment and depletion with highest values of about 94 meq/L (SO_4/Cl equivalent ratio = 0.173) and 44 meq/L respectively, and SO_4/Cl ratios as low as 0.075 accompanying the lowest Mg/Ca ratios. Fig. 2C shows the relationship between Na/Cl and Ca/Cl equivalent ratios: all waters differ from modern seawater for higher Ca/Cl ratios. As to the Na/Cl ratio, salt waters show values either higher or lower than seawater. Minor constituents (Li and Sr) are mostly enriched with respect to present sea water. Sr increases to about 20 mg/L in SCW salt waters (Fig. 2D). Lithium reaches 420 μg/L in the salt water with the highest TDS (48 g/L).

Fig. 3 shows a series of $^{87}Sr/^{86}Sr$ ratio - Sr mixing curves (Barbieri et al.1999) estimated according to Banner et al. (1994) on the basis of the $^{87}Sr/^{86}Sr$ ratios measured on freshwater (9 samples), seawater, and brackish coastal spring waters (19 samples) belonging to the karstic aquifers of the Apulian Region. The freshwater end-member is identified by the average values of fresh waters (Sr = 0.2 mg/L and $^{87}Sr/^{86}Sr$ ratio = 0.70788). Modern seawater is the salt end-member for the uppermost mixing curve ($^{87}Sr/^{86}Sr$ = 0.709198 ± 0.000020, De Paolo and Ingram 1985; Sr = 9.2 mg/L); the intermediate mixing curves intercept the values measured for coastal springs ($^{87}Sr/^{86}Sr$ from 0.70761 to 0.70889, Sr from 1 to 5 mg/L).The lowest mixing curve is driven by the lowest value of $^{87}Sr/^{86}Sr$ (0.70760) and by Sr of about 4 mg/L (Tara spring). The hyperbola connecting the mixing curves at their right ends led Barbieri et al (1999) to assume the existence in Apulia of aged salt waters featuring, with respect to seawater, progressively lower strontium isotope ratios, and higher Sr concentrations. The $^{87}Sr/^{86}Sr$ ratio and Sr concentration of the most evolved salt end-member were respectively 0.70760 and 50 mg/L.

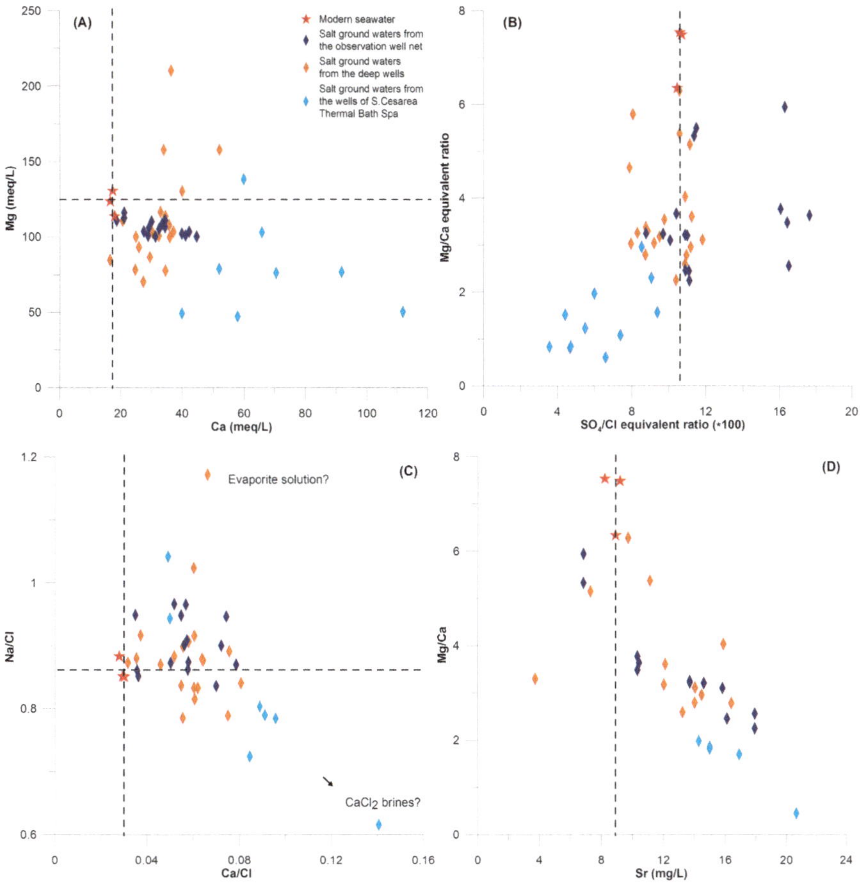

Fig. 2. Salt ground waters in the Salento aquifer: (A) Mg vs. Ca relationship; (B) Mg/Ca equivalent ratio vs. SO_4/Cl equivalent ratio; (C) Na/Cl equivalent ratio vs. Ca/Cl equivalent ratio; (D) Mg/Ca equivalent ratio vs. Sr concentration. Dashed lines mark modern seawater ratios/concentrations.

The $^{87}Sr/^{86}Sr$ ratio of salt ground waters sampled in the DWs and SCWs ranges from 0.70911 to 0.70825. Data almost perfectly adjust along the assumed hyperbola: the matching of real measured data to those hypothesized by extrapolation validates the approach maintained by Barbieri et al. (1999). If salt ground waters were in equilibrium with rocks, their $^{87}Sr/^{86}Sr$ ratio would vary between 0.70720 (lowest limit of Lower Cretaceous) and 0.70780 (highest limit of Upper Cretaceous). However, neither salt waters nor the (assumedly) most evolved salt end-member reach such low values. In fact Tara spring, which drives the lowest mixing curve, does not discharge from the Salento aquifer: its related end-member may be in the Murgia aquifer, in the basement sunk under the Plio-Pleistocene clay covers of Bradanic Trough. Yet data unequivocally indicate that salt waters are very ancient, as do the few historical and new data available on ^{14}C (as low as 2.2 pmc).

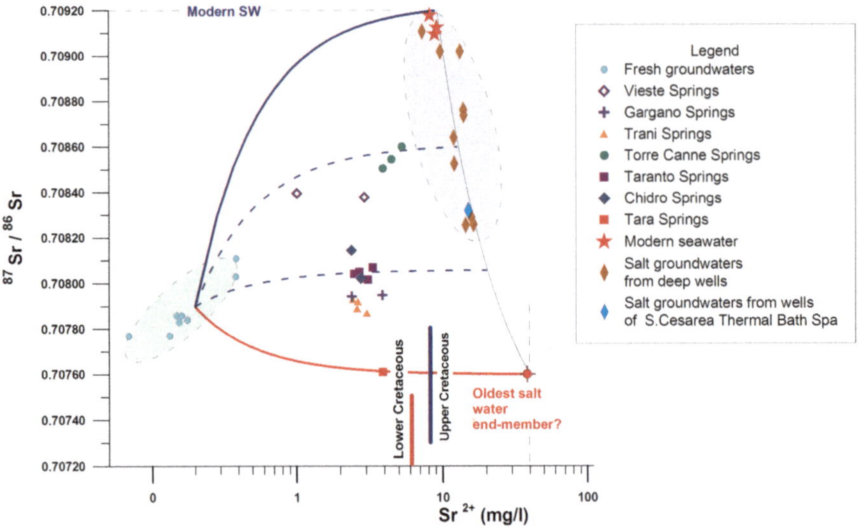

Fig. 3. $^{87}Sr/^{86}Sr$ ratio vs. Sr concentration for seawater (Salento), salt waters (Salento aquifer) and coastal brackish springs (Murgia and Gargano aquifers). The dotted areas respectively gather fresh waters and salt waters (from Barbieri et al. 1999, modified).

4 Discussion and conclusions

Major reactions that could explain all the examined variations are: Ca^{2+}-Mg^{2+} exchange due to dolomitization, Na^{+}-Ca^{2+} or Mg^{2+} base-exchange, SO_4^{2-} reduction, solution of evaporite salts, and mixing with high TDS Ca-Cl_2 brines.

The concurrent Ca enrichments and Mg depletions found in the salt waters of Apulia's karstic coastal aquifers (Fidelibus and Tulipano 1991; Tulipano and Fidelibus 1984) were previously explained as due to dolomitization, dissolution and precipitation of carbonate rocks. The low SO_4/Cl ratios, in a few cases similar to those found in subsurface waters of ancient basins, indicate sulfate reduction, which can also occur due to thermo-chemical reduction at high temperatures. This could explain the very low SO_4/Cl ratios found in the salt thermal waters of SCWs. Reduction processes produce CO_2 and H_2S: thermal salt waters of SCWs, as well as some salt waters from DWs, show bicarbonate contents up to 5.3 meq/L (higher than seawater) and sulfides up to 14.8 mg/L in connection with temperatures of about 30°C. The reduction process produces either high or lower pH: a reductive environment and a pH decrease favor the dissolution of carbonate minerals: this way sulfate reduction can contribute to Mg/Ca ratio variations. The feasibility of Na^+-Ca^{2+} or Na^+-Mg^{2+} base exchange in coastal carbonate aquifers has been demonstrated by Pascual and Custodio (1990) and Nadler et al. (1980). In the study area, the appreciable depletion of Na (low Na/Cl ratios) in some salt waters could be ascribed to the presence of *terra rossa* or, locally, to the influence of clay formations lying on the basement top (Fig. 4). A Na/Cl ratio exceeding seawater value points to the solution of evaporite salts of marine origin. On the other hand, Ca/Cl ratios highly exceeding the seawater value in SCW waters point to the influence of very old brines deeply modified. Sr is abundant in almost all salt waters, and Sr/Cl ratio is higher than that of modern seawater: carbonate rock-water interaction is the main source of Sr. Li, generally enriched in Salento salt waters, shows very high contents in SCW thermal waters: this occurrence confirms that Li is a useful pathfinder for hydrothermal systems (Brondi et al. 1973).

Figure 4 shows a geological section, outlined between a deep well along the Ionian coast and S. Cesarea thermal area: it can be representative of the general structural features of Southern Salento. If we consider its averagely low topographic elevations, it is easy to imagine how the last glaciation and the following Flandrian transgression may have changed the position and thickness of fresh-, brackish- and salt-waters within the rock mass (Calò et al. 2005). To the E (left of section) the covers have acted as a barrier for further exchange with the open sea, blocking seawater entered during transgression: such setting has the chance to favor reductive conditions, whose effects have been recognized in some salt water samples. To the W (right of section) the structural reconstruction indicates the presence of an important fault that could be the entryway for thermal brines present in the bottom sediments of the Adriatic Sea.

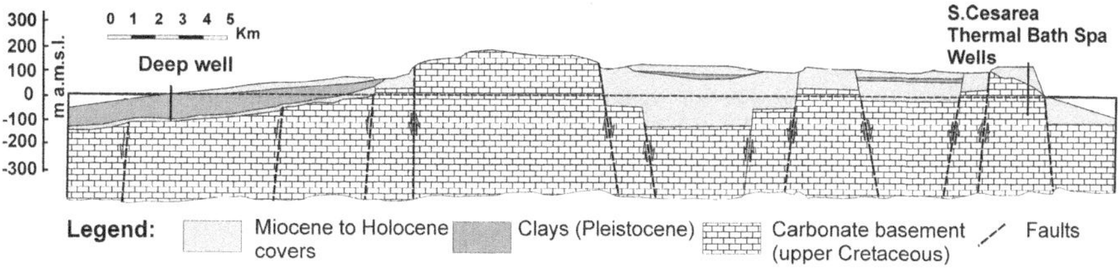

Fig. 4. Schematic geological section (trace in Fig. 1) of Southern Salento.

In conclusion, the chemical and isotope variations shown by salt ground waters can be mainly interpreted as caused by water-rock interactions acting during a progressive aging of original seawater. The new data on S. Cesarea thermal waters (Adriatic coast) reveal the presence of a new salt end-member ($CaCl_2$ brine?), whose genesis is surely different from that of other end-members. This occurrence opens new scenarios for the interpretation of the features of salt waters located near the Adriatic border that can only be explored with further purposed surveys.

References

Banner JL, Musgrove M, Capo R (1994) Tracing ground-water evolution in a limestone aquifer using Sr isotopes: Effects of multiple sources of dissolved ions and mineral-solution reactions. Geology 22, 687-690

Barbieri Mr, Barbieri Mz, Fidelibus MD, Morotti M, Sappa G, Tulipano L (1999). First results of isotopic ratio $^{87}Sr/^{86}Sr$ in the characterization of seawater intrusion in the coastal karstic aquifer of Murgia (Southern Italy). In Procs. 15th SWIM, Natuurwet. Tijdschr., 79, 132-139

Barbecot F, Marlin C, Gibert E, Dever L (2000) Hydrochemical and isotopic characterisation of the Bathonian and Bajocian coastal aquifer of the Caen area (Northern France). Appl. Geochem. 15, 791-805

Brondi M, Dall'Aglio M, Vitrani F (1973) Lithium as pathfinder element in the large scale hydrochemical exploration for hydrothermal systems. Geothermics, 2 (3), 142-153

Calò GC, Tinelli R, Lucrezio D, Stani M (2005) Riscontri delle oscillazioni eustatiche Flandriane nelle acque profonde degli acquiferi Salentini (Puglia). Giornale di Geologia Applicata 2, 341-347 (in Italian)

Cotecchia V, Tazioli GS, Magri G (1974) Isotopic measurements in research of seawater ingression in the carbonate aquifer of the Salentine Peninsula, Southern Italy. IAEA-SM-182/38 1, IAEA, pp 445–463, Vienna

Darling WG, Edmunds WM, Smedley PL (1997) Isotopic evidence for palaeowaters in the British Isles. Applied Geochemistry 12, 813-829

De Paolo DJ, Ingram BL (1985) High resolution stratigraphy with strontium. Science 227, 938-941

Fidelibus MD, Tulipano L (1991) Mixing phenomena owing to seawater intrusion for the interpretation of chemical and isotopic data of discharge waters in the Apulian coastal carbonate aquifer (Southern Italy). In: De Breuck W (Ed), Hydrogeology of Salt Water Intrusion, A Selection of SWIM Papers, Int. Contr. to Hydrog., 11, 317-327, Hannover: Heise

Fidelibus MD, Tulipano L (1996) Regional flow of intruding sea water in the carbonate aquifers of Apulia (Southern Italy). Proc. 14th SWIM, Malmo, Sweden, Rapporter och meddelanden, Geological Survey of Sweden, 87

Han D, Kohfahl C, Song X, Xiao G., Yang J (2011) Geochemical and isotopic evidence for palaeo-seawater intrusion into the south coast aquifer of Laizhou Bay, China. Appl. Geochem., doi:10.1016/j.apgeochem.2011.02.007

Hanor JS (1983) Fifty years of development of thought on the origin and evolution of subsurface sedimentary brines. In: Boardman ST (Ed.) Revolution in the Earth Sciences, advances in the past half century. Lendall/Hunt Publ. Co

Jørgensen NO, Banoeng-Yakubo BK (2001) Environmental isotopes (18O, ^2H, and $^{87}Sr/^{86}Sr$) as a tool in groundwater investigations in the Keta Basin, Ghana. Hydrogeol. J.9, 190–201

Kapelj S, Lambrakis N, Morell I, Petalas C (2003) Sources of aquifer salinization. In Tulipano et al (Eds.) Final Report of Action COST 621, Groundwater management of karst coastal aquifers, Part II (3.4):152-157. EUR 21366, ISBN 92-898-0015-1, Luxembourg

Land LS, Prezbindowski DR (1985) Chemical constraints and origins of four groups of Gulf Coast reservoirs fluids. Discussion. Am Assoc Petrol Geol Bull 69, 119-121

Nadler A, Magaritz M, Mazor E. (1980) Chemical reactions of seawater with rocks and freshwater: experimental and field observations on brackish waters in Israel. Geoch. et Cosmoch. Acta, 44. pp. 879-886

Ng KC, Jones B (1995) Hydrogeochemistry of Grand Cayman, British West Indies: implications for carbonate diagenetic studies. J. Hydrol. 164,193-216

Oetting GC, Banner JL, Sharp Jr. JM (1996) Regional controls on the geochemical evolution of saline groundwaters in the Edwards aquifer, central Texas. J. Hydrol. 181, 251-283

Pascual M, Custodio E (1990) Geochemical observations in a continuously seawater intruded area: Garraf, Catalonia (Spain). In: Kozerski and Sadurski (Eds.) Proc. of 11th SWIM, Gdansk, Poland

Richter BC, Kreitler CW (1993) Geochemical techniques for Identifying Sources of Ground-Water Salinization. CRC Press. Inv. FL (USA)

Tulipano L, Fidelibus MD (1984). Geochemical characteristics of Apulian coastal springs (Southern Italy) related to mixing processes of groundwaters with sea water having different residence time into the aquifer. In: Tsakiris G (Ed.), Proc. 5th Int Conf on Water Resources Planning and Management, Water in the year 2000, Athens, 1984, 2.55-2.67

Yechieli Y, Kafri U, Sivan O (2009) The inter-relationship between coastal sub-aquifers and the Mediterranean Sea, deduced from radioactive isotopes analysis. Hydrogeol. J. 17(2), 265-274

An oceanographic survey for the detection of a possible Submarine Groundwater Discharge in the coastal zone of Campo de Dalias, SE Spain

M.A. Díaz-Puga, A. Vallejos, L. Daniele[1], F. Sola, D. Rodríguez-Delgado[2], L. Molina, A. Pulido-Bosch

Department of Hydrogeology, University of Almería, 04120 Almería, Spain.
mdpuga@ual.es, avallejo@ual.es, fesola@ual.es, apulido@ual.es, lmolina@ual.es

[1]Department of Geology, Universitat Autònoma de Barcelona, Spain. linda.daniele@uab.cat

[2]Ciencia e Tecnoloxía Ambiental, S.L.N.E. Departamento de Oceanografía (Ourense, Spain)

Abstract The Campo de Dalías, in south-eastern Spain, is an area of important economic activity linked to agriculture and tourism, both of which have exacted fierce exploitation of aquifer water. The recovery of one of these aquifers in recent years has even triggered fresh water discharges into the sea. An oceanographic survey was undertaken along the coastline in order to detect possible Submarine Groundwater Discharge (SGD). Salinity and temperature data were collected on the seasurface, as well as in 51 vertical profiles. These results suggest the existence of a thermohaline anomaly in the area of the port of Aguadulce possibly due to submarine groundwater discharges.

1 Introduction

Submarine Groundwater Discharge (SGD) can be defined as the flow of water through continental margins from the seabed into the ocean, independent of the flow mechanism or chemical composition (Burnett et al. 2003; Moore 2010; Johannesson et al. 2011). In this context, SGD can derive both from fresh water originated from meteoric recharge to terrestrial aquifers (due to differences in hydraulic head), and from salt water derived from recirculation of seawater in coastal aquifers (due to tides, ocean swell, density and geothermal gradients (Taniguchi et al. 2002; Kim and Swarzenski 2010).

The main methods for detecting SGD are infiltration counters, hydrogeological models, natural tracers (Ra, Rn, CH_4 and salinity), numerical models and airborne or satellite thermal imaging (Burnett et al. 2006; Charette et al. 2001; Kim and Hwang 2002) and marine remote sensing techniques (Christodoulou et al 1993, Karageorgis et al 2011). Additionally, recent studies have demonstrated that the spatial distribution of SGD can also be studied using sediment resistivity profiles

(Viso et al. 2010; Henderson et al. 2010). Each of the above mentioned methods has its advantages and disadvantages. Thermal imaging, resistivity profiles and marine remote sensing techniques can not quantify the discharge, but they do identify it and provide information for its spatial distribution.

The specific hydrological situation of the Southern Basin in Spain, with its unfavourable water balance, makes it necessary to understand all water resources of the area in detail, including SGD. For this reason, since the 1970s, numerous studies have been carried out to develop techniques for locating and quantifying upwelling of groundwater water through the seabed (Espejo et al. 1988). In 1988, the Confederación Hidrográfica del Sur de España, in collaboration with Empresa Nacional Adaro de Investigaciones Mineras, S.A., used infrared remote sensing of electromagnetic radiation to determine surface thermal anomalies that could be attributable to SGD along the coast of the Southern Basin. Moreover, this survey was complemented by meteorological, oceanographic and hydrogeological data. This multidisciplinary survey showed the existence of a thermal and salinity anomaly in the marine environment of Aguadulce which is probably related to a SGD.

This aquifer has been subject to fierce exploitation, with the consequent decline in piezometric levels. The visible springs close to Aguadulce port dried up and the wells in the area began to be affected by what seemed to be marine intrusion (Molina et al. 2002, Morell et al. 2008). Water levels in the aquifer are now recovering as a result of the abandonment of some salinized wells, and the decrease in the groundwater abstractions in this sector due to deteriorating water quality, and improved regulation of these water resources.

Although in the study area has been a previous study (Espejo et al. 1988), this was carried out at the regional scale and mainly using remote sensing techniques. This paper focuses on a particular sector of the Andalusian coast using data from an oceanographic survey and it aims to verify possible SGD in the Aguadulce sector, by means of a survey of temperature and conductivity of coastal waters.

2 Hydrogeological framework

The Campo de Dalías coastal plain is situated in the extreme south-east of Spain, to the west of the bay of Almería. It covers an area of about 330 km^2 and traces the coastline for about 50 km. Its northern limit is in the foothills of the Sierra de Gádor, whilst the other borders are coastal (Fig. 1). The importance of intense agriculture practised in this area is fundamental for the local economy. Campo de Dalías includes about 20,000 ha of cultivated ground. In addition, the population has grown from 8,000 inhabitants in 1950 to more than 120,000 at present. Tourism also plays an important role in the local economy. Together, these factors have lead to an alarming growth in the water demand in this part of Almería.

Three hydrogeological units are distinguished: Balerma-Las Marinas, Balanegra and Aguadulce (Fig. 1). The Aguadulce aquifer unit has a complex geometry as a consequence of the lithological diversity and structure of the area (Pulido et al. 2000). The carbonate strata of the Felix nappe are highly developed and overlie the thicker Gádor nappe. A layer of phyllites separates the two units, making two distinct aquifer layers. Miocene calcarenites overlie the carbonate layers of both nappes and are in hydraulic connection with these; Pliocene calcarenites and other more recent detritic sediments can also be found, which also behave as aquifers.

Thus, three aquifers occur in the area: an upper aquifer consisting of Pliocene calcarenites together with Plioquaternary detritic materials and two carbonate aquifers: one corresponding to the Felix carbonate strata and the other comprising the Gádor carbonate rocks, which extend to depth. Overall, there is a sequence of formations separated by layers of low permeability. The intense fracturing favours hydraulic connection between the various aquifer levels, which, under a natural regime, would show the same piezometric level. However, overexploitation of the unit has given rise to a difference in levels, depending on the cross-section.

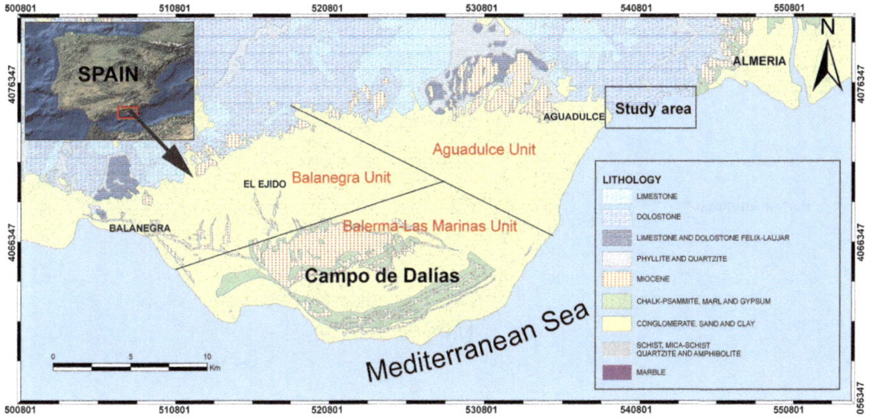

Fig. 1. Hydrogeological map of the study area.

3 Methods

In December 2010, an oceanographic survey was undertaken by the company CITECAM and the Water Resources and Environmental Geology Research Group of the University of Almería. The survey was designed to measure physical characteristics of the seawater in order to map the thermohaline structure of a zone from the port of Aguadulce to a point that is 5.6 Km eastward to the port (Fig. 2). Surface measurements were made using a SBE 45 Thermosalinograph, while conductivity, temperature and pressure measurements through the water column were made using a SBE 25 Sealogger CTD.

The survey transects have a total length of 23 km, and the surface thermoaline data was collected at a constant depth of 20-30 cm every 10 seconds with a velocity of 4 knots. Fifty-one (51) vertical profiles of temperature and conductivity from the water surface to the sea bed were carried out using CTD.

Data treatment was consisted of an initial calculation of statistical parameters in order to check goodness of fit and confidence intervals. Then the data were analysed using the Ocean Data View (Schlitzer 2007) and Arcgis 9.3 software. Furthermore, the data was treated statistically by cluster analysis.

Additionally, we used other derived variables, such as potential temperature and potential density. Potential temperature is used to compare waters taken from different depths or when vertical movements during the measurement procedure have to be taken into account, so avoiding the influence of compressibility of the water. Potential density allows a better estimate of the density difference between two water types. Using the Ocean Data View software, vertical temperature and salinity profiles were produced for all the sampling points, as well as T-S diagrams.

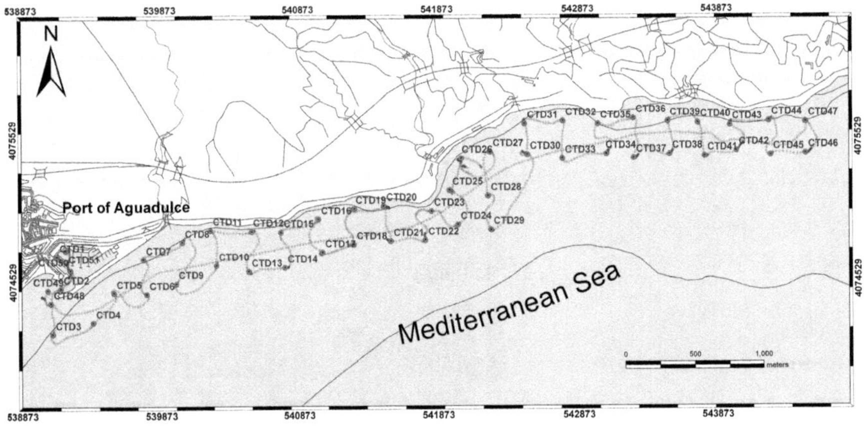

Fig. 2. Map showing the location of sampling stations.

4 Results and Discussion

Surface water analysis

The cluster analysis based on seawater temperature and salinity values showed that five seawater classes can be defined in the study area (Table 1).

Table 1. Statistics of each seawater class identified by cluster analysis of thermohaline data.

Class	Code	N° of cells	Average Temperature	Average Salinity	Covariance
1		11	16.04	36.11	0.00628
2		57	15.68	37.01	-0.00045
3		64	15.89	37.02	0.00024
4		60	16.04	37.00	0.00010
5		32	16.19	36.99	0.00022

Class 1 is characterised by much lower values of salinity than the other groups, and correspond to the water inside the port of Aguadulce (Fig. 3). Classes 2 to 5 differ from each other mainly on the basis of temperature; a clear and continual increase of temperature was observed from the port of Aguadulce to the east (Fig. 3). The relationship between the variables is more significant in Class 1. In the other classes, the two variables are statistically independent, which allow us to suggest that the temperature changes were not linked to salinity changes, but rather to the diurnal pattern of insolation.

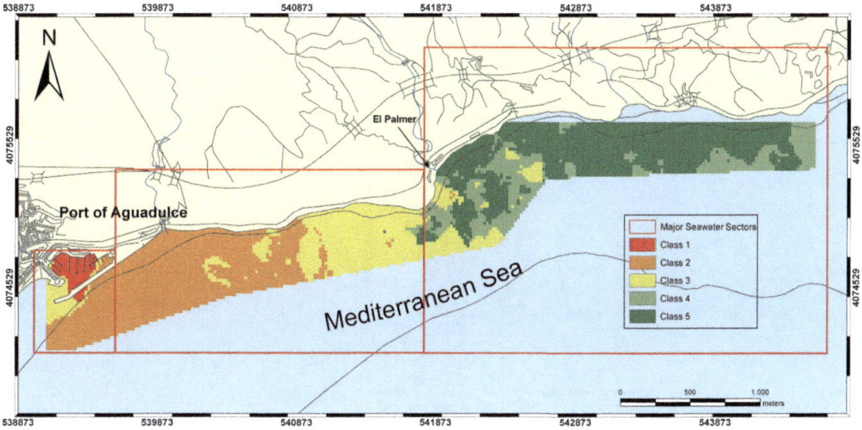

Fig. 3. Map showing the five (5) seawater classes produced by the cluster analysis on the basis of temperature and salinity values.

This interpretation leads to a simplified classification scheme consisting of three major seawater sectors in the survey area (Fig. 3). The first sector corresponds to Aguadulce port and is dominated by water belonging to class 1, whose salinity is lower than at the remaining sites. Class 1 salinity does not correspond to typical values for seawater in the Mediterranean at this time of year, and this suggests that there is an input of continental water within the confines of the port. The second sector extends from outside the port to El Palmer and is dominated by water classified in classes 2 and 3. The third sector is dominated by classes 4 and 5 and corresponds to the easternmost zone of the survey area. In this sector, only isolated anomalies were detected at certain points, which could be due to some continental inflows, possibly from urban wastewater discharges along the coast (Fig. 3).

The data exhibited a thermosaline anomaly in the area of Aguadulce port. This accords with the presence of a SGD, since the date of the survey (December 2010), the seawater temperature was around 15.5 ºC, which was below the 21.5 ºC measured in springs and boreholes in the aquifer. The salinity measurements also showed an anomaly.

Water column analysis (CTD)

The interpretation of the CTD data was done taking into account the three major sectors defined by the cluster analysis. CTD data showed that the stations located outside from the port exhibit almost similar temperature and salinity profiles. In these stations the seawater was significantly colder and more saline than the stations inside the port. This can be also observed in spatial distribution of salinity at the seasurface (Fig. 4d).

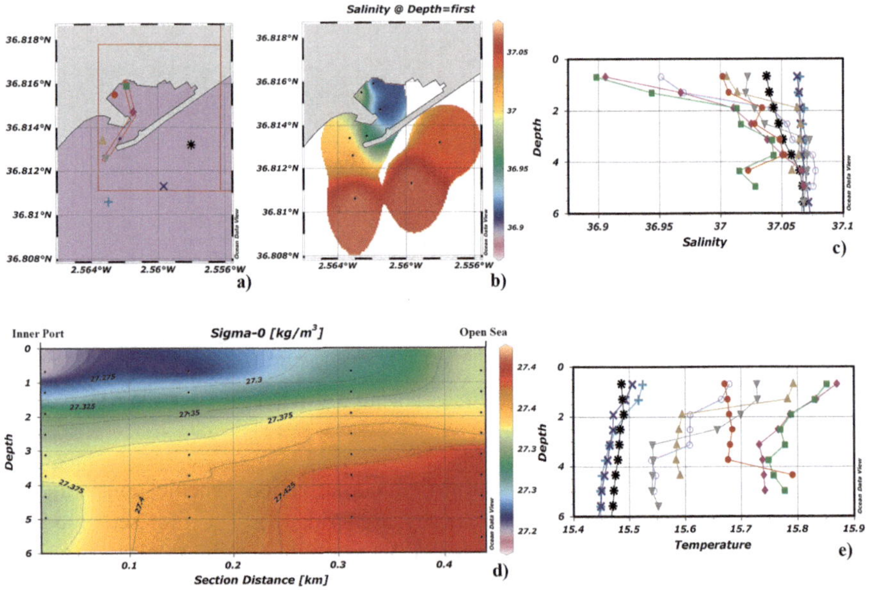

Fig. 4. Graphs resulting from CTD data. a) location of sampling stations; b) temperature and salinity profiles c); d) surface salinity map; e) potential density cross-section.

The potential density profile along a transect from the inner part of the port to the open sea clearly shows a band of lower density inside the port with the lowest values at the sea surface (1m) (Fig. 4e). This band of lower density is typical of seawater that is receiving inflows of continental fresh water (Stewart 2009).

The sectors 2 and 3, identified on Fig. 3, show values in-keeping with what is expected in the Mediterranean Sea at this time of year. There are only small anomalies at depth, which may be attributable to anthropogenic inputs as uncontrolled water waste discharges from the land, although submarine freshwater discharges cannot be ruled out.

5 Conclusions

The oceanographic survey undertaken over the 5.6 km between Aguadulce port and the city of Almería deduced a thermohaline anomaly in the vicinity of Aguadulce port, consisting of markedly lower salinity and higher temperature than for seawater. The temperature and salinity values obtained are typical of coastal waters that receive inflows of continental water. This discharge also appears in 1988, but during the nineties and early 2000, the entry of seawater, cooler and more saline waters into the aquifer were clearly detected by means of groundwater temperature and electrical conductivity logs.

The thermohaline anomaly occurred the last years may have been established due to the recent recovery of water levels in the aquifer, following a decline in the quantity of pumped abstractions as a result of worsening water quality, and because a new desalination plant now covers part of the water demand for the Campo de Dalias.

Acknowledgments This research was funded by Spanish Research Project MICINN CGL2007-63450/HID, and Andalusian Research Project PO6-RNM-01696. We are grateful to Cristina Pérez (INTECSA-INARSA) for their contribution during the oceanographic survey.

References

Burnett W, Bokuniewicz H, Huettel M, Moore W, Taniguchi M (2003) Groundwater and pore water inputs to the coastal zone. Biogeochemistry 66, 3-33

Burnett W, Aggarwal P, Aureli A, Bokuniewicz H, Cable J, Charette M (2006) Quantifying submarine groundwater discharge in the coastal zone via multiple methods. Science of The Total Environment 367, 498-543

Charette M, Buesseler K, Andrews J (2001) Utility of radium isotopes for evaluating the input and transport of groundwater-derived nitrogen to a Cape Cod estuary. Limnology and Oceanography 46, 46—470

Christodoulou D, Papatheodorou G, Ferentinos G, Masson M (2003): Active seepage in two contrasting pockmark fields in Patras and Corinth Gulfs, Greece. Geo-Marine Letters 23, 194-199

Espejo JA, Fernandez MC, Linares L (1988) Inventario de surgencias de aguas de origen continental en el litoral mediterráneo del sur de España, mediante utilización de sensores aeroportados con apoyo de técnicas oceanográficas. TIAC'88, I: 191-228

Henderson RD, Day-Lewis FD, Abarca E, Charles F, Harvey CF, Karam HN, Liu L, Lane JW (2010) Marine electrical resistivity imaging of submarine groundwater discharge: sensitivity analysis and application in Waquoit Bay, Massachusetts, USA. Hydrogeology Journal 18, 173-185

Johannesson KH, Darren A, Chevis DA, Burdige DJ, Cable JE, Martin JB, Roy M (2011) Submarine groundwater discharge is an important net source of light and middle REEs to coastal waters of the Indian River Lagoon, Florida, USA. Geochimica et Cosmochimica Acta 75, 825-843

Karageorgis AP, Papadopoulos V, Rousakis G, Kannelopoulos Th, Georgopoulos D (2011) Submarine groundwater discharges in Kalogria Bay, Messinia-Greece: geophysical investigation and one-year high resolution monitoring of hydrological parameters. Proceedings book. 9th International Hydrogeological Congress. Kalavrita, Greece

Kim G, Hwang D (2002) Tidal pumping of groundwater into the coastal ocean revealed from submarine 222Rn and CH4 monitoring. Geoph. Res. Lett. 29, 1678 doi:1610.1029/2002GL015093

Kim G, Swarzenski PW (2010) Submarine groundwater discharge (SGD) and associated nutrient fluxes to the coastal ocean. In: Quiñones, L. A., Talaue- McManus (eds.) Carbon and nutrient fluxes in continental margins: A global synthesis, part III pp. 529–538. New York: Springer Berlin Heidelberg

Molina L, Vallejos A, Pulido-Bosch A and Sánchez-Martos F (2002). Water temperature and conductivity variability as indicators of groundwater behavior in complex systems in the south-east of Spain. Hydrological Processes, 16(17), 3365-3378

Moore WS (2010) The effect of submarine groundwater discharge on the ocean. Annu. Rev. Mar. Sci. 2, 59-88

Morell I, Pulido-Bosch A, Sánchez Martos F, Vallejos A, Daniele A, Molina L, Calaforra JM, Francesc Roig A, Renau A (2008) Characterization of the salinisation processes in aquifers using boron isotopes; application to South-Eastern Spain. Water Air Soil Pollut. 187: 65-80

Pulido-Bosch A, Pulido-Leboeuf P, Molina L, Vallejos A and Martín-Rosales W (2000). Intensive agriculture, wetlands, quarries and water management. A case study (Campo de Dalías, SE Spain). Env. Geol: 40 (1-2), 163-168

Schlitzer R (2007) Ocean Data View, http://odv.awi.de

Stewart RH (2009) Introduction to Physical Oceanography. Texas A & M University. 353 pages (http://oceanworld.tamu.edu/ocean410/ocng410_text_book.html)

Taniguchi M, Burnett W C, Cable J E, Turner JV (2002) Investigation of submarine groundwater discharge. Hydrological Processes 16, 2115–2129

Viso R, McCoy C, Gayes P, Quafisi D (2010) Geological controls on submarine groundwater discharge in Long Bay, South Carolina (USA). Continental Shelf Research 30, 335-341

Aquifer systems of Epirus, Greece: An overview

E. Nikolaou[1], S. Pavlidou[1], K. Katsanou[2]

[1] Institute of Geology and Mineral Exploration (IGME). Greece, email: nikolaou@igmeprv.gr

[2] Laboratory of Hydrogeology, Department of Geology, University of Patras

Abstract The Epirus region covers an area of 9385 km^2 and displays one of the highest surpluses in water regime of the country. The annual renewable water supplies of the major aquifer systems are $2880 x 10^6$ m^3 and the total volume of the available groundwater $3220 x 10^6$ m^3. $2390 x 10^6$ m^3 are in storage in the karst aquifers and cover 75% of the Epirus water demand. Major elements analyses performed on 2310 groundwater samples from the study area showed that they are mainly of good quality and can be classified into three major groups. It should be noted that a significant percentage of the available groundwater is degraded, due to high SO_4^{2-} concentrations. In the coastal aquifer systems either porous or karst ones that are in hydraulic connection with the sea, occur active and passive seawater intrusion respectively. In those areas a small percentage of the groundwater is brackish. Elevated NO_3^- concentrations ranging from 80 to 150 mg/L were observed in many boreholes located in Preveza peninsula. Groundwater also from several springs and boreholes displays locally high nitrate content that varies from 25 to50 mg/L.

1 Introduction

This paper outlines the hydrogeological and hydrochemical conditions that prevail in the Epirus region. It is based both on a long term study that was carried out from the Institute of Geology and Mineral Exploration since 1985, and the findings of the project "Groundwater resources" evaluation and development funded by the 3rd Community Support Framework Programme: Competitiveness during 2004-2008.

The Epirus region is located in the NW Greece and covers 9385 km^2. It includes Aoos (2140 km^2), Arachtos (2160 km^2), Kalamas (1270 km^2), Louros (926 km^2), Acheron (762 km^2) Ioannina (508 km^2) watersheds as well as some other smaller ones. The aforementioned rivers along with Pamvotis Lake are the main hydrographic features of the study area. Physiographycally, the region is predominantly mountainous (74%). These mountains rise to 2500 m a.s.l. Considering the physiography, the geological structure, the annual precipitation which varies between 1000 and 1800 mm, and the hydrogeologic characteristics (55%) of the rocks are essentially impermeable), it can be concluded that the Epirus region is among the few regions of Greece displaying groundwater surplus and therefore there is a need for groundwater management.

In order to determine the quantitative and qualitative status of groundwater in the Epirus region, within the frame of this study, 605 groundwater samples from 311 springs and 294 boreholes were collected from 25 aquifer systems (primary and secondary) and local aquifers. The samples were analyzed for major elements. Also, discharge and water level measurements were carried out on a monthly basis. Moreover, 182 contaminating sources were also noted.

2 Geological and hydrogeological setting

The Epirus region is composed of formations of the Ionian, Gavrovo-Tripolitsa and Olonos-Pindos external geotectonic zones (Karakitsios 1992). Moreover, in the northeast part of the region (Metsovo-Grammos) the tectonic nappe of the Pelagonian zone is developed.

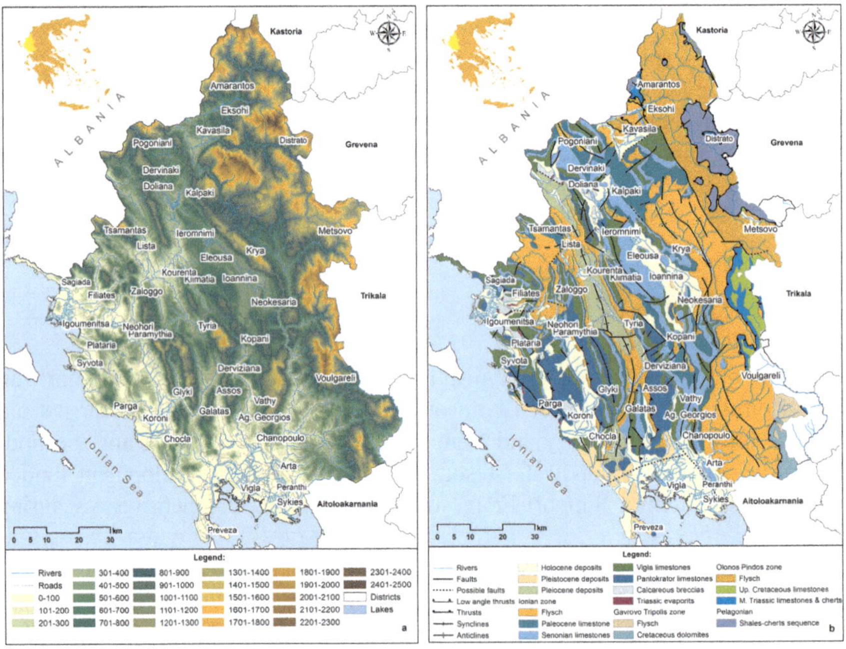

Fig. 1. a. Morphological and **b.** geological map of the study area.

The study area is mainly composed of formations of the Ionian zone. The lithostratigraphy of the zone determines the hydrogeological setting of the region. The older formations consist of Triassic evaporites overlain by a thick sequence of carbonate and clastic rocks of Late Triassic to Upper Eocene age which in turn are overlain by the clastic sequence of the Oligocene flysch (Skourtsis-Coroneou et al. 1995). The cores of the large anticlines in the internal Ionian zone consists of evaporites and no outcrops can be observed, while in the central and external part they irrupt along reverse faults in the thrusts (Nicolaou and Paschos 1999).The Cretaceous limestones and the flysch are the dominant formations in the Gavrovo-Tripolitsa zone. The Olonos-Pindos zone is composed of deep-sea sedimentary rocks of Triassic to Palaeocene age (Skourlis and Doutsos 2003). In the tectonic nape of Pelagonian age, the carbonate sequence and the ophiolithic complex are present. These formations are overlain by Neogene, Pleistocene and Holocene fluvial deposits, mainly developed in the Plains of Arta and Preveza, in Ioannina polje and in the Konitsa basin.

Based on their hydrolithological behavior the geological formations can be grouped into the following categories: a) Porous-fractured formations of Neogene to Quaternary age composed of alluvial deposits, conglomerates and unconsolidated lacustrine or terrestrial deposits of variable composition and particle size. Extended aquifer systems of low to high potential are formed within these rocks. The aquifers formed in the porous deposits of Arta-Preveza are of high potential, while the aquifer formed in the sedimentary filling of the Ioannina polje displays low to moderate potential. b) Karstified-fissured formations of all geological zones, comprise the most important hydrogeological unit and display moderate to high potential. It is worth mentioning that limestone with intercalations of chert layers (Vigla), constitute a formation of variable permeability (impermeable-permeable) which plays an important role in the configuration of the aquifer's geometrical characteristics. c) Fissured, non karstified formations consisting of ophiolitic rocks (basalts, serpentines, dolerites, gabbros etc) that form a significant discontinuous aquifer in their upper zone. d) Fissured or locally porous formations consisting of evaporitic breccias and evaporites of Triassic age control to a great degree the hydrogeologic and hydrochemical conditions of the study area. e) Practically impermeable formations, such as flysch, blue marls, shales, etc. The impermeable formations caver entirely 4223 km^2 (45%), the fissured karst formations 3590 km^2 (38.25%), the fissured rocks of the ophiolitic complex 869 km^2 (9.26%) and the porous-fissured formations of Neogene-Quaternary age 703 km^2 (7.49%).

Base level is an important control that affects the evolving porosity and permeability of carbonate rocks. Base level defines the altitude of the hydraulic drain from the flow systems. Four types of base-level control the hydrogeology. The four types are: (1) The water level of Pamvotis Lake (+469 m), in Ioannina city. (2) The perennial streams of Louros, Aoos, Kalamas, and Acheron, as they affects the water table and groundwater recharge (120-450 m), (3) The impervious formations at the base of the aquifer systems as (a) the large faults, from south to north, of Ziros, Petousi and Doliana respectively, (b) the contact along thrusts between limestone and flysch in Olonos-Pindos and Ionian zone, (c) the level of the base of the alluvial deposits of Arta-Preveza plain (4) The sea-level position as it affects the water table or artesian pressure and groundwater discharge in the coastal karst systems.

3 Aquifer systems of the Epirus region

The hydrogeological structure of Epirus as well as the major aquifer systems is presented in Figure 2 a, b. From east to west the hydrolitholig setting of the region comprises of: (1) the extended karst systems formed in the carbonate formations of Gavrovo, Pindos and Pelagonian geotectonic zones at altitudes ranging from 700 to2500 m. Lithologic and stratigraphic controls (e.g. the interbedded chert layers between the carbonate formations as well as the contacts between limestones and turbidite deposits in the major thrusts) play an important role in concentrating flow within specific zones. (2) The Louros, Ioannina, Tymfi, and Pogoni large karst systems of the central Ionian zone, which extend from the Greek-Albanian borders on the south to the Ziros transverse fault on the north.

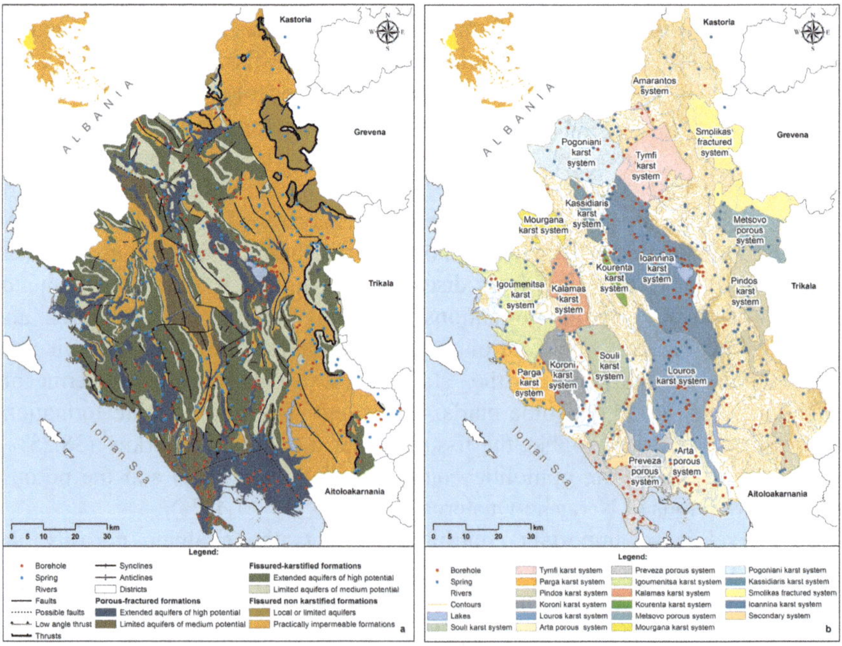

Fig. 2. a. Hydrolithological map, **b.** Map with the major aquifer systems of the study area.

This fault brings into contact limestones and alluvial deposits of Arta's plain. The lack of well defined impermeable boundaries between the limestones and alluvial deposits reinforces the point of view that they constitute a huge uniform karst system. Within this karst system two distinct aquifers are formed: the first is

formed in the Upper Senonian-Eocene limestones and the second in the Liassic limestones of Pantokrator series. They display three levels of discharge: (a) the uppermost one, determined by the water level of Pamvotis Lake (469 m), (b) the intermediate one determined by the Arachtos, Aoos, Kalamas, Louros perennial streams (260-460 m), and (c) the lower one at the altitudes of 10 to 40 m.
(3) The Filiates-Igoumenitsa and Parga large coastal karst systems developed within the external Ionian zone, at the coastal part of the Epirus region, extending from the estuaries of Kalamas River (on the north) to those of Acheron River (on the south), (4) The unconfined and confined alluvial aquifers formed in the Arta-Preveza Plain (Nikolaou and Tzouli 2000). Within the Arta's plain alluvial deposits which are estimated to be 250 m thick, an artesian aquifer of high potential is developed. This aquifer is composed of vertical alternations of permeable sand and gravel and impermeable clay marls. The Arta's plain include the following aquifers: (a) the upper shallow unconfined aquifer which occurs in the top 5 to 15 m, (b) the middle confined aquifer which occurs at depths 40 to 60 m, (c) the lower confined aquifer at depths of 60 to 90 m and (d) the lowermost confined aquifer at depths greater than 100 m. (5) The homonymous aquifer systems developed within the fissured rocks of the ophiolithic complex along Metsovo-Smolikas mountainous area. In many aquifer systems such as Amarantos, Grammos etc. the intense tectonics led to hydraulic integration among serpentines and limestones. Except from the major ones, many other of local interest or limited in extent aquifer systems, such as the aquifer formed in the sandstones of the turbidite deposits of Arachtos' watershed, etc. are also present.

Meteorological and hydrological data for the last decades were used in order to estimate the hydrological budget of the study area according to Thornthwaite and Mather (1957) (Nikolaou, 2001). The annual precipitation ranges from 400 to 785 mm (or $0.5x10^6$ $m^3/km^2/y$). Out of them 46% is the mean annual infiltration and 55% the mean annual runoff. The annual renewable groundwater supplies of Epirus totals $2880x10^6$ m^3. Moreover, $2880x10^6$ m^3 (about 10%) of the total groundwater is groundwater discharging from the fractured-karst formations. $62x10^6$ m^3 (2.16%) of groundwater are produced from the porous formations of Neogene and Quaternary age. $140x10^6$ m^3 (4.86%) of groundwater are in storage within the fractured non-karstified or locally porous formations, such as Triassic breccias.

The specific yield of the individual karst units as calculated by using natural Lysimeters ranges from 13.4 to 18.5 $L/s/km^2$ for Paleocene-Eocene limestones, from 17.4 to23 $L/s/km^2$ for Upper Senonian limestones, from 20 to25 $L/s/km^2$ for Pantokrator limestones and from 15 to17 $L/s/km^2$ for dolomites.

Moreover, in order to determine the hydraulic properties of the different hydrogeological units, several pumping tests were also carried out. The results regarding hydraulic conductivity, transmissivity and storativity (according to Cooper Jacobs's method) are shown in Table 1.

Table 1. Hydraulic properties of the hydrogeological units.

Formation	K (m/s)	T (m^2/s)	S (%)
Pantoktator limestones	$3.2 \times 10^{-4} < K < 4.6 \times 10^{-2}$	$3.7 \times 10^{-3} < T < 1.5 \times 10^{-2}$	2-3
Up. Senonian-Eocene limestones	$9.7 \times 10^{-4} < K < 3.25 \times 10^{-2}$	$7.5 \times 10^{-4} < T < 1.1 \times 10^{-1}$	1.5-6.28
Neogene-Quaternary deposits	$10^{-4} < K < 10^{-3}$	$10^{-4} < T < 10^{-3}$	5-10

4 Hydrochemistry and contamination problems

In total, 2310 groundwater samples were collected from 580 sampling sites. The unstable parameters (Ec, pH, Tw) were measured in situ. The samples were stored in 1L plastic bottles and analyzed in IGME laboratories for major and trace elements.

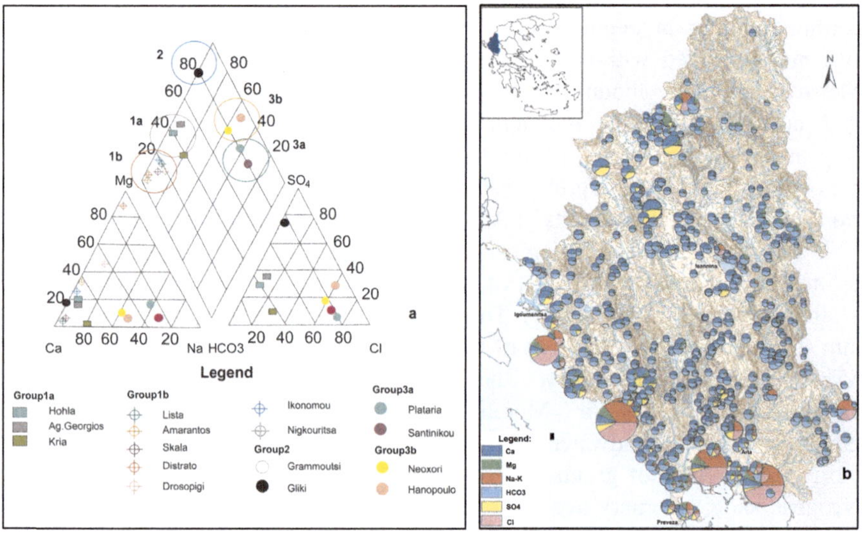

Fig. 3. a. Piper diagram, **b.** Hydrochemical map (Nikolaou 2010).

The classification of the samples was based on the Piper diagram of Figure 3a, showing the most representative ones for all groups (samples from springs with discharge rates exceeding 1 m^3/s. They can be classified into three major groups.

The first one includes samples from karst, fractured and porous aquifers of the study area. The second and the third one include the rest of the samples from the karst aquifer systems and particularly those of the Ionian zone. The first group is characterized by Ca-(Mg)-HCO_3 groundwater type and low water temperature values. It is the most abundant groundwater type representing fresh samples (recharging waters) and can be divided into two subgroups. This group (Fig. 1a) displays low TDS values (140-400 mg/L) and corresponds to groundwater samples from the karst systems and the porous aquifers of Arta-Preveza Plain. The samples from the Skala spring, which discharges the limestone aquifer of Senonian age also belongs to previous group (Fig. 3a). The samples of the second subgroup (1b) are enriched in Mg^{2+} which concentration varies between 26 and 45 mg/L. These are fresh groundwater with low TDS values (140-160 mg/L) corresponding to the aquifer systems in the ophiolitic complex and in the dolomites. Distrato spring (Fig 3b, 1a) is classified in this group.

The second group representing 20% of the samples is of Ca-HCO_3-SO_4 groundwater type. They display average water temperature values (Tw) of about 15°C and are mainly from aquifers in Pantokrator limestones. They display enrichment in SO_4^{2-} content with values that range between 250 and 900 mg/L. The samples of this group are connected to the evaporites and the evaporitic breccias, of the lower formations of the Ionian zone. Evaporites and the evaporitic breccias often outcrop in extensive areas in the immediate or wider catchment area of the large karst springs. Some typical springs of this group are the big sulfur springs of Kalamas River, Ieromnimi, Rogozi, and Mana Nerou of Pogoni karst system; Gliki springs of Souli karst system (Fig. 3a, 1b), Anakoli spring of Filiates karst system, Vathy spring in Louros karst system, Santonikou spring of Mitsikeli and Vouvos springs in Tymfi karst system.

The third group includes the samples that are enriched in Cl^- and belong to the Na-Cl-(SO_4) water type. They display increased salt content compared to those of the second and the third group and can be further divided into two subgroups. Chloride content for both 3a and b exceed the drinking water standards (250 mg/L). Group 3a includes groundwater samples from the karst aquifers which are in hydraulic communication with the sea in the west part of the Epirus coast inthe Ionian Sea, and is subjected to passive saline water encroachment and groundwater samples from the overexploited S-SW coastal part of the aquifer system of Preveza Plain which is also subjected to active seawater intrusion. In this area, the seawater intrusion affected many boreholes, where the Cl^- content ranges between 300 and1600 mg/L. A number of samples indicate sea water intrusion mainly in the northwest part were the coastal karst system of Parga is in hydraulic communication with the sea. The brackish zone extends along the northern part of Plataria up to Syvota village, where Cl^- values range between 415 and600 mg/L. A zone with brakish groundwater is also observed in the northern coastal area of Amvrakikos region (Vigla, Kommeno) and in the northern part of Parga in the Acheron and Kalamas watersheds (Plataria-Syvota karst system). Samples of group 3b display elevated Tw values (up to 33°C) and are enriched in H_2S. They exhibit excess in SO_4^{2-} (Fig. 3b) representing deep groundwater from the Hanopoulo and Peranthi springs of Xerovouni, Neles spring of Tymfi karst system and Kavasila spring in Konitsa, and are connected to the salt domes in the cores of anticlines.

The problems of NO_3^- contamination occur mainly in the Preveza plain. The samples from the porous confined and unconfined aquifers display high NO_3^- content varying between 100 and 200 mg/L. The high values of NO_3^- are attributed to the excessive and uncontrolled use of fertilizers. Typical cases of NO_3^- contamination can be observed in Assos, Galatas (27 mg/L) and Kopani (34 mg/L) springs of the Louros karst system and Eleousa spring (36 mg/L) of the Ioannina karst system.

5 Discussion-Conclusions

The renewable supplies of the major aquifer systems were estimated to be about 2880x10^6 m^3. Significant groundwater resources are also in storage in the low-permeability formations (discontinuous local aquifers). The geological setting illustrates characteristic alternations of limestone anticlines and flysch synclines of NW-SE trend which coincide with the Ionian coastline. This setting acts as a natural barrier preventing any potential seawater intrusion.

The total annual renewable groundwater resources of the Epirus region totals 3220x10^6 m^3 and exceed the annual water demands, which is approximately 400x10^6 m^3 (60x10^6 m^3 serve needs for drinking water, while 330x10^6 m^3 and 10x10^6 m^3 of water serve the irrigation and industrial needs respectively). The aquifers exploitation is low and estimated to be 25% of the total available groundwater. 75% of the total water use is produced from the karst aquifers.

The overexploitation of the aquifer in the Preveza peninsula by numerous boreholes that serve irrigation needs originated locally seawater intrusion problems.

The interactions between surface/groundwater are common, either through sinkholes or by direct infiltration along the river beds. The water table variations among wet and dry season depend on the pattern of annual precipitation, the hydraulic properties and the abstraction rate of the aquifers. The greater variations ranging between 14.8 and 35.1 m were recorded boreholes of the Ioannina karst system that serve drinking water needs. Moreover, great variations in groundwater level have also been observed in inadequately recharged karstic subsystems and hydrogeological units, such as the subsystem of Klimatia in Ioannina (54.4 m) and in areas where the aquifer systems are recharged through surface water, such as the upstream part of Louros' watershed (50.8 m). The groundwater level varies slightly in the aquifer systems of Pindos Mountains, where there is a better distribution of precipitation (3.5-10 m) and in the porous aquifer of Arta (1.5-3.5 m) and Preveza (0.9-9.4 m) Plains. Problems due to groundwater deficit occur during the peak season of water demand (August to September), primarily in Ioannina watershed. This watershed is characterized by the presence of intense human activities (dense population, industrial and agricultural activity).

Groundwater is mostly of very good quality and display low salinity, although approximately 20% of the water resources display high SO_4^{2-} concentrations, due to contact with the Triassic evaporites. A small percentage of groundwater is brackish due to seawater intrusion, either caused by the overexploitation of the porous aquifers or by passive seawater intrusion into the coastal karst aquifer systems. High NO_3^- values (80-150 mg/L) attributed to intense anthropogenic activities were observed locally in boreholes located in Preveza peninsula.

References

Katsikatsos G (1992) Geology of Greece, 451 pp. University of Patras (In Greek)

Nikolaou E, Paschos P (1999) The karst aquifers of Epirus and the reasons of their water quality degradation. Proc. 5th Hydrogeol. Congress, pp. 347-362, Nicosia, Cyprus (1999) (in Greek)

Nikolaou E, Tzouli Ch (2000) Hydrogeological study of Preveza Plain. IGME 33 pp. (In Greek)

Nikolaou E (2001) Study on the groundwater regime of Epirus karst aquifers. 2nd E.U. Framework project, IGME (in Greek)

Nikolaou E, Angelopoulos Ch, Paschos P (2002) The karst system of Ioannina and the connection of the aquifer to the lake. Proc. 6th Hydrogeol. Congress, pp. 131-142, Xanthi, Greece (2002) (in Greek)

Nikolaou E (2010) Update of Epirus groundwater data. Hydrogeology Report (V). 3rd E.U. Framework project IGME, pp.192 (In Greek)

Skourtsis-Coroneou V, Solakius N and Constantinidis I (1995) Cretaceous stratigraphy of the Ionian Zone, Hellenides, Western Greece. Cretaceous Res. 16: 539-558

Skourlis K, Doutsos T (2003) The Pindos Fold-and-thrust belt (Greece): inversion kinematics of a passive continental margin. Int. J. Earth Sci. 92, 903

Thornthwaite CW and Mather JR (1957) Instructions and tables for computing potential evapotranspiration and the water balance. Publ. Climatol. 10(3):311

Application of stochastic models to rational management of water resources at the Damasi Titanos karstic aquifer in Thessaly Greece

A. Manakos[1], P. Georgiou[2], I. Mouratidis[3]

[1] IGME, 1 Fragon, 54626 Thessaloniki, amanakos@thes.igme.gr
[2] Department of Hydraulics, Soil Science and Agricultural Engineering, Faculty of Agriculture, A.U.TH, 54124 Thessaloniki, pantaz@agro.auth.gr
[3] Voulgaroktonou , 50100 Kozani, ilmouratid@gmail.com

Abstract Several stochastic models, known as Box and Jenkins or SARIMA (Seasonal Autoregressive Integrated Moving Average) have been used in the past for forecasting hydrological time series in general and stream flow or spring discharge time series in particular. SARIMA models became very popular because of their simple mathematical structure, convenient representation of data in terms of a relatively small number of parameters and their applicability to stationary as well as nonstationary process. The application of SARIMA model to the Mati spring's monthly discharge time series for the period 1974-2007 at Damasi Titanos karst system yielded the following results. The stationary is obtained by logarithmic transformation and the suitable model $(2,0,0)(0,1,1)_{12}$ is selected by different criteria. This type of model is suitable for the Damasi Titanos karst aquifer simulation and can be utilised as a tool to forecast monthly discharge values at Mati spring for at least a 4 year period. SARIMA model seem to be capable of simulating both runoff and groundwater flow conditions on a karst system and also easily adapt to their natural conditions.

1 Introduction

Stochastic models, known as Box-Jenkins or Seasonal and nonseasonal ARIMA (AutoRegressive Integrated Moving Average) models were applied in scientific, economic and engineering applications to time series analysis and forecasting (Box and Jenkins, 1976). Some useful applications of these models can be found in the literature for the simulation of time series of flow data, groundwater head, water quality parameters (Box and Jenkins 1976, Hipel and McLeod 1994, Ahn and Salas 1997, Manakos 1999, Papamichail et al. 2000, Papamichail and Georgiou 2001, Manakos and Dimopoulos 2004, Manakos and Georgiou 2009, Voudouris et al 2010). SARIMA models became very popular because of their simple mathematical structure, convenient representation of data in terms of a relatively small number of parameters and their applicability to stationary as well as nonstationary process. In this paper the SARIMA models were used to simulate the discrete time series of monthly discharge values at Mati spring (1974-2003) for the Damasi Titanos karst aquifer and was investigated the suitability to forecast monthly discharge values for the rational management of water resources of the aquifer. The suitable SARIMA model was used to forecast monthly discharge values at Mati spring for a 4 year period (2004-2007).

2 SARIMA models

For the description of the seasonal SARIMA models (Seasonal Autoregressive Integrated Moving Average) are used the notations (p, d, q) (P, D, Q)$_S$. The SARIMA model has the general form:

$$\varphi(B)\Phi(B^S)(1-B)^d(1-B^S)^D(Z_t-\mu)=\theta(B)\Theta(B^S)e_t \quad (1)$$

The application of the SARIMA models requires stationarity of time series data obtained by different transformations. The logarithmic transformation is chosen to stabilize the variance of the series and to transform the usually skewed distribution into a normal distribution.

The construction of SARIMA models involves various stages which are the identification, the estimation and the diagnostic checking (Box and Jenkins 1976, Hipel and McLeod 1994, Mohan and Vedula 1995). The purpose of the identification stage is to determine the differencing required to produce stationarity and to estimate the order of both the seasonal and nonseasonal AR and MA operators (p, d, q, P, D, Q) for the stationary series. The *identification* is examined by the cumulative periodogram, the autocorrelation function (ACF) and the partial autocorrelation function (PACF) (Box and Jenkins, 1976). This information is then used to determine the general form of the univariate model to be fit.

The estimation stage involves the estimation of the time series model parameters. This estimation is obtained by the residual sum of squares minimization using the Marquardt optimization algorithm for non-linear least squares. The value of the parameters is associated standard errors, T-values and P-values.

Finally, the *diagnostic checking* involves examination of the residuals fitted model, and this can prove or not the model inadequacy and inform about the model improvement. This determination can be achieved by using the identification stage tests and furthermore the Portmanteau test (Box and Pierce, 1970), the Akaike Information Criterion (AIC) (Akaike 1974), the Schwarz Bayesian Criterion (SBC) (Schwarz 1978), the Hannan and Quinn Criterion (HQC) (Hannan and

Quinn 1979) and the residual variance (σ_e^2). For the log transformed series the AIC, SBC and HQC criterions are given, respectively, by:

$$\mathrm{AIC} = N \ln \sigma_e^2 + 2n + 2N\overline{\ln Z_t} \tag{2}$$

$$\mathrm{SBC} = N \ln \sigma_e^2 + n \bullet \ln(N) + 2N\overline{\ln Z_t} \tag{3}$$

$$\mathrm{HQC} = N \ln \sigma_e^2 + 2n \bullet \ln\left[\ln(N)\right] + 2N\overline{\ln Z_t} \tag{4}$$

where σ_e^2 is the estimated residual variance, $\overline{\ln Z_t}$ is the sample mean of the logarithms of measured values and n is the number of parameters.

For a SARIMA model the number of parameters n is given by:

$$n = p + P + q + Q + pP + qQ + 1 \tag{5}$$

The selection rule in the AIC, SBC and HQC criterions is to select the model with the lowest AIC, SBC and HQC value. The best model chosen is used for forecasting.

3 Geological-hydrogeological setting

The Damasi Titanos karstic aquifer extends along the west border of the east Thessaly plain at a distance of 2 Km west of the city of Tyrnavos. Geotectonically belongs to the wider area of the Pelagonian zone and it develops within a continuous carbonate series of Triassic-Jurassic age.

A simplified geological map of the aquifer is shown in Figure 1. The karst system, with an areal extent of 180 km^2, represents a high productive aquifer and supplies water mainly for irrigation and drinking purpose. Mean multi annual recharge of the karst aquifer was estimated to be over 80 hm^3/year predominantly originating from direct infiltration in the marbles of both precipitation and overland flow of the Titarisios river. Three karstic springs (Mati, Agia Anna and Amygdalia), located at the east margin, discharge the aquifer, in an altitude between 73 and 75 m a.s.l. The large spring of Mati is the main discharge point of the aquifer. A schematic geological profile through the Mati spring is added in Fig. 1. Its annual spring discharge, for the year 2007, was measured to be 70.55 hm^3, whilst the mean annual water temperature was 14.5 ^{o}C. Intensive pumping in the karst area reaches to groundwater table decline and leads to large changes in storage and discharge rate of the springs.

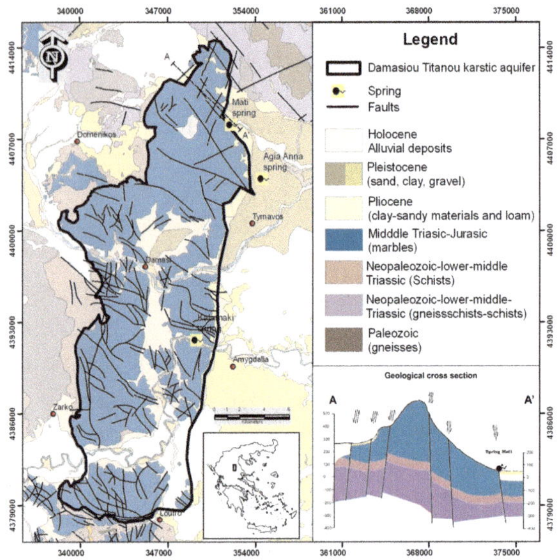

Fig. 1. Geological map of Damasi Titanos karstic aquifer.

4 Results and discussion

In this paper, the SARIMA models are used for the simulation of discrete time series of historical monthly discharge at Mati spring for the Damasi Titanos karst aquifer. The available historical monthly data cover 34 years (1/1974-12/2007) and is shown in Figure 2.

Observed discharge measurements were obtained from the Land Reclamation Service of Larissa and Institute of Geology and Mineral Exploration. From the time series we used the period 1974-2003 to construct a suitable SARIMA model,

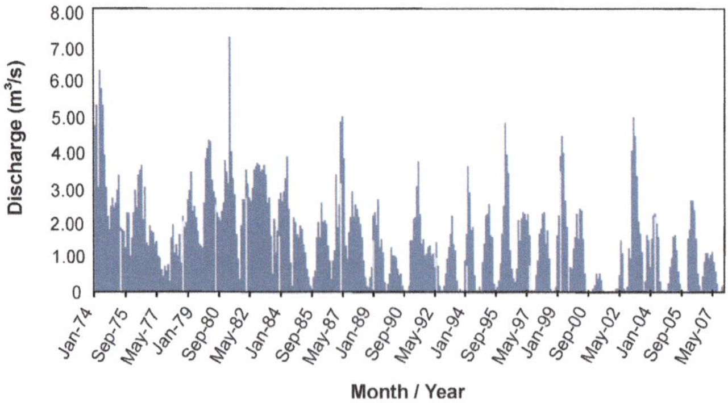

Fig. 2. Monthly discharge at Mati spring.

while the period 2004-2007 was used as a forecasting period to check the reliability of the model

SARIMA models take into account both the seasonality, which was introduced by the length of seasons S, and the persistence of time series. The stationarity of time series data was obtained by logarithmic transformation. For the logarithmic transformed time series of monthly discharge (1974-2003) at Mati spring was computed the cumulative periodogram [$C(f_k)$], the autocorrelation function (ACF) and the partial autocorrelation function (PACF) (Fig. 3).

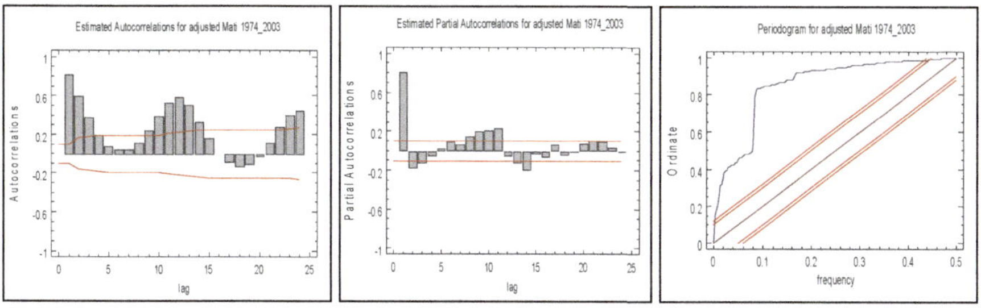

Fig. 3. a) Autocorrelation Function (ACF) b) Partial Autocorrelation Function (PACF) c) Cumulative Periodogram [$C(f_k)$] of the logarithmic transformed time series of monthly discharge (1974-2003) at Mati Tirnavou spring.

Figure 3 shows that the logarithmic transformed time series of monthly discharge isn't white noise, the length of seasons S is equaled to 12 and the model with structure $(3,0,0)(0,1,1)_{12}$ is a candidate model. To allow for possible identification errors, a set of six models (Table 1) were considered.

Table 1. Number of parameters (n), residual variance (σ_e^2) Akaike information criterion (AIC), Schwarz bayesian criterion (SBC), Hannan and Quinn criterion (HQC) and Portmanteau test (Q_p) of the logarithmic transformed time series of monthly discharge (1974-2003) at Mati spring.

Model Structure	n	σ_e^2	AIC	SBC	HQC	Q_p
$(1,0,0)(0,1,1)_{12}$	*3*	0.703477	-208.8034	*-197.1451*	-204.1678	25.6558
$(1,0,1)(1,1,1)_{12}$	7	0.694439	-205.4585	-178.2558	-194.6422	17.1416
$(1,0,1)(0,1,1)_{12}$	5	*0.693329*	-210.0344	-190.6039	-202.3084	17.1950
$(2,1,1)(0,1,1)_{12}$	6	0.713248	-197.8376	-174.5209	-188.5664	*16.5782*
$(2,0,0)(0,1,1)_{12}$	4	0.694606	*-211.3719*	-195.8275	*-205.1912*	17.6178
$(3,0,0)(0,1,1)_{12}$	5	0.695109	-209.1113	-189.6808	-201.3854	17.6567

The number of parameters (n), the residual variance (σ_e^2), the Akaike information criterion (AIC), the Schwarz bayesian criterion (SBC), the Hannan and Quinn criterion (HQC) and the Portmanteau test (Q_p) were applied to select the best model (Table 1).

The model $(2,0,0)(0,1,1)_{12}$ has the lowest AIC and HQC and was selected to simulate the logarithmic transformed time series of monthly discharge. The parameters of the model (2,0,0)(0,1,1)$_{12}$ were estimated using the Marquardt algorithm. For the above model the parameters with their confidence limits, standard errors, T-values and P-values are given in Table 2.

Table 2. Parameter estimates and their 95% confidence limits, standard errors, T-values, and P-values for the (2,0,0)(0,1,1)$_{12}$ of the logarithmic transformed time series of monthly discharge (1974-2003) at Mati spring.

Parameter	Estimate	95% Conf. limits	Stnd. Error	T-value	P- value
φ_1	0.654883	(0.725028,0.584738)	0.053958	12.137	0.0000000
φ_2	0.127369	(0.197372,0.057366)	0.053848	2.3653	0.0185660
Θ_1	0.816183	(0.859633,0.772733)	0.033423	24.420	0.0000000

The selected model (2,0,0)(0,1,1)$_{12}$ has to be validated for its suitability through the diagnostic checking. The first check is that if the residuals are white-noise. For this reason were computed the cumulative periodogram [C(f_k)], the autocorrelation function (ACF) and the partial autocorrelation function (PACF) of the residuals which are shown in Figure 4. It is observed that the all values are well within the 95% confidence limits confirming the residuals are white noise.

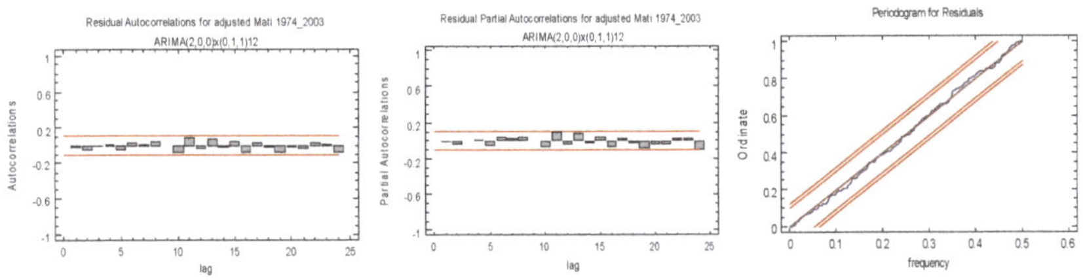

Fig. 4. a) Autocorrelation Function (ACF) b) Partial Autocorrelation Function (PACF) c) Cumulative Periodogram [C(f_k)] of the residuals for the (2,0,0)(0,1,1,)$_{12}$ model.

The second check is referred to examine the confidence limits of the parameters. The hypothesis is that if the confidence limits include the value zero, then the corresponding parameter needs to be discarded and if they include 1, then replacement of the associated term by a difference operator (1-B) is necessary. From Table 2 it can be seen that the confidence limits of the parameters of (2,0,0)(0,1,1)$_{12}$ model do not include the value zero or 1 confirming the suitability of the model.

Therefore, the SARIMA model (2,0,0)(0,1,1)$_{12}$ can be used to forecast the monthly discharge for one or more time steps. This was done for the monthly discharge of period 2004-2007. The observed and forecasted values of discharge with (2,0,0)(0,1,1)$_{12}$ model for the period 2004-2007 at Mati spring are shown in Figure 5. It can be seen the ability of the SARIMA model (2,0,0)(0,1,1)$_{12}$ to forecast the monthly discharge of Mati spring with sufficient accuracy.

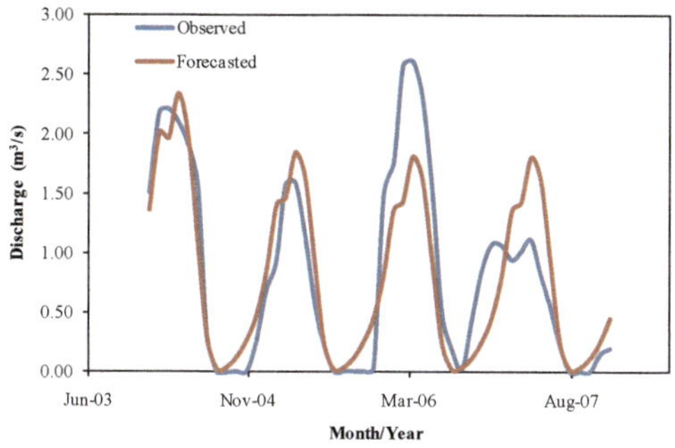

Fig. 5. Comparison between observed and forecasted values of discharge with $(2,0,0)(0,1,1)_{12}$ model for the period 2004-2007 at Mati spring.

Some basic statistical properties of the observed and forecasted data of discharge with $(2,0,0)(0,1,1)_{12}$ model for the period 2004-2007 at Mati spring are given in Table 3. Using t-test for the means and F-test for the variances (Yurekli et al 2005) to determine whether there is significant difference, the mean and variance values of the observed and forecasted data were tested.

The hypothesis that the mean and variance values of the forecasted data are not significantly different from those of the observed data can be accepted at 5% significance level (Table 3). Thus, the results show that the forecasted data preserve the basic statistical properties of the observed data.

Table 3. Statistical properties of observed and forecasted values of discharge with $(2,0,0)(0,1,1)_{12}$ model for the period 2004-2007 at Mati Tirnavou spring.

	Mean	$t_{cal}<t_{table}$	Standard deviation	$F_{cal}<F_{table}$
Observed	0.845	0.12<1.66	0.814	1.31<1.62
Forecasted	0.826		0.710	

From the above analysis results that the SARIMA models can be used successfully for short term forecasting of the discharge values at Mati spring and can also generate sequences of monthly synthetic time series of unlimited duration. This offers the possibility for decision making for integrated protection and management of groundwater and lead to the right decisions for the programming of various management plans.

5 Conclusions

Application of the seasonal stochastic model SARIMA to the Mati spring's monthly discharge time series for the period 1974-2007 in Damasi Titanos karst system yielded the following results. Logarithms of the monthly spring discharge time series can be simulated on aSARIMA model $(2,0,0)(0,1,1)_{12}$. This model is suitable for the Damasi Titanos karst aquifer simulation and can be utilised as a tool to predict monthly discharge values at Mati spring for at least a 4 year period. Seasonal stochastic models SARIMA seem to be capable of simulating both run-off and groundwater flow conditions on a karst system and also easily adapt to their natural conditions. Adapting the proper stochastic model to the karst groundwater flow conditions offers the possibility to obtain accurate short term predictions, thus contributing to rational groundwater resources exploitation and management planning.

References

Ahn H, Salas JD, (1997) Groundwater head sampling based on stochastic analysis. Water Resour. Res. 33(12), 2769-2780

Akaike H, (1974) A new look at the statistical model identification. IEEE Trans. Autom. Control. AC-19(6), 716-723

Box GEP, Jenkins GM, (1976) Time Series Analysis: Forecasting and Control. Revised Edition. Holden Day, Inc., San Francisco, Calif., 532 p

Box GEP, Pierce DA, (1970) Distribution of autocorrelations in autoregressive integrated moving average time series models. J. Amer. Stat. Assoc, 180

Hannan EL, Quinn BG, (1979) The determination of the order of an autoregression. J. Roy. Stat. Soc. B., 41, 190-195

Hipel KW, McLeod AI, (1994) Time Series Modelling of Water Resources and Environmental Systems. Elsevier Science B.V., Developments in Water Science, No 45, 1013

Manakos A, (1999) Hydrogeological behavior and stochastic simulation of Krania Elassona karstic aquifer Thessaly. PhD Thesis, Aristotle University of Thessaloniki, 214 p. (in Greek)

Manakos A, Dimopoulos G (2004) Contribution of stochastic models to the sustainable water management. The example of Krania Elassona karstic aquifer in Thessaly. Proceedings of the 10th International Conference of Greek Geol. Society, April 15-17, Thessaloniki, 361-368 (in Greek)

Manakos A, Georgiou P, (2009) Time series modeling of groundwater head using seasonal stochastic models SARIMA. Proceedings of the common Conference of the 11th Hellenic Hydrotechnical Society and of the 7th Conference of the Hellenic Committee of Water Management, 27-30 May, Volos, 709-716 (in Greek)

Mohan S, Vedula S, (1995) Multiplicative seasonal arima model for longterm forecasting of inflows. Water Res Manag 9, 115-126

Papamichail DM, Antonopoulos VZ, Georgiou PE (2000) Stochastic models for Strymon river flow and water quality parameters. Proceedings of the International Conference "Protection and Restoration of the Environment V", Thassos, Greece, Vol. I, 219-226

Papamichail DM, Georgiou PE, (2001) Seasonal ARIMA inflow models for reservoir sizing. J. Amer. Water Res. Assoc. 37(4), 877-885

Schwartz G, (1978) Estimating the dimension of a model. Ann. Statist, 6, 461-464

Yurekli K, Kurunc A, Oztrurk F (2005) Application of linear stochastic models to monthly flow data of Kelkit Stream. Ecol. Mod. 183, 67-75

Voudouris K, Georgiou P, Stiakakis E, Monopolis D, (2010) Comparative analysis of stochastic models for simulation of discharge and chloride concentration in Almyros Karstic Spring in Greece. e-Proceedings of the 14th Annual Conference of the International Association of Mathematical Geosciences, IAMG 2010, Budapest, Hungary, 15 p

Solution of operation and exploitation issues of the Almiros (Heraklion Crete) brackish karst spring through its simulation with the MODKARST model

A. Maramathas

National Technical University of Athens, School of Chemical Engineering, Zografos Campus GR-15780 Athens, Greece, thamar@chemeng.ntua.gr

Abstract A simulation is presented of the operation of the Almiros brackish karst spring at Heraklion Crete Greece with the MODKARST deterministic mathematical model. Through this simulation, major issues regarding the operation mechanisms as well as the sustainable development of the spring are addressed and analyzed. Of concern are the following aspects: (a) The fractal characteristics of the Almiros karst formation, (b) the sea intrusion mechanism for the Almiros spring and (c) the choice of a suitable method for the sustainable development of the Almiros spring. It was found that the seawater intrusion in the Almiros karst-spring reservoir follows a power law, the exponent of which provides the fractal dimension of the system. Regarding sea intrusion, the spring becomes brackish due to the crossing of the tube, which carries the fresh water to the spring, with other tubes, carrying saltwater from the sea. MODKARST simulates the hydrograph along with the variation with time of the chloride concentration and correlates the discharges from the fresh water and the seawater tubes. The outcome is the relative importance of the two potential mechanisms for seawater intrusion, namely the density difference between fresh water and sea water and the venturi effect; it is found that the density difference dominates in the Almiros case. An important conclusion of the analysis is that raising the Almiros-spring water outlet could prevent the seawater intrusion; the location of the elevated outlet and the fresh water loss to the sea are estimated.

1 Introduction

The MODKARST model is a deterministic mathematical model that simulates the operation of brackish karst springs (Maramathas et al. 2003). The input to the model is the rainfall data; the output is the hydrograph and the variation with time of the chloride concentration of the spring water. The model is based on a hydrodynamic analogue, consisting of a number of reservoirs properly connected. Each of the reservoirs is discharged to the tube that abuts to the spring. The governing equations of the model are mass and mechanical energy macroscopic balances. More details are available in Maramathas et al. 2003.

The model is used to simulate the operation of the Almiros karst spring at Heraklion Crete in Greece. The Almiros spring is located 10 km west of the city of Heraklion, at an elevation of about 4m and at a distance of 1km away from the sea. It is a periodically brackish karst spring, the discharge of which fluctuates between $4m^3/sec$ in the summer and 70-80 m^3/sec in the winter. During almost 90% of the year the spring water is brackish due to seawater entering the spring's reservoir. The ground-water reservoir of the spring is structured by Mesozoic limestone with preferential permeability and small porosity.

Representative results of the Almiros-spring simulation are shown in Figures 1A and 1B; the model predictions for the hydrograph and for the water chloride concentration versus time are compared with measurements.

Below are presented the three major results drawn from the simulation, with regard to the fractal characteristics of the Almiros karst formation, the sea intrusion mechanism and the sustainable development of the Almiros spring.

2 Measurement of the fractal characteristics of the Almiros karst formation

Although predictions and measurements of the Almiros-spring hydrograph appear to be in good agreement (cf. Fig. 1B), the corresponding comparisons of earlier predictions with measurements for chloride concentration revealed a discrepancy during the depletion periods. The discrepancy is shown in Fig. 2A; it is not shown in Figure 1B because the corresponding theoretical predictions are obtained with an improved model – see below.

The spring exhibits an unexpected behaviour during depletion periods when there is no rain at all in Crete. The slope of the real spring water chloride concentration curve decreases, that is, the rate of increase of the chloride concentration of the spring water decreases during depletion periods. Moreover, in some cases, the chloride concentration decreases without any prior rainfall (Fig. 2B).

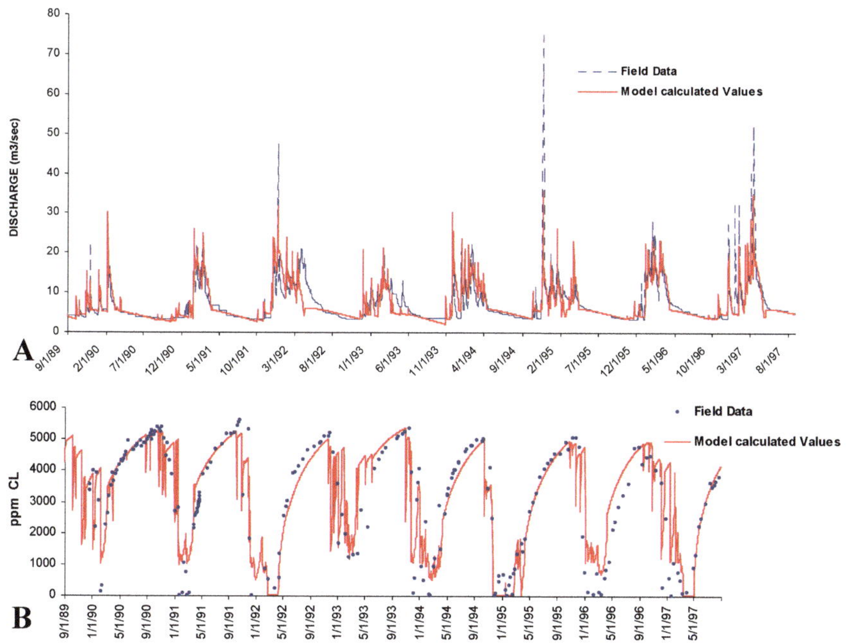

Fig. 1. A. Comparison between the model-predicted hydrograph and field measurements (field data) for the Almiros spring, **B.** Comparison between the model-predicted water chloride concentration versus time and field measurements (field data) for the Almiros spring.

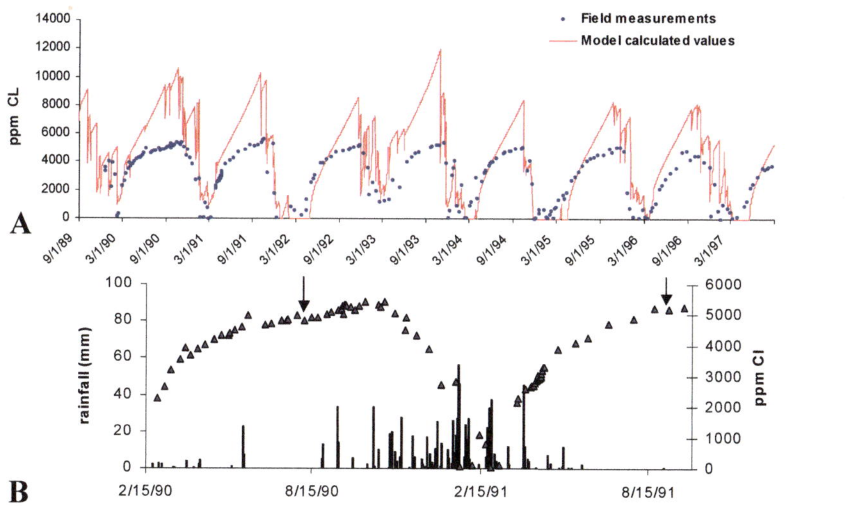

Fig. 2. A. Comparison between predictions and measurements of water chloride concentration before the use of the power law. A discrepancy is observed during depletion periods. **B.** Rainfall and chloride concentration versus time for the Almiros spring. In two cases (arrows) the chloride concentration decreases without any prior rainfall.

The key to explain the behaviour of the spring with regard to chloride concentration during the depletion period is in the seawater intrusion into the karst system. For seawater to intrude, the pressure in the freshwater tubes must be lower than the pressure in the tubes carrying the saltwater from the sea. This is the case during the depletion periods and as the water level in the karst system reservoir decreases, due to the absence of rainfall, the flow ceases in an increasing number of freshwater tubes. As a result, the saltwater flow in the tubes, which come from the sea and meet the freshwater tubes, ceases too. Consequently, the rate of increase of the saltwater discharge to the system decreases and the rate of increase of the spring water chloride concentration decreases too.

To account for these effects, we adopt the reasonable assumption that the effective seawater outflow tube cross-section (S_3) depends on the saturated volume of the karst reservoir and consequently on the water level (H_f) in it. It comes out, through model fitting, that the best agreement between theoretical predictions and measurements is obtained with the power law $S_3 = aH_f^b$. From further fitting it was found that b = 0.59. This value coincides with the fractal dimension of the system, which has been estimated following the analysis by Maramathas and Boudouvis 2006.

3 Sea intrusion mechanism

From the hydrodynamic point of view, two are the potential mechanisms of seawater intrusion (Maramathas et al. 2006, Breznic 1973):

1. The fresh water tube that abuts to the spring intersects another tube that comes from the sea and, as a result, the freshwater becomes brackish at the intersection, when the pressure in the freshwater tube is lower than the pressure in the seawater one – cf. Figure 3B. This becomes possible if the intersection is at large enough depth below the sea level, since the seawater density is larger than that of the freshwater.
2. The pressure in the freshwater tube at the intersection is less than that in the seawater due to geometry variations. In other words, the freshwater tube is quite narrow at the intersection and the pressure decreases due to the venturi effect – cf. Figure 3A.

Actually, even though both mechanisms are active in any case, it is important to find the dominant one; this is valuable in deciding which the suitable spring development method is.

Deciding about the prevailing seawater intrusion mechanism amounts to determining the way in which the freshwater discharge (Q_1 in Figures 3B, 3A) affects the seawater discharge (Q_2 in Figures 3B, 3A). Decreasing the water level over the discharge point of the spring's reservoir decreases freshwater discharge and the imbalance between saltwater and freshwater pressure increases; consequently, the seawater discharge increases. In this case the dominant intrusion mechanism is the difference between freshwater and seawater density.

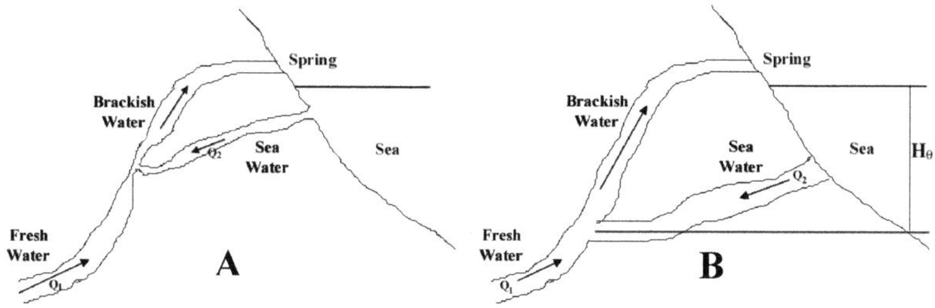

Fig. 3. A Sea intrusion mechanism of the spring based on the venturi effect. **B.** Sea intrusion mechanism of the spring based on the difference between the freshwater density and the seawater one.

On the other hand, there is a mutual augmentation between the seawater and freshwater discharge when the mechanism caused by the venturi effect prevails. This happens due to the intensification of the venturi effect, as a result of the increase of the freshwater velocity at the intersection between the seawater and the freshwater tube which thus provokes the increase of the freshwater discharge.

MODKARST calculates Q_1 and Q_2 in every run and uses them as intermediate variables to calculate the spring discharge and the chloride concentration of the spring water. Thus, a relationship between freshwater and saltwater discharge is obtained, as shown in Figure 4A for the Almiros spring. As it can be observed, the seawater discharge decreases when the freshwater discharge increases leading to the conclusion that the dominant sea intrusion mechanism for the Almiros spring is the difference between the seawater and freshwater density.

4 The development of the Almiros spring

In some brackish springs it is possible for sea intrusion to be blocked by raising the water outlet point of the spring, through the construction of a dam in front of it as, in this way, the freshwater level in the spring reservoir will be raised and the resulting pressure increase will prevent sea intrusion (Maramathas 2004, Breznic 1973). This is an old but not very popular method since it raises two difficulties to be dealt with. First, the determination of the necessary vertical movement of the outlet and, second, the estimation of the freshwater loss (discharge) to the sea; the latter is caused since the tubes, that previously brought seawater to the spring reservoir, will bring fresh water to the sea after the elevation of the spring outlet which increased the freshwater pressure. MODKARST is a valuable tool in dealing with these difficulties.

One of the parameters of the model is the elevation of the spring's water outlet point. Consequently, the model can simulate the operation of the spring at different spring elevations. This has been applied to the Almiros spring for the years from 1989 to 1997 and the results are presented in Figures 4B, 4C and in Table 1. In Figures 4B and 4C, the model predicted chloride variation versus time for the 10-meters and 20-meters spring elevation respectively, is compared to the real one at the real elevation (4m). As it can be observed, the quality of the water improves as the elevation of the water outlet point increases. In Figure 5, the simulated percentage of the days throughout the year with fresh water versus the spring elevation for the hydrologic years 1989-90 to 1996-97 is presented. It comes out that the periods during which the spring offers fresh water become longer as the spring elevation increases. Moreover, the spring offers fresh water during the entire year when the elevation reaches the 22m. Finally, in Table 1, the model-calculated freshwater loss versus the spring elevation for the depletion period and the entire period, respectively, is presented.

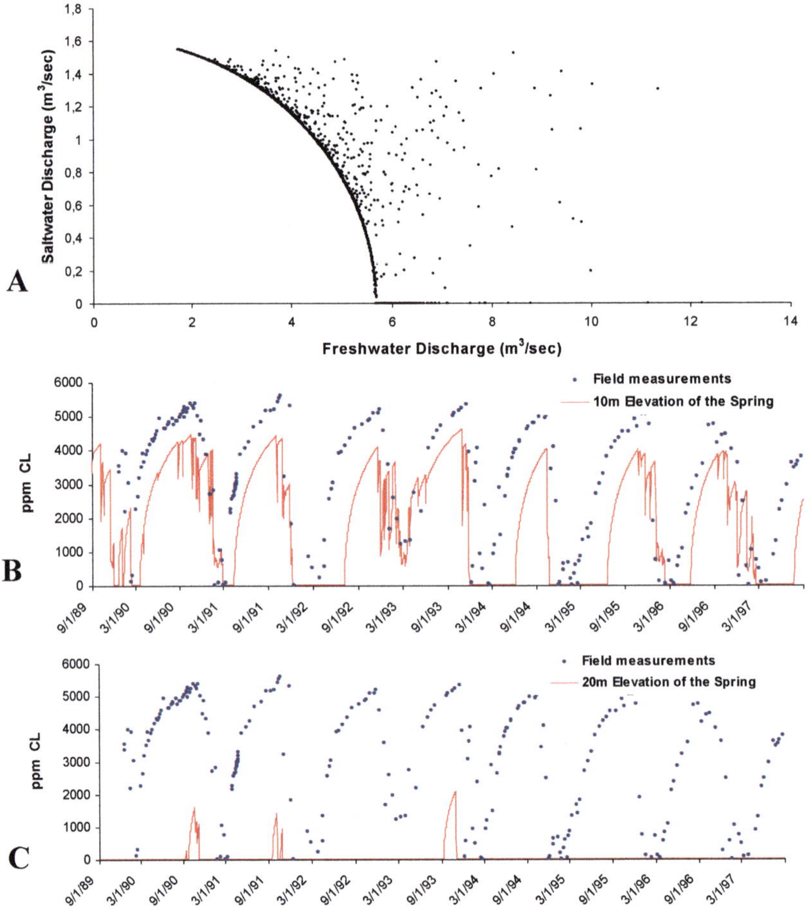

Fig. 4. A. Saltwater discharge versus freshwater discharge for the Almiros spring. Saltwater discharge increases when freshwater discharge decreases. **B.** Comparison between the observed chloride concentration versus time at the real spring elevation point (field measurements) and the simulated one at 10m **C.** Comparison between the observed chloride concentration versus time at the real spring elevation point (field measurements) and the simulated one at 20m.

Table 1. Spring freshwater loss (%).

Elevation	Whole period	Depletion period
4	0.00	0.00
5	0.03	0.00
6	0.81	0.00
7	1.70	0.50
8	2.96	1.02
9	4.24	2.32
10	5.55	4.06
11	6.99	6.39
12	8.53	8.93
13	9.98	11.63
14	11.50	13.91
15	13.15	17.05
16	14.68	20.83
17	16.21	24.26
18	17.68	27.75
19	18.97	30.38
20	19.96	33.22
21	20.74	34.75
22	21.23	35.33

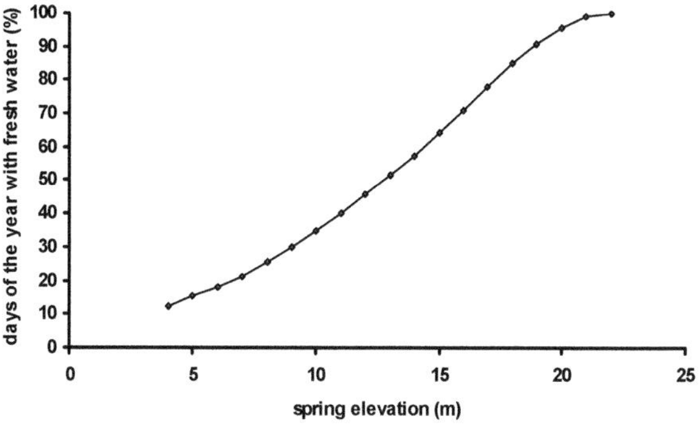

Fig. 5. Days during the year (%) with fresh water versus spring elevation.

5 Conclusions

The simulation of the operation of brackish karst springs with the deterministic MODKARST model provides clear physical insights to the karst system. The Almiros spring simulation has given information about the fractal nature of the karst system, the seawater intrusion mechanism and the choice of a suitable development method. The dominant sea intrusion mechanism for the Almiros spring is the difference between the freshwater and seawater densities. The sea intrusion could be prevented by raising the spring water outlet location, through the construction of a small dam; an estimated elevation of 22m appears to be the best choice.

References

Breznic M (1973) The origin of brackish Karst springs and their development Col. Razpr. In Por. 16 Knjig, 83-186

Maramathas A, Pergialiotis P, Gialamas I (2006) Estimation of sea intrusion mechanism of brackish karst springs by their simulation with the "MODKARST" deterministic model. Hydrogeology Journal 14, 657-662

Maramathas A, Boudouvis A (2006) Manifestation and measurement of the fractal characteristics of karst hydrogeological formations. Advances in Water Resources 29(1), 112-116

Maramathas A, Maroulis Z, Marinos-Kouris D (2003) A Brackish Karstic Springs Model. Application on Almiros Crete Greece. Ground Water 41(5), 608-620

Maramathas A (2006) A new approach for the development and management of brackish karst springs. Hydrogeology Journal 14, 1360-1366

The hydrodynamic behaviour of the coastal karst aquifer system of Zarakas - Parnon (Southeastern Peloponissos)

I. Lappas[1], P. Sabatakakis[1], M. Stefouli[2]

[1] Institute of Geology and Mineral Exploration, Sector of Water Resources and Environment, Department of Hydrogeology, 13677 Acharnai, Athens, Greece, ilappas@igme.gr, sampa013@yahoo.gr
[2] Institute of Geology and Mineral Exploration, Sector of Basic and Applied Geology, Department of General Geology and Geological Mapping, 13677 Acharnai, Athens, Greece, stefouli@igme.gr

Abstract The aquifer system of Zarakas is developed at central Parnon area and presents significant water potential having also hydraulic contact with the sea. The area is geologically structured limestones and dolomitic limestones of Tripolis Zone and of Tyros beds. The hydrogeological boundaries have been determined through the implementation of the following combined scientific works: geophysical soundings, aero-magnetic measurements, remote sensing, water - boreholes, hydro-chemical analyses and hydraulic measurements of the aquifers. The total extent of the area is approximately 385km^2 and the average annual efficient infiltration is 92x10^{6}m^3/yr of water. Based on the measurements of the hydrostatic level at the western part of the area where water wells have been drilled, the average hydraulic head is 0.32‰ while the groundwater flow direction is towards SE. According to the drilling data, the depth of karstification presents irregular development due to the type of the carbonate formations, resulting in groundwater circulation at specific zones. The seawater intrusion front under nonpumping conditions covers an extent of ~230km^2, in which Cl^- concentrations exceed 200 ppm.

1 Introduction

The accurate delineation of groundwater recharge / discharge areas in coastal karst aquifer systems is essential for their effective long-term development and management. Zarakas coastal karst aquifer system (Fig. 1) has been selected as a study area because it constitutes a hydrogeological environment of exceptional interest as a) it represents an extensive carbonate massif, b) there is direct / significant infiltration of meteoric water, c) there is lack of inland water springs except for Kremasti springs while there are large number of coastal springs and d) there has been water shortage for years in the settlements of Zarakas region, while drilling efforts proved to be unsuccessful.

The study area is covered by the 1:50,000 topographical/geological map sheets of MOLAI-RIKHEA. Zarakas is a part of Parnon mountain range, which is located in the south-eastern Peloponissos. There are (3) municipalities and 10 settlements, for which there is an urgent need to cover their water supply needs.

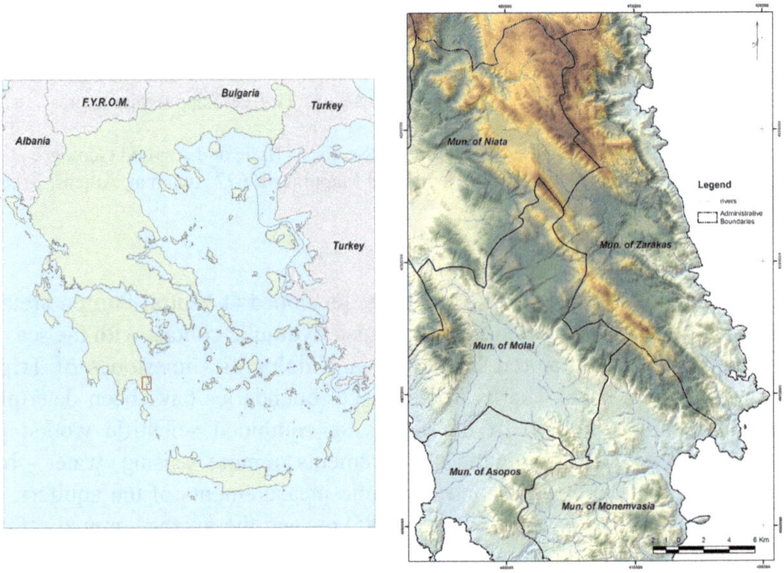

Fig. 1. The geomorphologic relief of the investigated area, as revealed by the Digital Elevation Model (25m resolution), showing the intensive, steep and rocky mountainous topography. The administrative boundaries of Zarakas region are also shown.

The study's objective is to apply a multi-level approach using various types of data and technologies, so as to identify recoverable groundwater in appropriate locations and furthermore to contribute to the region's hydro-economy.

2 Data and Techniques

All known methods have been evaluated like: remote sensing techniques, geological mapping, aero-magnetic measurements, geo-electrical soundings, conductivity and temperature coastal records, well drilling, pumping data and tracers. Remote sensing techniques have been applied in a previous hyrogeological study (Sabatakakis and Aggelopoulos 1999). The Landsat image has been analyzed for assessing surface seawater temperatures using the thermal bands of the system so as to map water outflows.

In order to define the aquifer's hydrogeological boundaries previous research data and studies have been evaluated such as geophysical soundings, salinization and groundwater flow maps and well data. In particular the groundwater flow map produced by Sabatakakis and Makris (1993) has been used to define the groundwater flow along the west margin which appeared to cover both carbonate formations and granular deposits.

During the Hydrogeological Research Program of the 2nd European Union Support Framework (conducted in 1996), the Institute of Geology and Mineral Exploration (I.G.M.E.) has accomplished the geoelectrical soundings in order to detect the depth that Tyros beds lay beneath carbonate series. The production wells of previous years have been drilled in locations that are close to the western margin of carbonate formations. In two of those boreholes three - stages pumping tests have taken place while controlling the Cl^- ions concentration. Many boreholes were drilled and pumping tests took place in previous years so as to analyse the impact of seawater intrusion. Two new deep wells have been drilled since 2008 at Eiconostasi (Moutzouri) by I.G.M.E and Niata locations with absolute elevation (a.s.l.) of 410m and 325m respectively, so as to analyze the stratigraphy of the area.

Isotopic analyses of Oxygen - 18 (^{18}O) and Tritium (T) have been done by the Isotope Hydrology Laboratory of the National Centre of Physical Sciences "Demokritos" at Parnon region and Molai catchment (Ghikas et al. 1983). Three coastal discharges have been selected from which samples have been taken. Additional sampling and isotopic analyses of $\delta^{18}O$ took place in four (4) different hydropoints which are affected by seawater intrusion.

Using this integrative data approach groundwater catchments are delineated associated with surface water catchments contributing to spring flow.

3 The Geomorphologic, geological and hydrogeological setting of the study area

Zarakas aquifer is highly faulted having a steep, rocky mountainous topography with altitudes from sea level up to 1200m and is as follows:

- Main tectonic discontinuities of Zarakas area are of NNW – SSE direction. Coastlines have the same NNW-SSE direction which is also consistent with the average strike of carbonate formations. This general direction is interrupted by E-W fault lines that coincide with the coast lines.
- There are distinctive karst geomorphological features like sinkholes, uvalas, dolines which are mainly located at Koupia area.
- Eastern coasts are steep enough due to recent fault activity and it is considered that some of these faults' steps is more than 250m, for instance in the section of Kyparissi - Avlaki coastline. Recent tectonic activity is documented by the absence of Holocene deposits along the coastline. The presence of such

deposits at Avlaki and Kyparissi bay and at several other creeks is due to the activity of local streams. Also the small thickness and rough structure demonstrate their recent geological formation.

The geology in Zarakas region is as follows:

- Based on I.G.M.E. geological mapping (Exindavelonis and Taktikos 1979, Stamatis 1979), the stratigraphical thickness of this series is more than 1200m.
- The Tyros volcanic - sedimentary series lay beneath the carbonate series.

According to I.G.M.E. drilling data of 1980s in Gaganias area (north of Molai), Tyros series thickness is approximately estimated to 530m (Aggelopoulos and Constantinides 1988). This formation appears in Northwest margin of study area with an average dip towards ESE and 300 degrees strike. In Zarakas region, except for the northwest margin, Tyros series are displayed at Kremasti area (Fig. 3). None of the wells under investigation (over 400m drilling) which have taken place at Labokampos, Rikhea and Koupia sites have cut Tyros series.

The geophysical measurements available showed that the depth at which Tyros beds extend follow the geomorphologic relief. Moreover, Tyros formations are in altitude above sea level even in low altitude areas. Specifically in Aghios Dimitrios region upper parts of Tyros beds are located in an altitude of 60m (a.s.l.). In Niata region both geoelectrical soundings and research boreholes have shown that Tyros formations are not located at sea level, but at least deeper than 420m (b.s.l.).

Irrigation pumping of ground water commenced in the mid-1980s, which caused gradual groundwater level decline. Continued regional groundwater resource development may further decrease spring discharge.

4 Data analysis and Re-interpretation of Groundwater flow

Possible recharge mechanisms in the area include (1) direct recharge on fractured carbonate mountain highlands, (2) limited recharge of mountain-front runoff on calcified alluvial fans, (3) recharge in valley floors and (4) interbasin flow.

The main directions of faulting are of NW-SE and E-NE directions while the compiled map shows water outflows of coastal springs or rivers (Fig.2). The locations of water outflows are related to the faulting system and the tectonic discontinuities that prevail in the main tectonic blocks. These discontinuities probably act as conduits for the subsurface water flow. The surface area of the karstifed limestones of Parnon system for the pilot project area is 385km^2. Karst geomorphologic features develop along the fault lines as they have the same NE-SW directions. Lineaments that are related to faulting can be also interpreted on the 15 meters spatial resolution image product (Fig. 2).

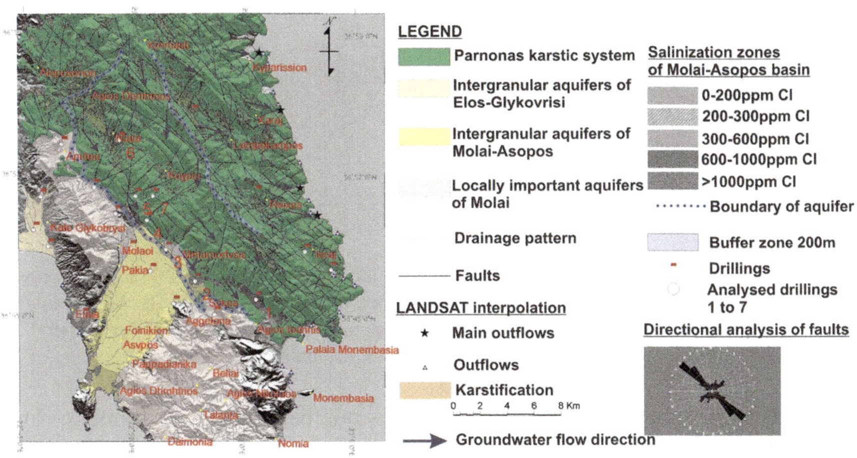

Fig. 2. Combined map of hydrogeological information. The terrain with surface karstification and the outflows have been extracted from the satellite image. The dark-blue arrows indicate the groundwater flow direction within the karstic aquifer. The rose diagram shows the main directions of tectonics on the 1:50.000 scale geologic maps, in the pilot project area.

The western boundary of Zarakas aquifer is bounded by the Molai - Asopos Neogene basin which is defined by both Carbonate and Tyros formations that are covered by Pliocene - Pleistocene - Holocene granular deposits. Groundwater flow within the carbonate formations appears to have a slight hydraulic gradient towards SSE, along the coastline of Palaia Monemvasia - Ormos Kremmidi. The hydraulic gradient determined during the nonpumping period, inside the carbonate series, is 0.55‰ at the Northwestern locations and 0.20‰ at the Southeastern ones (Sabatakakis and Aggelopoulos 1997). The hydraulic head increases towards SE and is consistent with the groundwater flow direction as it is confirmed by the salinization map (Sabatakakis and Makris 1993).

The northern hydrogeological boundary is defined by Tyros beds which follow the geomorphologic relief according to geoelectrical soundings along Aghios Dimitrios - Kremasti line. The uplift of Tyros beds defines the aquifer's groundwater drainage towards North direction (Fig.3). Also, in northern locations far away from this hydrogeological limit, research boreholes confirm the uplift based on the carbonate formation's thickness.

The aquifer's eastern part according to aero-magnetic measurements is delineated by the mountains of Gaidourovouni - Tourla - Kalogerovouni - Koulochera while Tyros beds are found to be located considerably above the sea level. These data have not been taken into account when trying to define the aquifers' limits in previous years and hence the limits have been set at 385 km^2. In this way the previous defined area which has regarded the aquifer as a whole till the eastern coastline, should be divided into two subareas, western and eastern towards Kremasti – Koulochera axis.

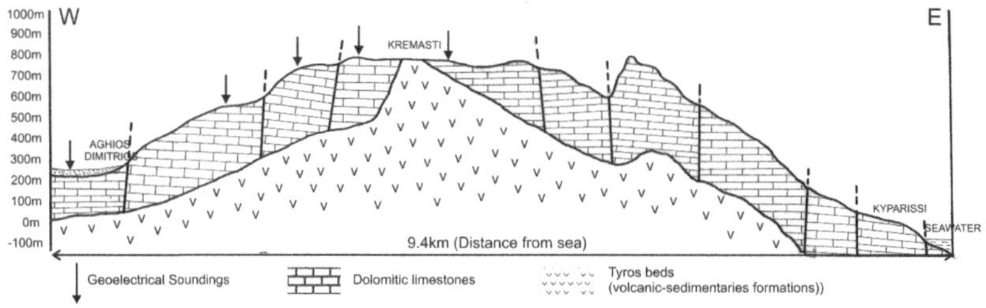

Fig. 3. Geological section in W – E direction showing the uplift of Tyros beds. The black arrows indicate the positions where geoelectrical soundings took place in order to specify the absolute altitude of Tyros beds. In the eastern side of Kremasti area, Tyros beds' depth has been determined based on stratigraphical, tectonic and hydrogeological data.

The eastern subarea is drained through a number of coastal saline springs (>3500ppm Cl^-). Some drilling efforts that have been made in the past at Aghios Ioannis and Rikhea areas, have yielded highly saline water (2500ppm Cl^-) with pumping rate of 3-4m^3/h. This is because the groundwater has extensive front with the sea so seawater intrusion is quite extensive due to weak fresh water hydraulic head. It should be noted that significant coastal discharges are mainly located within major bays which in turn, have tectonic origin as mentioned above (Tavitian 1981).

Step Drawdown tests: According to pumping tests data no change of salinity has been observed during the step drawdown test at Koprisses borehole which means that the equivalent groundwater flowlines are not disrupted and maintain subhorizontal slope towards the shoreline (southeast). Thus, the interface lies deeper and intersects the impermeable basement in southeastern locations.

Differences in hydraulic gradient from NW (Niata) to SE (Ormos Kremmidi and Palaia Monemvasia coastline) such as 0.55‰ - 0.31‰ - 0.22‰ - 0.20‰, suggest according to Darcy's law gradual groundwater flow decrease which is consistent with the density alteration in compared with the increasing salinity. The hydraulic gradient suggests small differences which mean that at macro level the groundwater flow can be considered laminar.

During the step drawdown test a slight salinity increase and soon after stabilization has occurred while the discharge rate changes. This means that Zarakas aquifer has great capacity and requires long term monitoring. After a 15 year exploitation of Smirtiza's borehole exploitation, salinity concentration has not differed significantly between 2007 and 2008 droughts and Cl^- ions levels have been in the order of 78ppm (Foukria-1). In the southeast close to the coastline (Foukria) many boreholes that have been drilled, show that there is seawater intrusion (Cl^- concentration exceeds 3500ppm).

Deep research wells: The deep wells at Eiconostasi (7) and Niata (6) show that the static water level is at 8.2m (a.s.l.). The well's yield during the pumping test was up to 7m^3/h with static water level at 458m below surface (Karydakis 2009). Niata well with static water level at 13.40m (a.s.l.) yielded 5m^3/h approximately during the pumping test.

The Eiconostasi boreholes' limited yields may be attributed to one of the two following causes: the karstification zone does not probably exceed the depth of about 300m (below surface) or the fact that the two wells are not located near tectonic structures. This results in poor groundwater circulation.

Table 1. Drilling data with mean values of Cl^- concentration.

Boreholes	Niata	Eiconostasi	Gaganias	Metamorphosi	Monoporo
Drilling depth	362m	510m	302m	137m	110m
Formation	Dolomitic limestones	Dolomitic limestones	Dolomitic limestones	Dolomitic lime-stones	Dolomitic limestones
Hydraulic Head	+13.4m	+8.2m	+6.2m	+2.8m	+1.3m
Pumping rate	$5m^3/h$	$10m^3/h$	$50m^3/h$	$50m^3/h$	$50m^3/h$
Cl^- (pumping)	45ppm	42ppm	75ppm	420ppm	940ppm
Cl^- (non-pumping)	38ppm	42ppm	50ppm	370ppm	750ppm

Triassic limestones dekarstification: Based on drilling data from the above mentioned wells it seems that Triassic limestones dekarstification is very limited below 340m±20m (below surface). That's why the wells at Niata and Koupia sites have limited yield of about $7\text{-}8m^3/h$.

Isotopic analyses (environmental isotopes): The analyses of the three coastal discharges of the springs of Gerakas, Palaia Monemvasia and Ormos Kremmidi show that according to $\delta^{18}O$ values, the highest recharge area of coastal discharges is 550, 500 and 650m respectively.

The $\delta^{18}O$ values in Gerakas discharge is -7.418 (November 2005) and -7.022 (March 2006). Although the water quality of coastal discharge is affected by salinization, $\delta^{18}O$ values indicate that recharge area comes from groundwater infiltrated at elevations up to 800m approximately. On the contrary, $\delta^{18}O$ values of Metamorfosi (3), Monoporo (2) and Foukria (1) wells indicate that infiltration areas of meteoric or surface water come from elevations which are below 500m.

5 Conclusions – Remarks

An integrative data approach has been adopted in this research work and the conclusions can be drawn as follows:

- The groundwater flow on the western margin towards the SE coastline represents the aquifer zone recharged by meteoric water at elevations above 500m with significant hydraulic conductivity.
- At higher locations the karstification of the carbonate formations does not exceed a depth of approximately 320m (below surface) and therefore this area

represents the groundwater drainage which is confirmed by the two deep wells of Niata and Eiconostasi (Moutzouri) drilled in 2008.

- The "foot" of saline – fresh water interface has been defined between Smirtiza and Koprisses wells, with front direction towards NE - SW which is vertical to groundwater flow, as revealed by salinization and groundwater flow maps.

Therefore, the optimal positions of exploitation wells should be along the axis of Smirtiza's stream (SW – NE). The current well total discharges have been estimated to reach $450m^3/h$ during the summer and have remained stable for over 15 years. The fact that the saline – fresh water front has not moved during that period means that the aquifer capacity has not yet reached its limits, not to mention that the well yields are almost continuous during the summer.

This study develops procedures to delineate regional groundwater catchments in coastal karst aquifers using various types of hydrogeological data. The approach is tested in the Zarakas region and the procedures may be used to understand the spatial extent and to quantify recharge processes in similar terrains.

The authors would like to thank the Isotope Hydrology Laboratory of the National Centre of Physical Sciences "Demokritos" for the isotopic analyses of Oxygen -18 (^{18}O) and Tritium (T) and also the Mayor and the board of Zarakas for the full support during the well drilling period.

References

Aggelopoulos K, Constadinides D (1988) Zn – Ar – Pb ore deposits of Molai – Lakonia area. Bulletin of Greek Geological Society, Vol. XX/2, pp. 305-320

Aggelopoulos A, Sabatakakis P (1995) Geoelectrical applications for seawater intrusion control. Zarakas N. Karvali case studies. 3rd Greek Hydrogeological Congress, Iraklio, Crete

Doutsos Th, Koukouvelas I (1986) Structure analysis of Molai ore deposits. Mineral Wealth, No 40, pp. 7-16

Exidavelonis P, Taktikos S (1979) Molai geological map, scale 1:50.000, I.G.M.E., Athens

Ghikas I., Kruseman, Leontiadis, Wozab (1983) Flow pattern of a karst aquifer in the Molai Area, Greece. Groundwater, Vol. 21, No 4

Karydakis G. (2009) Economical – technical data of hydrogeological drilling projects (Municipality of Zarakas borehole). I.G.M.E., Athens

Marakis G (1965) The Paleozoic volcanic rocks of Lakonia area. Phd, University of Athens

Sabatakakis P, Makris A (1993) Salinization case and rational management capabilities to coastal aquifers of SW and SE Peloponnese. 2nd Greek Hydrogeological Congress, Patras

Sabatakakis P, Aggelopoulos A (1999) Hydrogeological and geophysical study of coastal karst aquifer in Zarakas area, Lakonia. Proceeidings of 4th National Congress, Greek Committee of Water Resources Management, Vol. 2, Volos

Stamatis A (1979) Rikxea geological map, scale 1:50.000, I.G.M.E. publication, Athens

Tavitian, I (1981) Hydrogeological conditions of Neocene karst aquifers of Molai – Asopos basin in Lakonia area. Phd, University of Patras

Thiebault E, Kozur H (1979) Precisions sur l'age de la formation Tyros (Paleozoique superieur – Carnien). Peloponnèse meridional Grèce. C.R.Acad. Sc. Paris 228 D, pp. 23-26, Paris

Application of tracer method and hydrochemical analyses regarding the investigation of the coastal karstic springs and the submarine spring (Anavalos) in Stoupa Bay (W. Mani Peninsula)

G. Stamatis, G. Migiros, A. Kontari, E. Dikarou, D. Gamvroula

Agricultural University of Athens, Laboratory of Mineralogy-Geology, Iera Odos 75, GR-118 55 Athens, stamatis@aua.gr .

Abstract The study of the SW Taygetos Mt. karst system, based on the implementation of tracer test and hydrochemical analyses, is aimed at the investigation of the geohydraulic characteristics and the water quality of the coastal and submarine springs at Stoupa region (Peloponnese). The study area is formed by intensively karstified carbonate rocks, flysch, Neogene sediments and Quaternary deposits. The high potential karstic aquifer is been discharged via coastal and submarine springs in Messinian Gulf. The submarine spring (Anavalos) is brackish to saline (EC: 0.023-5.438 S/m and Sal: 0.14-39.12), periodically acquires freshwater characteristics and it is not suitable for exploitation for human consumption. Among the four coastal springs, two of them having permanent flow, appear brackish, while the other two are of freshwater and flow periodically during the wet period. The tracer test pronounced a SW direction of groundwater flow and relatively high velocities, up to 150m/h, revealing the intense karstification of the carbonate formations.

1 Introduction

The coastal karstic aquifers are an important part of water resources and sometimes the only available to be used in a region, although their exploitation tend to be very difficult mainly due to technical problems. The coastal karstic systems discharge through coastal and/or submarine springs. However, most of the submarine discharges remain unexploited primarily because of their direct contact to seawater. This fact leads to water quality degradation. In most cases, the submarine springs are characterized by high discharge rates. The karstic system of SW

Taygetos (Fig. 1) is consisted of intensive karstified Cretaceous and Triassic-Jurassic carbonate rocks. The aquifer presents high potentiality and is been discharged through coastal springs which appear in the bay between Kardamily-Stoupa- Ag. Nikolaos. One of the greatest submarine discharge is the Anavalos spring (Stoupa-Kalogria bay), which was also the main subject of the hydrogeological investigation. The municipality of Lefktron has to drill with the deficiency of water resources as only a part of the water supply needs are covered through wells and boreholes. However, the intensive withdrawal from wells and boreholes along the coastal area induces intrusion of saline water towards coastal groundwater intakes, especially during the dry period. The marine investigation was carried out by a research group of HCMR (Hellenic Centre for Marine Research) the period between June 2009 and June 2010 and was focused on the submarine spring discharge function, and the monitoring of its quantitative and qualitative characteristics (HCMR 2010). On the other hand, the terrestrial survey was conducted by a research team of the Mineralogy-Geology Laboratory of the Agriculture University of Athens aiming the investigation of: a) the geological and hydrogeological conditions in the area of SW Taygetos, b) the functioning of the karst system and c) the estimation of the adequacy of the Anavalos discharges for human consumption (Stamatis 2009). The ultimate purpose of this combined research was to take further decisions on the use and exploitation of the submarine spring (Anavalos) in order to cover the water supply needs of the region. The research was funded by the municipality of Lefktron and Messinia Prefecture.

2 Study area

Geological setting: The study area is part of the Mani geotectonic unit, which appears as a tectonic window (Fig. 1). Above the Mani unit there is a sequence of thrusts with different nappes resting on top including successively from the base to the top the Arna unit, the Tripolis unit and the Pindos unit (northwards) respectively. The Mani unit has a thick large carbonate series, with frequent siliceous intercalations that end up with the metaflysch. Above the Mani unit within the study area, the overthrust series of the Tripoli geotectonic unit is resting on the top with the carbonate formations of Cretaceous that end up with the (Oligocene?) flysch. At the east part of the Messinian gulf Neogene marine formations (Pliocene) are encountered consisting of marls, marly limestones and conglomerates while Quaternary deposits resting above are represented by Pleistocene terraces, old and recent fans, scree as well as by alluvial deposits. The major part of the study area is formed by intensively karstified carbonate rocks, in that a significant aquifer is developed. The wide spread of the carbonate sequences combined with the neotectonic deformation of the area, have led to the formation of a complex and extensive surficial, but mostly underground karst (Papanikolaou and Skarpelis 1987; Migiros et al. 2008).

Hydrogeology: Within the karstified carbonate rocks a series of coastal springs and submarine discharges in the gulf of Stoupa drainage the karstic aquifer. The underground water, following the trend of the major structures has initially a N-S direction flow, which turns to be E-W when it comes along with the transverse faults within the area. The discharge rates both of the coastal springs and of the submarine springs are characterized by strong seasonal fluctuations. Especially the coastal springs Sc2 and Sc3 appear only during the wet period after excessive rain and act as regulators of the karstic aquifer relief.

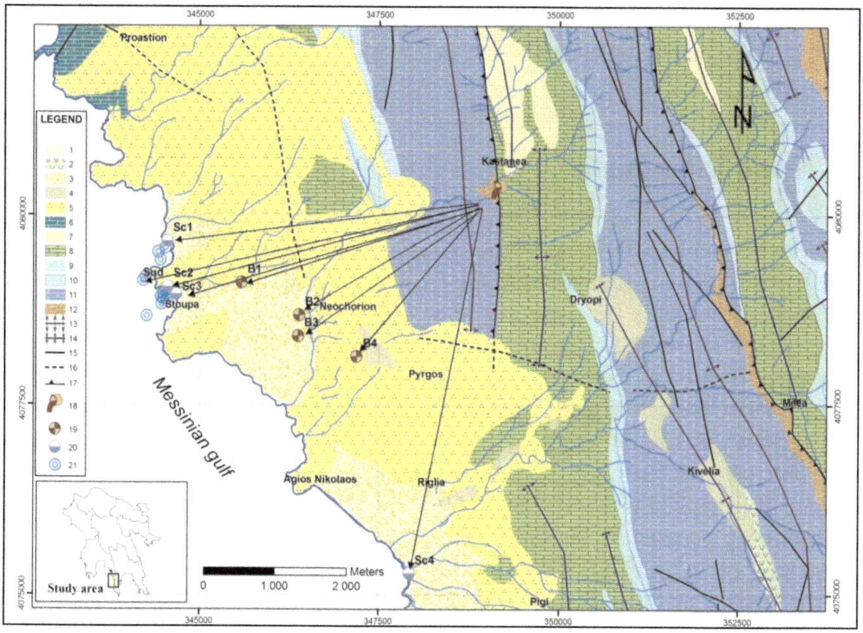

Fig. 1. A simplified geological map of the study area (Katsikatsos et al. 1983); 1: Alluvial deposits (Holocene); 2: Scree (Holocene); 3: Old & recent fans Holocene); 4: Terraces (Pleistocene); 5: Marine formations (Pliocene); 6: Limestones, dolomitic limestones (Cret-U.Jurassic); 7: Flysch (U.Eocene-Oligocene); 8: Limestones (U.Senonian- U.Eocene); 9: Crystalline limestones (U.Jurassic-Cretaceous); 10: Sillicate schists (L.-M.Jurassic); 11: Dolomitic limestones- dolomites (M.-L. Triassic); 12: Phyllitic- quartizic series (Permian-L.Triassic); 13: Antcline; 14: Syncline; 15: Normal; 16: Probable; 17: Overthrust; 18: Injection point (Uranine dye); 19: Borehole; 20: Coastal spring; 21: Submarine spring.

According to the HCMR measurements the estimated discharge rate of the Anavalos varies between 720 m^3/h and 4500 m^3/h (HCMR 2010), while for the coastal springs Sc2 and Sc3 varied between 0-80m^3/h and 0-250m^3/h respectively. The recharge area of the submarine karstic springs and of the groundwater aquifer in general, is located in eastern and northern area of Stoupa, between the hydrological basins of Ridomo, Biros and Ag. Nikolaos. The springs appear scattered in the mountain of Taygetos following the tectonic structures and the fracture zones.

These are contact springs appearing to the contact between the U.Cretaceous-Eocene limestones and the L.Cretaceous- M.Jurassic multimict formation (carbonate- siliceous- clastic), as well as at the contact of the above quoted limestones and the flysch. The springs appearing at the semi-mountainous and mountainous region are related to the Plio-Pleistocene formations and show low discharge rates. Aquifers of limited efficiency are found within the Neogene sediments and Quaternary deposits developed in lowland. Their recharge is obtained through precipitation and mainly via lateral seepage by the karstic aquifer. The aquifers within the plain area are of intense exploitation by a large number of wells and boreholes for water supply and irrigation. The discharge rate of the wells is estimated at 10-15m^3/h, while on boreholes at 30-40m^3/h.

3 Groundwater quality

Between September 2009 and October 2010 several samplings of local boreholes (B1, B2, B3, B4), the submarine spring (S_{ud}) and of coastal springs (Sc1, Sc2, Sc3, Sc4) have been held. The sampling aimed to the investigation of the qualitative composition of groundwater in the region, especially of the Anavalos spring. Measured in situ with portable instruments: discharge rate of springs Q L/s (Flow Probe FP101/Global Water), temperature T (°C), electrical conductivity EC (mS/cm) (WTW/LF-330) and pH (WTW/330i). At the laboratory were identified by using titration methods, spectrophotometry (HACH, DR/3000), flame photometry (Flamephotometer INTECH/420) and atomic absorption (GBC/908AAS) the following parameters: Hardness, Ca^{2+}, Mg^{2+}, Na^+, K^+, HCO_3^-, Cl^-, SO_4^{2-}, NO_3^-, NO_2^-, NH_4^+, PO_4^{3-}, Br^-, SiO_2, and the heavy metals, Fe_{tot}, Mn, Cu, Cd και Zn. The hydrochemical analyses took place on the Mineralogy- Geology Laboratory of the Agriculture University of Athens (Table 1). Boreholes B2 and B3 provide very good water quality, with seasonal fluctuations (EC: 510-810μS/cm, TDS: 429-773 mg/L, Na^+: 12.9-110.3 mg/L, Cl^-: 17.7-248.2 mg/L and SO_4^{2-}: 5.9-19.2 mg/L). However, boreholes B1 and B4 reveal intense seawater intrusion (EC: 1465-3360 μS/cm, TDS: 415-2254 mg/L, Na^+: 15.6-436 mg/L, Cl^-: 17.7-1170.2 mg/L and SO_4^{2-}:5.9-106.2 mg/L). The intrusion dominates mostly during the dry period when water demands are increased. Seasonal springs Sc2 and Sc3, show variable values of electrical conductivity (782-953μS/cm), and TDS (591-659 mg/L). According to the concentration of Na^+ (Sc2: 124.1mg/L and Sc3: 96.1 mg/L) and of Cl^- (Sc2: 216.3 mg/L and Sc3:190.1mg/L), it is obvious that both springs are slightly affected by the seawater. Sc1 spring shows continuous flow, and is characterized by low temperature between 12.2 and 13.8 □C and by seasonal fluctuation as far the electrical conductivity is concerned (EC:585- 2980 μS/cm), while during the tracer test period it was 1.145 μS/cm. The permanent flow spring Sc4 shows high intrusion during the whole year (EC: 5800-6970 μS/cm, TDS: 3302-3873 mg/L, Na^+: 949-1.095mg/L, Cl^-: 1800-2180mg/L and SO_4^{2-}: 98-280mg/L).

Table 1. Results of hydrochemical analyses.

Sampling point	Scl	Sc2	Sc3	Sc4	Sud	B1	B2	B3	B4
Tw °C	12.3-12.8	12.3	12.3	16-16.7	-	16.4-22.9	14.2-20.4	15.4-20.3	15.7-21.6
Ec (µS/cm)	1145-4115	782	953	5800-6970	19020-45900	1465-2300	510-748	648-810	513-3360
pH	7.7	7.7	7.7	7.4-8	7.8-8	7.8-8.3	7.3-8	7-7.9	7.3-7.9
Eh (mV)	116-167	230	232	43.3-206	156	46.4-200	53.6-240	43.4-217	38.3-263
TDS (mg·L^{-1})	378-2390	659	591	3302-3873	10949-28222	1254-1779	429-551	526-773	415-2254
Hardness (°dH)									
Total	10.7-30.9	12.6	12.3	48.9-51.2	128.1-314.6	27.3-18.9	12.6-16.5	15.2-17.8	12.6-54.3
Temporal	8.1-9	8.4	8.1	10.9-14	8.2-8.4	11.8-14	8.4-14	10.6-14	3.5-12.6
Permanent	1.9-21.9	4.2	4.2	34.9-40	119.7-306.2	7.1-13.3	0-5.1	1.2-7.1	4.5-45.5
Major elements (mg·L^{-1})									
Ca^{2+}	23.2-96	24	23.2	32-124	24-380	79.2-111.9	73.6-84.8	30.4-97.6	73.6-316
Mg^{2+}	9.9-45.5	40.1	39.6	148.7-203	149.5-1136.6	33.9-50.8	9.9-20.2	10-58.7	5.1-54.3
Na^{+}	35.8-747.2	124.1	96.1	949.4-1095.2	3198.4-8620.7	165-436	12.9-72.6	35.4-110.3	15.6-424.1
K^{+}	2.7-18.2	14.6	5	26.8-84	290-320	7.4-7.8	1-3.1	17.2-60	0.6-6.6

Table 1. Continue.

Sampling point	Scl	Sc2	Sc3	Sc4	Sud	B1	B2	B3	B4
Major elements ($mg \cdot L^{-1}$)-continue									
HCO_3^-	176.9-195.1	183	176.9	237.9-304.9	183-244	256.2-304.9	182-350	231.8-305	18.3-274.5
Cl^-	21.3-1205.7	216.3	190.1	1800-2180.8	6596-17376	74.7-344	17.7-141.8	102.8-248.2	17.7-1170.2
SO_4^{2-}	13.8-100.2	35.7	35.3	98.4-280	110.2-139.5	30.1-97.7	5.9-13	3.4-19.2	5.9-106.2
NO_3^-	3.5-6.2	7.5	6.2	7.5-10.6	4.8-5.3	16.7-17.6	13.6-33.4	11.4-28.6	11.9-19.8
NH_4^+	0.03-0.23	0.06	0.1	0.58-0.96	0.01-0.97	0.11-0.46	0.01-0.65	0-0.32	0.02-0.95
NO_2^-	<0.01	<0.01	-	-	-	0.01	-	-	-
PO_4^{3-}	0.04-0.13	0.06	0.07	0.09-0.17	0.05-0.11	0.15-0.17	0.19-0.24	0-0.23	0.11-0.25
Br^-	0.01-0.03	0.1	-	0.01-0.13	-	0.03-0.17	0.01-0.1	0.07-0.08	0.06-0.25
SiO_2	4.3-12.6	12.8	18.2	6.6-16.5	-	8.3-12.8	11.7-18.8	11.8-22.6	12.6-13.9
Trace elements (ppm)									
Cd^{2+}	0.007-0.064	0.047	0.047	0.006-0.05	0.2	0.051-0.053	0.023-0.038	<0.001-0.004	0.001-0.007
Sn^{2+}	<0.001	<0.001	-	<0.001	<0.001	<0.001	<0.001	0.001	<0.001
Co^{2+}	<0.001	<0.001	<0.001	<0.001	<0.001	0.001	0.001	0.001	-
Mn^{2+}	<0.001	<0.001	0.054	0.008-0.097	0.044	0.001-0.018	0.009-0.047	0.001	0,011
Pb^{2+}	0.034-0.037	0.066	0.042	0.08-0.176	0.064	0.032-0.416	0.008-0.152	0.001-0.289	0.001-0.184
Fe_{tot}	0,011	0.185	0.112	0.001-0.002	<0.001	0.001-0.004	<0.001	0.001-0.006	0.001-0.004
Cu^{2+}	0.014-0.059	0.079	0.069	0.001-0.073	0.053	0.003-0.084	0.044-0.096	0.069-0.07	0.005-0.069
Zn^{2+}	0.02-0.07	0.009	0.001	0.012-0.016	0.014	0.008-0.053	0.001-0.078	0.038-0.078	0.078-0.688

All samples collected by the submarine Anavalos spring, show very high intrusion (EC: 19020-45900μS/cm, TDS: 10949-28222mg/L, Na^+: 3198-8620mg/L, Cl^- : 6596-17376mg/L and SO_4^{2-} :110-140mg/L) revealing the mixture of karstic water with seawater. It is difficult to conclude whether the mixture takes place into the karstic aquifer or in the discharge point, mostly due to technical reasons.

The specialized team of HCMR placed advanced equipment, as closest as possible to the Anavalos discharge point and recorded several series of measurements during the investigation. The results showed that the lowest chemical values appear during the wet period, after long term precipitations, while the highest values during the dry period (T: 12.1-26.1 °C, EC: 0.023-5.438S/m and Sal: 0.14-39.12 (HCMR 2010). According to this evidence we concluded that the submarine spring water quality is inappropriate for human consumption, so its utilization is not indicated. Generally, the overexploitation of the coastal aquifers, especially during the dry period causes locally salinity effects, which are limited during the wet period.

4 Tracer test

On 15.01.2010 a tracer test took place on the SW slopes of Taygetos Mt. in order to investigate the direction and the velocity of groundwater flow and the hydraulic connection between the surface water of SW Taygetos Mt. and the karstic groundwater. The injection point was sited near Kastanea village, within the Megalo Lagadi stream. The tracer test began in 23:00pm during excessive rainfall, which had already begun an hour before. As trace element was used 10 kg Uranine dye (Uranine, Na-Fluorescein $C_{20}H_{10}O_5Na_2$). The dye was dissolved in approximately 50 L plastic bottles of water. The monitoring points selected were two of the water supplying boreholes of the municipality B_1 and B_2, four coastal karst springs Sc_1, Sc_2, Sc_3 and Sc_4 and the Anavalos spring (S_{ud}). The sampling began on 16.01.2010, at 06:00 and finished on 23.01.2010, at 16:00. Samples were taken every two to three hours from the monitoring points.

In all 325 water samples was taken. The water samples were collected into 100ml polythene plastic vials, and maintained in darkness until their analysis. The Uranine was detected in all samples. The detection of the Uranine dye took place in the Laboratory of Mineralogy- Geology of Agriculture University of Athens, with the use of PERKIN/ELMER LS3 Spectrofluorometer. Figure 2 demonstrates the breakthrough curve for Uranine in Sc_1 spring as well as the curve of recovered Uranine. The initial arrival of the dye was 8 hours after the injection, while the maximum concentration revealed 34 hours later. The appearance of other characteristic peaks with high dye concentration, after the gap, has to do with the excessive rainfall during the next days, which has probably flushed the dye from the karstic conduits in the karst system. The representative times after injection and the perspective flow velocities were recorded in all monitoring points.

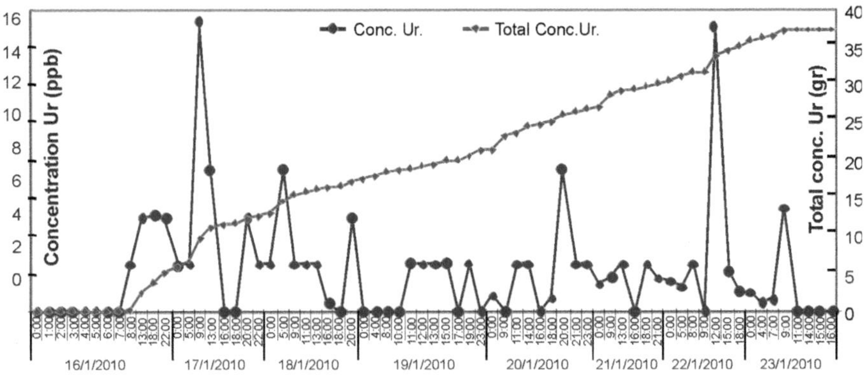

Fig. 2. Breakthrough curve for Uranine, Sc_1 spring (Printzipas) and the respective cast line.

In Table 2 representative results are presented. The maximal flow velocity (V_{max}), which is taken as the velocity that corresponds to the first appearance of the tracer, has a mean value of 497.6m/h.

Table 2. Time appearance of the Uranine and groundwater flow velocities.

Monitoring points		Sc1	Sc3	Sc4	Mean Value
Total distance from injection point	(m)	5200	5250	5800	54175
Initial arrival of Uranine dye	t_1 (h)	8	8	31	15.6
	V_{max} (m/h)	650	656	187	497.6
Maximal concentration	t_2 (h)	34	34	58	42
	V_{dom} (m/h)	153	155	100	136
Last appearance	t_4 (h)	72	96	88	85.3
	V_{min} (m/h)	72	55	66	64.3
Recovered Uranine	Conc. (gr)	38	3095	68	T=3201 (32%)

The dominant velocity (V_{dom}), which is defined as the time that corresponds to passage time of the maximal concentration, has a mean value of 136 m/h. Finally, minimum velocity (V_{min}) corresponding to the last appearance of the Uranine dye has a mean value of 64.3 m/h. The recovered amount at the sampling springs was 3.201 gr that represents 32% of the total injected amount of Uranine, which was 10kg. Apparently, a great amount of the Uranine injected, was been discharged by the submarine spring within Stoupa-Kalogria bay, which was impossible to measure. Another explanation could be that an amount of the dye was trapped in the fissures of the karst system. Generally the flow velocities of the karst aquifer are very high due to the intensive karstification. The values of the flow velocities of SW Taygetos karst system are comparable with the values of the flow velocitites derived from tracer tests in other karst systems (Morfis and Zojer 1986; Stamatis and Zagana 2004; Stamatis 2010).

Conclusions

Within the karstified carbonate formations, an aquifer of high exploitation potential has been developed. Numerous hydrochemical analyses of the coastal springs and the submarine (Anavalos) spring water, took place generating the following conclusions: a) the submarine spring is brackish, while periodically obtains characteristics of freshwater, but it is still inappropriate for human consumption, b) among the four coastal springs, two of them having permanent flow, appear brackish while the other two are of freshwater and flow periodically during wet periods. Thus, they cannot contribute to the solution of the water supply deficiency of the area. According to the results of the tracer test, we came up with the following conclusions: i) there is a direct hydraulic connection between the injection point of the Uranine dye, the coastal springs and the submarine Anavalos spring, ii) the groundwater flow has a SW general direction, in accordance with the slope direction of the geological layers and the fault structures and iii) the mean velocity of the groundwater flow is estimated at about 150m/h, revealing the intense karstification of the carbonate rocks and obviously the cavernous regime that dominates in the area.

References

HCMR (Hellenic Centre for Marine Research) (2009) Integrated marine and terrestrial study to the investigation of the quantity, quality and exploitation of the submarine spring (Anavalos), Stoupa, Messinia Prefecture. Progress report II. Unpublished report, pp 140

Migiros G, Psomiadis E, Papanikolaou I, Karamousalis T, Stamatis G (2008) Groundwater coastal discharge of the karstic system of Mani Peninsula, southern Peloponnesus- Greece, Proceedings of the 8th Int. Hydrogeological Congrss of Creece and 3rd MEMWorkshop on Fissured rocks Hydrology, Athens 2008, 1:317-326

Morfis A and Zojer, H (1986) Karst Hydrogeology of the Central and Eastern Peloponnisos, (Greece) 78-91, 5th International Symposium on Underground Water Tracing, 1986. – Steir. Beitr. Hydrogeologie, B.37/38:301 S., Graz

Papanikolaou D and Skarpelis N (1987) The blueschists in the external metamorphic belt of the Hellenides: composition, structures and geotectonic significance of the Arna Unit. Geol Pays Hellen. 33:47-68

Stamatis G and Zagana E (2004) Hydrochemical analysis and application of the tracer method regarding the investigation of the N. Ossa karstic system (NE Thessaly), Proceedings of the 10th Int. Congress of the Geological Society of Greece, Vol XXXVI, Patras, May 2004

Stamatis G (2009). Integrated marine and terrestrial study to the investigation of the quantity, quality and exploitation of the submarine spring (Anavalos), Stoupa, Messinia Prefecture. Progress report II. Unpublished report, pp 140

Stamatis G (2010) Groundwater Quality of the Ag. Paraskevi/Tempi valley karstic springs-Application of a tracing test for research of the microbial pollution (Kato Olympos/NE Thessaly), 12th Int. Cong., Bulletin of the Geological Society of Greece, Vol XLIII, No 4, p.1868-1877, 2010

Submarine groundwater discharges in Kalogria Bay, Messinia-Greece: geophysical investigation and one-year high resolution monitoring of hydrological parameters

A.P. Karageorgis, V. Papadopoulos, G. Rousakis, Th. Kanellopoulos, D. Georgopoulos

Institute of Oceanography, Hellenic Centre for Marine Research GR 19 013, Greece.
ak@ath.hcmr.gr

Abstract Submarine groundwater discharges (SGD) in the SW Peloponnesus are ubiquitous, but the Kalogria Bay underwater spring is surprisingly permanent, and strong, according to previous information. During 2006, preliminary investigations revealed the presence of SGDs characterized by high discharges of freshwater (salinity <2). The possibility of potential use of the SGD water for drinking purposes motivated a multi-disciplinary study of the spring from July 2009 to July 2010. The spring discharges at ~26 m depth, at the section of two faults, which expose the limestone bedrock to the surface of the seafloor. Monitoring of flow velocity, temperature, and salinity every 30 minutes was achieved by the deployment of a lander, equipped with various sensors, designed and modified accordingly for the purposes of this study. The results have shown high flow velocities (>125 cm s^{-1}), which correspond to a discharge estimate of 4500 m^3 h^{-1}. Salinity variations were high, but minimum values ~1-2 were constantly recorded since mid-October 2009, and until the beginning of May 2010, where abruptly salinity increased to values higher than 12. The complex morphology of the seabed at the SGD point, which was covered by a big rock, is probably creating a rather fast mixing of freshwater and ambient high salinity water. Under these conditions, the potential exploitation of the spring water was not justified.

1 Introduction

The general scarcity of drinking water worldwide has attracted the attention of scientists to study in more detail submarine groundwater discharges (SGD), which could provide water for irrigation and in some cases for drinking purposes (Taniguchi et al. 2002). SGDs are known since the Roman geographer, Strabo, who lived from 63 B.C. to 21 A.D., mentioned a submarine spring (fresh groundwater) four kilometers from Latakia, Syria. Water from this spring was collected from a boat, utilizing a lead funnel and leather tube, and transported to the city as a source of fresh water (Taniguchi et al., 2002; UNESCO, 2004). Another particularly spectacular example of such use involved the construction of dams in the sea near the southeastern coast of Greece. The resulting "fence" allowed the formation of a fresh water lake in the sea that was then used for irrigation on the adjacent coastal lands (Zektser 1996; Burnett et al. 2003). So far, this is the only SGD exploited continuously since the 1970's for irrigation in Greece.

The main goal of this study is to present, for the first time, one-year high resolution monitoring of hydrological data for the Kalogria Bay SGD, near Stoupa town, in SW Peloponnissos (Figs. 1a). Bathymetric and geophysical surveys were also carried out in order to unravel the complexity of the underwater features of the area.

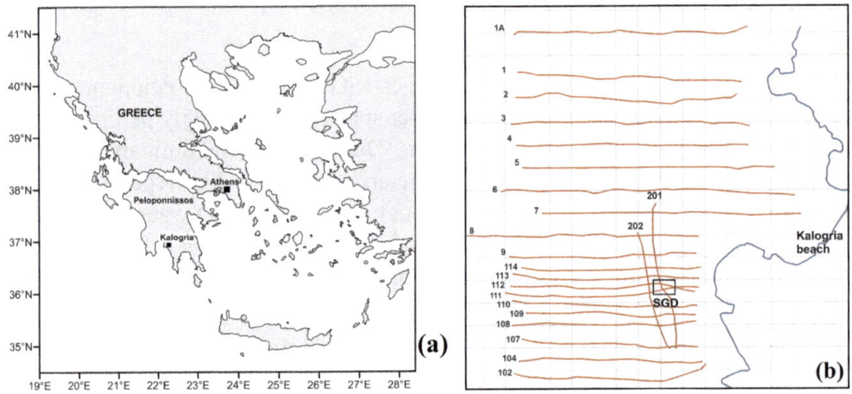

Fig. 1. Study area location (a) and geophysics tracks and SGD location (b).

2 Materials and methods

Bathymetry

The bathymetric measurements were conducted along 50-m distance tracks, and near the SGD 20-m distance, and orientation W-E, utilizing the ELAC SEABEAM 1180, in combination with the Inertial Positioning and Altitude system CODA and the Sound Velocity Probe CTD 048 M. The R/V 'Orion' position was controlled by DGPS systems Trimble R8 L1/L2, Trimble 5800 L1/L2, and GNSS radio for HEPOS RTK and DGPS corrections. The same instrumentation was used for the topographic measurement of the coastline.

Geophysics

Sub-bottom profiling tracks (Fig. 1b) were conducted with BOOMER of Geoacoustics Ltd., which is suitable for penetrating coarse sediments at frequencies 1-3 kHz.

The surficial morphology of the seabed was recorded with side scan sonar by Geoacoustics Ltd., operating at 500 kHz.

Time series

Sensordata 6000 current meters specially modified (curved blades) for measuring vertical water flow were used to monitor the velocity of the underwater spring's freshwater/brackish flow. Conductivity and temperature were recorded with SBE-37 by Sea-Bird Electronics Inc. Sampling was set to 30 min. An autonomous underwater γ-ray spectrometer (KATERINA) was deployed simultaneously with the other sensors to measure, among other radionucleides, ^{222}Rn activity; details on the sensor principles and preliminary results are given by Tsabaris et al. (2010). All sensors were assembled on a lander designed and constructed for the specific needs of the SGD measurements. The lander was modified during the survey according to different requirements; e.g. due to failure of the sensors, mainly by sand and pebble blocking, it was decided to use pairs of sensors, instead of a single one. Placement of the lander the nearest possible to the spring outflow was performed by experienced divers.

3 Bottom morphology and SGD features

The morphology of the seabed is characterized by depths 20-40 m with smooth slopes in the NW sector. At the central sector, a crest with orientation NE-SW and relatively sharp slopes dominates; it extends ~500 m to the NE and then fades at shallow depths (10-11 m). The SGD depression appears to the SE sector of the area in a short distance (~150 m) from the coast (Fig. 1b). According to the geophysical recordings, the bedrock (most probably limestone) ascends to the surface of the seabed at the area where two fault lines converge (Fig. 2). The fault surfaces are vertical and covered by rich flora. The underwater feature is oriented W-E and its approximate length is 50-60 m, where as the N-S width is about 20 m (Fig. 3). During autumn and winter four single point springs of varying intensity were observed by the divers, which however did not operate throughout the survey period.

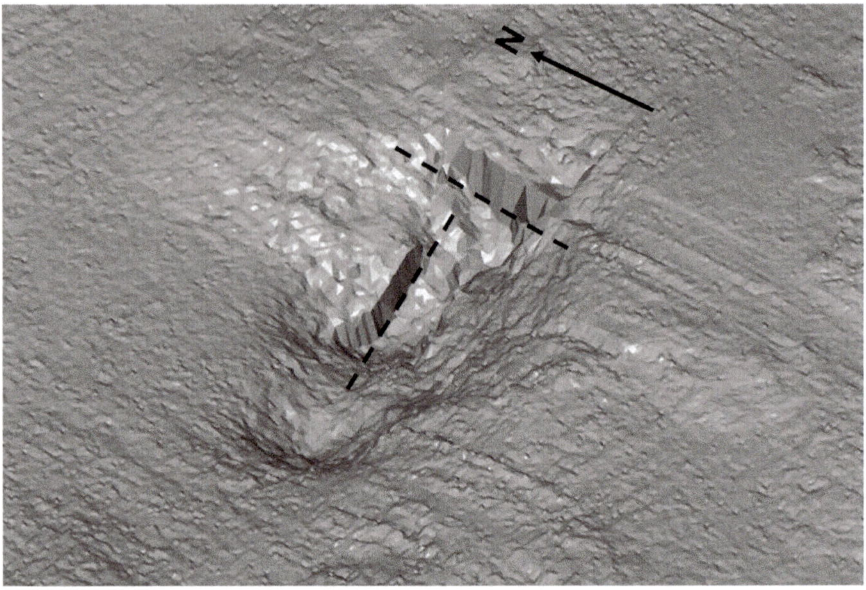

Fig. 2. 3-D representation of the SGD depression. Dashed lines: faults.

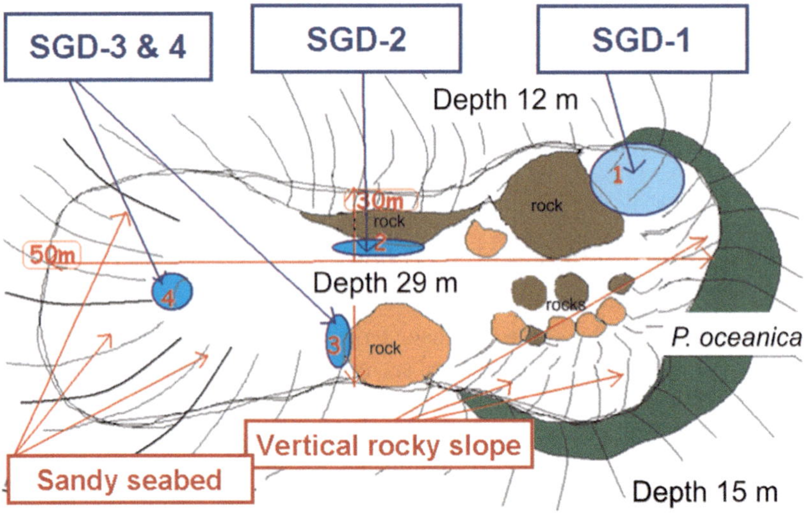

Fig. 3. Sketch of the SGD depression.

The only permanent and most impressive was SGD-1, emanating at 25-26 m depth (Fig. 3). The SGD-1 was covered by a limestone rock (dimensions ~5x4x1 m), thus direct placement of the lander exactly on the main outflow was impossible. Except for the apparent morphological obstacles, another problem was the safety of divers, as the water jet was flowing vertically towards the surface with high velocity. At times of high discharge, the divers could hardly reach the SGD without the danger of abrupt ascending, which is extremely dangerous. On the other hand, they described blasts of smaller water jets from all around the rock, thus hampering their underwater activities (lander placement and water sampling). It should be noted that the seabed around the main SGD was covered by rock debris, coarse sand, and after every visit (9 in total) the underwater landscape was slightly different, indicating strong flows and continuous movement of small and larger rock fragments. At the surface, the outflowing water creates a gyre with variable diameter, from 25 to 60 m, visible from long distance.

4 Hydrological parameters

Flow velocity

During the first deployments in summer 2009, and until the first days of October 2009, the flow of the main SGD was stable ~20 cm s^{-1} (Fig. 4). A major peak exceeding 120 cm s^{-1} was recorded in late October 2009, and another one in mid-January 2010, when the lander was positioned to the south side of the limestone rock and therefore not at the core of the SGD jet. The divers were not able to reach the site due to extreme outflow of freshwater. Relatively high flow velocities were recorded during the entire spring season (~60-80 cm s^{-1}) and until June-July 2010, when they decreased gradually from 40 to 20 cm s^{-1}.

An important outcome from time series study is the phenomenon of sharp gaps between successive deployments. This shows that the environment is exceptionally dynamic and variable, thus even a few 10s of centimeters of difference in the lander location resulted in great variability of the hydrological measurements. The presence of the limestone rock probably plays an important role in the geometry of the SGD, as it disrupts the normal upward and vertical flow, it generates increased turbulence. In addition, it may favor *in-situ* rapid mixing of groundwater and seawater.

Another issue that deserves particular attention is the difficulty in the estimation of the area of the spring which actively emanates water (in m^2). Therefore, the calculation of discharge can be only based on the divers' observations. It seems likely that an area of at least 1 m^2 was actively discharging groundwater, but this is probably a conservative estimation. According to this value, the discharge of the main SGD ranged from 720 to 4500 m^3 h^{-1} during the survey period. Such high fluxes are commonly expected in karstic and fractured formations (Gallardo and Marui 2006). According to Taniguchi et al. (2002) the areal extent of offshore SGDs and the groundwater flux are largely unknown. In the present study, the only way to estimate accurately the flux would be to remove all rocks blocking the access to the SGD and to measure the discharge area of the SGD. However, this operation would be rather costly.

Finally, it is noteworthy that during the survey the current meters suffered twice blocking of the rotor by falling pebbles, and one time the entire blade part was lost. Therefore the use of twin sensors proved to be the best solution to ensure data recovery from the hostile (for the instruments) environment of the SGD.

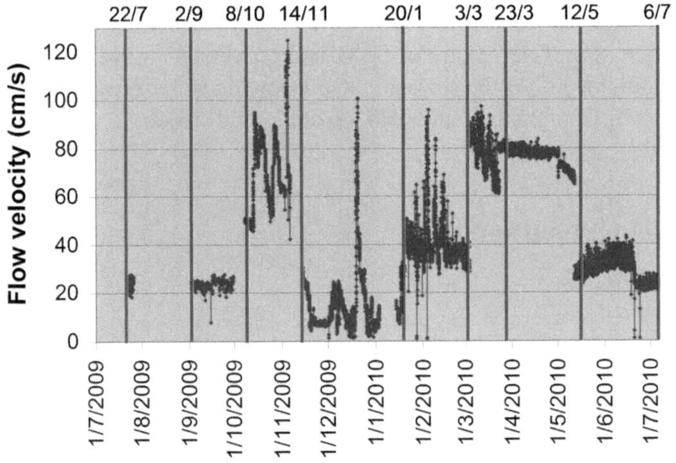

Fig. 4. Time series of flow velocity during the survey period. Red lines indicate deployment dates. From the 14th of November 2009 to January 20th 2010 the lander was positioned a few meters away from the main discharge point due to extremely intense freshwater outflow making access impossible.

Salinity and temperature

Considering the main scope of the survey, i.e. to verify if freshwater would be constantly outflowing from the SGD, salinity was probably the most critical parameter to monitor, together with groundwater volume. From the first deployment during July 2009, it became obvious that brackish waters were prevailing until October 2009, when salinity decreased to low values (Fig. 5). It is remarkable that during the deployment on the 9th of September 2009, the salinity increases from ~20 to values greater than 30. Similarly, during other recoveries and re-deployments salinity pattern exhibits substantial differences. This can be explained only by the specific conditions that prevail in the SGD area, and particularly to high turbulence. The synergy of explicitly complex underwater topography, the presence of the limestone rock on top of the main spring discharge point result in high variability of all measured parameters even during the same deployment period (e.g. 20/1-3/3 2010). In general, low salinity values representative of freshwater inputs were recorded from October 2009 to May 2010, indicating that the SGD emanated freshwater, which was rapidly mixed with ambient seawater of high salinity (~39). This period coincide with rainfall and snow melting, which supply the aquifers with freshwater.

Nevertheless, during the deployment 23/3-12/5 2010 it appears that without moving the lander to another position, there is a sudden increase in salinity, maintained until the end of the measurements. This is probably a critical point where freshwater discharges decrease (see also Fig. 4) to a threshold that causes salinization of the submarine groundwater (recharge) and triggers the recirculation and discharge of brackish waters (Taniguchi 2002). This phenomenon may be intensified by human intervention, since May is usually the month of the year when irrigation commences, and the operation of numerous private drill holes apparently pumps significant amounts of water from the aquifers feeding the SGD system.

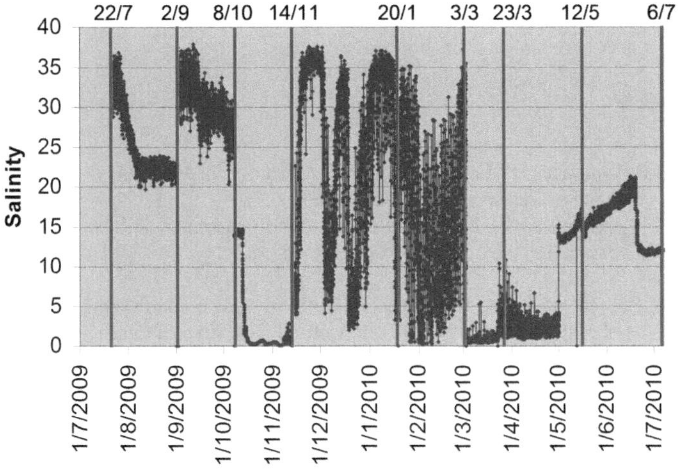

Fig. 5. Time series of salinity during the survey period. Red lines indicate deployment dates.

Temperature records may be separated in two periods: (i) May to October, with temperature mostly regulated by the ambient seawater temperature (18-26 °C); and (ii) November to April, when freshwater spring discharges have a clear signal ~12.2 °C, often recorded by the sensors. In general, temperature and salinity exhibit similar trends and represent the influence of submarine groundwater inputs in the area.

5 Conclusions

The Kalogria SGD in Messinia, Greece, is a unique site that currently concentrates a wealth of information, far more than any other Greek SGDs. The initial target of the study was to ascertain that potable water could be available for the local community. Our survey proved that the SGD is largely unsuitable for exploitation.

During the autumn and winter freshwater emanates from the spring, however strong turbulence causes fast mixing with seawater of high salinity. At the same time, the flow velocity is generally high, and water discharge was more than 4500 $m^3 h^{-1}$. On the contrary, during spring and summer the spring's water was brackish, and accompanied by relatively low discharge (~720 $m^3 h^{-1}$).

The high variability of salinity, velocity (and temperature) measurements show clearly that the SGD is a highly dynamic environment. The presence of a limestone rock, which covers the primary SGD, makes it difficult to obtain more accurate measurements and also to estimate actual dimensions (area) of the spring. Additional work in needed to identify the exact mechanism that enhances rapid mixing, and possibly the hydraulic conditions, which favor the salinization of the aquifer during spring and summer.

References

Burnett WC, Bokuniewicz H, Huettel M, Moore WS, Taniguchi M (2003) Groundwater and pore water inputs to the coastal zone. Biogeochem. 66, 3–33

Gallardo AH, Marui A (2006) Submarine groundwater discharge: an outlook of recent advancesand current knowledge. Geo-Mar. Lett. 26, 102–113. DOI 10.1007/s00367-006-0021-7

Taniguchi M, Burnett WC Cable JE Turner JV (2002) Investigation of submarine groundwater discharge. Hydrol. Process. 16, 2115–2129. DOI: 10.1002/hyp.1145

Tsabaris C, Scholten J, Karageorgis AP, Comanducci J.-F, Georgopoulos D, Liong Wee Kwong L, Patiris DL, Papathanassiou E (2010) Underwater in situ measurements of radionuclides in selected submarine groundwater springs, Mediterranean Sea. Radiat. Prot. Dosim. 1-9. doi:10.1093/rpd/ncq190

UNESCO (2004) Submarine groundwater discharge: Management implications, measurements and effects. IHP-VI, Series on groundwater No. 5, IOC Manuals and Guides No. 44

Zektser IS (1996) Groundwater discharge into the seas and oceans: state of the art. In: Buddemeier RW (ed.), Groundwater discharge in the coastal zone: Proceedings of an International Symposium. LOICZ/R&S/96-8, iv+179 pp. IGBP, LOICZ, Texel, The Netherlands, 122–126

Water tracing test of the Ag. Taxiarches spring (South Achaia, Peloponnese, Greece). Infiltration of the Olonos-Pindos geotectonic unit, Upper Cretaceous-Paleocenic carbonate rocks

N. Tsoukalas[2], K. Papaspyropoulos[1], R. Koutsi[3]

[1]Department of Geology and Geoenvironment, National and Kapodistrian University of Athens, GR 157 84, Panepistimiopolis, Athens, Greece. kpapaspy@geol.uoa.gr
[2] MSc Geologist, Ano Vlassia - Achaia, niktsouk@hotmail.com
[3] Dr Geologist, Athens, reginakoutsi@yahoo.gr

Abstract Ag. Taxiarches is one of the several springs that discharge Erymanthos mountain chain underground water in South Achaia (Peloponnese). The prevailing geotectonical unit is Olonos-Pindos, constituted mainly of deep-sea sediments consisted of carbonate, siliceous and clastic rocks. In order to realize the type and the expansion of the karstification into Upper Cretaceous-Paleocenic carbonate rocks, we carried out an experiment based on the implementation of tracer tests. The tracer we used was Uranine, and the whole procedure was completed in a day time. During this procedure, the velocity of the dissolved Uranine on the surface stream was also calculated. Furthermore, the mean underground water flow was determined, concerning to the karstic network system. In order to test the hydraulic conductivity, water samples were collected both from the physical exit of the spring and the interior of the adjacent underground river. The calculated mean velocity for both surficial and ground water flow velocity is about 160m/h, revealing the expected expansion of karstification. Thus, it is obvious that the Ag. Taxiarches discharging water is highly vulnerable in any water dissolved pollutant.

1 Introduction – Purpose

The area belongs to Regional Unit of Achaia, in Northern Peloponnese, 30 km SW of Kalavryta city. Ano and Kato Vlassia are the closest villages, 3 and 4 km respectively to the north. The surrounding area is located on the NE parts of Erymanthos Mountain chain in the Selinous river hydrological basin, where altitudes range from 750 m up to 2169 m a.s.l.; the second highest summit after Olonos (Fig. 1). Thus, Ag. Taxiarches is one of the several springs which discharge significant water volumes, flowing northeastwards to the Corinth Gulf, and forming the southernmost catchment area of Selinous river. Nearby to spring, there is an extensive underground karst system, developed in multiple levels.

The main objective of this study is to test hypothesis of direct infiltration of surficial stream water into the Upper Cretaceous-Paleocenic carbonate rocks, to define the hydraulic connection of the penetrated water into subsurface karstic river and finally the adjacent Ag. Taxiarches spring.

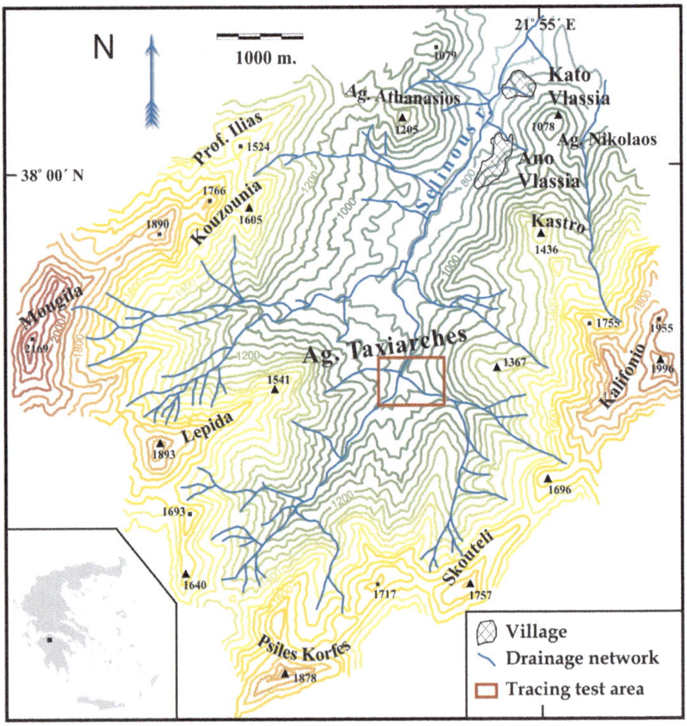

Fig. 1. Topographical map of the wider region, showing the tracing test area (based on HMGS).

2 Geological setting

Geotectonic units from both the Internal and External Hellenides are present in Peloponnese (Jacobshagen 1979, Papanikolaou 1997). The Internal Hellenides crop out only the Argolis area of eastern Peloponnese, whereas the rest of it is mostly composed by the geotectonic units of External Hellenides. The tectono-stratigraphically subjacent unit is the Ionian, in the northwestern part of Peloponnese, as well as the metamorphosed equivalent of Mani – Crete (Plattenkalk) to the southeast. Upwards, there is the Phyllite - Quartzite unit in central to south-eastern. Above is the Tripolis unit with the largest coverage. Lastly, the Olonos –

Pindos geotectonic unit is thrust on Tripolis and has widespread exposure too, in northern, central and southwestern Peloponnese; moreover, it constitutes the rocks of the study area. Significant areas mainly in western and north Peloponnese are covered by post alpine sediments very thick in some cases.

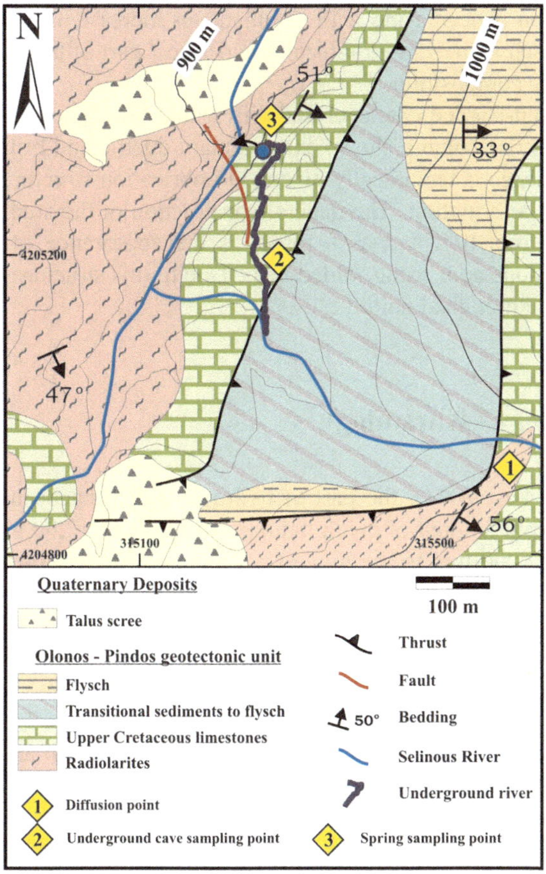

Fig. 2. Geological map of the tracing test area (based on Tsoukalas 2006).

The Alpine formations of Olonos – Pindos represent one of the most characteristic tectonic nappes in Hellenides. The NE Erymanthos Mountain is formed by the whole sedimentary succession (Tsoukalas 2006) (Fig. 2). The upper Triassic clastic sediments compose the basal lithological unit of the stratigraphic sequence, containing alternations of sandstones and pelites with minor carbonate intercalations. The following formation is the Triassic "Drymos limestones". The maximum thickness of this lithology is around 60 m, sporadically including interbedded chert. This is overlain by an at least 100 m thick radiolaritic sequence. Besides the siliceous part, there are also intercalations of pelites and limestones, mostly in the base and top of the succession. Moving upwards, the relatively thick Upper Cretaceous carbonate successions are ubiquitous, outcropping significant part of the area and mostly the peaks of the Erymanthos horst. This outstanding lithological unit has a total thickness of 200 – 250 m, where platy limestone is predominant. The next unit contains the 120 m transitional sediments to flysch. The carbonate proportion is decreasing with simultaneous increase of marls, black cherts and pelites. Flysch is the stratigraphically uppermost part of the succession with a sandstone-pelitic composition.

Due to the thin-bedded character and differential lithofacies of each individual formation, the entire column shows extraordinary ductile and brittle deformation. The whole sequence is tectonically placed in imbrication structure, caused by the Alpine tectonic deformation. The general construction of the studied area follows this type of deformation and as a result, is the appearance of successive upthrusting formations and strong folded beds. Due to this structure we have the repentance of the lithological units.

3 Hydrogeology - Karstology

According to Nikas (2004), the part of hydrogeological basin in the study area, receives significant amount of precipitation, varying from 1,200 mm to the lowest altitudes, up to 2,000 mm to the highest per annum. Furthermore, the calculation of the karst springs discharging rates in Ano Vouraikos basin -a few kilometers to the east- in the same Upper Cretaceous limestones, yielded an infiltration coefficient of 47% (Giannatos 1999). This number indicates high permeability of the carbonate rocks.

By the aforementioned formations, “Drymos” limestones, Upper Cretaceous limestones and the superjacent transitional sediments to flysch are permeable and consequently important for the overall hydrological conditions. On the contrary, the rest three lithologies comprise the impermeable rocks. In our case, radiolarites constitute the major impervious rock, whereas in agreement with the stratigraphically above Upper Cretaceous limestones, the latter are able to accommodate and release via springs, remarkable groundwater volumes (Fig 3a). Furthermore, it is considered that the underground water flowing follows not only the expanded tectonic discontinuities and faults, but also the stratigraphy which arise by the siliceous rock intercalations.

Consequently, the obvious reason of water infiltration and underground water flow, remains the karstification of carbonate rocks (Fig 3b). In the surrounding area of few tens of square kilometers, the Olonos – Pindos geotectonic unit contains remarkable surface karst landforms such as karrens, shallow dolines, gorges and dry karst valleys are developed. Moreover, subsurface karstification is present with caves, potholes, underground river with karstic conduits, classifying it to the Transitional type of karst (Tsoukalas and Papadopoulou 2004).

Fig. 3. a: The Ag. Taxiarches spring (in the front); the natural exit of the subsurface karstic river with emerging water (in the back) and **b:** the interior of the karstic river.

It is consisted of several levels of karstification, in an overall height of 100 m, creating corresponding caves. The most significant of them is an inflow cave with a total length of 215 m. The general direction is from south to north, following often the strike direction (NNE-SSW) of the Upper Cretaceous limestones as well as faults and fractures of similar or perpendicular direction. This underground river is active during the whole year. In wet seasons, water comes out from the natural exit, whereas when the flowing volume decreases, the water is completely cached by at least two karstic conduits and gradually remains one, 158 m deeply in the rock mass. The above spring is situated ten meters southwards from the exit of the cave, at 900 m a.s.l., in a steep slope. Regarding to the operational mechanism the spring is classified as overflow spring (Soulios 2010). The annual mean discharge volume of the spring is about10 L/s, according to field measurements of first author, during 2002.

4 Tracing procedure and experiment

The unraveling of the underground karst drainage system, its geometric parameters and water provenance, the relationship with the surface runoff, and the significance of karst water in total, excites the interest of the geoscientists to accomplish ground-water tracing study. Thus, a few decades ago Morfis and Zojer (1986) carried out pioneering tracing projects in eastern Peloponnese, extracting important hydrogeological and karstological implications by qualitative and quantitative point of view.

In our research, the tracing test deviated from the typical procedure where the dye is injected in a point of absolute infiltration. The test modified in order to explore the ability of the thin-bedded carbonate rocks to absorb gradually and spatially the surficial water from the riverbed. In this way, on the 4th of December 2004 at 11:10 a.m., Uranine tracer was dissolved and diffused in the stream at 960 m of altitude, following the flow direction of the current northwestwards approaching the main stream which is Selinous River.

The sampling positions were two. The one was the Ag. Taxiarches spring, north-westwards from the tracing point. The latter was also in the same direction inside the underground river, right before the position of the active karstic conduit. The distance (path) from the tracing point to the spring is about 700 m, and 550 m respectively from the underground river sampling point. The first samples were received in 20 min from the cave sampling point and in 50 min from the spring. Sampling was continued with time intervals of 30 min to 1 h for the first hours, whereas the entire procedure lasted for almost 23 hours. Uranine dye was slightly able to be seen macroscopically flushing from the spring after 4 h and 20 min; in agreement with the maximum concentration indicated from the breakthrough curve (Fig. 4a, b). The maximum concentration in the subsurface river appears almost an hour before. Furthermore, the first arriving of the dissolved substance in the spring is after 3 h and 50 min, whereas in the cave in almost 3 h respectively.

The curve of Uranine concentration illustrates that 6 h 30 min after the beginning and 5 h 50 min respectively in the spring and the cave, most of the material has been washed out.

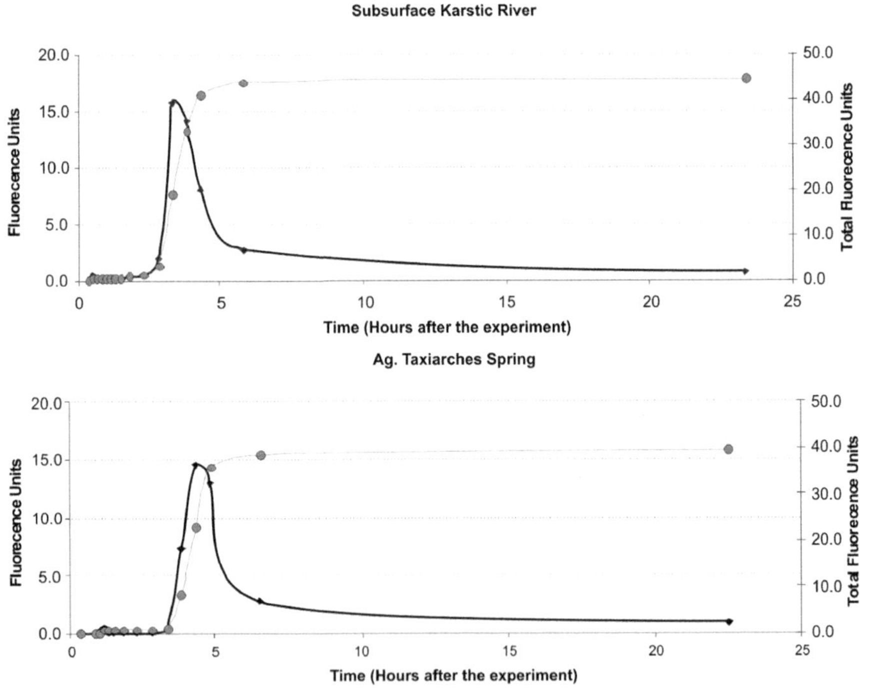

Fig. 4. a: Uranine breakthrough curve from the sampling point of the subsurface karstic river, **b:** Uranine breakthrough curve from the sampling point of the Ag. Taxiarches spring.

5 Conclusions – Results

The tracing experiment took place in Ag. Taxiarches karst spring, followed a relatively different procedure, where the Uranine tracer was dissolved and diffused in the stream. The sampling positions were two; the Ag. Taxiarches spring, northwestwards from the tracing point and the underground river in the same direction. The ultimate goal was the exploration of the ability of the thin-bedded carbonate rocks to absorb gradually and spatially the surficial water from the riverbed.

Taking into consideration the relative distance from the throwing point to those of collecting, the dominant flow speed according to the maximum concentration of the Uranine was calculated for both, spring and cave, as 162 m/h and 165 m/h respectively. Additionally, the calculated current speed, based on the surficial tracer propagation in the river stream is about 340m/h, as it was measured in the field. It is obvious that, analyzing the breakthrough concentration curves, the rapid culmination and the following exponential decrement reveals a karstic aquifer with considerable karstification (Koutsi and Stournaras 2005; Stamatis 2010).

Acknowledgments We would like to thank Ass. Prof. Papadopoulou – Vrynioti K. for the initial idea of the experiment. We would also like to acknowledge Dr. Ioannou P. for the lab assistance she provided.

References

Giannatos G (1999) Spring's hydrodynamic analysis from rocks in which the fissure type permeability prevails. The wider area of the Ano Bouraikos catchment in the Pindos geotectonic rock unit, PhD thesis, National & Kapodistrian Univ. of Athens, p.141

Jacobshagen V (1979) Structure and geotectonic evolution of the Hellenides. In: VI Colloq. Geol.Aegean Region 3, Athens, 1355-1367

Koutsi R, Stournaras G (2005) Tracer test at Loussi Polje Karst system (Kalavryta - North Peloponnesus, Hellas). 7th Hellenic Hydrogeological Conference, 2nd MEM Workshop of Fissured Rocks Hydrology 2, p.241-250

Morfis A, Zojer H (1986) Karst Hydrogeology of the Central and Eastern Peloponnesus (Greece). 5th International Symposium on Underground Water Tracing, Athens, 1986. – Steir. Beitr. Hydrogeologie, B.37/38: 301 S., Graz

Nikas K (2004) Hydrogeological condititions in NE part of Achaia Prefecture, PhD Thesis, University of Patras, 331p (in greek)

Papanikolaou D (1997) The Tectonostratigraphic Terranes of the Hellenides. Ann. Géol. des Pays Héll., vol. XXXVII, p. 495-514, Athens

Soulios G (2010) Springs (classification, function, capturing). Bull. Geol. Soc. Greece, Vol XLIII/1, p. 196-215

Stamatis, G (2010) Groundwater quality of the Ag. Paraskevi/Tempi valley karstic springs - application of a tracing test for research of the microbial pollution (Kato Olympos/NE Thessaly). Bull. Geol. Soc. Greece, Vol XLIII/4, p 1868-1877

Tsoukalas N, Papadopoulou K (2004) The karst in Ano Vlassia Achaia (Greece) and the contribution of Olonos – Pindos geotectonic unit on its formation, Bull. Geol. Soc. Greece, XXXVI/2, 1068-1076 (in Greek)

Tsoukalas N (2006) Geological structure and brittle-ductile deformation in the eastern part of the Olonos mountain chain (Achaia – Greece). Gaia, V. 14(2), p. 837-852, Athens (in Greek)

Effective infiltration assessment in Kourtaliotis karstic basin (S. Crete)

E. Steiakakis[1], D. Monopolis[†], D. Vavadakis[1], N. Lambrakis[2]

[1] Department of Mineral Resources Engineering, Technical University of Crete, Chania, GR 73 100. stiakaki@mred.tuc.gr

[2] Department of Geology, University of Patras, Rio-Patras, GR 26 504, Greece.

Abstract The objective of this work is to determine the coefficient of effective infiltration in Kourtaliotis karstic basin (S. Crete). Initially, the geological structure of the catchment basin that drains through Kourtaliotis springs was investigated and its boundaries were defined. Following, the coefficient of effective infiltration was determined by hydrologic balance calculations. The estimated value can be used for the determination of hydrologic balance of other basins in Crete with similar geology, physiographic and climatic conditions, contributing in the rational exploitation of the aquifers.

1 Introduction

Kourtaliotis hydrogeologic basin is in the southern part of Rethymno Prefecture and is formed of carbonates rocks (Tripolis and Pindos zones).

The basin discharges through a series of springs located in various altitudes and sites. Kourtaliotis spring comprise the main discharge outlet with mean annual yield 38.8×10^6 m^3.

The hydrogeologic basin of Kourtaliotis was evaluated as exemplary for the assessment of the effective infiltration in carbonate formations (Tripolis zone). The calculations were based on the groundwater balance of the karst spring.

2 Geology - Hydrogeology boundaries of Kourtaliotis spring catchment

In the southern part of Rethymno Prefecture, the catchment of Kourtaliotis spring is formed of limestone and dolomite formations (Tripolis zone) that are widely thrust onto Phyllites-Quartzites series.

The surface geological formations in the area are displayed in Figure 1 (Karakitsios 1982, Bonneau 1982) and the geological section AA' (Fig. 2) presents the geological structure in depth.

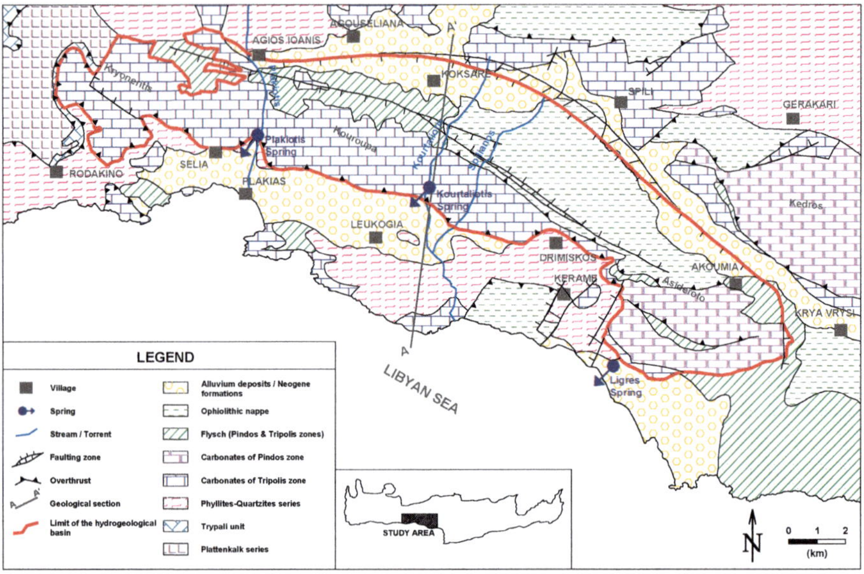

Fig. 1. Geological map of the study area.

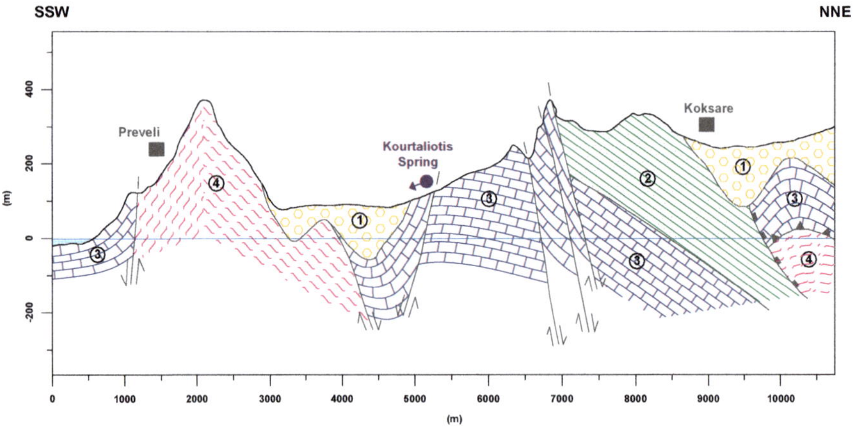

Fig. 2. Cross geological section of the Kourtaliotis basin along line AA' (Fig. 1). 1: Alluvium deposits / Neogene formations, 2: Flysch of Tripolis zone, 3: Carbonates of Tripolis zone, 4: Phyllite-Quartzite series.

Kourtaliotis basin is a calcareous karstic, hydrogeological system with well-defined permeability boundaries and forms a separate hydrogeological entity.

To the south-southwest, the permeability boundary is marked by a tectonic contact between the calcareous units of the aquifer and the phyllite - quartzite series which act as a barrier to the coastal area.

The northern boundaries are clearly marked by a syncline structure filled with formations of low permeability (flysh, flyschoide and other formations of similar composition).

This carbonate confining unit which is developed in a direction WNW-ESE, along the mountains range of Kryoneritis, Kouroupa and Asideroto (Fig. 1), constitutes the hydrogeological basin that feeds exclusively the Kourtaliotis spring.

3 Essential components of Kourtaliotis spring catchment

The water balance of the hydrogeological "isolated" Kourtaliotis basin it is relatively easy to be estimated given that the main inflows to the system are the recharge from rainfall (direct infiltration) and the leakage from the streams/ torrents which are flowing across the basin from north to south (secondary recharge). These streams flow across the carbonate formations and loses flow to the aquifer, contributing locally to water recharge (losses), while in locations where karstic springs emerge along their beds, streams gain yield.

The main streams that flow across the hydrogeologic basin (Fig. 1) are (a) Kourtaliotis, (b) Spilianos (Myxorouma) and (c) Plakiotis (Kotsifos).

The unknown coefficient of infiltration (percentage of primary precipitation that is absorbed by the ground and leads to the underground reservoir), can be estimated from the total amount of the springs yield, decreased by the quantity of water that precipitates through the stream beds. This quantity of water is supplied by hinterland basins, beyond the boundaries of Kourtaliotis basin.

Next, comparing the estimated infiltration with the volume of rainfall that falls in the hydrogeologic basin, the effective infiltration in Kourtaliotis karstic basin will be estimated.

In following, the calculation of the essential parameters and the estimation of the effective infiltration are presented in detail.

3.1 Discharge of the springs

The estimation of the mean annual yield of the springs (Table 1), was based on available records over a relatively long and continuous time spam (20 years) (YEB 2006), individual measurements, estimations and information gathered in the field during research time.

It should be noted that measurements of the rainfall and the spring discharge were non concurrent, during the investigation period. However, it was realized that the mean value of rainfall during the time period when discharge measurements are available, is identified with the corresponding value of the mean annual value of rainfall. Therefore, it is expected that the available measurements of spring's outlet reflect satisfactorily the mean annual spring discharge for the same period.

Table 1. Mean annual yield of the springs.

Springs	Kourtaliotis	Plakiotis	Ligres	Total Amount
Mean annual disharge ($\times 10^6$ m^3)	38.8	3.5	2.4	44.7

Furthermore, because the resolution that was realized, refers to measurements taken in a time interval of 20 years (1967-1987), it was judged essential to complete the time series data in the stations where they are missing, so that to exist a common reference time period for all the stations.

3.2 Secondary recharge - Streambed infiltration

Catchment of Kourtaliotis springs, apart from the direct infiltration, it is also recharged, from the streambed infiltration (Table 2). Analytically, the attendance of each one of the streams in the basin recharge is as follow:

Table 2. Water balance of streams catchment in the study area.

Catchment(Stream/Torrent)	Area	Precipitation	Runoff	Secondary
	km^2	$\times 10^6$ m^3/y	$\times 10^6$ m^3/y	$\times 10^6$ m^3/y
Kourtaliotis	55.75	60.95	15.5	3.10
Spilianos	42.45	55.85	14.7	4.59
Plakiotis	10.25	12.16	1.22	0.40
	Total secondary recharge			8.09

Kourtaliotis: The stream drains a basin of 55.75 km^2. It flows across Kourtaliotis basin and loses flow to the aquifer, through permeable carbonate outcrops.

Upstream of the Kourtaliotis spring, the stream was observed to lose a part of its flow to the aquifer, while during summer the entire stream flow is completely lost to the aquifer.

Flow in the stream is gauged at a location, upstream of the spring (Fig. 1), after the stream has flowed a length of 1 km across the basin, along the homonym gorge. The mean annual flow rate estimated 15.5×10^6 m^3. By observations (since no data are available to assess the magnitude of this flow at different sites), it was realized that the surface flow is decreased in the interval between the entry of the stream in the gorge and the point of measurement of flow, upstream of the springs.

Hence, from upstream side and for about 1 km, it behaves as influent stream, while downstream the spring (that outlets with discharge rate 0.8 to 3 m^3/sec), it behaves as effluent stream.

The recharge to the aquifer from the streambed infiltration was estimated to be 3.1×10^6 m^3/y (20% of the measured flow rate). This quantity contributes to the springs yield and it should be subtracted from the observed yield of the Kourtaliotis spring, since it concerns an allogenic recharge; this quantity of water emanates from basins beyond the study catchment.

Spilianos: It is an alloctonous stream that drains a catchment basin of 42.45 m^2. It flows across Kourtaliotis basin and loses flow to this aquifer, through permeable outcrops.

The average runoff, measured at a site after the exit of the stream from the Kourtaliotis basin, was estimated 14.7×10^6 m^3/y.

According to sporadic measurements of the stream flow, comparative data, and assuming similar conditions between Spilianos and Kourtaliotis streambeds, the infiltration was estimated to be equal to 25% the gauged flow.

This percentage was assumed higher than the adjust Kourtaliotis stream, because the flow length of Spilianos stream through the carbonate outcrops is more extended, and mainly because it entertains aestival yields higher than Kourtaliotis.

Based on the above mentioned estimation, (e.g. 25% of the surface runoff contributes to the yield of Kourtaliotis aquifer and the homonym spring), the secondary recharge to the underground basin from Spilianos stream is calculated in 4.59×10^6 m^3/y.

Plakiotis: It is a torrent that drains a catchment of 10.25 km^2. Runoff, after the exit of the torrent from the Kourtaliotis basin, was determined in 1.22×10^6 m^3/y. Taking into consideration that the streambed infiltration is 30% of the runoff, the water that Kourtaliotis basin receives from the Plakiotis torrent, is calculated in 0.4×10^6 m^3/y.

By summing all the above mentioned streambeds infiltration, the recharge to the Kourtaliotis aquifer, it is estimated to be 8.1×10^6 m^3/y.

3.3 Rainfall recharge

The annual rainfall volume in the study area was calculated from the time series of monthly rainfall values measured in the stations located in the area. Rainfall data from six precipitation stations which are found under comparable climatic conditions (stations of Agia Galini, Leykogeia, Vyzari, Spili, Melampes and Gerakari) were used.

A 20-year time series (1967-1987) of precipitation data were initially checked for continuity and consistency by the double-mass curve technique. The deduced figures / curves indicate that the time series of the stations have no inconsistencies.

However, the time series (20 years long) have shown gaps in short periods, resulting from the absence of concurrent data. Hence, it was judged essential to complete (extent) the observations in the stations where the existing measurements were insufficient and non concurrent. The gap was filled using the double-mass curve technique. Missing data were estimated using data of neighboring stations.

Afterwards the check of the observations and the completion of missing values, the mean annual precipitation during the specified 20 years of records, was calculated. Then, the average rainfall value of the stations was estimated and was correlated to their altitude.

A straight-line relationship established, giving a general relationship:

$$H=1.14h+582.6 \tag{2}$$

where H the rainfall in mm and h the altitude of station in m.

The precipitation map of the study area (Fig. 3) was constructed based on the above mentioned equation.

Based on the relation of rainfall - altitude (equation 1), the topographic map of the region (GYS 1972) and the use of Surfer software (2003), isohyets were drawn (Fig. 3) and the volume of precipitation falling per year to the basin (Table 1) was estimated. The annual rainfall volume in the study area was calculated in 71.32×10^6 m^3/y.

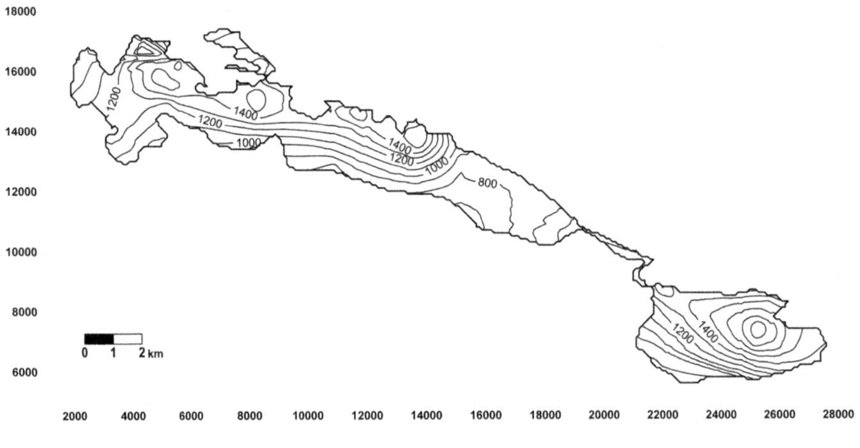

Fig. 3. Isohyets in Kourtaliotis basin (Interval: 100 mm of rainfall).

4 Coefficient of effective infiltration

In Table 3 the components of groundwater recharge to the basin are presented. They concern the yield of the springs, the mean annual precipitation and the streambed infiltration that accepts the hydrogeologic basin. According to the results, the annual yield of springs that outlet in the hydrogeologic basin is calculated in 44.7×10^6 m^3/y, the secondary recharge is estimated in 8.1×10^6 m^3/y, and the "estimated" direct precipitation (as it results from the difference between the springs yield and secondary recharge) it amounts in 36.6×10^6 m^3/y.

Table 3. Annual groundwater balance of Kourtaliotis basin.

Annual spring discharge	44.7×10^6 m^3
Leakage from the streams and torrents (secondary recharge)	8.1×10^6 m^3
Estimated direct infiltration (Annual spring discharge – secondary recharge)	36.6×10^6 m^3
Precipitation	71.32×10^6 m^3
Coefficient of effective infiltration	51.3 %

This quantity compared to the volume of rainfalls (71.32×10^6 m^3/y) that accepts annually the Kourtaliotis hydrogeologic basin, leads to the conclusion that the factor of infiltration in the carbonate rocks of the basin is 51.3%.

Similar values have been determined in other areas of Greece with same lithology and intense of karstification. More concretely, the values that were estimated in the Parnassos, Giona and various other regions in Central and Northern Greece using the method of discharge via springs, it varies between 45.6 and 60% (Aronis et al 1963; Mastoris 1968; Monopolis 1971; Soulios 1984).

5 Conclusions

Based on the geological structure of the region and the hydrogeological survey in the field, it is concluded that Kourtaliotis spring catchment constitutes an "exemplary" karstic basin and its hydrogeologic parameters can be applied in other basins with similar geology, physiographic and climatic conditions, and insufficient data, to calculate their balance.

The Kourtaliotis basin extents for 54.45 km^2, receives rainfall 71.32×10^6 m^3/y, while the direct infiltration in the carbonate formations of the area was calculated in 36.6×10^6 m^3/y.

As a result, the coefficient of effective infiltration in the carbonate formations of the basin was estimated to 51.3%. This value is similar to infiltration coefficient determined in other basins in the Para- Mediterranean region.

The value was achieved for first time in karstic limestones of Crete and it is of paramount importance to the groundwater balance estimation, and the management of both surface and groundwater resources in the island.

References

Aronis G, Burdon DJ, Zeris K (1961) Development of a karst limestone spring in Greece. Intern. Assoc. Sci. Hydrol., Groundwater in Arid Zones, 57, 564-585

Bonneau M (1982) Geological map of Greece, sheet Melampes: Athens, Institute for Geology and Subsurface Research

Golden Software, Inc (2003) Surfer v. 8.04 Manual

GYS (Hellenic Military Geographical Service) (1972) Topographic maps – Sellia and Melampes sheets, Scale 1:50.000

Karakitsios V (1982) Geological map of Greece, sheet Sellia: Athens, Institute for Geology and Subsurface Research

Mastoris K (1968) Hydrogeological investigation in the karstic limestone area of south Ghiona. PhD dissertation, NTUA, Athens (in Greek)

Monopolis D (1971) Hydrogeological study of the karstic carbonate rocks in the Mt Paranssos complex. PhD dissertation, NTUA, Athens (in Greek)

Soulios G (1984) Infiltration efficace dans le karst hellénique. J. of Hydrology, 75, pp. 343-356

Soulios G (1985) A contribution to the hydrogeological study of the karstic systems of the Greek territory. Scientific Annuals of the Faculty of Sciences, Aristotle University, Vol 3, No27. Thessaloniki (in Greek)

Soulios G (1985) Recherches sur l'unité des systèmes aquifères karstiques d'après des exemples du karst hellénique. Journal of Hydrology, 81, pp. 333-354

Technical University Crete (1990) Development of exploitation methods of the underground water in West Crete (PENED 87ED53: Program for the researchers support Ministry of Industry, Research and Technology/ General Secretariat for Research and Technology GGET, Progress Report, Chania), (unpublished, in Greek)

YEB (Land Reclamation Service) (2006) Hydro-meteorological data of Crete. Heraklio (in Greek)

The use of hydrographs in the study of the water regime of the Louros watershed karst formations

K. Katsanou[1], A. Maramathas[2], N. Lambrakis[1]

[1] University of Patras, Department of Geology, Laboratory of Hydrogeology, Rio-Patras, GR 26 504, Greece
[2] National Technical University of Athens, School of Chemical Engineering, Zographos Campus, GR 15780 Athens, Greece
Corresponding author: katsanou@upatras.gr

Abstract Louros watershed hosts several karst springs with discharge rates that vary between 0.08-3.14 m^3/s. In order to study the groundwater regime of the study area, the hydrographs of the springs were simulated with MODKARST model for a five-year period. Moreover, the classified discharge curves and the recession coefficient values were also processed. Based on the results three different systems can be distinguished. The first consists of Viros spring outflowing at altitudes of 300 m a.s.l. The second consists of Agios Georgios, Omorfada, Vathy and Kerasovo springs in the central part of the watershed at altitudes of 150 m a.s.l. And a third one that is formed at lower altitudes (30-50 m a.s.l.) and consists of Skala and Chanopoulo springs.

1 Introduction

The objective of this work is the study of Louros karst watershed, which geographically lies in the province of Epirus in North-West Greece (Fig. 1). Numerous studies have been carried out in karst, aiming at the investigation of groundwater regime. The term of karst aquifer system has occasionally different expressions, due to the complexity of karst conduits network that create diversity at all levels of observation (Margat 1971; Mangin 1975; Kallergis 2001). In the present study, the definition adopted considers a karst aquifer system as an area limited by well defined boundaries, that prevent the spread of any water quantity out of the system while inside, any part of the system is in relation with the others (Margat 1971). According to this definition, any action, e.g. recharge, discharge, within any part of the system affects the whole and never the outer part of the system.

The spring hydrographs of the karst aquifer system can reflect the groundwater regime and consequently, the study could be based on them (Mangin 1984). Except hydrographs, other analyses based on statistical methods have also been proposed (Padilla et al. 1994; Padilla and Pulido-Bosch 1995; Lambrakis et al. 2000; Panagopoulos and Lambrakis 2006).

In order to investigate the water regime of the different karst units of Louros' karst watershed, the hydrographs of nine major springs were simulated with the MODKARST model and the conclusions of this simulation were examined in combination with conclusions from the springs' classified discharge curves obtained by field measurements and Maillet recession coefficient values.

2 Geology and hydrogeology of Louros watershed

Geologically, the study area belongs to the formations of the Ionian geotectonic zone (Katsikatsos 1992). The older formations consist of evaporitic layers of Early to Middle Triassic age, which outcrop at the western and southern part of Ziros Lake (Karakitsios and Pomoni-Papaioannou 1998) (Fig. 1), overlain by a thick sequence of carbonate and clastic rocks reflecting a continuous sedimentation from Late Triassic to Upper Eocene (Skourtis- Coroneou and Solakius 1999). Carbonate formations are the major units constituting the highlands of the study area. These units include dolomites, the Pantokrator limestone, the Vigla limestone, Late Senonian limestone, Paleocene-Eocene limestone, and Posidonian chert (Skourtis-Coroneou et.al. 1995; Skourtis- Coroneou and Manacos 1995, Rigakis and Karakitsos 1998). Oligocene turbidite deposits outcrop at the margins of Louros' watershed, whereas at the lower part of it these formations overlain by Neogene and Pleistocene sediments as well as Holocene fluvial deposits (Macleod and Vita-Finzi 1982).

Within the Ionian Zone several major thrusts have been described as well as other important tectonic features like E-W trending strike-slip fault systems (IGRS-IFP 1966). Most of the thrusts show a normal westward dip (Louros, Paramithia, Margariti and Parga thrusts) and similarly, most of the folds are asymmetric with east-dipping axial planes. Nevertheless important tectonic structures with an opposite dip also exist (Xerovouni and Tomaros back-thrusts).

Since all the alpidic edifice of the Hellenides was created by west-dipping thrusting, the east-dipping thrusts are referred as back-thrusts. Moreover, the large E-W fault zones, like the Souli strike-slip fault system, also known as Petoussi fault, play an important role in the Alpine tectonic evolution of the area. Indeed, on the two sides of these faults the amount of displacement and the style of the NNW-SSE trending shortening structures often differ (IGRS-IFP 1966). The cross sections of Figure 1.b present the complex geological setting of the study area.

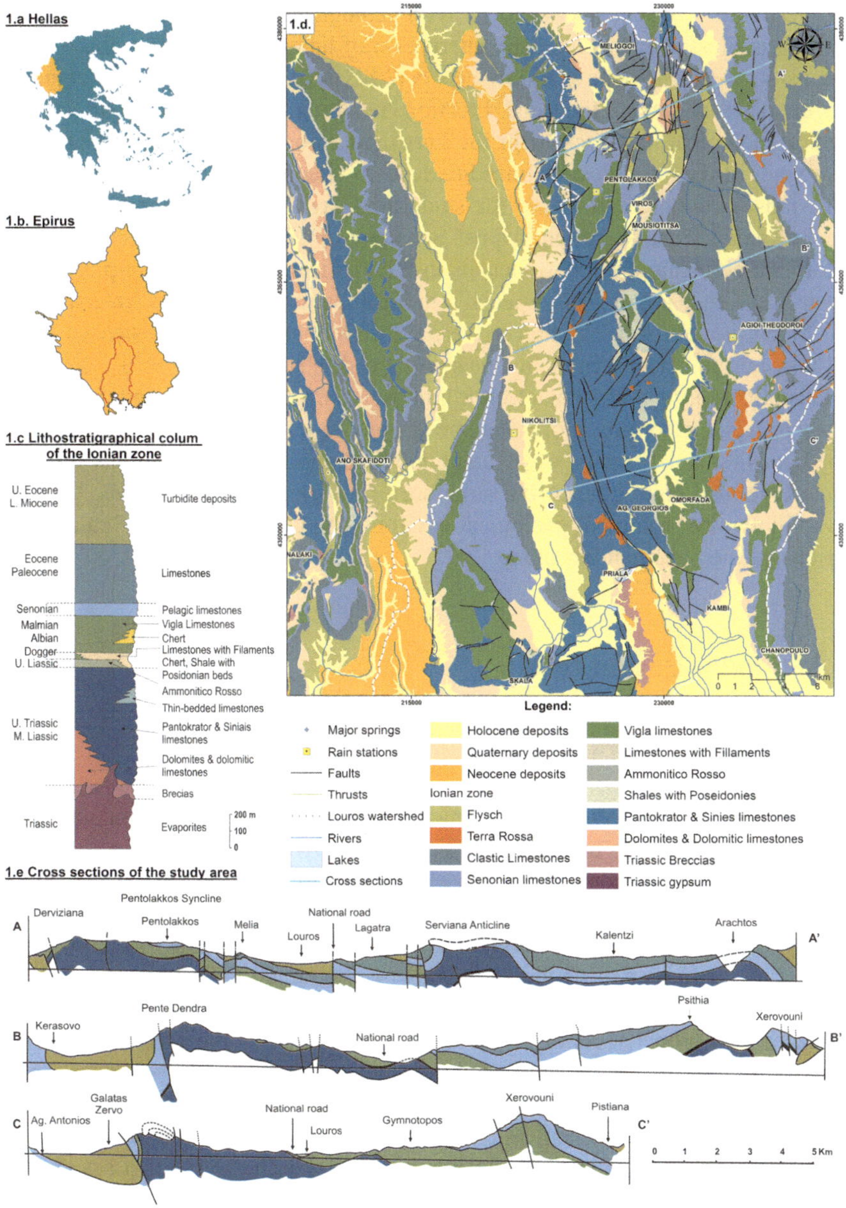

Fig. 1. a, b. Location of Louros watershed (red line) **c.** Stratigraphical column, **d.** Geological map. **e.** Cross sections of the study area (simplified map of the study area scale 1:50.000, IGME).

The formations present different hydrogeological properties. The geological conditions, the lithostratigraphical diversity, and the complicated tectonics in combination with the geomorphological conditions have resulted to the formation of independent or semi-independent hydrogeological units. The geological setting of Epirus illustrates characteristic alternation of limestone anticlines and turbidite deposits syncline structures following a NW-SE general axis direction. Groundwater circulates following the same directions resulting in the formation of extensive hydrolithological units hosted in the karstified carbonate anticlines separated by synclines of the impermeable turbidite deposits (Leontiadis and Smyrniotis 1986). The most important aquifers in the study area have been developed in carbonate rocks (mainly Pantokrator and Upper Senonian limestone). These formations present high permeability due to their intense karstification and fracture porosity and their occurrence is extended especially in the upper parts of Louros watershed.

Along the left bank of Louros river, at different altitudes and distances from it (Table 1), the clastic senonian limestone are discharging through the Viros and Mousiotitsa springs. The Pantokrator limestones are being discharged through Kerasovo, Agios Georgios and Priala springs on the left bank of the river, while on the opposite bank through Vathy and Omorfada springs. It is likely that the Vigla limestone also contribute to the structure of karstic units. At a lower level and even in a more prominent geological environment, Chanopoulo and Skala springs discharge the Senonian limestone.

3 Methodology and results

3.1 MODKARST model simulations

As mentioned above MODKARST model was used in order to simulate the hydrographs of the study area karst springs. MODKARST is a deterministic mathematical model which simulates the hydrograph (discharge versus time) of a karst spring (Maramathas et al. 2003). This model also, simulates the chloride concentration curve (chloride concentration of the spring water versus time), in case of the spring is brackish or periodically brackish (Maramathas and Boudouvis 2006). It is based on a hydrodynamic analogue of a karst spring that consists of a number of reservoirs properly connected with conduits. Each of these reservoirs is discharged to the conduit that abuts to the spring. Consequently, the spring discharge is given by the sum of these discharges. The macroscopic mass balance equation and the macroscopic mechanical energy balance equation in a control volume and in the respective control surface have been used in each reservoir to calculate the corresponding discharge (Maramathas et al. 2003).

Input information of the model is the rainfall of the recharge area while output information is the hydrograph (discharge versus time). With MODKARST model the hydrographs of nine springs that discharge discrete units of the Louros watershed have been simulated. In Figure 2 the model predicted hydrograph in comparison with discharge field measurements for the Skala spring and the Omorfada spring is presented.

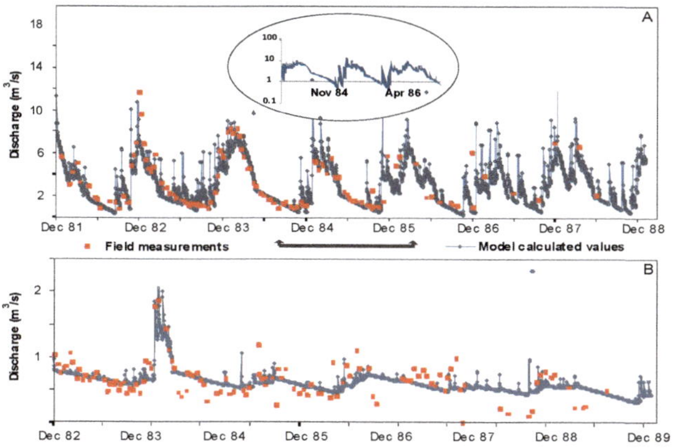

Fig. 2. Comparison between the model-predicted hydrograph and field measurements for a. Skala, b. Omorfada spring.

One could confirm in this Figure (Fig. 2) the satisfactory coincidence between the model calculated discharge values and the real ones. From model fitting the values of parameters that characterize the karst system of the spring, e.g. recharge area and coefficient of storage, was determined (Table 1).

Table 1. Natural features and results of the MODKARST and MAILLET simulation.

Spring	Recharge area (km^2)	Spring's altitude (m a.s.l.)	Coefficient of storage	Recession period	Mean discharge Q (m^3/s)	Recession Coefficient (1/s)	r^2
Ag. Georgios	151.0	100	$3.4x10^{-4}$, $2.5x10^{-6}$	22/2-30/10/86	3.14	$1.8x10^{-3}$	0.93
Kerasovo	22.0	140	$1.4x10^{-4}$	30/5- 20/11/84	0.48	$7.6x10^{-3}$	0.98
Mousiotitsa	18.0	300	$5.0x10^{-2}$	14/7-30/11/85	0.47	$5.1x10^{-3}$	0.98
Omorfada	31.5	120	$2.0x10^{-3}$	04/3-15/10/83	0.67	$1.1x10^{-3}$	0.86
Priala	26.0		$2.0x10^{-3}$	15/6-18/8/82	0.08	$4.6x10^{-3}$	1
Skala	152.0		$5.9x10^{-3}$, $4.2x10^{-7}$	20/5-15/9/84	2.00	$6.0x10^{-3}$	0.99
Vathy	79.0	138	$6.0x10^{-3}$	15/3-28/8/84	1.89	$3.3x10^{-3}$	0.98
Viros	28.2	300	$4.0x10^{-2}$	18/4- 15/10/84	0.92	$8.1x10^{-3}$	0.96
Chanopoulo	179.0		$4.8x10^{-3}$, $2.5x10^{-7}$	20/4-20/8/85	3.00	$8.5x10^{-3}$	0.98

Moreover, the simulation demonstrated that some of the springs, namely Agios Georgios, Skala and Chanopoulo drain karst reservoirs, where channels are present. In the springs' discharge an additional dense net of fractures is contributing and therefore this model gives two different storage coefficients (see Table 1). In other words, the karst reservoirs of the above-mentioned springs' present dual porosity.

Viros spring seems to discharge an isolated karst system. A two months time lag was obtained between rainfall occurrence and spring's discharge response, after a long dry season. Similar results were obtained for Mousiotitsa spring which also presents hysteresis in relation to rainfall. These springs are at an altitude of 300 m a.s.l and present similar storage coefficients values ($4x10^{-2}$, $5x10^{-2}$ respectively). At a lower altitude (110-150 m a.s.l.) the Kerasovo, Vathy, Omorfada and Agios Georgios springs are laid. Vathy and Omorfada present similar storage coefficient values ($6x10^{-3}$, $2x10^{-3}$ respectively). The rest of the springs are characterized by storage coefficients values of an order of magnitude lower.

At an even lower altitude of 30-50 m a.s.l, Skala and Chanopoulo springs present similar mean discharge and average storage coefficient values.

3.2 Recession curves and Maillet's recession coefficient values

The recession coefficient values (s) were calculated from the application of the Maillet's equation to the spring's discharge data. They have physical meaning and therefore can be used for the study of the springs. As it is known, the recession coefficient is proportional to the hydraulic conductivity and inversely proportional to storativity and the karst aquifer extent. It shows the rate of the decrease of the karst unit's discharge during recession. Thus, high coefficient values correspond to rapid recession and low values to slow recession.

According to the study of hydrographs, two of which are shown in Figure 2, it can be concluded that the separation of the absolute recession period is not easy due to the continuous rainfall throughout the year in Louros watershed. Nevertheless, the recession coefficients were calculated taking into account the part of the curves that correspond to the longer time period without rainfall.

3.3 Curves of the classified discharges

The curves of the classified discharges are used to investigate anomalies in recharge or discharge of a karst aquifer system (Mangin 1971; Soulios 1985). In order to construct these curves, discharge values on a daily basis for a five years period are used in a probabilistic sheet (Fig.3). An isolated karst system presents a normal distribution. In the opposite case any perturbation in the system deforms the curves. The classified discharge distributions were plotted for the nine springs

(Fig. 3a-i). Viros spring (Fig. 3h) displays a relatively normal distribution, typical of a relatively isolated karst unit which does not seem to be discharged from other springs. Mousiotitsa spring (Fig. 3c) shows similar behavior. Nevertheless, in this case obviously, there is a supplementary recharge contribution to the karst system for discharge values that exceed 0.45 m^3/s and 0.8 m^3/s respectively, which displays relatively low probability of emergence (cumulative probability below 95%).

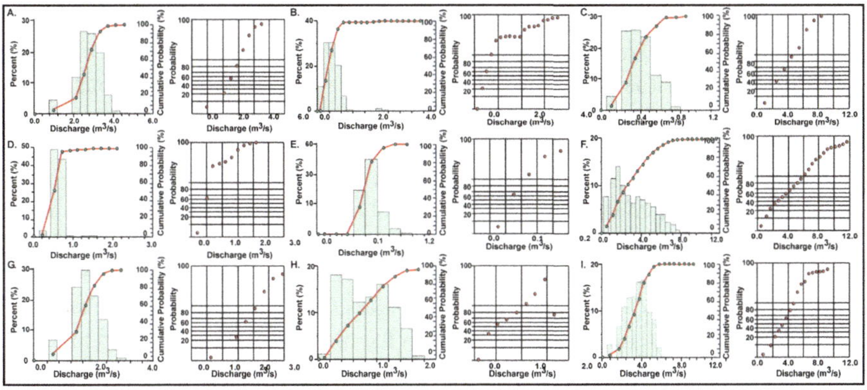

Fig. 3. Classified discharge curves for **a.** Agios Georgios, **b.** Kerasovo, **c.** Mousiotitsa, **d.** Omorfada, **e.** Priala, **f.** Skala, **g.** Vathy, **h.** Viros, **i.** Chanopoulo springs.

Kerasovo, Omorfada, Vathy, Agios Georgios and Priala springs correspond to complex karst systems. Kerasovo (Fig. 3b) spring shows a supplementary recharge component contribution of relatively low discharge values ranging between 1-2 m^3/s, while during increased rainfall intensities additional conduits are recharging the system with larger amounts of water and the discharge of the spring reaches higher values (2.5-3.5 m^3/s). The rest springs of this group present similar behavior with lower mean discharge values.

4 Discussion and Conclusions

The carbonate formations of Louros hydrogeological basin include very important karst springs that correspond to semi autonomous, complex karst aquifers that in many cases are in hydraulic contact.

The curves of classified discharge (Fig. 3) clearly show that except Viros, all other springs have a supplementary recharge component. The most significant springs, in terms of mean daily discharge values, namely Vathy, Agios Georgios, Skala and Chanopoulo show certain regularity in the distribution of daily discharge that could be explained by considering a very large catchment area that acts as regulator in the quantities of inflows and outflows of the karst systems. However, the storage coefficients vary depending on geological formation. Thus, Agios Georgios spring which drains the Pantokrator limestone presents storage coefficient values of an order of magnitude lower than the other three springs that drain Pantokrator and Senonian limestone (Table 1). The complexity of karst systems is more evident in springs of lower discharge such as Priala, Kerasovo and Omorfada. The base level of the karst seems to vary among the studied springs and therefore is considered to play a key role.

The recession coefficient values do not show considerable variation among different springs and particularly different geological formations. Springs of higher altitudes such as Viros and Mousiotitsa that discharge different formations (Senonian and Pantokrator limestone respectively) display higher storage coefficient values compared with those of lower altitudes.

Thus, for the springs of higher altitudes, Viros and Mousiotitsa and for those hosted in the center of the basin the base level of the karst system is driven a few meters below the level of Louros dam. The aquifer hosted in the unconsolidated and carbonate formations, brings the Pantokrator and the Senonian limestone in hydraulic connection. The karst system of Agios Georgios, Omorfada, Vathy and Kerasovo springs interacts and thus justifies the high overall discharge in total catchment area of relatively limited extent. Another base level of the karst is formed at lower altitudes and hosts Skala and Chanopoulo springs.

References

IGRS-IFP (Institut de Geologie et de Recherches du Sous-sol et Institut Francaise du Petrole) (1966) Etude Geologique de l'Epire (Grece nord-occidentale), Technip, Paris, 306 pp

Kallergis G (2001) Applied environmental hydrogeology 2nd ed., Vol. 3, 432 pp. Athens: Technical Chamber of Greece

Karakitsios V and Pomoni-Papaioannou F (1998) Sedimentological Study of the Triassic Solution-collapse Breccias of the Ionian zone (NW Greece). Carbonates and Evaporites 13(2):207-218

Katsikatsos G, (1992) The Geology of Greece, 451 pp. University of Patras

Lambrakis N, Andreou AS, Polydoropoulos P, Georgopoulos E, Bountis T (2000) Non-linear analysis and forecasting of a brackish karstic spring. Water Resour. Res. 36(4):875-884

Leontiadis I, Smyrniotis, Ch, (1986) Isotope hydrology study of the Louros Riverplain area. Proc. of 5th International Symposium on underground water tracing, Athens, 75-90

Macleod DA and Vita-Finzi C (1982) Environment and provenance in the development of recent alluvial deposits in Epirus, NW Greece. Earth surface processes and landforms, 7:29-43

Mangin A, (1971) Etude des debits classes d' exutoires karstiques portant sur un cycle hydrologique. Ann. peleol., 26(2): 283-339

Mangin A (1984) Pour une meilleure connaissance des systemes hydrologiques a´ partir des analyses corre´latoire et spectrales. Journal of Hydrology 67(1-4):25-43

Maramathas A, Maroulis Z, Marinos-Kouris D (2003) A Brackish Karstic Springs Model: Application on Almiros Crete Greece. Ground Water 41(5), 608-620

Maramathas A, Boudouvis A (2006) Manifestation and measurement of the fractal characteristics of karst hydrogeological formations. Advances in Water Resources 29(1), 112-116
Margat J (1971) Terminologie hydrogeologique. Bull. B.R.G.M., Sect. III, No, 1, pp. 155-168
Padilla A, Pulido-Bosch A, Mangin A (1994) Relative importance of baseflow and quickflow from hydrographs of karst spring. Ground Water 32 (2):267-277
Padilla A, Pulido-Bosch A (1995) Study of hydrographs of karstic aquifers by means of correlation and cross-spectral analysis. Journal of Hydrology 168:73-89
Panagopoulos G, Lambrakis N (2006) The contribution of time series analysis to the study of the hydrodynamic characteristics of the karst systems: Application on two typical karst aquifers of Greece (Trifilia, Almyros Crete). Journal of Hydrology 329: 368-376
Rigakis N and Karakitsios V (1998) The source rock horizons of the Ionian Basin (NW Greece). Marine Petrol. Geol. 15:593-617
Soulios G (1985) Recherches sur l'unite des systems aquifers karstiques d' après des examples du karst Hellenique. Journal of Hydrology, 81 : 333-354
Soulios G (1991) Contribution à l'étude des courbes de récession des sources karstiques: Exemples du pays Hellénique. Hydrol. 127:29-42
Skourtsis-Coroneou V and Manacos C (1995) New micropaleontological data on the age of the onset of the deposition of the Vigla Limestone Formation. Special Publication of the Geological Society of Greece 4:269-274
Skourtsis-Coroneou V and Solakius N (1999) Calpionellid zonation at the Jurassic/Cretaceous boundary within the Vigla Limestone Formation (Ionian Zone, western Greece) and carbon isotope analysis. Cretaceous Research 20: 583-595
Skourtsis-Coroneou V, Solakius N and Constantinidis I (1995) Cretaceous stratigraphy of the Ionian Zone, Hellenides, Western Greece. Cretaceous Research 16: 539-558

Hydrogeological conditions of the coastal area of the Hydrological basin Almyros, Prefecture Magnesia, Greece

Ch. Myriounis[1], G. Dimopoulos[2], A. Manakos[3]

[1]Dr. Geologist - Rural and Surveyor Engineer - Igoumenitsa P. Tsaldari 17 46100 Greece
cmyriounis@teemail.gr
[2]School of Geology, Aristotle University of Thessaloniki, Thessaloniki, Greece
[3]I.G.M.E. Department of Central Macedonia, Greece

Abstract In this paper are analyzed the hydrogeological conditions of the coastal area of the hydrological basin Almyros which belongs in the Prefecrure of Magnesia, Greece. At first, the geological, tectonic and stratigraphic conditions of the coastal area are described. Then, the genesis mechanism of Kefalosi spring that emerges at the southern margin of the region and has an important role in shaping the quality and quantity characteristics of the regional groundwater resources regime is analyzed. A reference is made to the particular conditions that characterise the recharge conditions of the regional aquifers. Based on the results produced by the conducted geophysical survey, the lithostratighraphic setup that controls the documented seawater intrusion is presented. An overall strong correlation of the geological, lithostratigraphic and tectonic characteristics of the regional aquifers to their hydrogeological properties is noted.

1 Introduction

Almyros basin is located in Magnesia, Thessaly Greece, south of the city of Volos. The study area covers a total area of 134.32 km^2 and encompasses the coastal part of Almyros basin. To the north is bounded by the fault zone of Nea Anchialos, to the south reaches the southern part of the village Sourpi while it is confined to the west in the coastal basin (Fig. 1). It belongs to the Almyros Municipality (NSSG 2001).

Almyros basin is characterized by the existence of Kefalosi karstic spring, at south, which has an important role in shaping the quality and quantity characteristics of the regional groundwater resources.

The present study aims to investigate the hydrogeological conditions of the coastal area of the hydrological basin Almyros using analyzing the geological, tectonic, stratigraphic, geophysical, and hydrological data.

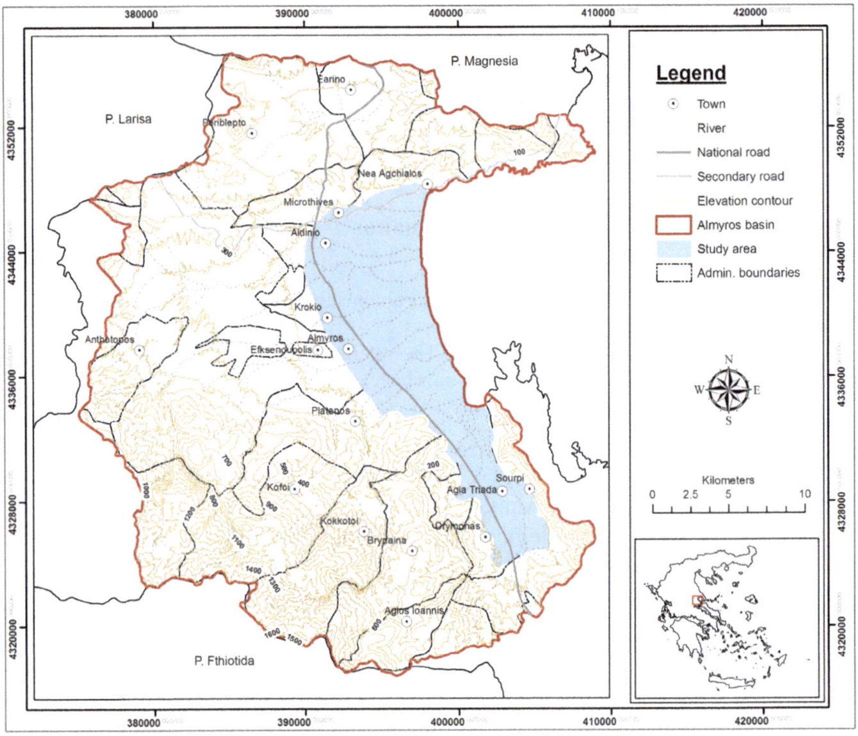

Fig. 1. Area of interest.

2 Geomorphological, geological and tectonic setting

The Almyros basin is part of Almyros-Pagasitikos uniform graben. The 2/3 of the graben are covered by eastern Pagasitikos Gulf, while 1/3 to the west is the hilly area and the low-relief land area, with altitudes up to 200m, which is the study area. The basin geomorphologically is distinguished in the Sourpi and Efxenoupoli sub-basins (Galanakis 1997).

According to Galanakis (1997), the relief of the Almyros basin is mainly due to tectonic activity of active faults that are active in the Pliocene until today, with reactivation of many of them during the Quaternary, and secondarily to climate conditions and differential erosion of rocks. The main feature of Almyros basin is the relative sinking of the eastern part of the basin compared to the west.

Alpine sediments present in the Almyros basin area belong to Pelagonian and Sub-Pelagonian zone (Fig. 2). Summarily, in the broader area of research formations of Pelagonian zone encountered are the pre-alpine background, slates, schists and ultrabasic rocks. Sub-Pelagonian formations found in the area are Paleozoic rocks of Permian, Triassic crystalline limestone, Jurassic, the formation of shales and ophiolites, the transgressive limestones and flysch.

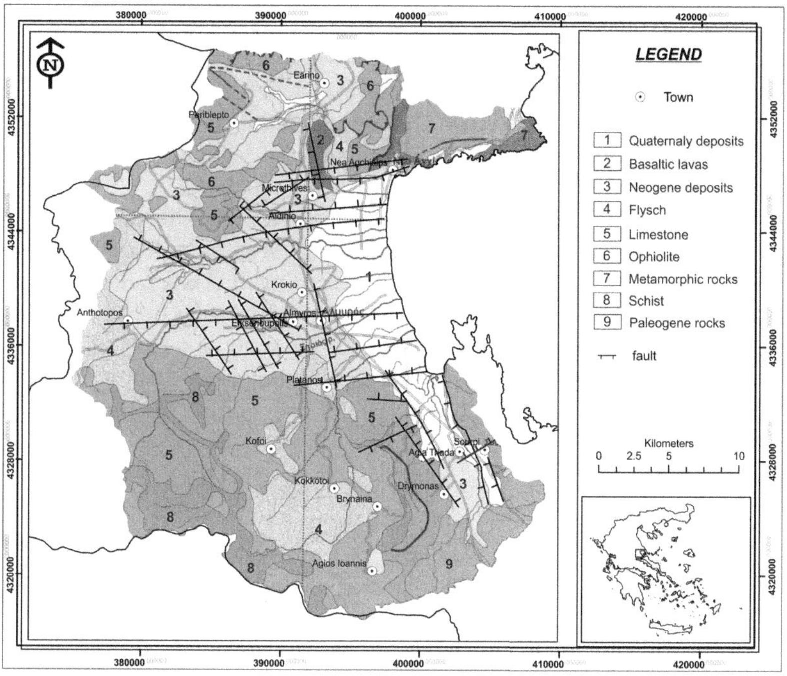

Fig. 2. Geological Map of Almyros Basin from IGME and modified by Myriounis (2008).

Regarding the post alpine stratigraphic structure, there can be found Neogene deposits with lignite deposits (loam) which are of lacustrine origin and of Lower Pliocene age. During the Upper Pliocene takes place deposition of yellowish clay marls, interrupted by short periods of semi-arid and continental facies. Finally, during Pleistocene clastic sedimentation is dominant, mainly fluvial and continental origin (Galanakis 1997).

3 Hydrogeological setting in the coastal area

The hydrogeological properties of geologic formations are controlled by the lithological composition and tectonic activity. Thus, the distinguished karst system of Northeastern Othris, containing the spring of brackish water Kefalosi, is formed, which is responsible for the hydrogeological conditions in the southern part of the study area. The above karst system shows great capacity. In the area of the same name exists Sourpi karst system of dolomitic and limestone composition. In this karst system, groundwater resources have better quality characteristics than those of Othris. In the northern region (N. Ankhialos) is found the Mavrorahi karst system. This system consists of limestones of large capacity. A feature of this system is the intense effects of seawater intrusions, due to the hydraulic communication between the carbonate blocks and the sea.

The systems of the Neogene sediments consist of alterations of coarse and fine-grained material that create free and pressurized aquifers. Moving westward aquifers of Neogene sediments have reduced capacity because of the lithological composition and their refreshing conditions. Quaternary sediments in the area cover most of the region and are generally impermeable formations.

The Kefalosi spring is located within the karst rocks of Othris, in the southern part of Efxenoupoli sub-basin. The spring occurs at an altitude of 26.5 m above sea level and 3250 m away from the coast. According to Dimopoulos (1983, 1984) it presents brackish water that comes from mixing of seawater and freshwater. Both discharge of the spring and the content of the water in Cl ions presents sharp fluctuations. This is attributed to the highly complex mechanism of operation of the spring (Fig. 3a.b.).

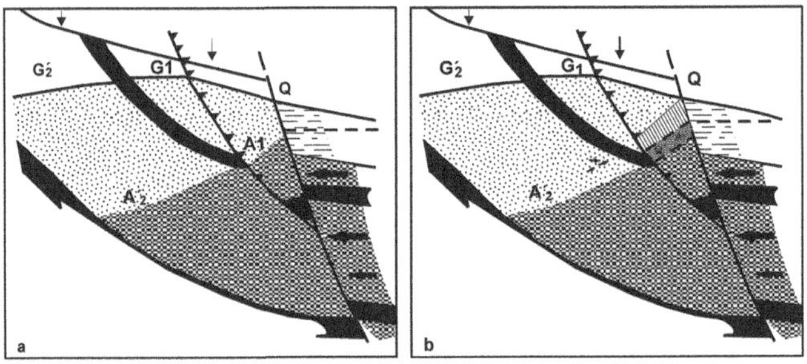

Fig. 3. Mechanism of operation of Kefalosi spring after Dimopoulos (1983, 1984) (1., 2. karstic aquifer, $G_{1.}$, $G_{2.}$, groundwater level, Q. Kefalosi spring, O. point of communication)

According to Dimopoulos (1983, 1984) and Myriounis (2008), special geological, tectonic and stratigraphic conditions prevailing in the region have brought together two independent karstic aquifers (1) and (2) who in particular times and hydrological conditions, communicate.

At the end of winder hydrological period (Fig 3.a) a rising of the groundwater level (G1) on the aquifer (1) is occurred, due to refreshing phenomena, while in aquifer (2) the groundwater level is in position (G2'), lower than (G1), due to long distance between the karstic aquifer (2) and the spring Kefalosi. At this stage the aquifers don't communicate. When water from aquifer (2) refreshes the groundwater resources of the area (Fig 3.b), groundwater in aquifer (1) is lower than before (reduce of refreshing), and the two aquifers are communicating (point O, Fig. 3.b).

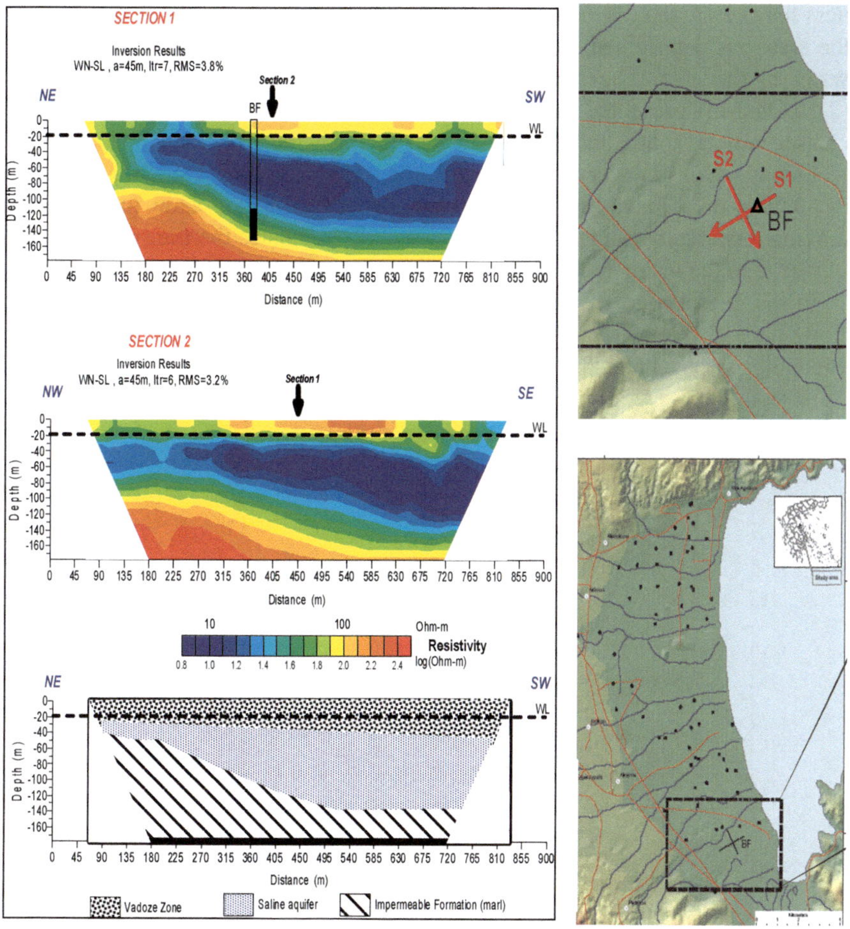

Fig. 4. Positions of geoelectric sections in Almyros basin, after Myriounis et al. (2006)

Through this communication, water from aquifer (2) leaves the spring, dragging brackish water from aquifer (1). As a result, the spring presents brackish character in an altitude of 26.5 m above sea level.

In order to investigate hydrogeological conditions in the area east of the Kefalosi spring, two geophysical surveys (Fig. 4) were made. The aim of these geophysical surveys was to determine the stratigraphy of the area and the location of the salty wedge (Aracil et. al. 2004, Goldman and Karfi 2004, Myriounis and Theocharis 2010). The method used was the Wenner - Schlumberger. The average distance of the geophysical surveys was 900 m. The distance between the electrodes was 45 m, while each section contained 21 electrodes.

Using the geophysical surveys, initially was determined the thickness of the unsaturated zone, which in the area consists of loose formations and with a thickness between 5 and 30 m. Also, the depth and thickness of salty wedge shown in the area was determined, on the basis of physicochemical parameters of groundwater in the area. The salt wedge is found as a band of very low resistance (2-10 Ohm-m) presented at a depth up to 80m. The thickness of this zone increases moving south, and reaches the absolute elevation of-120m.

Finally, a third layer occurs in the region, characterized by high resistance (> 80 Ohm-m). This layer reaches depths greater than 160m below the sea surface moving towards the south. This layer essentially regulates the transmission of salty wedge to the inside of the interior land (Myriounis et. al. 2006).

In the coastal part of the study area artesian phenomena and submarine outflows have also been identified (Fig. 5). They occur in the central part of the Efxenoupoli sub-basin and also in the north, east of Aidino.

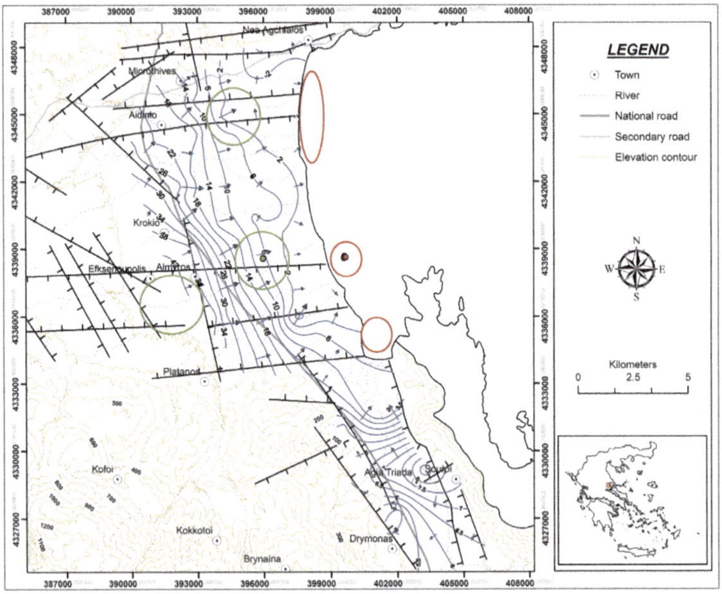

Fig. 5. Submarine outflows (red color) and artesian phenomena (green color) for Almyros basin area, and piezometric map for September 2005 after Myriounis (2008).

Artesian phenomena are attributed to faults with north-south direction, which have humbled the eastern parts of the aquifers in relation to the west, resulting in free-aquifers of the western regions to transitioning to pressurized, showing in places artesian phenomena moving eastward. This phenomenon occurs mainly during the winter months when the supply of the aquifer is greater.

Regarding submarine outflows, they are located in the northern coastline of the study area in Demetriada and south of the beach area of Almyros (Fig. 5). Generally the positions of vent are closely associated with the positions of faults with E-W direction present in the margins of the study area. Through these cracks groundwater find favorable corridors to discharge into the sea. In the Dimitriada the vents do not appear anymore, as the subsidence of the aquifer has led to their disappearance. In contrast, in Almyros beach vents are located even within 850 m from coast.

For the investigation of groundwater level in the coastal part of Almyros basin, groundwater level measurements took place during 2005, in May (147 wells) and September (93 wells), in a uniform aligned network of hydro-points (Myriounis 2008). All field data were collected in a database system (MS Access) and where analyzed using Geographical Information Systems (ArcInfo).

From the piezometric map of the region (Fig. 5) for September 2005, it is found that in the Efxenoupoli sub-basin are presented two axes of groundwater drainage which are identical to those in areas where there are submarine outflows. For the Sourpi sub-basin, in the area between the settlements Sourpi and Agia Triada, a local high of the piezmetric charge is present, attributed to local geological, stratigraphic and tectonic conditions in the area. More specifically, the presence of cross-faults with NNW-SSE and EW direction in the region has resulted in the supply of sediment aquifers from the underlying karst aquifer.

Finally, it should be emphasized that due to the existence of these faults, there is a diversity of the chemical characteristics of groundwater in the area of Sourpi sub-basin, and in the southern part of the site groundwater does not appear so aggravated by the phenomenon of seawater intrusion compared with waters of the north. This happens because the karst rocks of the southern part are hydraulically isolated compared with those of the north.

4 Results

From the hydrogeological investigation of the coastal part of the Almyros basin, it is found that the karst rocks of the region show strong seawater intrusion effects. This is attributed to hydraulic communication with the sea. The supply of alluvial aquifers in the area of karstic rocks leads to the qualitative degradation in the margins of Efxenoupoli sub-basin and in the coastal part of the Sourpi sub-basin.

The Kefalosi spring, which feeds the alluvium aquifers in the southern Efxenoupoli sub-basin presents brackish water that comes from mixing of seawater and freshwater. The salinity of the spring and her manifestation in this altitude is associated with local geological and tectonic conditions in the region.

The piezometric conditions in the region create conditions for submarine outflows and artesian phenomena. These phenomena are identified in both the southern and northern region. The presence of these phenomena depends on supply conditions of the aquifers and its geological and tectonic conditions in the region.

From the geophysical surveys in the area west of Kefalosi, is determined the depth and thickness of salty wedge shown in the area, determined also on the basis of physicochemical parameters of groundwater in the area. Also is specified the depth of an impervious formation, which essentially regulates the transmission of the salty wedge to the inside of the interior land, acting protectively and limiting the spread of the seawater intrusion front in the central part of the region, a fact confirmed also by the physicochemical parameters of groundwater waters of the region.

Because of the favorable refreshing of the alluvial aquifers, which has resulted in artesian phenomena and submarine outflows, as well as the specific lithostratigraphic conditions of the area, no seawater intrusion occurs in the central part of the Efxenoupoli sub-basin, despite the over-exploitation of groundwater.

In Sourpi sub-basin, recharge of the alluvial aquifers from the strained karst rocks in the area, combined with overexploitation of groundwater in the region has resulted in the intrusion of sea water in groundwater aquifers. At the same time the local geological and tectonic conditions in the region have resulted in modulation of chemical characteristics of groundwater, mainly in the southern part.

Acknowledgments The first author, Ch. Myriounis, would like to thank the State Scholarship Foundation for financial support under the 3998/2003-2004 program.

References

Aracil, E., Maruri, U., Vallés, J., Porres, A., Espinosa, A., Ibáñez, S., Martínez, P., (2004). Electrical resistivity tomography as a technique for studying and modelling saline water intrusion, 18 Salt Water Intrusion Metting, Cartagena, Spain, pp. 341-353

Galanakis D. (1997). Neotectonic structure and stratigraphy of Neogene-Quaternary sediments of the Almyros-Pagasitikos, Pilion, Oreon-Trikeri and Maliakos basins, PhD thesis, Thessaloniki pp. 258

Goldman, M., Karfi, U., (2004). The use of the time domain electromagnetic (Tdem) method to evaluate porosity of saline water saturated aquifers, 18 Salt Water Intrusion Metting, Cartagena, Spain, pp. 327-340

Dimopoulos, G (1983). Salzwasserfuhrende Karstquellen oderhalb der Meeresoberflche, Zeitschrift fur angewandle Geologie, Bd. 29, pp. 565-568

Dimopoulos, G (1984). Geohydrologische und hydrochemische Probleme bei der Untersuchung von Salzkarstquellen, DGM, 28 H. 2, pp. 49-53

Myriounis, C., Voudouris, K., Tsourlos, P., Soulios, G., Dimopoulos, G (2006). Hydrochemical and geophysical survey of the Almiros aquifer system, East Central Greece, 1st SWIM - SWICA, 19th Salt Water Intrusion Meeting, 3rd Salt Water Intrusion in Coastal Aquifers, Cagliari - Chia Launa, Italy, pp 221-227

Myriounis, Ch. (2008). Hydrogeological and hydrochemical conditions of groundwater in the coastal part of the hydrological basin of Almyros prefecture Magnesia. Phd Thesis, School of Geology, Thessaloniki, pp. 397

Myriounis, Ch. Theocharis, M., (2010). Quality characteristics of water resources, Environmental Impact assessment, Department of Crop Production, Technological Educational Institution of Epirus, pp. 333. (in Greek)

N.S.S.G., (2001). National Inventory of year 2000

Contribution on hydrogeological investigation of karstic systems in eastern Korinthia

K. Markantonis, J. Koumantakis

National Technical University of Athens. Scholl of Mining Engineering and Metallurgy, Section of Geological Studies, Laboratory of Engineering Geology and Hydrogeology, Iroon Polytechneiou 9, Zografou, Athens. Email: markantonis@metal.ntua.gr

Abstract The biggest part of Eastern Korinthia contains mesozoic carbonate rocks, which in the southern mountainous zone construct the northern section of Arachneon, while in the northern hilly zone, appear as tectonic horns inside the neogene sediments. The carbonate mass communicates hydraulically with the Saronikos gulf sea in the east for many kilometers. This fact, in combination with the plunge of base level of the karstic process tens of meters below the today sea level, as a result of the increase of the water level has led to the natural salinization of karstic water tables. This salinization is more enhanced in conditions of overexploitation.The average recharge of the karst systems because of the rainfall infiltration is around $90x10^6m^3$. Almost all karstic aquifers discharge through coastal and submarine springs in the Saronikos coast, from the Kechres gulf to Selonda gulf (Orea Eleni, Almiri, Korfos and Selonda springs). The exploitation of karstic aquifers with boreholes is very difficult and it is proven ineffective, because as exploitation continues the result is salinization of karstic aquifers. In the Koutalas - Spathovouni zone, in a 15km distance from Saronikos coast, there are 20 boreholes for about 20 years. The result of overexploitation of karstic aquifers was the increase of Cl^- concentration that is now 280 - 1246mg/L, when it was less than 100mg/L in the early 1990s. In a borehole in the Sofiko area (mountainous zone), in a 7km distance from Saronikos coast, Cl^- concentration is more than 1200mg/L (October 2007), when it was 240mg/L in 2003.

1 Geological conditions

The area of the eastern part of Corinth is included in Eastern Greece geotectonic unit, the formations of which are overthrusted on the flysch formation of "Tripolis", in the southwestern margin of research area (Dervenakia). The southern region is constructed by the extensive and continuous occurrence of carbonate rocks (northern sector Arachnaio) from the region of Dervenakia in the west to the shores of the Saronicos Gulf in the east (Barnert 1981, Gaitanaki et al. 1985). These formations are thick limestones and dolomites, of Triassic to Upper Jurassic age, that have been affected by the rift tectonics that formed the Corinthian Gulf in its present form and tectonic trenches and horns of the research area. E-W faults and fault zones are dominant in the area, along which grow strongly karstification phenomena, thereby regulating the flow direction of groundwater. St. Basil - Klenias and Athikia-Rito large fault zones are covered by lateral scree. These fault zones delimit the mountainous carbonate mass of Arachnaion the Neogene basin of Corinth, in the north (Papanikolaou et al. 1990). Along these fault zones there are small occurrences of schist-chert sediment formations with ophiolitic bodies. The same sediments occur along the EW overthrust in the region of Angelokastro (Tataris and Kallergis 1965; Tataris et al.1970). In the northern part of the research area, the continuation of the Neogene basin is interrupted by multiple occurrence of alpine background (Mapsos, Mavri Ora, Acrocorinthos, Onia). Plio-Plistocene post-alpine sediments consist of marls, sandstones and conglomerates. The conglomerates in the margins of the basins overlay on the carbonate formations of the background.

2 Karst hydrogeological units

In the limestones of the mountainous and hilly area of East Corinth, a total area of 300km^2, a karstic aquifer is developed. It is in direct contact and hydraulic communication with the sea water of the Saronicos Gulf. The supply of karst aquifers realised from direct infiltration of the limestone mass (45-50%) and indirect from conglomerates and lateral scree are also involved. These formations overlay directly on the limestones. The correlation of average annual rainfall values (1978-79 to 2009-10), seven (7) rainfall stations in the area, with the altitude, indicates the relation between the two sizes (P=0.3731xH+446.67). By applying this relation to the average altitude of altitudinal zones of 200m and the use of infiltration rates (45% limestone, conglomerates and lateral scree 15%), it was determined that the average annual recharge volume of karst groundwater supply, is around 90x10^{6}m^3. Almost all of these water quantities are discharged through coastal and submarine brackish karst springs into the Saronicos Gulf.

Hydrogeological units of limestones of the hilly zone. This area includes the limestone masses that emerge from the Neogene sediments in the form of tectonic horns. Fault making tectonics, expressed by east-west direction faults, have caused similar direction of these appearances and also similar main direction of movement of underground karst aquifer. The karstic aquifer of Onia mountains is discharged through the springs of Orea Eleni, while the Mavri Ora aquifer participates in the recharge of Almiri springs. As for the karstic aquifer of Mapsos limestone appearances, it has not been yet completely clarified, if there is direct contact with the Onia aquifer or the Mavri Ora aquifer.

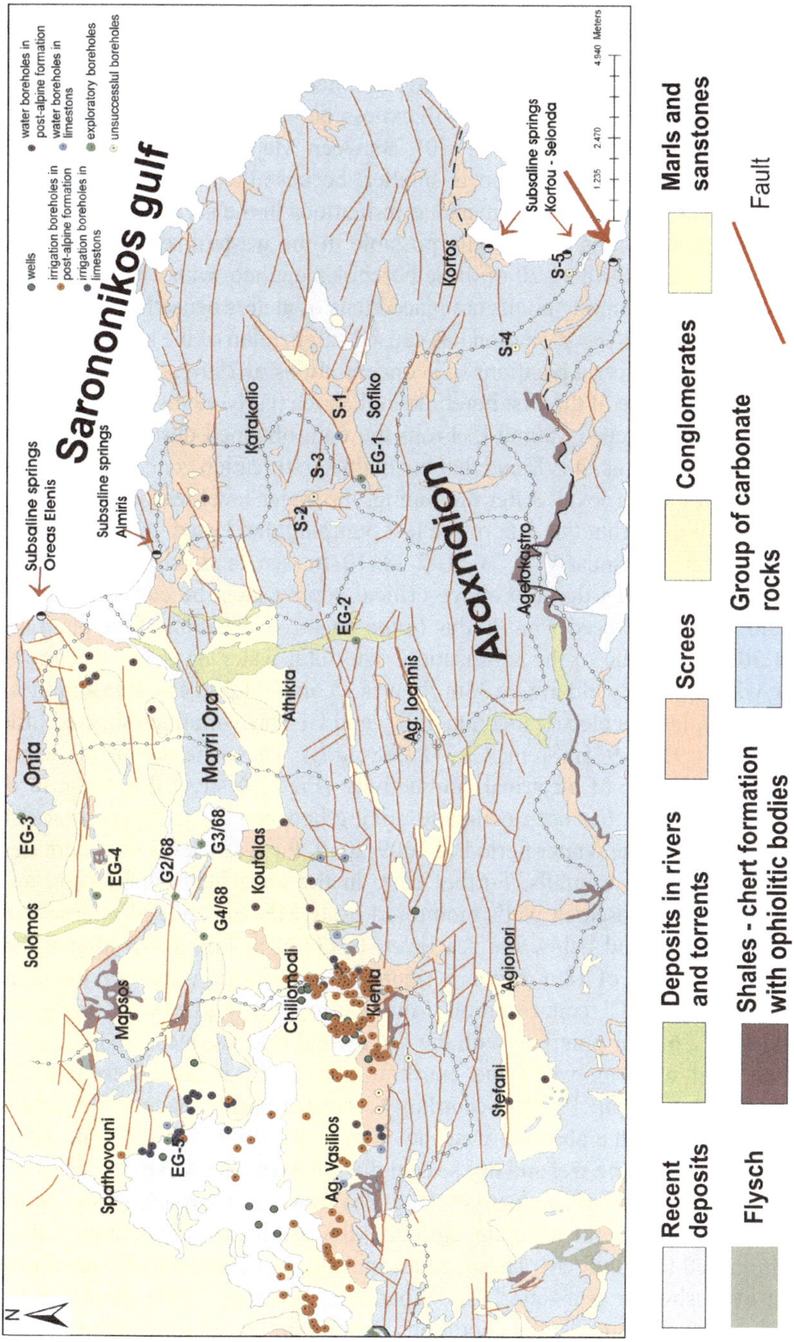

Fig. 1. Hydrogeological map of the research area.

It's certain, however, that it is in hydraulic communication with the sea water, as there has been salinization. The movement of karst waters towards Mavri Ora is more probable, as it is proven (with geophysical methods) that there are impermeable Neogene sediments of substantial thickness between Mapsos and Onia, beneath the sea level (Mastoris et al. 1970). Between Mapsos and Mavri Ora, the thickness of the Neogene formation is smaller, because their continuity is interrupted by successive small-scale limetsone formations in the surface. Exploitation of the karst aquifer in this area is only possible in the western sector of Mapsos limestone appearances, where 20 of these boreholes operate, with a yield between 40 and 70m^3/h. The pumping points are placed tens of meters beneath the sea level. The current status of overexploitation has caused salinization of the karst aquifer to varying degrees in different positions (Cl concentrations of 280-1,246mg/L), while during the start-up time of the first borehole in the area (early 1990s), chloride concentrations did not exceed 100mg/L. From the data obtained during exploratory drilling (EG5) in the area, it is found that the main karstification zone and aquifer is located beneath the sea level. After the piping, the water level reached a depth of 188.70m (+4.30 m). From the data of the test pumping that lasted 48 hours, it was determined that transmissivity is $T=2.38x10^{-3}m^2/sec$ and critical yield is $Q=23m^3/h$. It is obvious that the above critical supply must be corrected downwards, and must not exceed 15m^3/h, due to the negative elevation of the pumping level (-1.30 m) and due to the salinization. The total water level fall during the pumping with a steady yield of $Q = 33m^3/h$, was 15.6m (-11.3 m absolute altitude).

During pumping the electrical conductivity and Cl^- concentrations showed little variation (EC 1,910-1,935mS/cm, Cl 378-388mg/L). These measurements, combined with the profile of electrical conductivity (Fig. 2a) show that because of overexploitation, there has been complete mixing between the fresh and brackish water zones. During the winter period of 2009-2010, the water level was increased by 10.2m, due to the rainfalls. Further east, in the zone between Mapsou and Mavris Oras, in two research wells monitored by IGME, the zone of intense karstification is also found below the sea level. In G4/68 (Chiliomodi) after a 24h pumping with a yield of 18m^3/h, we measured a drop in the water level of 25m and increase of the Cl concentration from 74 to 308 mg/L, while in G3/68 (Athikia) after a 6 hour pumping with the same yield rate, we measured a drop in the water level of 21m, while the concentration of Cl increased from 35 to 650mg/L (Mastoris et al. 1971). According to measurements made in the period 2007-2011 in G3/68, the absolute water level ranges from +5.5 to +7.1 m, while the variation between the wet and dry season did not exceeding 0.6 m.

The profile of electrical conductivity (Fig. 2b) shows that throughout the investigated region (up to -6m absolute altitude), the electrical conductivity values ranged from 660 to 670μS/cm. These values indicate that under conditions of rest, the zone of freshwater (subzone D by Milanovic, Kallergis 2001), floats on the subzone E of saline water. Under conditions of pumping, we observe induced mixing of the two zones (fresh and saline water), as a result of the reduction of hydraulic load, the amount of which depends directly from the pumping supply.

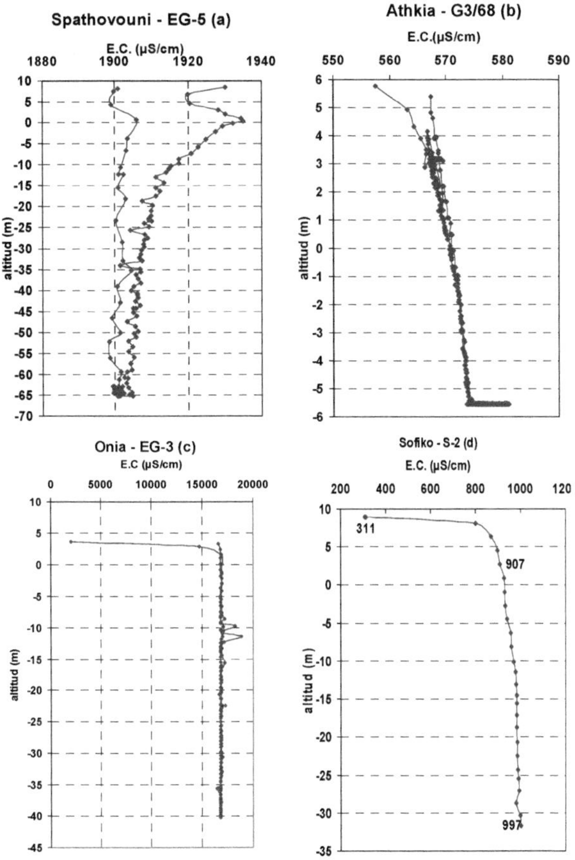

Fig. 2. Electrical conductivity profiles in the boreholes of the study area (hydrostatic conditions).

On the western edges of Onia mountains, as measured in the research well EG3, there was found little variation in the water table level, just above the sea level (+3 to +4.1m), for the period from 2009 to 2011. The profile of Ec (Fig. 2c) shows the complete salinization of the karstic aquifer of Oneion Mountains, which hinders the further development of proposals that have been made so far (Koumantakis 2005), in order to halt the phenomena of salinization. The values of electrical conductivity equal (16.8 mS/cm) the corresponding values of saline springs of Orea Eleni (20.0mS/cm), which is the point of discharge of karstic aquifers of Onia Mountains. Direct infiltration in limestone masses, owing to the intense karstification process, discharged quickly at the springs of Orea Eleni, and infiltrates the aquifer-transfusions of impeding conglomerates are slowly and is not likely to improve the quality of water in karst aquifers.

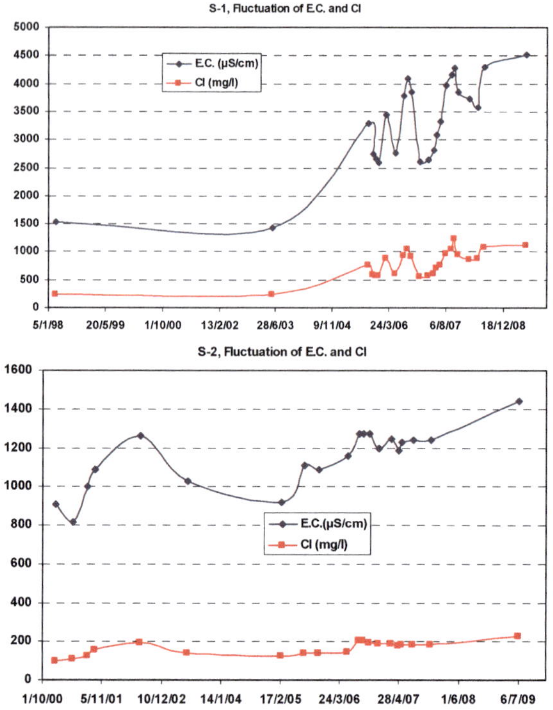

Fig. 3. Electrical conductivity fluctuations and Cl concentration in S-1 and S-2 Sofiko boreholes.

Hydrogeological unit of limestone masses of the mountainous area. This unit is part of the massive appearance of the limestones masses of the northern part of the mountain complex of Arachnaion stretching out from the flysch overthrust of the Tripoli zone, west of the area of Agiou Basileiou to the shores of the Saronicos Gulf. All along the coastline, from the bay of Almiri to the bay of Sofiko, limestone masses are found to exhibit hydraulic communication with the sea. From the north, the limestone masses are delineated by the long fault zone of Ag. Vasilios - Klenia - Rito - Katakalio. The southern boundary of the fault zone is difficult to be identified to its full length. Eastern of the village Lakes, the fault zone follows the overthurst of limestone formations upon the schist - chert formation, which having a direction East - West is visible almost to the bay of Selonda. The numerous and long faults, with directions East - West, which cross the limestone mass, have developed karstic structures with similar direction, through which the groundwater discharges to the sea of the Saronicos Gulf. The discharge of this karst aquifer is through coastal brackish karstic springs of very low absolute altitude (<1m) and many submarine ones. The springs of Almiri, which have higher yields, discharge directly from the limestone formations, through the tangential alluvial coastal zone, and through at least four distinct major submarine vents, located in the eastern areas at depths of 4-11m. Also the coastal springs of

Korfos in the area of Nisi Bloutsi, have large yields rate. Beyond these springs, substantial quantities of karstic water discharged along the steep shores of the Saronicos Gulf in the form of scattered discharge points (Fig. 1). The exploitation of the karstic aquifer through a well network is extremely limited due to intense lifting of the morphological terrain near the coastal zone. In many areas of the mountainous zone (Stefani, Agionori, Ag. Ioannis, Klenia, etc) several wells have been drilled at depths up to 350-450m without encountering the aquifer zone, as they didn't approached the sea level, near which the developed karstic aquifer is found.

In the area of Sofiko, the karstic aquifer is distributed in the karstified zone located approximately 30m below the sea level (exploratory drilling EG-1). It is about a "sunken karst ', formed as such, from the combination of tectonic action, and the rising of the sea level. Under conditions of rest the water table of the karstic aquifer, is formed in absolute altitude from +8 to +12m. During the test pumping in EG1 (July 2009), operated for 24h with a supply of 15m^3/h, the total drawdown was 1m, without substantial change in the concentration of chloride (81-85mg/L). By increasing the supply, the drawdown is higher and progressively increasing the concentration of chloride. The operation of two water wells (S1, S2) for more than 10 years, with 30 and 25m^3/h, supply rate respectively, has caused a significant increase in salination. In S1, the concentration of chloride has exceeded 1200mg/L, from 240 mg/L in 2003 (Fig. 3), while in S-2 exceeds 200 mg/L, with electrical conductivity greater than 1400mS/cm. From the profile of electrical conductivity in S-2 (Fig. 2d) it is founded that the electrical conductivity of the 310mS/cm at the first 30cm of water column, increases gradually to reach 1000mS/cm around the bottom of the well (-32m). From these values combined with the results of test pumping in EG1, that by reducing the supply rate, it is possible to keep the concentration of chloride within tolerable levels.

3 Quality characteristics of subsaline karstic springs

All coastal and submarine karstic discharges in the coast line of the Saronicos Gulf are subsaline, with little change in chloride concentration during the hydrological year. The water springs of Orea Eleni exhibit a smaller degree of salination (31% seawater), as the masses of the limestone of Onia mountains, communicates hydraulically with the sea in a small area, width of 120m. The concentration of chloride (analysis 31 samples period 2004-08), ranges from 6,290-6,900mg/L, with a mean value of 6,520mg/L. In Almiri springs (discharges from limestone masses) the concentration of chlorides (analyzes 33 samples, period 2004-08) ranges from 11,380-13,600mg/L, with a mean value of 12,524mg/L (58% seawater), while at the discharges from the alluvium that cover the coastal zone, the chlorides range from 9,900-11,400mg/L, with a mean value of 10,235mg/L. Regarding the springs in the bay of Korfos and the seasonal discharge of karstic waters in the area of Loutsa Tsigou, the concentration of chloride range about

8,000mg/L, while in the area of Nisi Bloutsi where high supply karstic waters are measured the concentration is of about 15,000mg/L. Finally the water springs of the bay of Seloda, where karstic water discharges are mixed with waste water from nearby fish farms, the concentration of chloride range about 4,000mg/L.

4 Conclusions

The karstic limestone aquifer of the eastern Corinthia is an important water resource for the region, despite the intense salination of discharges that propagated the shores of the Saronic Gulf. The average annual recharge of the karstic aquifers is about $90x10^6m^3$ and potential exploitation of these aquifers, would supply the region with a new water resource particularly important for economic and social development. Such a perspective is particularly difficult, as it will require major research projects to finalize the form of recovery works needed, the construction of which will require significant financial cost. Until now the use of the karstic aquifer via a well network, both in the area of Sofiko, and in the zone between Koutalas - Spathovouni in the hilly area, is made irrational, leading to the salination of the karstic aquifers. The yield rate of the wells, far exceed the critical yield, beyond which upsets the delicate balance between the band of fresh and saline water. In both cases it is vital to reduce the supply rate at the level of $15m^3/h$ and the installation of pumping stations to a smaller depth.

References

Bannert D (1981) Geological Map of Greece, scale. 1:50,000, Sheet Ligourio. Institute of Geology and Mineral Exploration, Athens

Bornovas J, Lalechos N, Filippakis N (1971). Geological Map of Greece, scale 1:50.000, Sheet Corinth. Institute of Geology and Mineral Exploration, Athens

Gaitanakis P, Mettos A, Fitikas M (1985) Geological Map of Greece, scale 1:50.000, Sheet Sofiko. Institute of Geology and Mineral Exploration, Athens

Kallergis C (2001) Applied - Environmental Hydrogeology Volume III, Athens

Koumantakis C (2005) Halting the salinization of aquifers with the use of underground dams. The case of the karst aquifer of the Orea Eleni springs in Corinth. Proceedings of the 7th National Hydrogeological Conference, pages 195-200, Athens

Mastoris K Monopolis D., Skagias S. (1971). Hydrogeological Research of the Corinth - Loutraki area. Institute of Geology and Mineral Exploration, Athens

Papanikolaou D, Logos E, Lozios S, Sideris C (1990) Comments on the kinematic and dynamic evolution of neotectonic basins of the Eastern Corinthia. Proceedings of the 5th Scientific Conference of the Greek Geological Society, XXV / 3, 177-191

Tataris A, Kallergis G (1965) The geology of the mountainous area of Trapezona and the Nafplion – Ligourio area. Institute of Geology and Mineral Exploration, Athens

Tataris A, Kallergis G, Kounis, G (1970). Geological Map of Greece, scale 1:50.000, Sheet Nafplion. Institute of Geology and Mineral Exploration, Athens

Contribution to the hydrogeological research of Western Crete

E. Manutsoglu, E. Steiakakis

Technical University of Crete, Department of Mineral Resources Engineering, University Campus, Akrotiri, 73100 Chania, Hellas. emanout@mred.tuc.gr; stiakaki@mred.tuc.gr

Abstract Hydrogeological research in western Crete, has led to the revision of the geological map of the area and the creation of a new lithological map in scale of 1:25.000. According to the investigation results, Trypali Unit forms a separate hydrogeological entity in the region that is extended widely. Folding and faulting of this unit combined with the widespread fault systems determine the ground water flow in the region.

1 Introduction

The Alpidic basement of Crete is composed of a number of fault bounded tectonic units with different lithologies, paleogeographical origins and metamorphic grade. Among others, a group of metamorphic rocks, or a part of it, composed mainly of phyllite and quartzite, has been described through time with variable names (Manutsoglu et al., 1995). This Group lies between Tripolitza nappe and "Crete-Mani Unit"/Plattenkalk Group or/and the Trypali Unit, and its maximum surface extension is located at Western Crete. The thickness of "Phyllite/Quartzite Series" (PQS) at Western Crete, which has been estimated at about 2150 – 2800 m (Cayeux 1902, Krahl et al. 1983), its stratigraphic position (normal or overturned), as well as the extension of the Trypali Unit and the emplacement of the large meta-evaporitic body in the region, still remains a matter of debate (Pomoni-Papaioannou and Karakitsios 2002, Krahl and Kauffman 2004).

In order to answer these main questions, which are of great importance considering the hydrogeological framework of the region, an area in the western Crete (belonging to Paleochora sheet, Manakos et al. 2002) was surveyed. The relation between the crystalline carbonate formations (Trypali Unit) and the overlying metaclastic formations (Phyllite/Quartzite Series) was investigated and a new lithological map was created in scale of 1:25.000. This map can be used to clarify the hydrogeological regime in the study area and to evaluate the groundwater resources.

2 Geological setting

The study area is located at the western side of the “White Ori” mountains range. The “Complex” of metamorphic rocks in the area has been described by Creutzburg and Seidel (1975) as “Phyllite/Quartzite Series”. It is dominated by phyllites of variable composition, quartzites of different thickness (meta) conglomerates, black platy dolomitic limestones, gypsum, rauhwake and basalts. This “Complex” corresponds to the “terrains metamorphique” Cayeux (1902), as well as to the formation of “Gypsum and Rauhwake” and to the “Phyllite-Quartzite-Shale Series” of Wurm (1950). Creutzburg and Seidel (1975) introduced also and defined the term “Trypali Unit” mainly to describe carbonaceous rocks (dolomites, dolomitic limestones, rarely pure limestones, carbonaceous breccia, rauhwake), as well as white, glace textured marbles. The mass of the aforementioned rocks has been subjected to different grade of recrystallization. The lithological evolution of this unit along with fossils (algae, corals, gastropods) led the researchers to accept that it was deposited rocks in a shallow water depositional environment. Gypsum and rauhwake occurrences in the western part of “White Ori” mountains range (between Stomio and Sougia settlement) were also included in Trypali Unit (Wurm 1950).

Regarding the transitional correlation between the carbonaceous formations of the Trypali Unit and the metaclastic formations of Phyllite/Quartzite Series, Creutzburg and Seidel (1975) came up with a surprising conclusion. Although in several positions (Kallikratis, south of Leukogia, east of Chosti, north and east of Kandanos, south of the Arkadi monastery) they notice that Trypali Unit seems to pass gradually to Phyllite/Quartzite Series, they question the Upper Triassic (Rhaetian) to Lower Jurassic (“Liassic”) age found for these rocks and claim that the contact with the underlying Plattenkalk is clearly tectonic. Considering the above they place Trypali Unit as underlying of the Phyllite/Quartzite Series, regarding it as a different structural unit. Further more, Krahl et al. (1982), studied biostratigraphical evidences in this unit and found that gypsum, dolomites and rauhwake, which until then were considered to consist the stratigraphically older horizons of the Trypali Unit, and they are in fact the youngest ones. Consequently they claimed that the Phyllite/Quartzite Series are in reverse position. However, this perspective was not adopted by all researchers (Kopp and Wernado 1983). In 1983 Krahl et al. provided new biostratigraphical data for the “Phyllite Group” of Western Crete. They dated the period between late Carboniferous to late Triassic, but they didn’t manage to cover the Middle Triassic period using fossils. Coming up with even more biostratigraphical data, they accepted that rocks of the Phyllite/Quartzite Series are isoclinal folded in a large scale. Also, they presented a series of geological sections, stratigraphical columns, a synthetic stratigraphical column and also the synthetic model for the structure of Phyllite/Quartzite Series as well as of Trypali Unit at western Crete. In this model meta-evaporites, that stratigraphically consist the youngest formations, are observed in the lowest tectonic position. For these rocks researchers accept that either consist a part of the Upper Triassic/Liassic meta-carboniferous formations of the "Phyllite Group" (and constitute the overturned part of the massive isoclinal fold) or they are a facial differentiation of the rocks of the Plattenkalk Group. The above claims were also supported by a paleogeographical evolution model of the External Hellenides during the Permian-Triassic period (Dornsiepen et al. 2001).

3 Methodology

A new geological map in scale 1:25.000 was created, using bibliographic data, data of the existing geological map, and data of the spatial distribution of relatively easy-mapped lithological units (such as quartzitic layers, black/graphitic schists - quartzite that enclose gypsum relicts), the new finding related to occurrences of marbles within phyllites and transitional formations (from phyllites to marbles and meta-evaporites).

The new lithological map covers an area of about 150 km^2. The geological information used to construct the new map, includes a simplification of the mapped units. Using the current geological map (Palaiochora sheet, Manakos et al. 2002) the large scale structures were checked and lithological sequences were grouped in order to construct three geological sections with E-W direction (Fig.1). The aspect that meta-evaporites are part of the Plattenkalk Group (Manakos et al. 2002) was not adopted, because none of the characteristic lithofacies elements of this Group was identified in the field. In order to proceed with the geological mapping, the limits for the upper non metamorphic units (Tripolitza and Pindos) and the limits of the Neogene and Quaternary formations were accepted as reliable. Moreover, according to the widely accepted model of the region, "Phyllite Group", was considered to be in a reverse position. Finally, the observation of characteristic formations such as black/graphitic schist-quartzite that enclose gypsum relics, it was the guide to reach in conclusions for the study area.

4 Results of the geological mapping

The tectonic structure of the region has been resulted by the combined action of thin-skinned deformation and fault tectonics. The structure of the study area has been defined by the existence of a large scale fault, striking NNW-SSE, which passes through Sfinari village. The geological survey led to the creation of a new lithological map of the study area (Fig.1). This map was based on lithological and interpreted biostratigraphical data from cited literature.

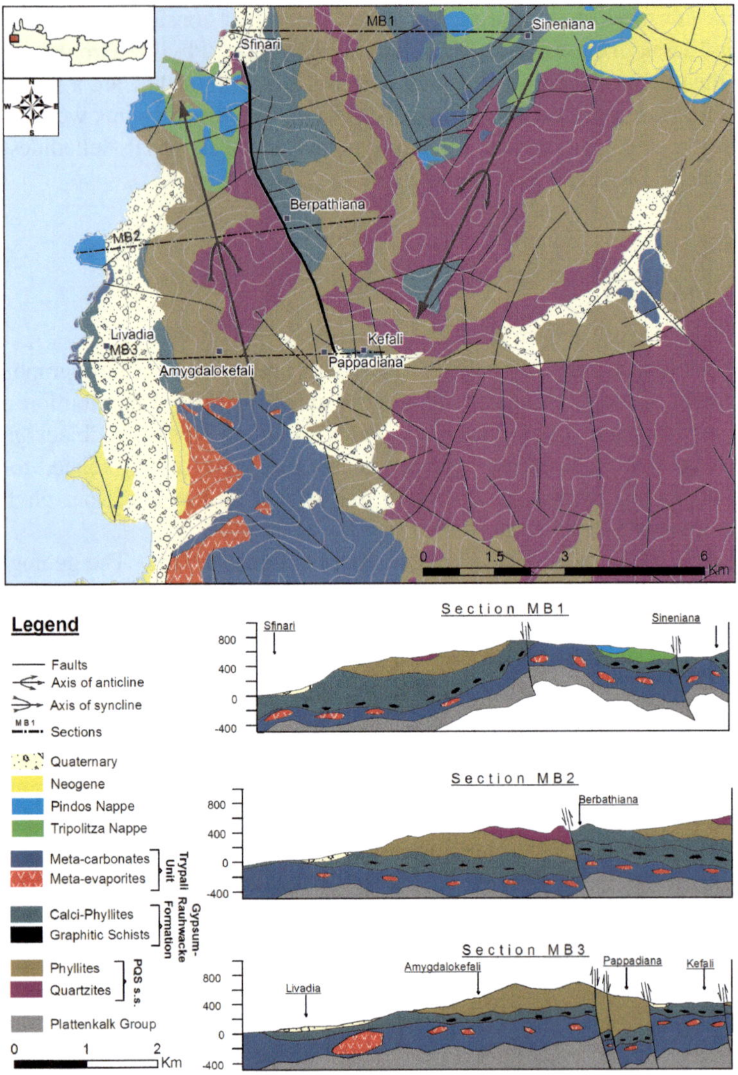

Fig. 1. New lithological map of the Sfinari area in Western Crete, Sheet Paleochora (after Manakos et al. (2002), modified).

The current work comes up for discussing the geological model of the area and they are presented in following:

- The meta-evaporitic body and meta-carbonate rocks of Upper Triassic (included in Ionian Zone according to Manakos et al. 2002), was regarded to be (together with the limestones of "Mana" formation) in Trypali Unit.
- Calci-phyllites and quartzites of "Mana" formation, the main part of "Sfinari" formation and "Poikilos" formation that consist of schists, re-

crystallized limestones and quartzites, (have been considered to be in Ionian Zone according to the existing geological map, Manakos et al. 2002), constitute a single formqation under the name of "Calci-phyllites". In this formation the existence of black graphitic schist-quartzite with gypsum intercalations is characteristic and outcrops in the whole study region. According to the current geological survey the "Calci-phyllites Unit" occurs stratigraphically between the clastic formations of Phyllite-Quartzite Series s.s and the carbonate formation of Trypali Unit. In the village Livadia, was observed intense karstification inside the aforementioned transitional layers.

- Taking into account the spatial distribution of the geological formations, the topography and the biostratigraphic data derived from bibliographic sources, the rest formations of Phyllities/Quartzite Series s.s. (schists, phyllites, quartzites) can be grouped into two stratigraphical successive formations: "Phyllites" and "Quartzites" formations.
- "Phyllites" formation is topographically lower, with extensive surface outcrop on both blocks of the fault. It is stated that "Varied Formation", "Stauros-Seli" schists and "Koutrouli" schists make up the "Phyllites" formation.
- The second one is topographically higher. The quartzites "Agios Dikaios" and the largest part of quartzites "Koutrouli" consists the "Quartzites" formation.
- The main fault is lifted the eastern part of the study area, preserving the NNE-SSW dip direction and the NNW and SSE direction of fold axes as well. The aforementioned directions were not preserved in the depressed western part. Instead, the general direction of the intensively folded rocks was changed after rotation to NNW-SSE. This rotation is evidenced by the presence of two additional tectonic clues. The first one is the existence of a large part of upper nappes (Tripolitza and Pindos) west of Sfinari and the second is the outcrop of tectonically reversed and stratigraphically younger meta-carbonate formations with anhydrite/gypsum in the proximity of Stomio area. Large scale anticline and syncline are observed in the eastern and western part, respectively. Apart from this large fault, several smaller faults run through the area with general direction NNW-SSE and NNE-SSW, as shown on the map.

5 Discussion and Conclusions

The authors have revised the lithostratigraphy in the area, and certain views are set out concerning:

- A large scale meta-evaporites body, with a meta-carbonate core, which (in contrast with the existing geological map which classifies it within the Plattenkalk Group), is assigned to the meta-carbonates formations of Trypali Unit. A large fault with NNW/SSE direction passes from Sfinari village and follows the general direction of the fold axes of the meta-carbonate rocks, with subduction either to NNW or to SSE.
- As it is shown in the three geological sections (Fig.1), there are three main large scale structures are formed over the metamorphic rocks of the Plattenkalk series:

  1. Trypali Unit which evolves into Calci-phyllites, Phyllites and Quartzites formations of the Phyllitic Nappe Group are overthrust on the preceding formations and stratigrapically are in a reversed position.
  2. The transition of Trypali Unit towards Phyllites/Quatzite Series s.s. (Phyllites and Quatzites Formations) is made gradually with transition beds of Calci-Phyllites. Based on the spatial distribution of Calci-Phyllites, it is concluded that apart from the anticlinic structure that exists in the eastern part of the area, an open synclinal structure is created to the west part that affects and is affected by the large fault striking the area. The general directions of the fold axes are N/NNW, while on the east block an N/NE direction is prevalent.

- The existence of the main fault, combined with the anticlinal and synclinal structures and synthetic faults that fracture the heterogeneous rocks in the area, determine the underground water flow.
- Small occasional aquifers and springs of low yield are developed in Phyllite/Quarzite Series s.s., that covers most of the surface of the study area and they are exploited for irrigation and water supplying needs.
- The overlaying carbonate rocks of Tripolitza Nappe are of great hydrogeological interest; however their extent is limited. They feed the spring of Agia Paraskevi - Sfinari at the N/NW of the study area, with average annual flow which reaches 1.9×106 m3/year.
- It is a contact spring (Lionis and Perleros 2001), without ruling out the existence of a small scale reservoir in negative altitudes. The quality of the water is good despite the fact that the spring discharges in the coastal zone (altitude +2.42 m). This is probably due to the narrow discharge front. We believe that the extent of the Tripolitza limestones ($\approx$ 2 km^2) and the estimated infiltration in the area, explain the average flow of the spring, indicating that its catcment is formed in the Tripolitza limestones (SW of the spring).
- Considering the macrotectonics of the area (formation of an open syncline with NNW/SSE axis), it is concluded that Trypali Unit is partially hydraulically isolated and it forms a hydrolithological entity of high interest. It is a hydrolithological entity extended widely and it forms a significantly long

synclinorium, with longitudinal axis in NNW/SSE direction. However, it should be noted that in the southern part of the study area the carbonate series include anhydrite-gypsum in variable sizes and consequently a high water concentration in sulfate ions should be expected.

It is suggested that an exporation program should be carried out in the area, by drilling the Phyllite/Quarzite Series s.s. Calci-Phyllites (with thickness up to 300 m) in order to reach the underlying Trypali Unit (0 to +100m asl) that is expected to contain a karstic aquifer. The final depth of the boreholes shall depend on the elevation of exploration site and it is estimated at about 500-600 m. It must be mentioned that due to lack of detailed mapping and hydrogeological-hydrochemical data, there is an increased uncertainty regarding the type and depth of the aquifer, as well as the depth where lithologies and groups might change.

References

Cayeux L (1902) Sur la composition et l'age des terrains metamorphiques de la Crete. C. R. Acad. Sc. Paris 134, 1116-1119

Creutzburg N, Seidel E (1975) Zum Stand der Geologie des Praneogens auf Kreta. Jb. Geol. Palaont. Abh. 149 (3), 363-83

Dornsiepen UF, Manutsoglu E, Mertmann D (2001) Permian-Triassic paleogeography of the External Hellenides. Palaeogeogr. Palaeoclim. Palaeoecol. 172, 327-338

Kopp KO, Wernado G (1983) Ueber eine intra-triadische Deckenbewegung auf Kreta. Geol. Rdsch 72, 895-909

Krahl J, Eberle P, Eickhoff J, Forster O, Kozur H (1982) Biostratigraphical Investigations in the Phyllite-Quartzite Group on Crete Island, Greece. Proc. of the International Symposium on the Hellenic Arc and Trench, 306-323, Athens

Krahl J, Kauffmann G, Kozur H, Richter D, Förster O, Heinritzi F. (1983) Neue Daten zur Biostratigraphie und zur Lagerung der Phyllit Gruppe und der Trypali Gruppe auf der Insel Kreta (Griechenland). Geol. Rdsch 72, 1147-1166

Krahl J, Kauffmann G (2004) New aspects for a palinspastic model of the External Hellenides on Crete. Proc. of the 5th International Symposium on Eastern Mediterranean Geology, 119-122, Thessaloniki

Lionis M, Perleros V. (2001) Hydrogeological Study of Kampos area, Chania (KA: 9481721). Ministry of Rural Development and Food, Department of Geology/Hydrology, Branch of hydrogeology/drills and mathematical models, Athens

Manakos K, Vidakis M, Kopp KO, Krahl J, Skourtzi - Koronaiou B (2002) Geological Map of Greece, scale 1:50.000, Paleochora Sheet. Inst Geol and Miner. Explor., Athens

Manutsoglu E, Mertmann D, Soujon A, Dornsiepen UF, Jacobshagen V (1995) Zur Nomenklatur der Metamorphite auf der Insel Kreta, Griechenland. Berliner geowiss. Abh E 16, 579-588

Pomoni-Papaioannou F, Karakitsios V (2002) Facies analysis of the Trypali carbonate unit (Upper Triassic) in central-western Crete (Greece): an evaporate formation transformed into solution-collapse breccias. Sedimentology 49, 1113-1132

Wurm A (1950) Zur Kenntnis des Metamorphikums der Insel Kreta. N. Jb. Geol. Palaont. Mh. 1950, 206-239

Karstic Aquifer Systems and relations of hydraulic communication with the Prespa Lakes in the Tri-national Prespa Basin

A. Stamos[1], A. Batsi[2], A. Xanthopoulou[3]

[1] Institute of Geology and Mineral Exploration, astamos@3175.syzefxis.gov.gr.

[2] M.Sc Geologist A.U.Th, Giouleka 1, 50300 Siatista, Kozani, annampatsi@gmail.com

[3] M.Sc Geologist A.U.Th, Aisopou 16, 54627, Thessaloniki, kaxanth@gmail.com

Abstract The Triassic-Jurassic limestones, which cover a large area in the region of Prespa, appear to be intensively karstic and create underground independent water reservoirs. It is estimated that the renewable water reserves of all the karstic aquifers that are located west of the basic watershed exceed $250*10^6$ m^3/yr. These water reserves are formed from the direct inflow of atmospheric precipitation without including the infiltration of water from the Lakes of Prespa. The lakes of Great Prespa and Small Prespa are surficially and hydraulically connected via an underground karstic water table; the water is discharged, northwest, to lake Ohrid either riparian or in the sub-surface via shallow holes in the limestones that exist between the two lakes Great and Small Prespa and between the lakes Great Prespa and Ohrid in FYROM and Albania. The sub-surface inflows of lake Great Prespa amount to $130*10^6$ m^3/yr and it is believed that there are more sub-surface inflows that have not been located.

1 Introduction

The lakes of Great Prespa and Small Prespa have been characterized as landscapes of particular natural beauty with the latter protected by the Ramsar Convention. They are wetlands of international importance with rare flora and fauna. Hydrologic and hydrogeologic studies were done in both lakes using modern methods of isotopic analyses, dye tracing as well as other methods, aiming to the full knowledge and understanding of the lakes' ecosystem characteristics and functionality, which would then become the basis for their protection, improvement, rational management and sustainable development without negative repercussions in the environment.

2 Geological Formations - Hydrolithology

The area of Prespa lakes geotectonically belongs in the Sub-Pelagonian zone and the geological formations that compose the area from newer to older, are summarised below:

Quaternary formations: Include all formations of Quaternary age, such as alluvial and fluvial deposits of sand and clay, conglomerates, talus cone etc. They can be mainly found in the plain of Korçë in Albania, in the plain of Resen in FYROM and in areas of lower altitude in the Greek territory. They are generally characterized as permeable but also as semi-permeable to impermeable formations.

Neogene formations: Consist of marls, clays, sandstones, conglomerates, marly limestones, occasionally with intercalations of sand and grit, with great thickness. They are characterized as non-permeable to locally semi-permeable formations. Generally they do not present active porosity and consequently they do not develop underground waters.

Molasse formations: These are very thick impermeable formations that are mainly occur between Korçë and Pogradec in Albania. They consist of sandstones, conglomerates and marls in alternations. Flysch and other related rocks such as conglomerates and sandstones, which have the same hydrologic behaviour, were included in this category. They are characterized as impermeable formations. Generally they do not present porosity and consequently they do not develop underground waters.

Trirassic-Jurassic limestones and marbles: They are widely distrubuted in the area, with thickness that reaches almost 1000 m. They consist of light grey, white and black limestones with various degrees of recrystallization, thick-bedded or low-bedded, white marbles, and dolomitic limestones. They are widely spread in the mountain range of Mali i Thate and Galicica in Albania and FYROM, between the lakes of Great Prespa and Ohrid, as well as in mount Triklari in the Greek side. They appear intensely tectonized and karstic with a decisive role in the hydrogeology of the region as it is described further down.

Granites: They are of considerable abundance and thickness in the Greek territory, while hydrologically they behave as impermeable formations.

Metamorphic rocks: They observed in the Greek side as well as in FYROM. They consist of phyllites, slates and gneiss, with hydrologically impermeable behaviour.

Ophiolites: They occupy a small area in the wider region and hydrologically are characterized as impermeable formations.

The factors of hydrologic coefficient for the aforementioned rocks are approximated in Table 1. The parameters in the table are the result of continuous measurements done for many years by the West Macedonia Department of the Institute of Geology and Mineral Exploration (IGME) in a natural lysimeter in the region of Kastoria, in the limestones and marbles of the Pelagonian zone as well as in other categories of rocks in other regions

Table 1. Hydrologic coefficients.

Geological Formations	Precipitation (P)	Surface flow (R)	Evapotranspiration (EVAP/TRANSPIR)	Infiltration (I)
Sedimentary rock formations	100%	24%	63%	13%
Metamporhic rocks	100%	34%	60%	6%
Limestones	100%	5%	35%	60%

3 Hydrogeological conditions of the Prespa hydrogeological basin

The Triassic-Jurassic limestones are those that mainly determine the hydrogeological regime in the area. Hydrogeologically they are significant because they develop in their mass a large secondary porosity. When they appear without admixtures of clay and other materials, the effect of the karst and dissolution is intense. The karst (dissolution) of the limestones is a complicated and multifunctional chemical activity. Rain water, enriched with CO_2 from the atmosphere or the CO_2-enriched ground air from decomposition of organic substances and root systems, forms the carbonic acid $CO_2 + H_2 \rightarrow H_2CO_3$ which acts on limestone and produces the bicarbonated calcium $CaCO_3 + H_2CO_3 \rightarrow Ca(HCO_3)_2$ which is soluble and is removed by water. The final result of this activity is dissolution of the interior of the limestones and the creation of all kinds of karstic holes.

Because of this process, all the karstic features (shallow holes, sinkholes, dolines, polges and caves) are observed on the surface as well as inside their mass. The inter-karst features were located during the water drills performed in the region by IGME. It is noted that in some water drills that penetrated the limestones in the Greek side, karstic sinkholes were found filled with Quaternary and Neogene sediments from fluvial deposits with rounded small diameter pebbles.

The corroded materials that were transferred by the streams in the region were selectively channelled in these karsts and mainly in the caverns because of their large opening. These limestones are anisotropic and non-homogeneous with regards to their permeability, which results in the underground water moving via channels. These karsts and the empty spaces are occupied by water and function as reservoirs of underground water. The percolation of atmospheric precipitation in the limestones amounts to roughly 60%.

The creation of karst is not enough to store underground waters. Essential is the existence of an underlying impermeable formation, which will regulate the flow and the storage of the underground water.

In our case, the underlying impermeable formation primarily consists of crystalline schist massif rocks and secondarily of the limestone itself, which at greater depths is not karstified. The karsts reach to a depth of greater than 500 m, as is shown by geophysical research done by IGME.

The axes of berm of this impermeable basement rock constitute the underground watersheds; the underground waters are stored on either sides of the underground watersheds and are led to springs and lakes, or overflow and feed other underground aquifers. Consequently, although the limestones are extensive in the three countries, they do not create one unified water table (Fig. 1).

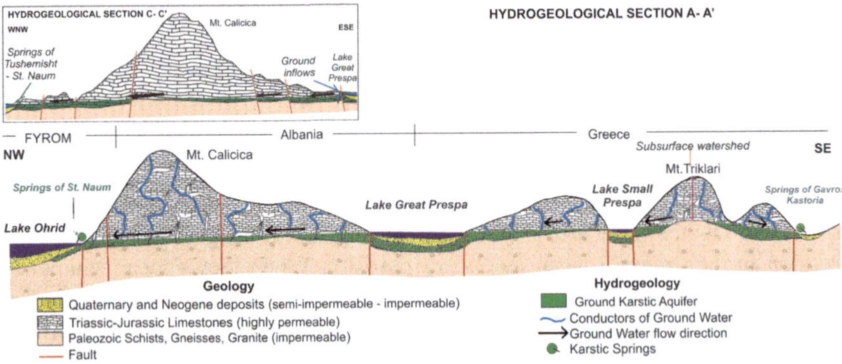

Fig. 1. Section A-A': Development of karstic aquifers and groundwater movement in the three neighbouring states (Stamos 2003). Section C-C': Sinkholes of Great Prespa Lake and karstic aquifers, which overflow undergrounds in succession the one in the other from east to west (Stamos 2010).

A number of underground watersheds that separate various independent aquifers were located using geological, hydrolithological, geophysical, drilling, dye tracing and isotopic analyses ^{18}O and many others that were done as part of the transnational research programmes for the region. The main underground watershed located in the eastern bank of Small Prespa (direction NE - SW) on the Greek side, divides the limestones' karstic waters in two main directions; to the east (Greece) and to the west (Albania and FYROM).

On the Greek side the underground karstic water moves southeast and is discharged in the big karstic springs of Gavro, Koromilia and Lefki in the area of Kastoria. On the west side the movement of karstic water is mainly northwestern, and creates independent aquifers. Thus, the karstic aquifer that is formed between the basic watershed and Small Prespa directs waters to Small Prespa. Thus a semi-independent aquifer is created in the limestones that separate the lakes Small Prespa and Great Prespa, with flow directed as much towards Great Prespa as to the main underlying limestones (active karst) to the northwest. In wet periods this aquifer has a higher water level - in altitude - than the level of Great Prespa (measurements of water level in drillings in the Greek territory) and then a portion of this water flows into the lake.

At the same time, a section of limestones located in the south develops a different independent aquifer that discharges its water to the springs of Proger and

Mancurisht in Albania. In these springs, water from Small Prespa is also discharged which flows into the shallow holes found at the lakeside as shown in Figure 2. This figure also shows the impermeable layer of schist with ophiolites in the limestones separating the underground karstic waters into independent aquifers (shown as yellow line in Fig. 3). This layer of schists in the limestones is only located in the south, and in some cases it creates a low hydrodynamic aquifer (Episkopi of Albania). In Figure 2, the schematic section B-D is shown in miniature with the point B' moved slightly to the northwest (point D), and in which the sub-surface watershed is shown with the layer of schist which separates the subsurface waters in different directions. All the remaining limestones northwest Great Prespa create the biggest limestone aquifer between Great Prespa and Lake Ohrid. This aquifer is enriched with the infiltration of big quantities of water from Great Prespa. There is likely to be lateral infiltration into the Quaternary deposits of the Korçë plain in Albania.

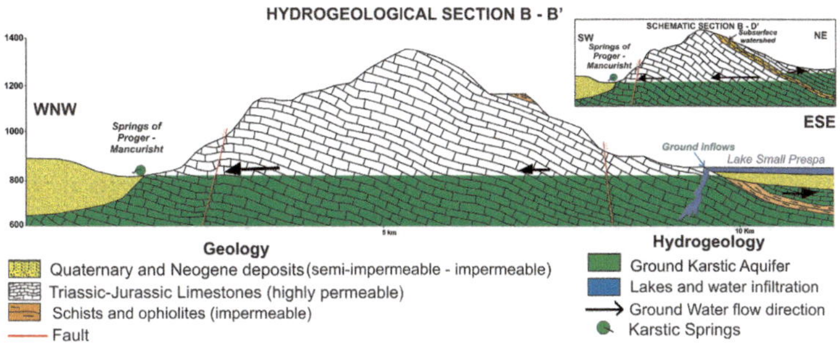

Fig. 2. Cross section showing the water infiltration of Small Prespa lake in to the karstic aquifer and the underground watershed with the impermeable layer of schists and ophiolites into the limestones (Stamos 2010).

As is shown in the hydrogeological map (Fig 3), water discharge of the aquifer happens via a large amount of water from the karstic springs at Tushemisht in the region of Pogradec in Albania and from St. Naum of Lake Ohrid in FYROM. These are more likely large fronts of overflow springs. The estimate of annual renewable water reservoirs of the aquifer amounts to over $190*10^6$ m^3. The estimate of annual renewable reserves of water from all of the above-mentioned karstic aquifer westwards of the main watershed is over $250*10^6$ m^3 (Stamos 2010).

The aforementioned reserves of water consist of quantities of water from the direct percolation of atmospheric precipitation in the karstic formations, without including the infiltration of waters from the Prespa lakes. Those were calculated based on specific output of karst (yield per km^2) measured in a natural lysimeter (Kefalari, Kastoria). Note that it was not possible to use the hydrologic and other measurements of three neighboring states, which were judged unreliable. On the map (Fig. 3) are also delimited the remaining water-bearing strata, as well as the movements of all their underground waters and discharge.

The determination of hydraulic parameters of karstic aquifers becomes very difficult because the aquifers are anisotropic and non-homogeneous. The hydraulic parameters which determined after a lot of pumping tests in boreholes that realized by IGME in the same limestones to the east and in Greek territory and were taken the mean, it is believed that they represent entire the karstic system of Prespa Lake. Only as for the Storage coefficient (S) exists reserve as in general for the all karstic aquifers. These parameters have as follows:
Transmission coefficient:

$$\mathbf{(T)} = 5.2*10^{-2}\ m^2/sec$$

Hydraulic conductivity coefficient:

$$\mathbf{(K)} = 2.6*10^{-4}\ m/sec$$

Storage coefficient:

(S) = 2% generally 0.1-10%.

4 The lake of Great Prespa and groundwater flow in the karstic aquifer

The catchment basin of the Prespa lakes is a closed basin with no surface drainage system, only underground. The sizes of the hydrological and hydrogeological basins of Prespa lakes are 1380 km^2 and 1630 km^2 respectively. The level of water in the lakes has been declining since 1987 and has dropped by approximately seven meters with a simultaneous reduction of its volume by 46% or $1750*10^6\ m^3$.

The Great Prespa lake is fed surficially by the overflowing waters of Small Prespa (via the channel Koula) and overland flow of the catchment, while in the subsurface is fed by karstic waters from the limestones that separate the two lakes. Mapping of the water balance of the lake and its subsurface outflow requires a lot of hydrologic and other data. These could result from a joint study done by the three countries to which the hydrogeologic and hydrologic basin belong.

Due to practical difficulties the volume of water outflows from the lakes cannot be directly calculated. Hence, an indirect method of calculation was followed using isotopic analyses. The technique takes advantage of natural tracers in the water resulting from changes in its isotopic composition throughout the hydrologic cycle, allowing monitoring of the movement of water mass. One of the isotopes found in water is the isotope ^{18}**O**, and the isotopic composition of water with regard to the heavy isotopes of oxygen and hydrogen can thus be considered as natural tracer for the investigation of the recharge areas of subsurface water, the study of the hydrologic balance and the dynamics of the lakes.

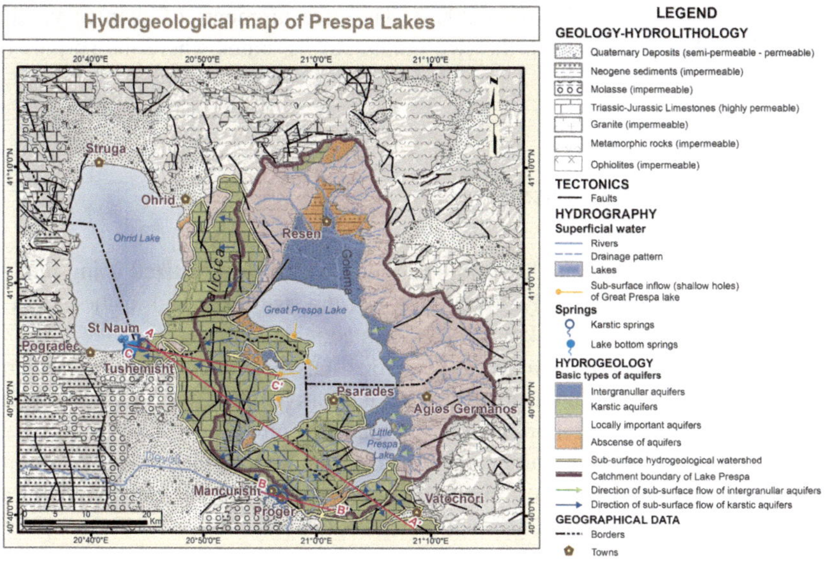

Fig. 3. Hydrogeological map of Prespa Lakes (Stamos and Stefouli 2010).

The method of calculation was as follows: first the relationship was established between the factor $\delta^{18}O$ of the underground water and the weighted average altitude of the recharge area (taking into consideration the surface and the level of rainfall). Secondly a sample of water was taken from the spring discharge system located in the area of Tushemisht in the region of Pogradec in Albania, St Naum of Ohrid in FYROM and in the lake of Great Prespa.

Table 2. Underground inflow of water of Great Prespa.

Springs	Average discharge (m^3/sec)	Participation of water from Great Prespa	
		(%)	(m^3/yr)
St. Naum	8.15	37.3	96.047.776
Tushemisht	1.885	54.4	32.338.275
		Total	128.386.051

The results of the isotopic analyses (which were done in collaboration with NCSR "Demokritos") determined the contribution of water from lake Great Prespa to the water from the excess discharge of springs of the karstic aquifer of Triassic-Jurassic limestones found in the mountain range of Galicica-Mali i Thate in FYROM and Albania (Table 2). Dye tracing with coloured substances was performed in the shallow holes of Great Prespa and Small Prespa, which confirmed the existence of water from Great Prespa in the springs located on the lakeside of Ohrid, the St. Naum and Tushemisht springs (FYROM and Albania) (Fig. 1 - section C-C') and from Small Prespa in the springs of Proger and Mancurisht in Albania (Amataj et al. 2003; IAEA 2003). The measurements of average yield of springs in Table 2 came from the two neighbouring states where the springs belong.

5 Conclusions - Discussion

After a hydrogeological and hydrological study which was realized using modern methods (isotopic analysis, dye tracing etc) in the two lakes and in the extent of hydrologic catchment (1,137 km^2), it has been concluded that:

Great Prespa Lake maintains its good qualitative situation; however it does show a dangerously declining trend in its quantitative situation. Small Prespa has similar characteristics but more intense qualitative problems.

In Triassic-Jurassic limestones of the wider region are independent karstic aquifers, which discharge mainly via a large reservoir of karstic springs which are located in the regions of Tushemisht (Albania) and St. Naum (FYROM). These are overflow springs and discharge from large riparian and lakeside areas of Lake Ohrid. They have secondary discharge in the two lakes of Prespa, which communicate mainly via surface flow as well as underground via the karstic aquifer of Mount Galicica-Mali i Thate between Great Prespa and Lake Ohrid. These aquifers are shown as yellow line on the hydrogeologic map.

The subsurface water outflows of Great Prespa are located along streams on the western side of the lake in Albanian territory, and amount to $130*10^6$ m^3/yr. Hypothetically there are also other outflows of groundwater in the lakebed, although they have not yet been located. It is imperative to complete an extensive and systematic study of the ecosystem of Prespa lakes and of the wider hydrologic and hydrogeologic basin of the two Prespa lakes and Lake Ohrid, by scientists from Greece, Albania and FYROM, taking into account the geomorphologic, hydrologic, hydrogeologic, geophysical, qualitative, quantitative, environmental and other parameters and characteristics of the system.

Reference

Amataj S, Anofski T, Leontiadis I, Stamos A (2003) Study of Prespa Lake Using Nuclear and Related Techniques (I.A.E.A)

Amataj S, Stamos A (2007) Tracer methods used to verify the hypothesis of Cvijic about the underground connection between Prespa and Ohrid lake

Stamos A (2010) Definition and Assessment of Hydrogeological Characteristics of Aquifer Systems of Greece, subproject 3: Water Balance Estimation of Upstream Aliakmonas, Vermio and Ptolemais (W.D 09), IGME

Flow geometry over a discharge measuring weir within inclined hydrogeological channels

J. Demetriou, D. Dimitriou, E. Retsinis

JD Research Hydrolab, 12th Polykarpou St., 17123, Athens, Greece. idimit@central.ntua.gr

Abstract When dealing with any hydrogeological issue, or the management of water environment, the water quality in agricultural constructions or even in hydrology, the first priority is to measure the water discharge. A suitable hydraulic structure, even in inclined watersheds within almost any runoff basin, is the construction of a weir and the experimental measurement of the corresponding water flow characteristics. Until now, the geometrical characteristics of such flows were limited in horizontal watersheds or channels, while in the present investigation an effort is made to extend the knowledge of such geometrical characteristics to weirs within inclined flows.
This experimental study concerns some interesting hydraulic characteristics of the flow around a thin weir within an inclined (angle φ) rectangular open channel. The length between the weir and the downstream contracted flow cross section is systematically investigated. The upstream to downstream depths′ ratio is also examined, while some interest has the ratio of the weir height to upstream hydraulic head - as a function of a reference Froude number. All results are accompanied by pertinent diagrams and empirical equations. The authors believe that the results are practical and may help to understand the thin weir overflows within inclined open channels or watersheds.

1 Introduction

The weirs are hydraulic structures mainly used to calculate relatively small flow discharges after measuring a convenient upstream water head (reference head) within an open channel. They usually consist of a thin and robust plate with the same width as the channel width (B), above which the water discharge q (per unit channel width) can be determined by the use of an appropriate equation produced through one-dimensional analysis.

Apart from q the thin weir overflow has some very interesting hydraulic characteristics such as the length L from weir to downstream flow reference cross section, the upstream to downstream depths' ratio, etc.

A large number of papers have been published in the past, mainly concerning the discharge coefficient of the vertical thin weir in horizontal open channel free flow. In the present paper the measurements of the main flow characteristics have been extended to flow in inclined open channels (angle φ, slope J_o=sinφ) and perpendicular (to channel bottom) thin weir.

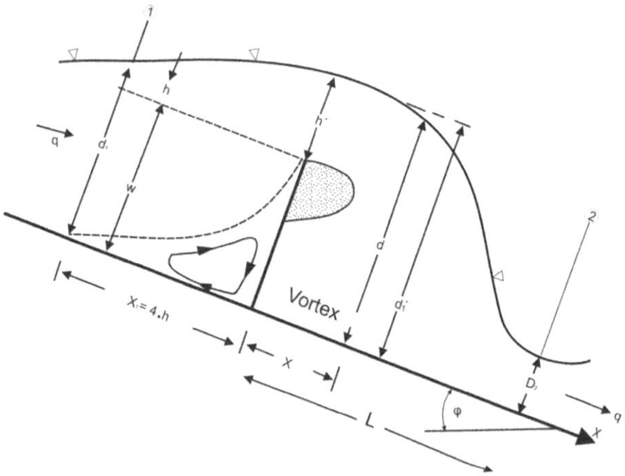

Fig. 1. Flow characteristics.

Figure 1 schematically shows the main free flow characteristics with the created air bubble-in touch with the weir's body - produced by itself. The weir height is w, the upstream head is h while the water depth at weir's crest is h'. The flow just upstream the weir is subcritical and gradually varied (almost horizontal free surface), above the weir is strongly curved-followed by a local descending flow. The control volume is included between cross sections 1 and 2, where corresponding water depths are d_1 and d_2. For d_1 measurements the same rule as in horizontal channels is followed, since d_1 is measured at distances $x_1 = 4 \cdot h$, where $h=d_1$-w. Thus, at cross section 1 the local pressures may consider as hydrostatic since they are unobstructed by any separation vortex effect. Since the free flow is ending here at a channel drop, cross section 2 is selected at a distance L where d_2 is the smallest depth, while local pressures at this cross section are also considered as hydrostatic.

Further, along the flow direction (x) the main flow parameters are the Froude numbers, $Fr_1 = q/g^{1/2} \cdot d_1^{3/2}$ and $Fr_2 = q/g^{1/2} \cdot d_2^{3/2}$, with

$$Fr_1 = Fr_2 \cdot (d_2/d_1)^{3/2} \tag{1}$$

To the best knowledge of the author there are no older experimental investigations concerning the flow length L and d_1/d_2 ratio within weir flows in inclined channels. Demetriou (2011) is presenting some first results of an experimental study concerning the dimensionless length and depths' ratio for the above flow geometry.

2 The Experiments

All experimental measurements were performed in the JD Research Hydrolab. A small tilting perspex flume was used with inclination angles φ=0°-3°-6°-9°-11°-14.5° and w/B=0.5-0.75-1.0-1.25. The discharges were volumetrically measured, and various depths were determined with a meter gauge. A number of 24 runs were organized and a large number of depths were recorded. Fr_2 varied between 2.6 and 20.2, Fr_1 between 0.09 and 0.54, Reynolds numbers Re_2 varied between 6,250 and 31,200, while w/h varied between 1 and 10.

At any angle φ and height w, q, d_1, d_2, h and L were recorded, while the rest of quantities were calculated after the end of measurements.

3 Results, Analysis and Discussion

Figure 2 (a-f) presents L/d_1' vs Fr_2 for the main angles φ=0°-3°-6°-9° -11° -14.5°, where it is clear that all experimental curves through the experimental points of measurements are descending when Fr_2 are increasing. If d_1/d_2 ratio is known then, from equation (1), Fr_1 may be used instead of Fr_2 when this is more convenient.

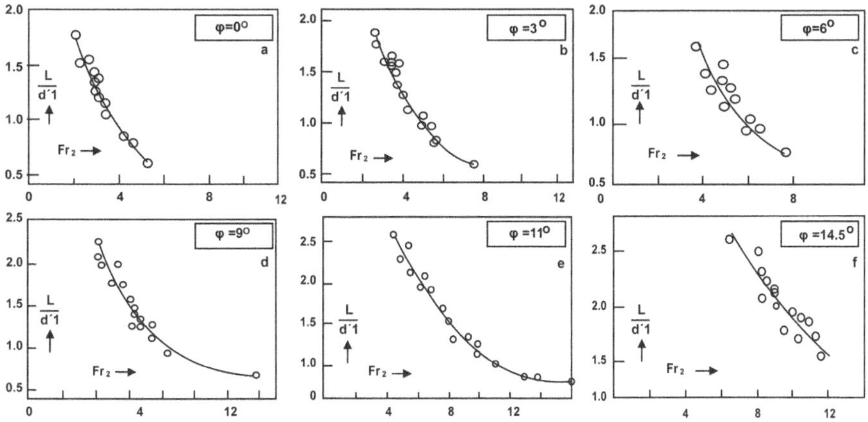

Fig. 2. a. L/d_1' vs Fr_2 for φ=0°, **b.** L/d_1' vs Fr_2 for φ=3°, **c.** L/d_1' vs Fr_2 for φ=6°, **d.** L/d_1' vs Fr_2 for φ=9°, e. L/d_1' vs Fr_2 for φ=11°, f. L/d_1' vs Fr_2 for φ=14.5°.

Figure 3 shows a comparison among all L/d_1' vs Fr_2 curves (solid lines) for angles φ, 0°≤φ≤14.5°. From this figure it is clear that all curves are systematic and for Fr_2=const. L/d_1' are increasing when angle φ is increasing, while for L/d_1'=const. this experimental value corresponds to larger Fr_2 values.

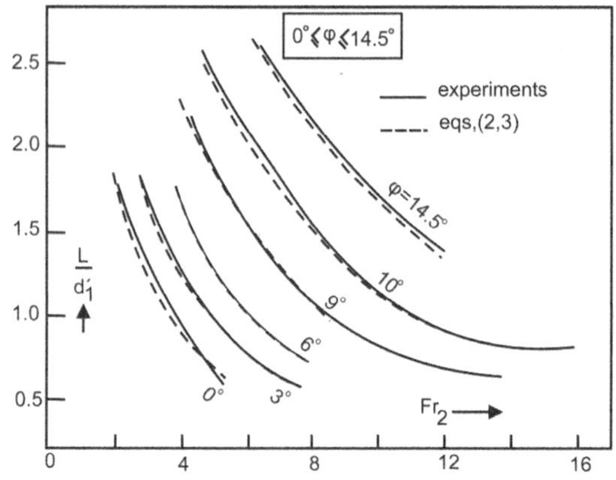

Fig. 3. L/d_1' vs Fr_2 for $0° \leq \varphi \leq 14.5°$.

In order to give an approximate and quantitative relationship between $L/d_1'$and Fr_2 for any angle φ, a number of mathematical equations - based on the previous experimental results - are produced with the following form:

$$L/d_1' = a + b \cdot Fr_2^{-1} \tag{2}$$

where a and b are arithmetic coefficients calculated from Fig. 2 (a-f). Table 1 presents all pertinent a and b coefficients for corresponding angles φ.

Table 1. Various arithmetic coefficients.

φ	a	b	a_1	b_1	c_1
0°	-0.086	3.861	-0.187	0.901	0.306
3°	-0.041	4.989	-0.746	2.357	0.510
6°	-0.289	7.856	0.841	-1.597	0.515
9°	-0.152	9.828	1.077	-2.117	0.537
11°	-0.047	12.268	1.260	-2.363	0.559
14.5°	0.10	15.854	-0.116	0.046	0.788

Equation (2) and Table 1 give a number of equations and pertinent curves - which are traced (dashed lines) in Figure 3, - while as one can see there is a very good agreement among experimental curves and curves representing equation (2) for $Fr_2>2$ and-even-$Fr_2 \leq 16$.

A very approximate and more general equation could also be deduced for L/d_1' if each one of a and b from Table 1 is correlated to corresponding angle φ. This was done in two separate diagrams (not shown here) with the results, where φ in degrees. Thus the set of equations (2) and (3) may be used in practice to give approximate values of L/d_1' vs angles φ and Fr_2.

$$\left.\begin{aligned} a &= -0.057 - 0.030\,\varphi^{1.5} + 0.009\cdot\varphi^2 \\ b &= 3.021 + 0.840\cdot\varphi \end{aligned}\right\} \quad (3)$$

Although the above ratio L/d_1' is rather small - since its minimum values remain between 0.6 and 1.4 while its maximum values are changing in the range 1.8 to 2.7 - in some cases it is necessary to know L as a function of d_1' and angle φ.

Figure 4 shows the relationship between d_1/d_2, angle φ and h/w ratio, for $0° \le \varphi \le 14.5°$. As one can see from this figure all experimental points are creating a cloud - without any systematic dependence on angle φ. A (solid) curve is statistically traced among all the experimental points and it is clear that - independently of angle φ - d_1/d_2 is reducing when h/w ratios are rising up.

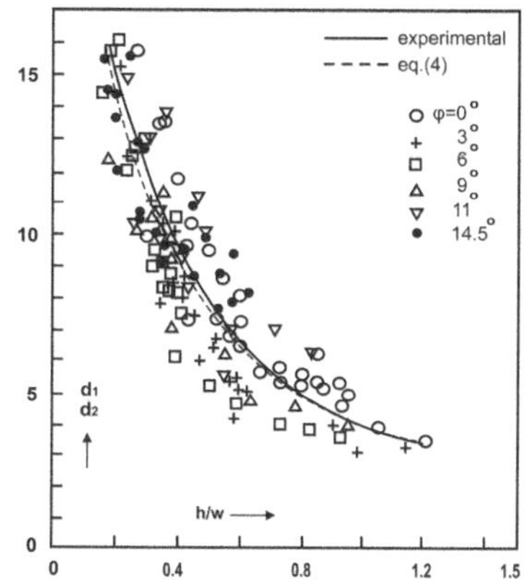

Fig. 4. d_1/d_2 vs h/w and angles φ.

A very good approximation of d_1/d_2 ratio vs h/w may be given by the empirical equation

$$s = d_1/d_2 = 1.353 + 2.871\cdot(h/w)^{-1} \quad (4)$$

showing that for $h/w \to 0$ $d_1/d_2 \to \infty$, for $h/w \to \infty$ d_1/d_2 tends to an asymptotic value around 1.35, while the characteristic value $d_1/d_2=4.224$ corresponds to h/w=1. The curve representing the above eq. (4) is also traced (dashed line) in Figure 4, and as one can see there is a good agreement with the experimental curve. In general when w/h is increasing d_1/d_2 is increasing rather fast, i.e. for d_1=const. d_2 is reducing and this may due to a lot of parameters such as the increase of the slope or the size of the air bubble.

The above ratio d_1/d_2 may also be connected to Fr_1 (or Fr_2-eq. 1) since w/h may be expressed as a function of angle φ and Fr_1. Figure 10 shows two typical experimental diagrams of w/h vs Fr_1 for φ=0° and φ=9°, while Figure 11 presents all pertinent curves of the present study for $0° \leq \varphi \leq 14.5°$.

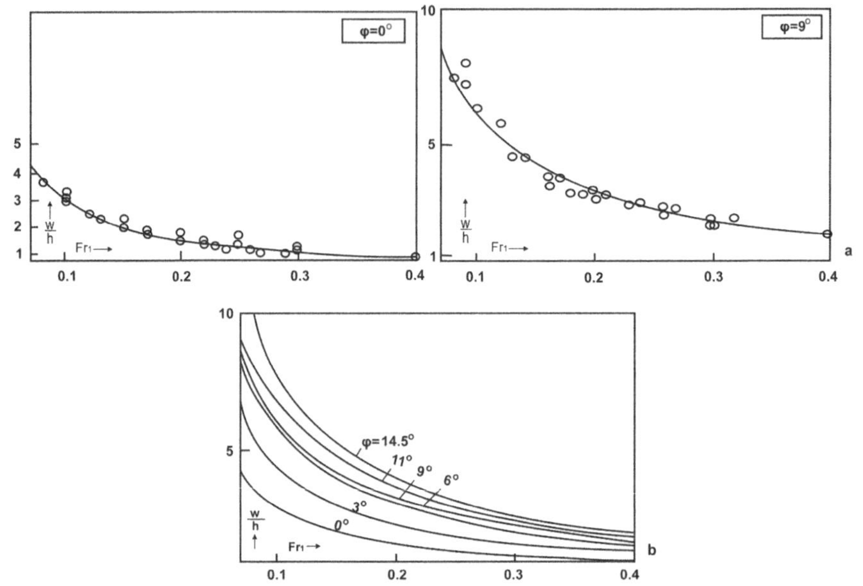

Fig. 5. a. w/h vs Fr_1 for φ=0° and φ=9°, **b.** w/h vs Fr_1 for $0° \leq \varphi \leq 14.5°$.

Further, an approximate empirical equation, may describe all experimental lines of Fig. 5b, where a_1, b_1 and c_1 are particular arithmetic coefficients taken from all present measurements and shown in Table 1 - for each angle φ.

$$w/h = a_1 + b_1 \cdot Fr_1 + c_1 \cdot Fr_1^{-1} \tag{5}$$

All a_1, b_1 and c_1 coefficients were also elaborated as functions of angle φ in separate diagrams (not shown here) with the following results, where φ in degrees.

$$\left.\begin{aligned} a_1 &= -0.021 \cdot \varphi^2 + 0.363 \cdot \varphi - 0.695 \\ b_1 &= 0.041 \cdot \varphi^2 - 0.783 \cdot \varphi + 2.059 \\ c_1 &= 0.0006 \cdot \varphi^2 + 0.018 \cdot \varphi + 0.362 \end{aligned}\right\} \tag{6}$$

The set of equations. (5) - (6) are represented by lines which are very close to the experimental (solid) lines of Fig. 5b.

Equations (1), (4) and (5), (6), also show that when Fr_1 is increasing, Fr_2 is decreasing. On the other hand if $d_2 = \lambda \cdot h$, where λ=suitable arithmetic coefficient then $s = d_1/d_2 = d_1/\lambda \cdot h$ or $d_1/h = \lambda \cdot s$, or $(h+w)/h = \lambda \cdot s$, $w/h = \lambda \cdot s - 1 = \lambda \cdot \left[1.353 + 2.871 \cdot (h/w)^{-1}\right] - 1$, or where h/w may be taken from

$$\lambda = \frac{(w/h)+1}{\left[1.353 + 2.871 \cdot (h/w)^{-1}\right]} \tag{7}$$

Equations (5) and (6). Moreover, since in thin weir flows $q = c_d \cdot (2/3) \cdot (2g)^{1/2} \cdot h^{3/2}$ where the discharge coefficient c_d=function of (angle φ and w/h), $Fr_2 = \left(2 \cdot \sqrt{2}/3\right) \cdot c_d \cdot (h/d_2)^{3/2}$ and since $d_2 = \lambda \cdot h$, $Fr_2 = \left(2 \cdot \sqrt{2}/3\right) \cdot c_d \cdot \lambda^{-3/2}$. Demetriou (2011) has given a number of equations for c_d in thin weir flows within inclined channels, while λ may be taken from eq. (7). Further, from eq. (1), $Fr_1 = Fr_2 \cdot \left[1.353 + 2.871 \cdot (h/w)^{-1}\right]^{-3/2}$, i.e. both Fr_1 and Fr_2 are functions of c_d and w/h.

4 Conclusions

This experimental study concerns some interesting hydraulic characteristics of the flow in the region around thin weir (w) within an inclined (angle φ) rectangular open channel, such as the dimensionless length between weir and a reference downstream cross section, or the upstream to downstream depth's ratio. Figure 1 presents the flow characteristics, Figure 2 (a-f) gives the above dimensionless lengths for φ=0°-3°-6°-9°-11°-14.5° and equations (2), (3) give mathematical expression for those lengths. The upstream to downstream depths ratio is given by Figure 4 and equation (4), while Figures 5a, b and equations (5) and (6) present the ratio w to upstream flow head h, as a function of Froude number and angle φ. The results are considered as practical and may help to understand the thin weir overflows in inclined open channels.

References

Demetriou J (2011) Experimental Measurements on Local Hydraulic Flows Within Inclined Open Channels. Vol II: Thin and Broad Crested Weirs-Sills. 146 pages, A publication by JD Research Hydrolab, Athens, Greece

The contribution of geomorphological mapping in the Ksiromero karstic region: land use and groundwater quality protection

M. Golubovic Deligianni[1], K. Pavlopoulos[1], G. Stournaras[2], K. Vouvalidis[3], G. Veni[4]

[1] Harokopio University, Faculty of Geography, 70 El. Venizelou Str., 17671 Athens, Greece miljanamakis@yahoo.com, kpavlop@hua.gr
[2] University of Athens, Faculty of Geology and Environment, Panepistimioupolis Zografou, 15784 Athens, Greece, stournaras@geol.uoa.gr
[3] Aristotle University, School of Geology, 54124 Thessaloniki, Greece, vouval@geo.auth.gr
[4] National Cave and Karst Research Institute, 400-1 Cascades Avenue, Carlsbad, New Mexico 88220, USA, gveni@nckri.org

Abstract Ksiromero is a karst region in the northeast section of the Prefecture of Aitoloakarnania, in western Greece. It has a total areal extent of 104.91 km^2 and is primarily developed on beds of Triassic carbonate breccia conglomerates and gypsum, and Upper Triassic to Early Jurassic limestones and dolomites. The region is normally dry, but streams flow after strong rains. A geographical digital geodatabase model was created to delineate its dense network of karst landforms, which include dolines, uvalas, poljes and stream sinks, and evaluate their impact on land use. Groundwater is poorly accessible to wells. Water for agriculture is mostly imported from adjacent regions and stored in earthen reservoirs, which include dolines and excavated depressions in terra rossa. Existing agricultural practices and the absence of sanitary landfills in this karst terrain may be resulting in groundwater pollution.

1 Introduction

Ksiromero is a karst region in the northeast section of the Prefecture of Aitoloakarnania in western Greece (Fig. 1). It is geographically bounded by the mountainous area of Akarnanika Ori to the west (Psili Koryfi, the highest peak is 1,157 m above sea level), the basin of Lake Ambrakia to the east, Amvrakikos Gulf to the north, and the southern watersheds whose rivers flow to the Ionian Sea through the Department of Akarnania. This paper describes the digital geographic mapping of the region and its application to land use and groundwater protection.

Fig. 1. Location of Ksiromero area (Akarnania, western Greece) with panorama of Stinadia polje.

2 Material and Methods

In this paper we present a practical application for digital geomorphological mapping. The study was conducted in five phases: literature search, field research, landform reconnaissance, laboratory analysis of water and rock samples, and field mapping and use of geographic information systems software ESRI ArcGIS® version 9.3.1. We stored, analyzed and displayed a detailed geodatabase of topographic, morphologic, hydrologic characteristics and establish a detailed geodatabase of karst landforms for the region to produce maps and reports. The digital database allows combination of the geodatabase's five different layers: topography, geology, hydrology, LandSat imagery, and karst features. For detailed visualization and analysis we digitalized 21 1:5,000 topographic sheets of the region at 4-m contour intervals and interpolated at 2 m. Digital geomorphological mapping combined the topographic data with landforms mapped with LandSat TM satellite images, acquired on 9 September 1992 at 09:25 UTC, and through the use of Principal Component Analysis and Tasseled Cap Transformation. We organized the karst landform database into sub-groups in different layers and calculated the surface area and volume each individual surface karst feature. The geological database was created by digitizing different layers of geological and topographical information on the 1:50,000 scale maps by the Institute of Geology and Mineral Exploration (1986, 1987).

3 Geology

Ksiromero is located in the Ionian geotectonic zone, part of the large External Hellenides Platform that extends west from mainland Greece and appears on the western edge of Peloponnesus and in some of the Dodecanese Islands. Primary tectonic features are long thrust faults that trend east-west and northeast-southwest, and long reversed and normal faults that trend northwest-southeast. Ksiromero is a tectonic basin that developed within this fault zone. Figure 2 illustrates the geologic units and karst geomorphic and hydrologic features. Lithologic units of interest are over 150 m of Triassic evaporites (2%) overlain by the 10-200 m thick Triassic Tryphos Formation carbonate breccia conglomerates (64%), followed by up to 200 m of dolomite (4%) and as much as 300 m of the Pantocrator Limestone (20%) into the Early Jurassic.

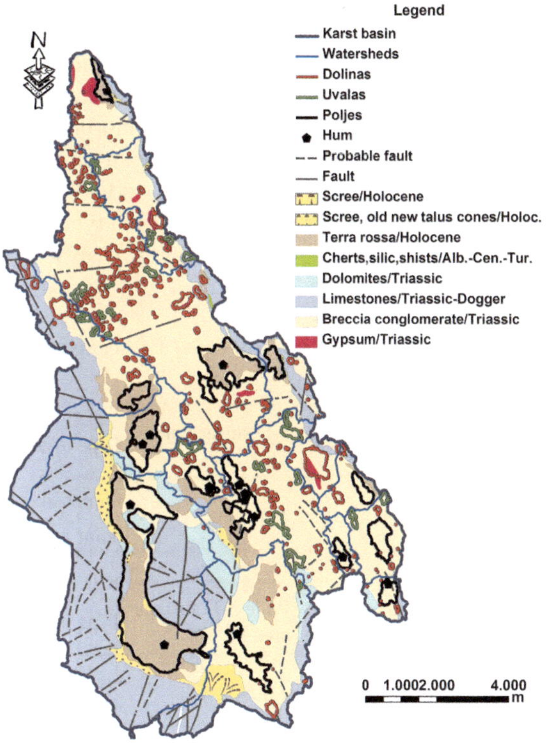

Fig. 2. Geological and geomorphological map of the Ksiromero region.

During Permo-Triassic time, the Ionian Zone was a shallow, restricted, marine basin which accumulated over 150 m of evaporites (Karakitsios 1992), of which gypsum is now most notably exposed. Their episodic deposition may have allowed the development of a paleointrastratal karst, as described in other regions by Bosák et al. (1989), although no direct evidence has been found. At that time, the shallow Thethos Sea covered a continental platform which extended throughout most of western Greece. During the Triassic epigenetic carbonate breccia conglomerates formed from major tectonic activity, diapiric deformation, and dissolution of underling evaporites. These conditions continued with small modifications to the end of the Jurassic. From the Pliocene to the Quaternary, more recent gypsum deformation occurred at the surface due to underlying diapiric movement along prominent faults (Underhill 1988).

4 Karst Geomorphology and GIS

Comparing the results of the karst geodatabase in combination with the fieldwork allowed us to identify Ksiromero as one big karst basin which includes 15 watersheds, 13 belonging to poljes and two to groups of dolines. Also, 16 residual hills occur in the poljes; they are comprised mostly of limestone intercalated with gypsum and are likely the result of lesser solubility than tectonic factors. We found 267 dolines with funnel, shallow, and a few collapse morphologies, and 23 uvalas (Fig. 2). Finally, we created a digital model of stream sinks and identified 229 sinks: 66 are in poljes, 30 in uvalas, 89 in dolines and 44 not within other karst features. Our database included different layers for each type of karst features, and using the DEM, we applied Spatial Analyst and 3D Analyst to calculate surface areas some results are summarized in Table 1.

Table 1. Morphometric and watershed characteristics of **a)** a major doline and **b)** a polje in Ksiromero.

a) Doline	Elevation (m a.m.s.l)	Depth (m)	3-D surface area (m^2)	Type of doline (Čar 2001)
Stadopigado	135.4 – 240	104.6	481,400	funnel-like, near-fault, point recharge

b) Polje	Elevation (m a.m.s.l)	Watershed 3-D surface area (m^2)	Polje 3-D surface area (m^2)	Type of polje	Hum
Fraksias	178.8 - 200	20,518,200	1,397,900	tectonic, evolved from dolines and uvalas	1

5 Regional Karst Hydrogeology

“Ksiromero” is the Greek word for “dry place.” Hydrogeologically, Ksiromero is a system of closed karst watersheds whose recharge characteristics are poorly defined. In the study area rainfall is typically 962-1,040 mm/year (NDBHMI 2010) and annual evaporation reaches 650 mm so the remaining 312-390 mm is available for runoff and groundwater recharge. The largest percentage infiltrates the carbonate breccia conglomerate which has the greatest karstic development. A smaller percentage infiltrates the Pantocrator Limestone due to its smaller surface area and steeper topography. Surface water flows in Ksiromero only after rainfall.

Domestically used water is supplied to Ksiromero mostly from a neighboring region. Water for agriculture is stored in 387 reservoirs, some natural dolines but most excavated in terra rossa soils of poljes and large dolines. Shallow groundwater occurs in some of the dolines and earthen reservoirs, as well as seasonally captured storm water, but much is pumped in from the other region or adjacent wells. Sea level at the Amvrakikos Gulf is the ultimate base level and the water of Lake Amvrakia suggests it is a major inland base level at 25 m above mean sea level. The permeability of the terra rossa soils seems generally low. The sparse well data suggest that groundwater occurs in two aquifers. A shallow, possibly perched aquifer is in terra rossa 4-15 m below the surface. The deeper aquifer is in the epikarstic zone, approximate 5-25 m deeper at an elevation of 216.65 m in well YΓ-3 (Fig. 2, Table 2); the elevation is 230 m at well ΠΓ-2 in the carbonate breccia conglomerate (Fig. 2). Groundwater is also found along the contact zone of the carbonate breccia conglomerate with the underlying gypsum (Stournaras et al. 1989). This contact ranges from 60-260 m in elevation with an approximate gradient of 4.5-5.5% near the villages Tryphos and Kompoti (Kalumenos 2002).

The quality of Ksiromero's water supplies is vulnerable to contamination. The shallow groundwater in the terra rossa is vulnerable to agricultural pollution, notably fertilizers and pesticides but also manure from livestock. The elevated nitrate in the well YΓ-3 (above U.S. drinking water standards) may result from the well's location near some large cultivated fields.

Table 2. Water chemistry of a well in the Spilja polje and a well near the village Aetos.

Parameter	Cave Doline Well	YΓ-3 Well	Parameter	Cave Doline Well	YΓ-3 Well
Ca^{++} (ppm)	39	125.3	$SO4^{=}$ (ppm)	--	3.36
Mg^{++} (ppm)	2	2.92	$NO3^{-}$ (ppm)	<1.6	13.02
Na^{+} (ppm)	3.8	5.96	pH	7.59	7.5
K+ (ppm)	--	0.39	TDS (ppm)	--	585.8
HCO_3^- (ppm)	2.03	395	Ec (μS cm^{-1})	227	760
Cl^- (ppm)	--	102.83	Total hardness	105	39.2

Groundwater near homes, especially in the villages, is vulnerable to human effluent in septic systems. The absence of sanitary landfills and other agricultural contaminants potentially threaten groundwater quality, but the impacts are not presently quantifiable due to insufficient aquifer characterization. Areas of greater and lesser vulnerability are currently undefined but related to changes in the terra rossa's heterogeneity and isotropic nature, and the presence and absence of underlying conduits. Ksiromero's limestone margins are steep with little agriculture or human presence, and while inherently vulnerable due to its high permeability, few contaminants are present. Garbage dumps are an exception. Ksiromero does not contain any sanitary landfills, and dumps are scattered in the region, putting underlying and adjacent water at risk (Fig. 3). The potential impacts of all of these activities are currently only theoretical due to insufficient aquifer and surface water characterization.

Fig. 3. a) Garbage dump on hillside near the village Kompoti, overlooking farms in a polje; **b)** Hillside and bottom of a ponded doline in the Spilja polje near the village of Pappadatos.

6 Land Use

Most land use in Ksiromero is farming, pasture and light agriculture. We supplemented the geodatabase with data using the program Corine Land Cover (2000) (version 13) and Table 3 and Figure 4 illustrate the results of this analysis. Slightly over 60% of the study area is used agriculturally and generally corresponds to level or hilly areas overlying the carbonate breccias conglomerate. Nearly all of the remaining area is covered by dense natural vegetation that closely correlates to steep, mountainous areas of limestone and dolomite. A small percentage (1.39%) corresponds to artificial surfaces such as roads and buildings.

Table 3. Summary distribution of land uses in the study area.

No.	Description	Surface area (km^2)	Percentage (%)
1	Urban	1.455	1.39
2	Natural vegetation	63.350	60.32
3	Agriculture and pasture	40.105	38.29
4	Ksiromero watershed	104.910	100.00

The smaller percentage of urban cover reflects the low population of 5,493 (Hellenic Statistic Authority 2010). The fact that a large part of the study area is used for agriculture leads to the conclusion that environmental degradation due to the use of pesticides and fertilizers is possible and likely for water stored in dolines and excavated reservoirs next to pastures and cultivated fields.

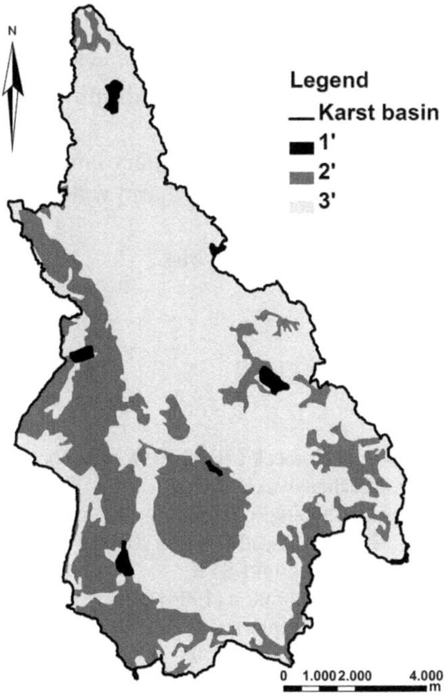

Fig. 4. Map of land uses in Ksiromero (keyed to Table 3).

7 Conclusions

Using Arc GIS 9.3.1 we made different geospatial layers with geodatabases of individual and grouped land features that can be easily updated and adapted to create specialized maps and analyses. For the Ksiromero region, we created a basic geographical digital geodatabase model as a matrix for the future exploration of relationships between the organization and origin of karst landforms, their influence on human activities, and the impact of those activities on karst. This paper provides an initial assessment of those karst features and their impacts.

Ksiromero is a region of topographically high carbonate karst surrounding lower elevations of richly developed fields of dramatic karst features, such as dolines, uvalas, polje's and rolling hills on a carbonate-evaporite breccia conglomerate and terra rossa, underlain by gypsum. Groundwater is poorly accessible to wells, although some groundwater is available through earthen reservoirs and some dolines. This water supply system is adequate for the area's modest demands, but insufficient for population or agricultural growth, and the locally-derived groundwater lacks adequate protection from contamination. Following are recommendations to increase water yield and to better protect the region's water supplies from pollution.

1. Most buildings are covered with ceramic tiles, which for relatively low cost can be modified to harvest clean runoff that needs minimal treatment relatively for domestic or agriculture use.
2. Clean-up all trash dumps and create a sanitary landfill for the region.
3. Regulate agricultural practices that may harm water quality in wells and surface reservoirs.
4. Controlled excavation of water reservoirs.

References

Bosák, P, Ford DC, Glazek J, and Horacek I (Eds.) (1989) Paleokarst - a systematic and regional view, Academia Prague, Czechoslovakia, 725 p

Čar J (2001) Structural bases for shaping of dolines. Acta Carsologica, 30(2), 239-256

Hellenic Statistic Authority (2010) accessed October 2010 at http//www.sttistics.gr/portl/page/portal/ESYE

Institute of Geology and Mineral Exploration (1986) Geological map of Greece: Astrakos Sheet. Institute of Geology and Mineral Exploration, 1:50.000

Institute of Geology and Mineral Exploration (1987) Geological map of Greece: Amphilochia Sheet. Institute of Geology and Mineral Exploration, 1:50.000

Kalumenos K (2002) Hydrogeological recognition of DD Katounas and Konopinas of municipality Medeonas, Aitoloakarnania, Institute of Geology and Mineral Exploitations, Division Hydrology, Athens, Greece

Karakitsos V (1992) Ouverture et Inversion Tectonique du Bassin Ionien (Epire, Grece): Annales Geologues des pays Helleniques, 35, 185-318

National Data Bank of Hydrological and Meteorological Information (NDBHMI) (2010) accessed September 2010 at http://ndbhmi.chi.civil.ntua.gr/en/index.html

Stournaras G, Panagopoulos A, Sotiropoulou K (1989) "La signification hydrogeologique, des conditions hydrochimiques et geomorphologiques d'un terrain gypseux. Les sources de Drymos", Geologie Mediterraneenne, XVI: 4, 311-320

Underhill, JR (1988) Triassic evaporates and Plio-Quaternary diapirism in western Greece, Journal of the Geological Society, 145:2, 269-282

The MEDYCYSS observatory, a Multi scalE observatory of flooD dYnamiCs and hYdrodynamicS in karSt (Mediterranean border Southern France)

H. Jourde, C. Batiot-Guilhe, V. Bailly-Comte, C. Bicalho, M. Blanc, V. Borrell, C. Bouvier, J.F. Boyer, P. Brunet, M. Cousteau, C. Dieulin, E. Gayrard, V. Guinot, F. Hernandez, L. Kong, A. Siou, A. Johannet, V. Leonardi, N. Mazzilli, P. Marchand, N. Patris, S. Pistre, J.L. Seidel, J.D. Taupin, S. Van-Exter

Laboratoire HydroSciences Montpellier UMR 5569, Université Montpellier 2, Place E. Bataillon CC MSE, 34095 Montpellier Cedex 5, France, herve.jourde@um2.fr

* : Ecole des Mines d'Alès, CMGD, 6 avenue de Clavières, 30319 Alès Cedex, France

Abstract In karst hydrosystems, the heterogeneity and the complexity of the transfer processes within the different compartments (soil, epikarst, unsaturated zone, saturated zone) control the recharge-discharge relationship. The MEDYCYSS observatory (Multi scalE observatory of flooD dYnamiCs and hYdrodynamicS in karSt) has been set up to better understand these transfers in a Mediterranean context. This observatory consists in a large karst hydrosystem located between the Hérault River, the Vidourle River and the Mediterranean Sea (Southern France, north and west from Montpellier). It comprises two main hydrogeological sites and three main hydrological sites where karst/river interactions occur. The challenges of the MEDYCYSS observatory are i) to better characterize the karst/river interactions for the risk assessment of floods and water resource contamination; ii) to discriminate and quantify flow distribution (rapid and slow transfers) between the recharge catchment and the outlets of the hydrosystem; iii) to understand the role of the unsaturated zone on the long term storage of the aquifer and quantify the water storage variation with time; iv) to characterize groundwater hydrodynamics in karst watersheds under climatic and anthropogenic forcing at nested observation scales; and v) to quantify the recharge, which is a key variable for the assessment and management of groundwater resources in karsts. Accordingly, a continuous monitoring of the hydrochemistry and hydrodynamics (wells, karst network, permanent and temporary springs), as well as of the river and spring discharges has been constructed to characterize transfer processes and parameterize local and regional scale models. Meteorological variables are also monitored at various points where measurements of soil humidity and evapotranspiration recently started.

1 Introduction

The number of damages related to flood has been increasing during the last years in the Mediterranean regions (Gaume et al. 2009). In these regions, most flood-prone streams and rivers flow on karstic rocks which are typical of the carbonates terrains of the Mediterranean circumference. Besides flood hazard, extreme rainfall events are also associated with the risk of groundwater resource contamination. Groundwater contamination can be related either to surface flow or groundwater flow because of the significant karst-river interactions. In all cases, a better assessment of these dynamics is a major stake for preserving the groundwater supply safety within the hydrosystem as groundwater management of these karsts is strongly dependent on the knowledge of their hydrodynamics. Moreover, because of the increased demand for water, the importance and awareness of karst aquifers in Mediterranean regions has grown.

For a better assessment of karst water resource, conceptual models are generally proposed to characterize the relationships between the recharge and the discharge at the outlets of the systems, based on hydrodynamics, hydrochemistry or isotopic studies (Bicalho et al. 2011 a,b,c; Fleury 2010; Mazzilli 2011; Borrell et al. 2008; Kong A Siou et al. 2011 c; Coustau et al. 2011; Bailly-Comte et al. 2011). The quantification (on the basis of long term time series) of these recharge-discharge relationships is a key parameter to estimate the sustainable yield of aquifers and assess the impacts of climatic and anthropogenic changes on the long term groundwater resources. However, an important limit of these approaches is related to the fact that it does not take into account the heterogeneity and the complexity of the transfers within each compartment (soil, epikarst/perched aquifer, unsaturated zone, saturated zone) of the karst aquifer. It is thus crucial to characterize and then consider these transfers in distributed models that account for each compartment.

Therefore, various experimental sites (Fig. 1) have been constructed to better understand these transfer processes. On these pilot sites, the challenges are i) to better characterize the karst/river interactions for the risk assessment of floods and water resource contamination; ii) to discriminate and quantify the distribution of the hydric fluxes (rapid and slow transfers) between the surface and the spring; iii) to quantify the water storage variation with time and understand the role of the unsaturated zone on the long term storage of the aquifer; and iv) to quantify the recharge as a function of field and climatic parameters, as it is a key variable for the assessment and management of groundwater resources.

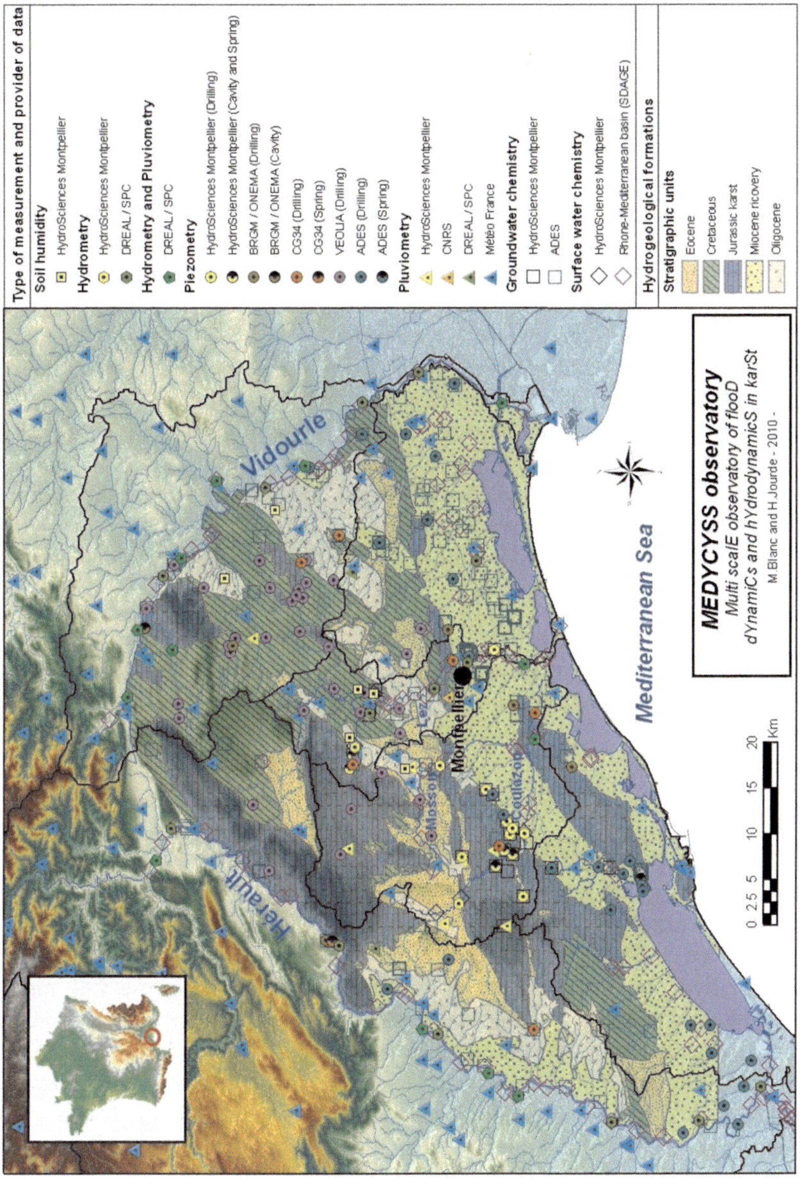

Fig. 1. Hydrogeological and hydrological contexts of the MEDYCYSS observatory (44° 00'N, 4°00'E - 43°17'N 3° 26'E).

2 Experimental sites of the MEDYCYSS Observatory

The MEDYCYSS observatory (Fig. 1) comprises three main watersheds where karst/river interactions occur: the Lez, the Mosson and the Coulazou river catchments. The major objectives of MEDYCYSS are:

- The observation and characterisation at local and regional scale of groundwater hydrodynamics in karst watersheds, under climatic and anthropogenic forcing;
- The observation and characterisation of the rapid hydrological processes during extreme flood events in Mediterranean watersheds, where interactions occurs between the river network and the karst aquifer;
- The observation and characterisation of groundwater resource contamination hazard, based on the coupled hydrochemistry /hydrodynamic monitoring;
- The assessment of the links between karst evolution over time (geology and geomorphology) and its current hydrodynamic behaviour.
- The characterisation and coupled modelling of the groundwater and surface water hydrodynamics in karst watersheds;
- The characterisation and modelling of the global change impact on the recharge/discharge relationships over long time period.

Long term time series (more than 40 years) of rain, water table level and discharge exist on the three experimental sites, which is a major asset for long term modelling and the characterization of global change impact on the long term groundwater resource.

2.1 Karst watersheds of the Coulazou and Mosson rivers

Since 2004, the experimental site of the Coulazou River allowed to study the modality of surface floods genesis (Jourde et al. 2007) and transfer (Bailly-Comte et al. 2008, Bailly-Comte et al. 2009) in karst terrains, at the scale of a well monitored small catchment. This site comprises a karst catchment, the Causse d'Aumelas, which is a Jurassic calcareous plateau crossed by the intermittent Coulazou River. This area is a binary karst system developed within the calcareous Jurassic plateau, in contact with an Oligocene impermeable basin in the North and a Miocene impermeable basin in the South.

In order to better understand the groundwater contribution to flash floods in the Coulazou river the water table in the karst aquifer is monitored in various locations (wells, karstic network, permanent and temporary springs), as well as the Coulazou river discharge at the entry and the exit of the karst system.

More than 15 karstic events, which can function alternately as sinkholes or springs, have been observed in this hydrosystem. In order to understand the interactions between surface water and groundwater, hydrodynamic monitoring is enriched with various chemical tracers monitoring (Major and trace elements, TOC:

Total Organic Carbon, stable isotopes of water and Carbon) on springs, boreholes, swallow holes, and on the *Coulazou* river. This "multiparameters" approach allows identifying the various types of water that contribute to flow, but also flow inversions between the karst system and the river. The understanding of these specific links between surface water and groundwater are fundamental to estimate the vulnerability of the aquifer and to protect water resources in this Mediterranean context (Batiot et al. 2005; Selles et al. 2011). Moreover, a regular monitoring (twice a month during two hydrological cycles) of CO_2 and ^{222}Rn allowed explaining the temporal variability of these elements (Batiot et al. 2007).

One of the major interest of this site is that, from upstream to downstream, the watershed evolution is characteristic of many Mediterranean rivers like Cèze, Hérault, Vidourle or Gardon in southern France, i.e: i) an upstream impervious watershed; ii) an intermediate watershed where karst/river interactions occur, with consequences on floods but also on groundwater contamination; iii) a downstream watershed in a alluvial plain, where flood hazard is generally more important.

This system thus constitutes a local scale experimental site to study karst/river interactions before focusing on larger systems where flood hazards have larger human and economic consequences.

On this site, some important experimental results characteristics of most Mediterranean karst watersheds and concerning hydrodynamics, groundwater/surface water interactions and transport have been highlighted:

- A hydrodynamic analysis coupled with a hydrological modelling showed that groundwater contribution from the karst to surface flood can reach one third of the total flood volume, and increase the flood peak discharge by a factor of two (Jourde et al. 2007).
- Times series analysis in time (correlation) and frequency (power spectral density) domains coupled with recession curves analysis showed that:
- karst groundwater flows may either attenuate or amplify the rapid surface flow transfers depending on the initial state (groundwater level) of the system (Bailly-Comte et al. 2008a, b).
- the initial groundwater level may be used to forecast the type of hydraulic connection between the river and the saturated zone during flood. This level is found to be also controlled by the limited discharge capacity of karst conduits. A conceptual model of karst/river interactions during flood has been proposed (Bailly-Comte et al. 2009).
- water exchange between matrix and conduits control the spring hydrograph (Bailly-Comte et al. 2010)
- A dye tracing experiment conducted in a small but highly transient tracing system interacting with the Coulazou River has shown that plurimodal tracer breakthrough curves only result from surface water/groundwater interactions, each flood pulse of the sinking stream inducing a tracer pulse at the outlet with low dilution and dispersion (Selles et al. 2011).

2.2 Karst watershed of the Lez river

The hydrological and hydrogeological catchments of the Lez River (North-West of the city of Montpellier) are remarkable as regards both their vulnerability to pollutions or flood hazards, and the active management of the groundwater resource. The objectives of the hydrodynamics monitoring are: (i) a better understanding of flash floods and contaminant transfers during extreme rain events, (ii) the characterization of the recharge and the transfer processes within each compartment of the karst system. To improve the understanding of the groundwater origins and its circulation dynamics within this karst system, regular sampling for hydrochemical and isotopical analysis is also carried out on the main springs and wells of the karst system. The application of a multi-tracer approach based on major ions, trace elements, isotopes and emerging tracers, helped determining water origins, water mixing proportions and levels of anthropogenic contaminations. Those results also allowed proposing a conceptual model of groundwater circulation for this aquifer (Batiot et al. 2008; Bicalho et al. 2009; Bicalho et al. 2010; Bicalho et al. submitted). High flow rate pumping (up to 6000 m^3/h) performed at the Lez spring for Montpellier water supply allows investigating the hydrodynamic properties of the karst aquifer at a regional scale. Besides, a small experimental site (Terrieu site) with 18 boreholes and 3 wells on a 500 m^2 area allows the characterisation of the hydrodynamic properties of the karst aquifer at a local scale (pumping up to 60 m^3/h). This configuration allows identifying the hydrodynamic behaviour of the karst system at nested observation scale and thus characterizing the scale effects in the assessment of hydrodynamic properties from pumping test analysis at local and regional scales (Jazayeri et al. 2011). Recent works focused on the karst drainage network development as a function of the hydrological, geodynamical and structural changes over time (Leonardi et al. 2011).

Recent studies also allowed proposing new results regarding hydrological modelling in karst watershed. For example, a distributed parsimonious event-based rainfall-runoff model was developed and applied on the Lez watershed upstream Montpellier (Coustau et al. 2011). On one hand, the model produces very satisfactory flash flood simulations, and on the other hand, its initialization is relatively easy thanks to a powerful relationship between the initial condition of the model and either the Hu2 index from the Meteo-France SIM model (soil humidity), or the piezometric levels. The interpretation of the model suggests that groundwater does not affect the first peaks of the flood, but can strongly impact subsequent peaks in the case of a multi-storm event. Because this kind of model is based on a limited amount of readily available data, it could be a crucial tool for flash floods real-time forecasting.

An original methodology based on neural networks was also developed to better understand the hydrodynamical behaviour of karst (Kong A Siou et al. 2011 a,b,c). A dedicated architecture allows describing the recharge of the system according to its geological heterogeneity. This model allows refining knowledge on the behaviour and the delimitation of its alimentation basin (Kong A Siou et al. 2011 c). Global modelling based on reservoir model that takes into account flow through both the saturated zone and the unsaturated infiltration zone was developed to characterize the functioning of this aquifer under active water management (Fleury et al. 2009; Mazzilli et al. 2011)

The Lez hydrosystem was also studied to assess the role of karst in flood dynamics, with a specific focus on the consequences of the active management of the groundwater resource, which offers some new insights in the integrated water resource/flood management on these hydrosystems with a high karstic component (Jourde et al. 2011

3 Conclusion and perspectives in terms of modelling and observation strategies

Flow modelling in karst hydrosystems implies various approaches to capture the spatio-temporal complexity of the studied site. Indeed groundwater hydrodynamics differs from surface water hydrodynamics, which involve referring to different conceptual models. Furthermore, depending on the considered temporal scale, the dominant processes within the hydrosystem will not be the same. For example, at the temporal scale of a flash flood, the processes will notably differ from the processes at the scale of one or more hydrological cycles.

The distributed measurement of humidity and evapotranspiration, together with piezometric monitoring in different compartments of the hydrosystems will enhance both hydrological modelling for flash floods forecasting and recharge estimation for hydrogeological modeling and water resource evolution with time.

Developing a multi-tracer approach based on major and trace elements and emerging tracers (isotopes, Rare Earth Elements, fluorescence of Dissolved Organic Matter), a better characterisation of the water origins, water mixing and of potential anthropogenic contaminations has been achieved (Batiot et al. 2008; Bicalho et al. 2009, Bicalho et al. 2010; Bicalho et al. submitted). The hydrochemical monitoring already shows that the chemical signatures vary all along the hydrological cycle but also during a flood event. This change will enhance a better understanding of the various processes occurring in different compartments of the karst system.

It will be useful to distinguish different types of pedological and landscape units, where infiltration and evapotranspiration will be determined. Defining a mean soil permeability and evapotranspiration for each units and characterizing the variability of these parameters within one unit will be possible. Based on these results, a GIS analysis taking into account landscape use, pedology and geology, will facilitate the drawing of a map with these distributed parameters.

The different perspectives in terms of modelling and observation strategies can be listed as follows:

- Couple distinct models for the integrated and distributed simulation of the hydrological, hydrogeological and hydraulic processes in karst hydrosystems at different time scales (flood event, hydrological cycle).
- Analyse the sensitivity and assess the uncertainty of the hydrological and hydrogeological models with regards to the main parameters (e.g. contrast between the hydrodynamic parameters of the different compartments, initial state of the system, soil humidity, recharge),
- Discriminate and quantify the distribution of the hydric fluxes (rapid and slow transfers) in each compartment of the hydrosystem.
- Characterize the transfer within each subsystem and develop distributed models taking into account the transfer within each compartment of the hydrosystem.
- Quantify the water storage variation with time and understand the role of the unsaturated zone on the long term storage of the aquifer.
- Quantify the recharge, which is a key variable for the assessment and management of groundwater resources in karsts
- Quantify the recharge-discharge relationships to estimate the sustainable yield of aquifers and determine the impacts of climatic change on the long term groundwater resources.
- Quantify the impact of anthropogenic forcing (pumping, active management) on the global water balance and hydrodynamic behaviour of the karst system.

References

Bailly-Comte V, Jourde H, Roesch A, Pistre S (2008) Mediterranean flash flood transfer through karstic area. Environmental Geology, 54(3), 605–614

Bailly-Comte V, Jourde H, Roesch A, Pistre S, Batiot-Guilhe C (2008) Time series analyses for Karst/River interactions assessment: Case of the Coulazou river (southern France). Journal of Hydrology, 349(1-2), 98–114

Bailly-Comte V, Jourde H, Pistre S (2009) Conceptualization and classification of groundwater-surface water hydrodynamic interactions in karst watersheds: Case of the karst watershed of the Coulazou River (Southern France). Journal of Hydrology, 376(3-4), 456–462

Bailly-Comte V, Borrell-Estupina V, Jourde H Pistre S (2011) Influence of karst aquifers on flood genesis and propagation in an ephemeral Mediterranean River: A semi-distributed conceptual model of the Coulazou River (Southern France). WRR, accepted

Batiot C, Desrousseaux C, Bailly-Comte V, Jourde H, Hébrard O, Pistre S (2005) Sink holes-springs study to underline flow inversion in a Mediterranean karst system partly supplied by an intermittent river (Causse d'Aumelas, southeastern France). Karst 2005, Water Resources and Environmental Problems in karst. International Conference and Field Seminar, Belgrade and Kotor, September 14-19, 2005

Batiot C, Seidel J L, Jourde H, Hébrard O, Bailly-Comte V (2007) Seasonal variations of CO2 and Radon 222 in a Mediterranean Cavity (Causse d'Aumelas, SE France), International journal of speleology, 36(1), 51–56

Batiot-Guilhe C`, Seidel J.L , Lafare A, Jourde H, Cordier M A, Van-Exter S, Rodier C (2008) Characterisation of underground flows in karstic aquifers by studying DOM fluorescence. Example of two Mediterranean systems (Lez and Causse d'Aumelas, Southeastern France), 13th IWRA World Water Congress, 1-4 septembre 2008, Montpellier

Bicalho C C, Batiot-Guilhe C, Seidel J.-L, Van-Exter S. Jourde H (2009) Hydrogeological functioning of a complex Mediterranean karst system by multivariable tracing. 37th IAH Congress, Hyderabad (India),6-12 Sept

Bicalho C C., Batiot C, Seidel J. L, Van Exter S, Jourde H (2010) Investigation of Groundwater dynamics in a Mediterranean karst system by using multiple geochemical tracers. Advances Res. Karst Media, 1, 157–162

Bicalho C C, Batiot-Guilhe C, Seidel J L, Van Exter S, Jourde H (2011) Geochemical evidence of water source characterisation and hydrodynamic responses in a karst aquifer. J. Hydrol., submitted

Bicalho C C, Batiot-Guilhe C, Seidel J. L, Taupin J D, Patris N, Van Exter S, Jourde H (2011) A conceptual model for groundwater circulation by using isotopic (δ18O, δ^2H and 87Sr/86Sr) and geochemical tracers in a Mediterranean karst system. H2Karst, 9th Conference on Limestone Hydrogeology, Besançon, 1-3 September

Bicalho C C, Batiot-Guilhe C, Seidel J L, Taupin J. D, Patris N, Van Exter S, Jourde H (2011) Isotopic (δ^{18}O, δ^2H, $\delta^{13}C_{TDIC}$ and ^{87}Sr/^{86}Sr) and geochemical monitoring of a Mediterranean karstic system: chemical evolution and hydrodynamics of groundwater, Chemical Geology, submitted

Borrell Estupina V, Coustau M, Jourde H, Bouvier C (2008) « Un modèle pluie-débit pour la prévision des crues sur un bassin karstique. Cas du bassin du Lez. » Colloque de la SHF Prévisions hydrométéorologiques. Lyon, 18 et 19 novembre 2008. Recueil des texts, pp.25–29

Coustau M., Borrell-Estupina V, Bouvier C (2011) Improvement of rainfall-runoff modelling with distributed radar rainfall data: a case study in the Lez French Mediterranean catchment. Weather Radar and Hydrology, IASH Pub, soumis

Coustau M, Bouvier C, Borrell-Estupina V, Jourde H (2011) Flood modelling with a distributed event-based parsimonious rainfall-runoff model: case of the karstic Lez river catchment. Journal of Hydrology, soumis

Fleury P, Ladouche B, Conroux Y, Jourde H, Dörfliger N. (2009) Modelling the hydrologic functions of a karst aquifer under active water management – the Lez spring

Gaume E, Bain V, Bernardara P, Newinger O, Barbuc M, Bateman A, Blaškovičová L, Blöschl G, Borga M, Dumitrescu A, Daliakopoulos I, Garcia J, Irimescu A, Kohnova S, Koutroulis A, Marchi L, Matreata S, Medina V, Preciso E, Sempere-Torres D, Stancalie G, Szolgay J, Tsanis J, Velasco D, Viglione A (2009) A collation of data on European flash floods. Journal of Hydrology. 367, 70–78

Jazayeri M, Jourde H, Massonnat G (2011) Influence of observation scale on the hydrodynamic analysis of well tests in a fractured reservoir. Case of the Lez karst system, Montpellier, France, Journal of Hydrology

Jourde H, Roesch A, Guinot, V, Bailly-Comte V (2007) Dynamics and contribution of karst groundwater to surface flow during Mediterranean flood. Environmental Geology Journal 51: 725–730

Jourde H, Lafarre A, Mazzilli N (2011) Flash flood mitigation as a positive consequence of anthropogenic forcing on the groundwater resource in a karst catchment. H2Karst, 9th Conference on Limestone Hydrogeology, Besançon, 1-3 September

Kong A Siou L, Johannet A, Borrell V, Pistre S (2011) Optimization of the generalization capability for rainfall-runoff modeling by neural networks: The case of the Lez aquifer (southern France). Environnemental Earth Sciences Advances in Karst Hydrogeology, soumis

Kong A Siou L, Johannet A, Borrell V, Pistre S (2011) Complexity selection of a neural network model for karst flood forecasting: The case of the Lez Basin (southern France). Journal of Hydrology, accepted

Kong A Siou L, Cros K, Johannet A, Borrell V, Pistre S (2011) Modélisation hydrodynamique des karsts par réseaux de neurones: comment dépasser la boîte noire ? - H2Karst, 1-3 Sept, Besançon, France

Kong A Siou Line, Cros K, Johannet A, Borrell Estupina V, Pistre S (2011) Utilization of Geology for the Hydrodynamical Modeling of a Karst Aquifer by Artificial Neural Networks. Case Study on the Lez Karstic Spring (southern France). Water Resources Research, soumis

Leonardi V, Tissier G, Jourde H (2011) Eléments de Genèse des karsts péri-méditerranéens : Impact de la tectonique sur l'évolution des drains karstiques (Karsts Nord-Montpelliérains). H2Karst, 9th Conference on Limestone Hydrogeology, Besançon, 1-3 Septembre

Mazzilli N, Jourde H, Guinot V, Bailly-Comte V, Fleury P (2011) Hydrological modelling of a karst aquifer under active groundwater management using a parsimonious conceptual model H2Karst, 9th Conference on Limestone Hydrogeology, Besançon, 1-3 September

Selles A, Leonardi V, Bailly-Comte V, Jourde H (2011) Influence des relations karst/rivière sur la restitution plurimodale d'un traçage artificiel: Cas du Causse d'Aumelas. H2Karst, 9th Conference on Limestone Hydrogeology, Besançon, 1-3 September

Hydrogeological research in Trypali carbonate Unit (NW Crete)

E. Steiakakis, D. Monopolis†, D. Vavadakis, E. Manutsoglu

Department of Mineral Resources Engineering, Technical University of Crete, Chania, GR73100, stiakaki@mred.tuc.gr

Abstract Trypali unit overlies the para-autochonous system of Plattenkalk in the NW Crete, and forms a separate hydrogeological entity. The objective of this work is to investigate the extent of Trypali unit in the region north and north-east of the Omalos plateau, and to evaluate the availability of groundwater to serve the water demands in the region. In the frame of this research, hydraulic properties of the aquifer were estimated by using pumping tests in Myloniana well-field. The transmissivity of the carbonate unit ranges between 0.1 to 1 m^2/sec which is equivalent to permeability between 10^{-3} and 10^{-2} m/sec.

1 Introduction

Geology of Crete can be summarized as tectonic units thrust into a pile of nappes. The Plattenkalk Series is the lowest tectonic unit and it is considered to be in a para-autochthonous position. Trypali unit overlies the system of Plattenkalk in the NW Crete, and forms a separate geological entity. The upper nappes consist of the Tripolis zone, the Pindos zone and the ophiolitic complex. Carbonate formations of these zones constitute the most extensive and productive water-bearing formations in the island.

The study area (north and north-west of the Omalos plateau) consists of limestones and dolomites (Trypali unit). At some places they look like the formations of Plattenkalk Series, but without cherts. The dark coloured parts of the unit are bituminous, and they have often a cellular texture, especially the dolomites, with dolomitic flour in their cells (Tataris and Christodoulou 1969).

Trypali unit forms a separate hydrogeological entity in the area with intensive karstification. It represents a productive aquifer and the Plattenkalk Series forms the accompanying basic layer.

Five major groups of springs near Agyia's village (Fig. 1) and more than ten pumping wells south of these springs comprise the main discharge outlets for the karstic aquifer.

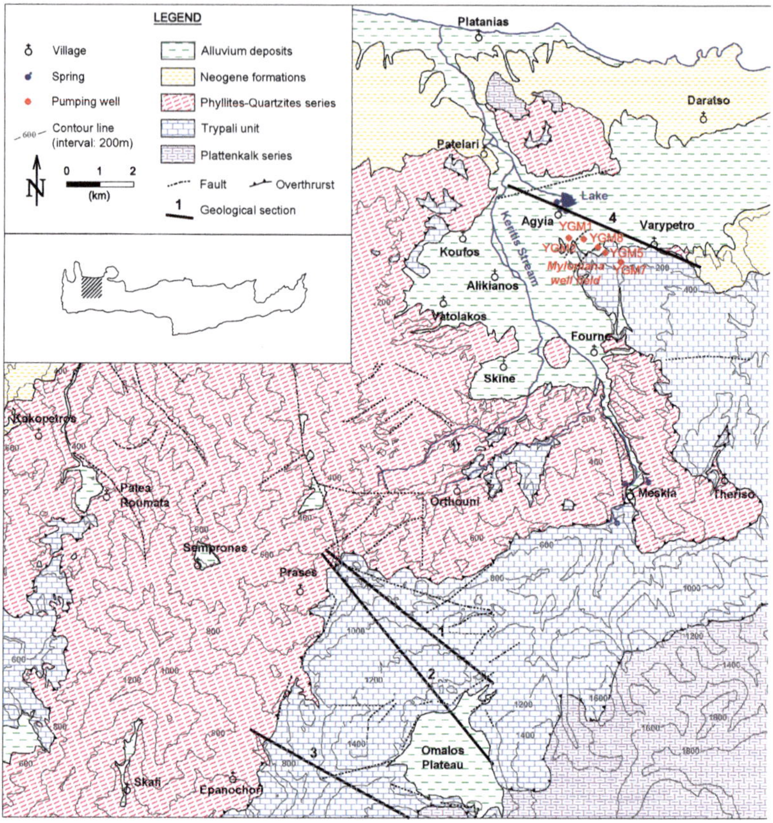

Fig. 1. Geological map of the study area.

The present work offers the possibility to evaluate the available groundwater potential in the area and to make suggestions for the exploration and exploitation of the groundwater resources.

2 Geology

The surface geological formations in the area north and north-east of the Omalos plateau are displayed in Figure 1. It was constructed with base the published geological maps (Tataris and Christodoulou 1969; Manakos et al. 1995) and the geological survey realized during the current research.

This map delineates the carbonate formations of Trypali unit in the region north and north-east of the Omalos plateau, as far as the Agyia's springs outlet.

Two different dominant opinions exist concerning the structural evolution of the study area.

- Initially, Tataris and Christodoulou (1964) based on the data obtained from the geological mapping of the Leuca Mountains area (Western Crete), suggest that the carbonate formation of Trypali forms a separate entity that was called temporarily as «Triassic trangressive system of Western Crete». It overlays trangressively the immediately preceding formations of Plattenkalk Series. In conformity with this view, Phyllitic nappe (made up of phyllites, rauwackes, gypsum, limestones etc) is overthrust on the preceding Trypali formation. Stratigraphically, the Plattenkalk unit, the carbonate formations of Trypali and the Phyllites-Quartzites series are regarded to be in a normal position.
- On the contrary, Krahl (1983) and Manakos et al. (1995) suggest that Trypali unit and Phyllitic nappe group form a separate tectonostratigraphic entity that is overthrust on the preceding Plattenkalk Series. Stratigraphically it is in a reversed position. The geological map of Figure 1 was drawn in conformity with this latter view.

Apart from the main structure lines, (derived from the published geologic mapping), it presents tectonic features identified in the field during the current research.

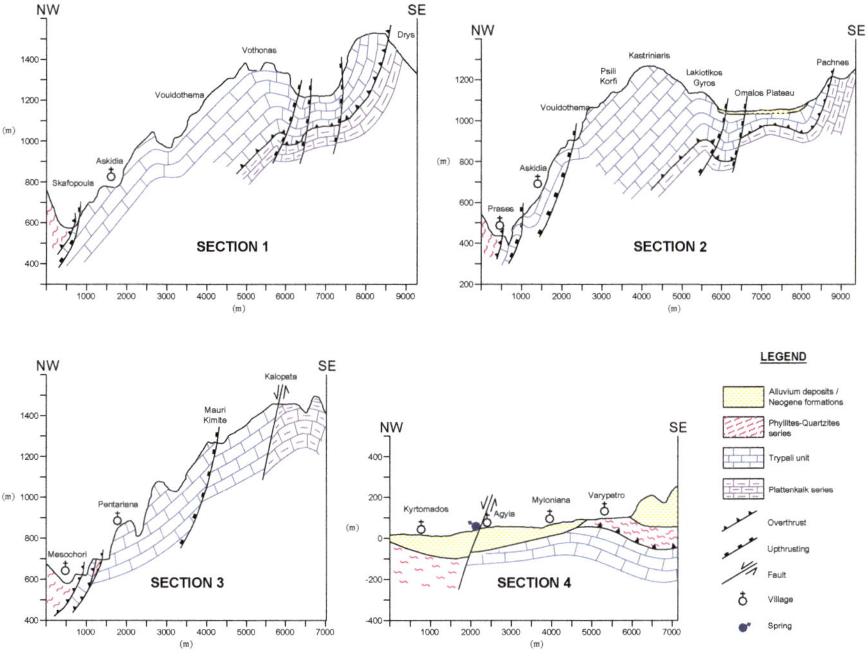

Fig. 2. Schematic geological sections. Their locations are defined in Figure 1.

Adopting the aspect that Trypali unit is overthrust onto the Plattenkalk Series, the large scale structures were checked in the field and four geological sections were constructed (Fig. 2). These sections are defined in Figure 1 and they present the geological structure in the region north and northeast of Omalos plateau, as far as the site where Agyia's springs discharge.

Based on the geological survey in the area, folds of various amplitude (tens to hundreds meters) were identified in the Trypali carbonate formations (Fig. 2).

Two groups of folds are mostly abundant: one group of NE-SW striking folds, and another one that strikes NW-SE. Both of them, verge slightly to N-NE and locally to south-southwestwards.

The overthrust tectonics (intense folding and dislodged slices) differentiates locally the thickness of the geological formations. In addition, further differentiation arises locally from the fault tectonics.

3 Hydrogeology

Widespread joint systems and localized fissures, induced by compressive forces and faults, have a pronounced effect on the solution patterns and the movement of water in Trypali carbonate formations.

Grooves of large scale, channels and gorges have been carved into the rock through flowing water and they are aligned mainly along the fault zones.

The surface water tends to concentrate into in the above mentioned features and following it infiltrates, recharging the groundwater.

It is worth noting that surface run off is very limited and surface drainage occurs only after strong rainfalls (for example Theriso's gorge). Pertinent studies show that 55% of the area precipitation infiltrates, recharging the groundwater (Lionis and Perleros 2001), while the runoff is negligible (about 5%).

Plattenkalk Series consists of several thousand meters of crystalline limestone and dolomite, interbedded with watertight horizons of chert and schists. These horizons restrict the downward movement of the water and influence strongly its movement and concentration in the overlying carbonate aquifer. This aspect is verified in many sites between Omalos, Laki and Meskla.

Five major groups of springs and more than ten pumping wells comprise the main discharge outlets for the karstic aquifer that is developed in Trypali unit (Fig. 1). The mean annual discharge of Agyia's springs is 76×10^6 m^3. Moreover, 12×10^6 m^3 of water are pumped during April to October, from Myloniana and Agyia well fields (located south of the springs) for public water supply and irrigation. It is appreciated that the ground water level ranges between 0 and +200 m a.s.l or even higher, depending on the local geological structure. Therefore, topography discourages (for financial and technical reasons) wells to be drilled for the exploitation of the water-bearing strata in high altitudes.

However, taking into account the geological structure of the area, the topography and the aspect that none important underground hydraulic conductivity exists between Trypali unit and the underlying Plattehkalk Series, it is possible to identify locally, even in high altitudes, optimal sites for water supply wells.

4 Transmissivity and permeability

Pumping test data were used to calculate transmissivity and permeability (hydraulic conductivity) of the aquifer.

The data derived from pumping tests in the wells YG1 and YG2 (Fig. 1). They have a depth up to 90 m and they are partially penetrating the karstic aquifer.

The test in well YG1 consisted of three phases at constant abstraction rates of 745, 764 and 972 m^3/h respectively. During pumping the levels in two nearby boreholes (P1 and P2), located in a distance of 7.3 and 27.1 m respectively from the abstraction well, were observed. The total drawdown of the water level after pumping was less than 1 m (0.91 m).

The test in YG2, consisted of two phases of pumping at a constant abstraction rate of 750 m^3/h. During the test the water levels in the well and in another borehole (P3) located in a distance of 14.25 m from YGM2, were measured.

Transmissivity and permeability were determined using Cooper and Jacob formula on drawdown data (Jacob method) and on recovery data (Recovery method).

Transmissivity (m^2/h) was estimated using the following non equilibrium equation:

$$T = \frac{0.18 \cdot Q}{c} \tag{1}$$

where Q the pumping rate (m^3/h), and c the difference in drawdown over one log cycle (m).

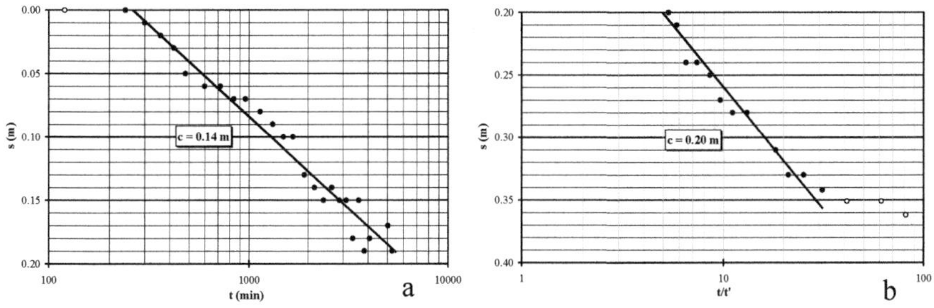

Fig. 3. a. Drawdown curve for YGM1, using observation well P1, plotted against pumping time (Jacob method), **b.** Residual drawdown for YGM1, using observation well P2, plotted against the ratio t/t' (Recovery method).

From the pumping test data, the drawdown s and the time t since pumping begun were plotted on semi logarithmic paper (Jacob method), as shown in Figure 3a. A straight line is fitted through the points and c = 0.14 m is read.

The water level recovery data were plotted on semi log diagram, with the residual drawdowns indicated on the vertical arithmetic scale, and the ratio t/t' on the horizontal logarithmic scale, where t the time since pumping begun and t' the time after pumping stops (Fig. 3b). A straight line was fitted through the points and c= 0.20 m was determined. The hydraulic conductivity was then calculated from the equation 2:

$$k = \frac{T}{H} \tag{2}$$

where H the drilled saturated thickness of the water-bearing formation.

For the abstraction and recovery measurements, the slopes of the curves indicate that the transmissivity ranges between 0.1 to 1 m^2/sec, which is equivalent to permeability between 10^{-3} and 10^{-2} m/sec. The variation in transmissivity can be attributed to the presence of extensive joints, fractures and karstification of Trypali formations.

It is worth noting that for the estimation of hydraulic conductivity (equation 2), the thickness of saturated water-bearing stratum (defined in the wells) was used, instead of the real but unknown thickness of the water-bearing formation (Todd 1980; Driscoll 1986). Consequently, the calculated values of permeability and transmissivity should be considered indicative.

Table 1. Analysis of pumping tests using the Jacob and Recovery method.

Pumping well	Pumping period	Duration of pumping t (h)	Pumping rate Q (m^3/h)	Analysis method	Observation well	T (m^2/sec)	k (m/sec)
YGM1	12/11/83	8	745	Jacob	P1	3.30×10^{-1}	5.49×10^{-3}
				Recovery	P1	2.69×10^{-1}	4.49×10^{-3}
	13/11/83	12	764	Jacob	P1	6.22×10^{-1}	1.04×10^{-2}
				Recovery	P1	6.97×10^{-1}	1.16×10^{-2}
	26/11/83 to 01/12/83	121	972	Jacob	P1	3.43×10^{-1}	5.72×10^{-3}
					P2	2.71×10^{-1}	4.52×10^{-3}
				Recovery	P1	1.70×10^{-1}	2.83×10^{-3}
					P2	2.52×10^{-1}	4.20×10^{-3}
YGM2	17/11/83	8	750	Jacob	P3	3.52×10^{-1}	8.81×10^{-3}
				Recovery	YGM2	7.39×10^{-1}	1.85×10^{-2}
					P3	1.84×10^{-0}	4.60×10^{-2}
	18/11/83 to 23/11/83	120	750	Jacob	P3	1.03×10^{-1}	2.57×10^{-3}
				Recovery	YGM2	2.54×10^{-1}	6.34×10^{-3}
					P3	8.04×10^{-1}	2.01×10^{-2}

The results of the test analysis are summarized in Table 1. Similar values have been reported by Soulios (1985) for the same aquifer. Also, the results are in good agreement with those estimated for other similar karstic aquifers in Greece (Monopolis et al. 1999). However, it should be noted that the permeability estimated at a well scale and it may not be applicable at the aquifer scale, because permeability in karstic aquifers is often scale dependent (Rovey 1994).

5 Conclusions

Based on pumping tests in Myloniana well field, the transmissibility of Trypali carbonate formations ranges between 0.1 to 1 m^2/sec which is equivalent to permeability between 10^{-3} and 10^{-2} m/sec.

The ground water level in the area ranges between 0 and +200 m a.s.l or even higher, depending on the local geological structure.

Investigating the geological structure of the area, and taking into account the topography, it is possible to identify optimal sites for water supply wells of small relatively depth, in sites with high elevation.

However, there is an increased uncertainty regarding the drilling results, because of the heterogeneity of the water-bearing stratum and the hydraulic system that runs through.

Acknowledgments The authors wish to thank YEB (Land Reclamation Service) and OADYK (Organization for the Development of Western Crete) for providing initial data for this research.

References

Driscoll FG (1986) Groundwater and Wells, Second Edition, Johnson Filtration Systems Inc., St. Paul, Minnesota

Krahl J, Kauffmann G, Kozur H, Richter D, Forster O, Heinritzi F (1983) Neue Daten zur Biostratigraphie und zur Lagerung der Phyllit Gruppe und der Trypali Gruppe auf der Insel Kreta (Griechenland).Geol. Rdsch, 72: 1147-1166

Lionis M, Perleros B (2001) Hydrogeological study for Chania area (KA 9481721). Ministry of Agricultural, Directorate of Hydrogeology. Athens (in Greek)

Manakos K, Vidakis M, Kopp KO, Krahl J, Skourtzi - Koronaiou B (2002) Geological Map of Greece, scale 1:50.000, Paleochora Sheet. Institute of Geology and Mineral Exploration

Monopolis D, Sofiou P, Steiakakis E, Kadianakis M, Vavadakis D, Kleidopoulou M (1999) Determination of hydraulic parameters of carbonate rocks (Methodology, Statistical analysis). 5th Hellenic Hydrogeological Congress, Nicosia, Cyprus, pp. 297-309 (in Greek)

Rovey, CW (1994). Assessing flow systems in carbonate aquifers using scale effects in hydraulic conductivity. Environ. Geol. 24 (4), 244-253

Soulios G (1985) A contribution to the hydrogeological study of the karstic systems of the Greek territory. Scientific Annual of the Faculty of Sciences, Aristotle University, Vol 23, No27. Thessaloniki (in Greek)

Tataris AA, Christodoulou GE (1964) The geological structure of Leuca Mountains (Western Crete). Bulletin of the Geological Society of Greece. Vol.6 No.2, 319-347

Tataris AA, Christodoulou GE (1969) Geological map of Greece, sheet Alikianou, scale 1:50,000. Institute for Geology and Subsurface Research

Todd DK (1980) Groundwater Hydrology, Second Edition, John Wiley and Sons Inc., New York

GPSR Compliance

The European Union's (EU) General Product Safety Regulation (GPSR) is a set of rules that requires consumer products to be safe and our obligations to ensure this.

If you have any concerns about our products, you can contact us on ProductSafety@springernature.com

In case Publisher is established outside the EU, the EU authorized representative is:

Springer Nature Customer Service Center GmbH
Europaplatz 3
69115 Heidelberg, Germany

Batch number: 10755310

Printed by Printforce, the Netherlands